Cell Mechanics and Cellular Engineering

Van C. Mow Farshid Guilak
Roger Tran-Son-Tay Robert M. Hochmuth
EDITORS

Cell Mechanics and Cellular Engineering

Springer-Verlag
New York Berlin Heidelberg London Paris
Tokyo Hong Kong Barcelona Budapest

Van C. Mow
Orthopaedic Research Laboratory
Columbia University
630 W. 168th Street
New York, NY 10032
USA

Farshid Guilak
Musculo-Skeletal Research Laboratory
Department of Orthopaedics
Health Science Center
State University of New York
Stony Brook, NY 11794
USA

Roger Tran-Son-Tay
Department of Aerospace Engineering,
Mechanics and Engineering Science
University of Florida
Gainesville, FL 32611
USA

Robert M. Hochmuth
Department of Mechanical Engineering
and Material Sciences
Duke University
Durham, NC 27708
USA

Library of Congress Cataloging-in-Publication Data
Cell mechanics and cellular engineering / Van C. Mow ... [et al.].
p. cm.
This volume is a record of papers presented at the Symposium on Cell Mechanics and Cellular Engineering that was held at the Second World Congress of Biomechanics in Amsterdam on July 10–15, 1994.
Includes bibliographical references.
ISBN 0-387-94307-2 (acid-free paper : New York). — ISBN 3-540-94307-2 (acid-free paper : Berlin)
1. Cells—Mechanical properties—Congresses. I. Mow, Van C. II. Symposium on Cell Mechanics and Cellular Engineering (1994 : Amsterdam, Netherlands)
QH645.5.C45 1994
547.87—dc20 94-11618

Printed on acid-free paper.

Production managed by Laura Carlson; manufacturing supervised by Vincent Scelta.
Camera-ready copy provided by the editors.
Printed and bound by Braun-Brumfield, Ann Arbor, MI.
Printed in the United States of America.

9 8 7 6 5 4 3 2 1

ISBN 0-387-94307-2 Springer-Verlag New York Berlin Heidelberg
ISBN 3-540-94307-2 Springer-Verlag Berlin Heidelberg New York

Preface

Cell mechanics and cellular engineering may be defined as the application of principles and methods of engineering and life sciences toward fundamental understanding of structure-function relationships in normal and pathological cells and the development of biological substitutes to restore cellular functions. This definition is derived from one developed for tissue engineering at a 1988 NSF workshop. The reader of this volume will see the definition being applied and stretched to study cell and tissue structure-function relationships.

The best way to define a field is really to let the investigators describe their areas of study. Perhaps cell mechanics could be compartmentalized by remembering how some of the earliest thinkers wrote about the effects of mechanics on growth. As early as 1638, Galileo hypothesized that gravity and mechanical forces place limits on the growth and architecture of living organisms. It seems only fitting that Robert Hooke, who gave us Hooke's law of elasticity, also gave us the word "cell" in his 1665 text, *Micrographiä*, to designate these elementary entities of life. Julius Wolff's 1899 treatise on the function and form of the trabecular architecture provided an incisive example of the relationship between the structure of the body and the mechanical load it bears. In 1917, D'Arcy Thompson's *On Growth and Form* revolutionized the analysis of biological processes by introducing cogent physical explanations of the relationships between the structure and function of cells and organisms.

In more recent times, investigators have used micropipet manipulation, laser tweezers, flow chambers, and rheometers to study cellular deformation and the material property relationships or "constitutive equations" that analyze cellular deformations. Complex deformations and motions can be studied using recent advances in computational mechanics. An overarching goal is to understand the relationship between mechanical forces and cellular functions in living organisms. Cellular engineering represents a newer field of investigation at the interface between cell and molecular biology, genetic engineering, and traditional engineering fields such as chemical and mechanical engineering. One emphasis in this field is on the molecular mechanisms that regulate cell differentiation, growth, shape, and locomotion. Genetic mechanisms that control the mechanical behavior of cells and the expression of adhesion proteins at the cell surface, for example, the selectins and integrins, are also being investigated through clever use of genetically engineered mutants and monoclonal antibodies to the adhesion proteins. These are only a small number of selected examples illustrating this broad new area of investigation.

In this volume, we have compiled a representative set of studies in seven different areas of cell mechanics and cellular engineering. The chapters in Part I deal with the constitutive modeling and mechanical properties of circulating cells, which have been objects of much of the pioneering work in the mechanical characterization of single cells. Part II presents recent work on flow-induced effects on cell morphology and function, with special emphasis on endothelial cell response. Part III examines the mechanics and biology of cell-substrate interactions, which addresses specific questions on the process of cell adhesion. Part IV describes the current knowledge of cell-matrix interactions and adhesion molecules. Part V focuses on the molecular and biophysical mechanisms of mechanical signal transduction in cells. Part VI utilizes articular cartilage as a paradigm for the physical regulation of tissue metabolic activity and examines how various mechanical factors may affect cellular response. Finally, the chapters in Part VII encompass the mechanics of cell motility and morphogenesis, and examine the mechanisms of force generation and mechanochemical coupling in cells.

The chapters in this book were written for the Symposium on Cell Mechanics and Cellular Engineering that was held at the Second World Congress of Biomechanics in Amsterdam on July 10–15, 1994. We are grateful to the organizers of the Second World Congress (e.g., Chairman Dr. Rik Huiskes) and to the Program Committee (Chairman Dr. Michel Y. Jaffrin and Co-Chairman Dr. Peter S. Walker) for inviting us to organize this symposium.

At the time of the invitation, we felt that this specialized area of research had gained sufficient scientific momentum to warrant such a symposium at the Second World Congress, and that a collection of works such as those included in this volume would serve a timely and important function for the field. Therefore, we accepted their invitation, and challenge, to develop the symposium. As our planning progressed, we were gratified at the overwhelming response from many investigators wishing not only to contribute to the program in Amsterdam, but also to contribute to this book. We fully appreciate that for busy researchers, particularly young investigators, writing a paper on short notice can be a significant challenge. Therefore, to the contributors of this volume, we express our sincere appreciation for your hard work. As the reader of this volume will appreciate, each chapter has been carefully written by experts in their area of specialization.

Finally we wish to thank Dr. Robert J. Foster of the Orthopaedic Research Laboratory of Columbia University for his effort in compilation of this volume, and Dr. Robert C. Garber and Ms. Jessica Downey of Springer-Verlag New York. Without their significant contributions, the publication of this volume would not have been possible.

Van C. Mow, Ph.D.
Columbia University

Farshid Guilak, Ph.D.
State University of New York, Stony Brook

Roger Tran-Son-Tay, D.Sc.
University of Florida, Gainesville

Robert M. Hochmuth, Ph.D.
Duke University

Contents

Contributors

Sadie Aznavoorian
Laboratory of Pathology
National Cancer Institute
National Institutes of Health
Bethesda, MD 20892, USA

Nathaniel M. Bachrach
Orthopaedic Research Laboratory
Department of Orthopaedic Surgery
Columbia University
New York, NY 10032, USA

Albert J. Banes
Plastic and Reconstructive Surgery
School of Medicine
University of North Carolina
Chapel Hill, NC 27599, USA

Victor H. Barocas
Department of Chemical Engineering
and Materials Science
University of Minnesota
Minneapolis, MN 55455, USA

Anne-Marie Benoliel
ISNERM U 387
Laboratoire d'Immunologie
Hopital de Saint-Marguerite
BP29 13274 Marseille Cedex 09
France

Avri Ben-Ze'ev
Department of Molecular Genetics
and Virology
The Weizmann Institute of Science
Rehovot 76100, Israel

Rena Bizios
Department of Biomedical Engineering
Rensselaer Polytechnic Institute
Troy, NY 12180-3590, USA

Frank A. Blumenstock
Department of Physiology
and Cell Biology
Albany Medical College
Albany, NY 12208, USA

Scott Boitano
Department of Cell Biology and Anatomy
University of California at Los Angeles
Los Angeles, CA 90024, USA

Pierre Bongrand
ISNERM U 387
Laboratoire d'Immunologie
Hopital de Saint-Marguerite
BP29 13274 Marseille Cedex 09
France

Brian Brigman
Plastic and Reconstructive Surgery
University of North Carolina
Chapel Hill, NC 27599, USA

Shu Chien
Institute for Biomedical Engineering
University of California, San Diego
La Jolla, CA 92093-0412, USA

Jean Delobel
Rhome-Merieux, Inc.
115 Transtech Drive
Athens, GA 30601, USA

Paul A. DiMilla
Department of Chemical Engineering and
Center for Light Microscope Imaging and
Biotechnology
Carnegie Mellon University
Pittsburgh, PA 15213, USA

Henry J. Donahue
Musculoskeletal Research Laboratory
Department of Orthopaedics
State University of New York
Stony Brook, NY 11794, USA

Cheng Dong
Department of Bioengineering
Pennsylvania State University
University Park, PA 16802, USA

Bard Ermentrout
Department of Mathematics
University of Pittsburgh
Pittsburgh, PA 15260, USA

Mark Fenster
Institute for Biomedical Engineering
University of California, San Diego
La Jolla, CA 92093-1412, USA

Jose Luis Rodriguez Fernandez
Department of Molecular Genetics
and Virology
The Weizmann Institute of Science
Rehovot 76100, Israel

Thomas Fischer
Department of Medicine
University of North Carolina
Chapel Hill, NC 27599, USA

John A. Frangos
Department of Chemical Engineering
Pennsylvania State University
University Park, PA 16802, USA

Benjamin Geiger
Chemical Immunology
The Weizmann Institute of Science
Rehovot 76100, Israel

Ursula Glück
Department of Molecular Genetics
and Virology
The Weizmann Institute of Science
Rehovot 76100, Israel

Douglas J. Goetz
School of Chemical Engineering
Cornell University
Ithaca, NY 14853, USA

Daniel A. Grande
Department of Orthopaedic Surgery
North Shore University Hospital
Manahasset, NY, USA

Siva R.P. Gudi
Department of Chemical Engineering
Pennsylvania State University
University Park, PA 16802, USA

Farshid Guilak
Musculoskeletal Research Laboratory
Department of Orthopaedics
State University of New York
Stony Brook, NY 11794, USA

Andrew C. Hall
Physiology Laboratory
Oxford University
Oxford, England, UK

Daniel A. Hammer
School of Chemical Engineering
Cornell University
Ithaca, NY 14853, USA

Heikki J. Helminen
Department of Anatomy
University of Kuopio
P.O. Box 1627, FIN 70211
Kuopio, Finland

Robert M. Hochmuth
Department of Mechanical Engineering
and Materials Science
Duke University
Durham, NC 27708-0300, USA

Peigi Hu
Plastic and Reconstructive Surgery
University of North Carolina
Chapel Hill, NC 27599, USA

Donald Ingber
Departments of Pathology and Surgery
Children's Hospital and
Harvard Medical School
Boston, MA 02115, USA

Raymond D. Iveson
Department of Biomedical Engineering
Rensselaer Polytechnic Institute
Troy, NY 12180-3590, USA

David A. Jones
Cox Laboratory for Biomedical Engineering
Rice University
Houston, TX 77251-1892, USA

Tomoatsu Kimura
Department of Orthopaedic Surgery
Osaka University Medical School
2-2 Yamada-oka, Suite 565
Japan

John Krupinski
Henry Hood Program for Research at the Weis Center for Research
Geisinger Clinic
100 N. Academy Avenue
Danville, PA 17822, USA

Mikko J. Lammi
Department of Anatomy
University of Kuopio
P.O. Box 1627, FIN 70211
Kuopio, Finland

W. Thomas Lawrence
Plastic and Reconstructive Surgery
University of North Carolina
Chapel Hill, NC 27599, USA

Shouyan Lee
Institute for Biomedical Engineering
University of California, San Diego
La Jolla, CA 92093-1412, USA

Lance A. Liotta
Laboratory of Pathology
National Cancer Institute
National Institutes of Health
Bethesda, MD 20892, USA

Larry V. McIntire
Institute of Biosciences and Bioengineering
Rice University
Houston, TX 77251-1892, USA

Kenneth J. McLeod
Musculoskeletal Research Laboratory
Department of Orthopaedics
State University of New York
Stony Brook, NY 11794, USA

Van C. Mow
Orthopaedic Research Laboratory
Departments of Orthopaedic Surgery and Mechanical Engineering
Columbia University
New York, NY 10032, USA

Ken Nakata
Department of Orthopaedic Surgery
Osaka University Medical School
2-2 Yamada-oka, Suite 565
Japan

Margaret K. Offermann
Winship Cancer Center
Emory University School of Medicine
Atlanta, GA 30322, USA

Norio Ohshima
Department of Biomedical Engineering
Institute of Basic Medical Sciences
University of Tsukuba
Tsukuba Ibaraki 305
Japan

Keiro Ono
Department of Orthopaedic Surgery
Osaka University Medical School
2-2 Yamada-oka, Suite 565
Japan

Keiko Ookawa
Department of Biomedical Engineering
Institute of Basic Medical Sciences
University of Tsukuba
Tsukuba Ibaraki 305
Japan

Michal Opas
Department of Anatomy and Cell Biology
University of Toronto
Toronto, Ontario
M5S 1A8 Canada

George Oster
Department of Molecular and Cellular Biology
University of California
Berkeley, CA 94720, USA

Jyrki J. Parkkinen
Department of Anatomy
University of Kuopio
P.O. Box 1627, FIN 70211
Kuopio, Finland

Charles Peskin
Courant Institute of Mathematical Sciences
251 Mercer Street
New York, NY 10012, USA

Anne Pierres
ISNERM U 387
Laboratoire d'Immunologie
Hopital de Saint-Marguerite
BP29 13274 Marseille Cedex 09
France

Clinton T. Rubin
Musculoskeletal Research Laboratory
Department of Orthopaedics
State University of New York
Stony Brook, NY 11794, USA

Frederick Sachs
SUNY Biophysical Sciences
118 Cary Hall
Buffalo, NY 14214, USA

Michael Sanderson
Department of Cell Biology and Anatomy
University of California at Los Angeles
Los Angeles, CA 90024, USA

Diane Savarese
Division of Hematology/Oncology
University of Massachusetts Medical School
Worcester, MA 01655, USA

Geert W. Schmid-Schoenbein
Institute for Biomedical Engineering
University of California, San Diego
La Jolla, CA 92093-1412, USA

Lori A. Setton
Orthopaedic Research Laboratory
Department of Orthopaedic Surgery
Columbia University
New York, NY 10032, USA

Hainsworth Y. Shin
Department of Biomedical Engineering
Rensselaer Polytechnic Institute
Troy, NY 12180-3590, USA

Richard Skalak
Institute for Biomedical Engineering
University of California, San Diego
La Jolla, CA 92093-0412, USA

Boguslaw A. Skierczynski
Institute for Biomedical Engineering
University of California, San Diego
La Jolla, CA 92093-0412, USA

C. Wayne Smith
Leukocyte Biology Laboratory
Baylor College of Medicine
Houston, TX 77030, USA

Don Sutton
Institute for Biomedical Engineering
University of California, San Diego
La Jolla, CA 92093-1412, USA

Markku I. Tammi
Department of Anatomy
University of Kuopio
P.O. Box 1627, FIN 70211
Kuopio, Finland

Linda A. Tempelman
Naval Research Laboratory
Center for Bio/Molecular Science and Engineering
Code 6900
Washington DC 20375-5000, USA

H. Ping Ting-Beall
Department of Mechanical Engineering and Materials Science
Duke University
Durham, NC 277089, USA

Aydin Tözeren
Department of Mechanical Engineering
The Catholic University of America
Washington, DC 20064, USA

Robert T. Tranquillo
Department of Chemical Engineering
and Materials Science
University of Minnesota
Minneapolis, MN 55455, USA

Mientao A. Tsai
Department of Biophysics
University of Rochester School
of Medicine and Dentistry
601 Elmwood Ave.
Rochester, NY 14642, USA

Noriyuki Tsumaki
Department of Orthopaedic Surgery
Osaka University Medical School
2-2 Yamada-oka, Suite 565
Japan

Mari Tsuzaki
Plastic and Reconstructive Surgery
University of North Carolina
Chapel Hill, NC 27599, USA

Jill P.G. Urban
Physiology Laboratory
Oxford University
Oxford, England, UK

Shunichi Usami
Institute for Biomedical Sciences
Academia Sinica
Taipai, 11529, Taiwan

Peter A. Watson
Henry Hood Program for Research at
the Weis Center for Research
Geisinger Clinic
100 N. Academy Avenue
Danville, PA 17822, USA

Richard E. Waugh
Department of Biophysics
University of Rochester School
of Medicine and Dentistry
601 Elmwood Ave.
Rochester, NY 14642, USA

Tom E. Williams
George W. Woodruff School
of Mechanical Engineering
Georgia Institute of Technology
Atlanta, GA 30332-0405, USA

Dongchun Xia
George W. Woodruff School
of Mechanical Engineering
Georgia Institute of Technology
Atlanta, GA 30332-0405, USA

Jun You
Department of Bioengineering
Pennsylvania State University
University Park, PA 16802, USA

Richard A. Zell
Musculoskeletal Research Laboratory
Department of Orthopaedics
State University of New York
Stony Brook, NY 11794, USA

Doncho V. Zhelev
Department of Mechanical Engineering
and Materials Science
Duke University
Durham, NC 27708, USA
Central Laboratory of Biophysics
Bulgarian Academy of Sciences
1113 Sofia, Bulgaria

Cheng Zhu
George W. Woodruff School of Mechanical
Engineering
Georgia Institute of Technology
Atlanta, GA 30332-0405, USA

Margot Zöller
Department of Radiology and
Pathophysiology
Germany Cancer Research Center
Heidelberg, Germany

Part I

Constitutive Modeling and Mechanical Properties of Circulating Cells

1
Human Neutrophils Under Mechanical Stress.

D. V. Zhelev and R. M. Hochmuth

Introduction

White blood cells play an important role in blood flow dynamics in the microcirculation because of their large volume and low deformability (Bagge et al. 1980). Among them, the neutrophil is of special interest, for it can activate and change its mechanical properties in seconds (Evans et al. 1993). Early studies of the mechanical properties of the passive neutrophil (Bagge et al. 1977) suggest that it behaves as a simple viscoelastic solid (represented as elastic and viscous elements in series with another elastic element). This model, known as the "standard solid model" (Schmidt-Schönbein et al. 1981), is based on small deformation experiments in which the viscosity and the two elasticities are considered as bulk properties of the cytoplasm. The results from large deformation experiments, however, cannot be explained by this model. Evans and Kukan (1984) proposed a model where the elastic resistance of the cell comes from a thin domain close to the cell surface (the "cortex"), while the cytoplasm interior is a liquid rather than a solid. According to this model there are two parameters: the cortical tension and the apparent cytoplasmic viscosity, which are sufficient for characterizing the rheology of the neutrophil. This model explains successfully the general behavior of the neutrophil in large deformation experiments (Evans and Yeung 1989) and in shape recovery experiments (Tran-Son-Tay et al. 1991). In both models it is assumed that the mechanical properties of the neutrophil do not depend on the applied stress or on the resulting strain during deformation (Chien et al. 1984, Evans and Yeung, 1989).

Even though the model of Evans and Yeung (1989) is relevant to neutrophil flow through small capillaries, it does not explain all of the experimental results. Zhelev et al. (1993) show that there is a bending resistance acting in synchrony with the isotropic cortical tension. Also Needham and Hochmuth (1990) found experimentally that the apparent cell viscosity depends on the suction pressure and Hochmuth et al. (1993a) found that it depends on the pipet size. The ratio of the maximum to the minimum measured apparent viscosity can be five or more (Hochmuth et al. 1993b).

The bending resistance of the neutrophil surface is an apparent characteristic of the neutral surface of the membrane-cortex domain (Zhelev et al. 1993). It has been shown by mechanical means that the membrane-cortex complex has a finite thickness on the order of 0.24 μm or less. A cortical meshwork with a thickness on the order of 0.05 μm to 0.2 μm was found by electron microscopy methods (Esaguy et al. 1989, Sheterline and Rickard 1989, Bray et al. 1986). The neutrophil cortex is made mostly of F-actin filaments (Bray et al. 1986). Then the apparent bending modulus depends on the cortical thickness, the density of F-actin filaments, their bending rigidity (Gittes et al. 1993) and the degree of their crosslinking (for a review of actin binding proteins see Hartwig and Kwiatkowski 1991).

Recent experiments show that there is a deviation from the model of Evans and Yeung (1989). The measured critical suction pressure for very small pipets is larger than that predicted by the law of Laplace and the measured apparent viscosity depends on the suction pressure and the pipet size. These results suggest that: there is a cortex with a finite thickness adjacent to the cell surface, the cell cytoplasm may not be a Newtonian liquid and the mechanical properties of the cell may change during the course of deformation. Therefore the objective of this work is to present and discuss the evidence for: 1) the existence of a membrane-associated cortex with a finite thickness, 2) the change of the neutrophil rheological properties during or after the application of a mechanical stress sufficient for deformation, and 3) the polymerization of F-actin induced by mechanical stimulation.

Materials and Methods

Cell Preparation

The preparation of neutrophils has been described elsewhere (Zhelev et al. 1993). Briefly, venous blood is drawn from healthy adult donors into vacutainers containing EDTA as an anticoagulant. The neutrophils are separated on a Ficoll-Hypaque gradient (Sigma Histopaque-1077 and 1119) at 800 g for 20 min. The cells are collected at the 1077/1119 interface and washed once with Ca^{2+} and Mg^{2+} free modified Hanks' balanced salt solution (HBSS, Sigma Chemical Co.). Finally the cells are resuspended in a 50% autologous plasma/HBSS solution. All the procedures and experiments are done at 23°C.

Micromanipulation

The experimental chamber is 2 mm thick and open from both sides to allow micromanipulation. The bottom of the chamber is covered with a glass coverslip and the top is covered with two parallel glass strips with a 1 mm thick gap. Thermostated water is allowed to flow in the gap for temperature control. The cells are observed with an inverted Leitz microscope with a 100x oil immersion objective. Micropipets are made from 0.75 mm capillary glass tubing pulled to a fine point with a vertical pipet puller and cut to the desired diameter as described elsewhere (Zhelev et al. 1993). The pipets are filled with the solution used in the experiments and are kept in the chamber for 15 min before starting the experiments in order to minimize cell adhesion to the glass surfaces (Evans and Kukan 1984). The pipets are connected to a manometer system, which allows the pipet-chamber pressure difference to be varied between 0.5 and 100 Pa using the micrometer driven displacement of a water filled reservoir, or to be increased to 40,000 Pa using a syringe. The pressures are measured with a differential pressure transducer (Validyne DP15-24). The pressure transducer readings are multiplexed on the recorded images together with a time counter (Vista Electronics, Model 401). Distances are measured with video calipers (Vista Electronics, Model 305).

Results

Static Experiments with Passive Cells

The passive (or resting) neutrophil has a spherical shape. When it is deformed into a pipet, the inside projection length increases simultaneously with increasing suction pressure until a hemispherical projection is formed (see Fig. 1). The suction pressure corresponding to the hemispherical projection is called the "critical pressure". After the hemispherical cap is formed, the cell flows continuously into the pipet at any pressure difference exceeding the critical one. For suction pressures smaller or equal to the critical pressure, the dependence of the projection length on the applied pressure difference is described by the law of Laplace. The resistance during the initial deformation of a neutrophil is provided by the isotropic stress in its surface, called the "cortical tension"

(Evans and Kukan 1984). The cortical tension has been measured in several experiments and found to be on the order of 0.024 to 0.038 mN/m (Evans and Yeung 1989, Needham and Hochmuth 1992, Zhelev et al. 1993). It has been shown also that the value for the cortical tension has a weak area dilation dependence, with an apparent area expansion modulus on the order of 0.04 mN/m (Needham and Hochmuth 1992).

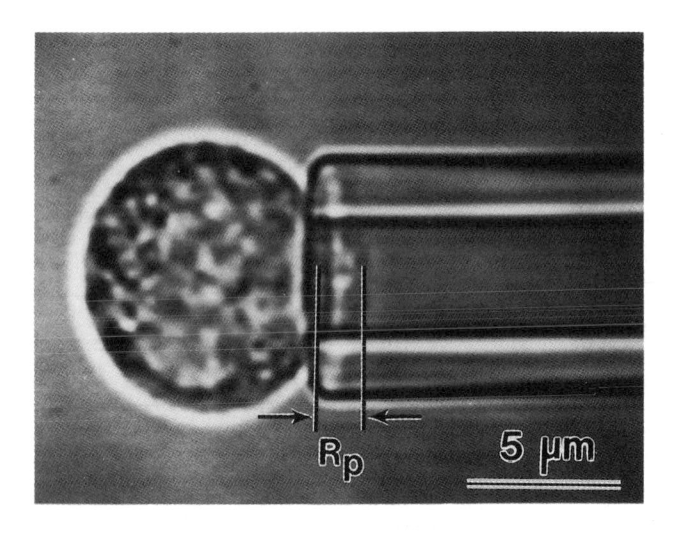

Fig. 1. Neutrophil held in a pipet with a suction pressure equal to the critical pressure. The projection length inside the pipet is equal to the pipet radius R_p.

The nature of the cortical tension is unknown. Evans and Kukan (1984) speculate that it arises from contractile elements tangent to the membrane surface. Zhelev et al. (1993) hypothesize that there is a structure with a finite thickness (cortex) associated with the cell membrane that determines the shape and the elastic resistance of the neutrophil. By measuring the critical pressure with increasingly smaller pipets (down to the optical resolution of the light microscope), they show experimentally the existence of this structure . In the proposed model by Zhelev et al. (1993) the cortical tension is defined as a free contraction energy per unit area of the cell surface. (In the model the cortical tension is an apparent characteristic of the neutral surface of the membrane-cortex complex.) They show that the membrane cortex has an apparent bending modulus on the order of $1\text{-}2\times10^{-18}$ J, which is an order of magnitude larger than the bending modulus of lipid-bilayer or red blood-cell membranes.

The law of Laplace for the neutral surface of a cortex with finite thickness has the form:

$$R_p = T \cdot \left(\frac{1 - \frac{R_p}{R_{out}}}{\frac{\Delta P_c}{2}} \right) + d \qquad (1)$$

where R_p is the pipet radius, R_{out} is the radius of the outside portion of the cell, T is the cortical tension, d is the distance from the neutral surface to the pipet wall and ΔP_c is the critical suction pressure related to the cortical tension. Then, when the experimental data (the pipet radius, the outside cell radius and the critical suction pressure related to the cortical tension) are plotted according to Eq. 1, the cortical tension and the distance of the neutral surface to the pipet surface are found from the slope and the intercept of the least-squares, straight-line fit.

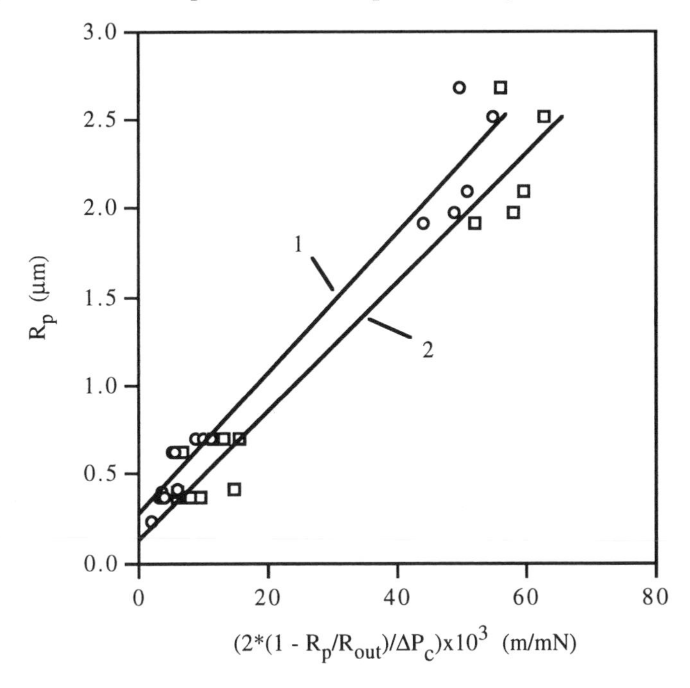

Fig. 2. Experimental data for the pipet radius and corresponding critical suction pressure presented in accordance with Eq. 1. Circles are for the measured critical pressures and squares are for the critical suction pressures corrected for the bending resistance (the apparent bending modulus is $2x10^{-18}$ J). The slopes and the intercepts for the least-squares lines are 0.038 mN/m and 0.3 µm for line 1 and 0.036 and 0.15 µm for line 2 respectively.

The measured critical suction pressures and the corresponding pipet radii using the coordinates given by Eq. 1 are shown with circles in Fig. 2. When the cell surface has also a bending resistance the critical suction pressure related to the cortical tension will be smaller than the measured one. The measured critical suction pressure is corrected for the bending resistance using the model described elsewhere (Zhelev at al. 1993). The data for the corrected critical suction pressures are shown in Fig. 2 with squares. The cortical tensions determined from the regression lines for both sets of data are similar, while the intercept, when the corrected critical suction pressure is used, is one half of the one found when the measured critical suction pressure is used. The membrane-cortex thickness estimated from the intercept of line 1 is on the order of 0.3 μm, which is larger than the cortical thickness of 0.05 to 0.1 μm measured by electron microscopy (Esaguy et al. 1989, Sheterline and Rickard 1989). The value for the membrane-cortex thickness found from line 2 is 0.15 μm. The intercept gives an average value for the cortical thickness. Its value will increase if the cell is activated. Then the value found from the intercept is an upper bound for the cortical thickness of an un-activated cell. The results in Fig. 2 show that there is a cortex with a finite thickness associated with the cell membrane.

Whole Cell Recovery

The neutrophil interior is filled with a cytoplasmic liquid and organelles. The nucleus, which occupies about 20% of the volume of the cytoplasm (Schmidt-Schönbein et al. 1980), is segmented and spread. Thus the cytoplasm is expected to be highly deformable. This is experimentally observed by Evans and Kukan (1984), and the neutrophil is modeled as a Newtonian liquid by Evans and Yeung (1989). Because of a persistent cortical tension and a liquid interior, the neutrophil returns to its spherical shape after deformation (see Fig. 3). The rate of recovery depends on the ratio of the cortical tension to the viscosity of the cytoplasm (Tran-Son-Tay et al. 1991). Although there is experimental evidence for the liquid behavior of the cytoplasm, the Newtonian model cannot explain all experimental data. The apparent viscosity of the cytoplasm determined from recovery experiments depends on the initial deformation. For large initial deformations it is on the order of 150 Pa·s (Tran-Son-Tay et al. 1991), whereas for small initial deformations it is on the order of 60 Pa·s (Hochmuth et al. 1993a). For both entry and recovery experiments, the value for the apparent viscosity of the neutrophil depends on the deformation ratio (the ratio of the maximum length of the deformed neutrophil to the diameter of the spherical cell), and it is smaller for smaller ratios (Hochmuth et al. 1993b).

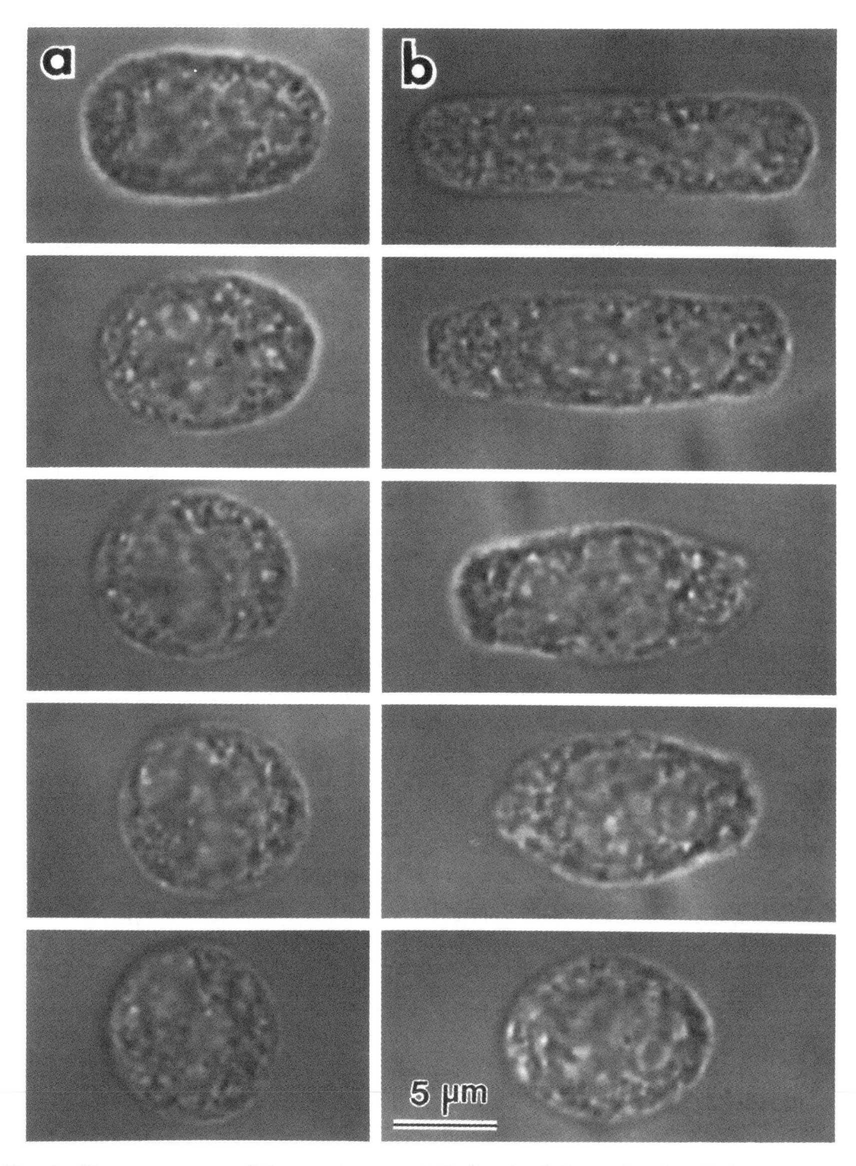

Fig. 3. Shape recovery of the same neutrophil after its deformation in a 3.4 μm radius pipet (a) and in 2.35 μm radius pipet (b). The times for the sequence (a) are 0 s, 10 s, 40 s, 60 s and 100 s after the cell is released from the pipet. The times for the sequence (b) are 0 s, 10 s, 30 s, 50 s and 95 s after the cell is released. The shape of the cell in the sequence (b) after 140 s (not shown) is essentially the same as that in the last photograph in sequence (a).

The experiments described above are for different cells from different donors. To study the dependence of the apparent viscosity on the initial deformation, the apparent viscosity is determined for the same cell for different initial deformations. (The optical system also is improved in order to follow precisely any shape changes and changes of the refractive index at the boundary of the cell. The activation in a local region of the neutrophil surface is observed first by the change of the refractive index. Later, a growing pseudopod is eventually observed in this region.) A single cell is aspirated into a large pipet (3.4 μm radius) and held for 8 to 12 s. Then the cell is released and allowed to recover. After its complete recovery the cell is aspirated into a smaller pipet (2.3 μm radius), held inside for the same time, and then allowed to recover again (Fig. 3). The shape characteristics (the cell length D and its width W) are measured (Fig. 4) and the ratio of the cortical tension to the apparent viscosity is determined using the model of Tran-Son-Tay et al. (1991). The ratio of the cortical tension to the apparent viscosity for five cells found from studying their recovery after small and large initial deformations is shown in Table I. After recovery from small and large initial deformations, cell #4 is deformed again in the larger pipet and the ratio of the cortical tension to the apparent viscosity is determined again (#4a).

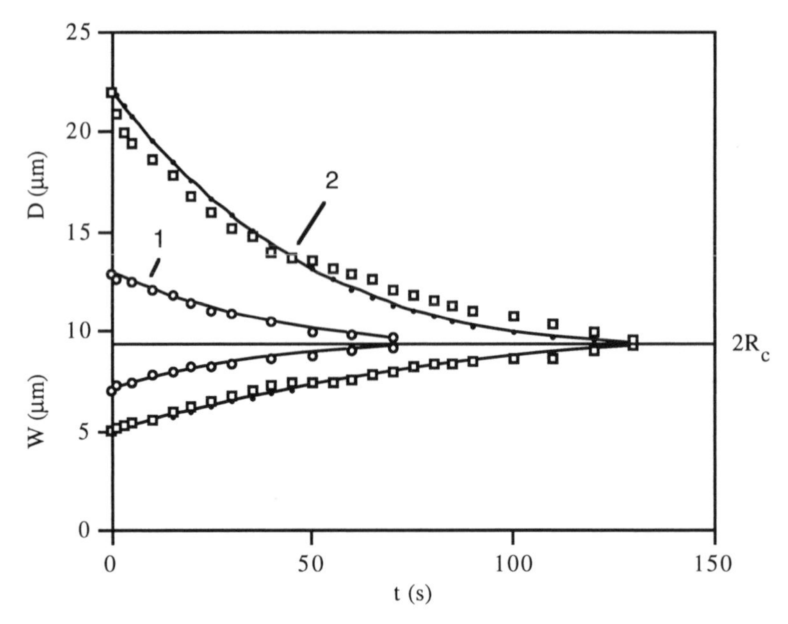

Fig. 4. Recovery of the same neutrophil after initial deformation into a 3.4 μm radius (1) and a 2.35 μm radius (2) pipet. The curves are predicted by the recovery model (Tran-Son-Tay et al. 1991). D is the cell length and W is its width. $2R_c$ is the resting diameter of the cell.

Table I. The ratio of the cortical tension to the apparent viscosity for five neutrophils determined from their recovery after initial deformation in a 3.4 μm radius and a 2.35 μm radius pipet. The apparent viscosities are calculated assuming a constant cortical tension of 0.03 mN/m.

Cell #	T/μ (μm/s) (from initial deformation in a 3.4 μm radius pipet)	μ (Pa·s) (from initial deformation in a 3.4 μm radius pipet)	T/μ (μm/s) (from initial deformation in a 2.35 μm radius pipet)	μ (Pa·s) (from initial deformation in a 2.35 μm radius pipet)
1	1.63	184	1.25	240
2	2.31	130	1.36	220
3	1.61	186	1.11	270
4	1.82	165	0.86	350
4a	1.58	190		
5	1.58	190	0.86	350

It is seen that the ratio of the cortical tension to the apparent viscosity for small initial deformations is consistently larger than that for large initial deformations (the corresponding apparent viscosities are smaller). This tendency is the same as the one reported by Hochmuth et al. (1993a) using different cells.

The shape changes corresponding to small and large initial deformations are seen in Fig. 3. The cell surface remains smooth during recovery after small initial deformations. After large initial deformations, however, there are irregularities at both ends of the cell. (In this experiment the cells do not stick to the pipet wall. They slipped along the inside of the pipet with pressure differences on the order of 4 to 5 Pa.) The irregular shape of the neutrophil surface is an indication of activation. Thus, after aspiration of a neutrophil into a 2.35 μm radius pipet, the cell is activated. The recovery model used to determine the ratio of the cortical tension to the apparent viscosity considers the neutrophil as a Newtonian liquid with a constant cortical tension. When the cell becomes activated, both the stress resultant along the cell surface (cortical tension) and the apparent viscosity can change. The higher apparent viscosities corresponding to smaller initial deformations can be due either to greater viscous dissipation inside the cell or to a smaller cortical tension or to both.

Cell Entry into Pipets

When a neutrophil is sucked into a pipet with a pressure difference above the critical pressure, it flows completely inside. The rate of cell entry is determined by the viscous dissipation in the cytoplasm core and the dissipation in the cortex. The dissipation in the cortex is shown to be negligible compared to that in the cytoplasm (Evans and Yeung 1989). The greatest dissipation occurs in a region close to the pipet orifice (M. Dembo, personal communication) while the dissipation in the region inside the pipet is negligible (the flow in this region resembles plug flow). The cell flows continuously into the pipet without any indication of approaching a static equilibrium when the suction pressure exceeds the critical one (Evans and Yeung 1989), which suggests that the shear elasticity (if it exists) is negligible.

The bulk apparent viscosity μ is found from (Evans and Yeung 1989)

$$\mu = \frac{\Delta P_{exc} \, R_p}{\left(\frac{dL}{dt}\right)} \cdot f\left(\frac{R_p}{R_{out}}, m\right) \tag{2}$$

where ΔP_{exc} is the excess suction pressure above the critical pressure, (dL/dt) is the rate of entry into the pipet and $f(R_p/R_{out}, m)$ is a function of the ratio R_p/R_{out} of the pipet radius R_p to the outside cell radius R_{out}. The coefficient m characterizes the ratio of dissipation in the cortex to that in the bulk phase. The exact expression for the functional term in Eq. 2 is rather complicated. Needham and Hochmuth (1990) showed that the numerical simulations of Evans and Yeung (1989) and the experimental results are in a good agreement with a simplified model for which the functional term has the form:

$$f\left(\frac{R_p}{R_{out}}, m\right) = \frac{1}{6\left(1 - \frac{R_p}{R_{out}}\right)}. \tag{3}$$

The value of *6* for the coefficient m corresponds to the ratio of cortical dissipation to bulk dissipation equal to 0.01. The assumptions made in the simplified model suggest that it will describe successfully the deformation of neutrophils into cylindrical pipets except when the outside cell radius approaches the pipet radius.

The models discussed in the preceding paragraph predict that the rate of neutrophil entry will increase monotonically with decreasing outside cell volume. In the experiments, however, it is commonly observed that the initial rate of entry is very rapid as can be seen in Fig. 5. Then, the rate of entry decreases, reaches a minimum, and increases again (see Fig. 5). The instantaneous apparent viscosity changes accordingly from its initial value, to the maximum value and then decreases again (see Fig. 6). The instantaneous viscosity in Fig. 6 is calculated from the simplified model (Eqs. 2 and 3). The rate of entry is found from the first derivative of the approximating least squares polynomial of the experimental data for the projection length verses time (Fig. 5). The experimental data are successfully approximated by a polynomial of fifth power (the correlation coefficient in this case is 0.999).

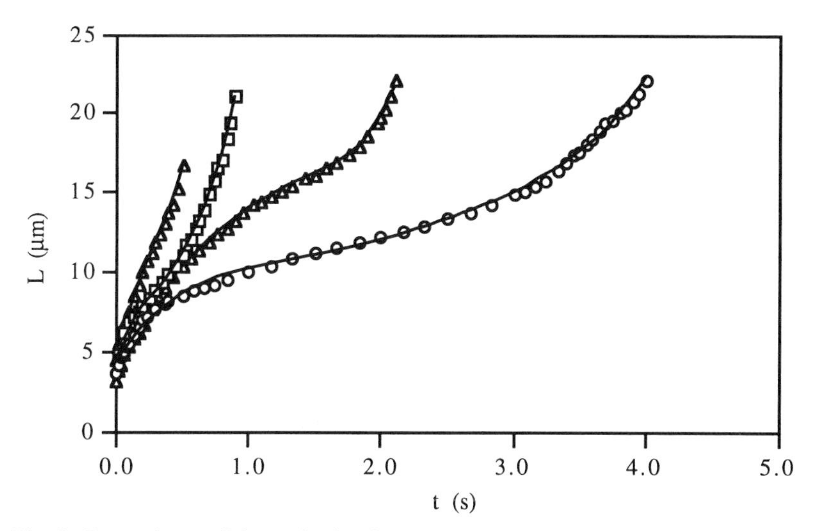

Fig. 5. Dependence of the projection length on time for entry of neutrophils into a large pipet at different suction pressures. Open symbols are for cell entry into a 2.0 μm radius pipet. The suction pressures are: circles - 800 Pa, squares - 1400 Pa and triangles - 2100 Pa. The filled triangles are for cell entry into 1.7 μm radius pipet at a suction pressure of 2000 Pa. The curves are generated by a fifth order polynomial that is used to calculate the instantaneous rate of entry *(dL/dt)*. All cells have the same resting radius of 4.6 μm.

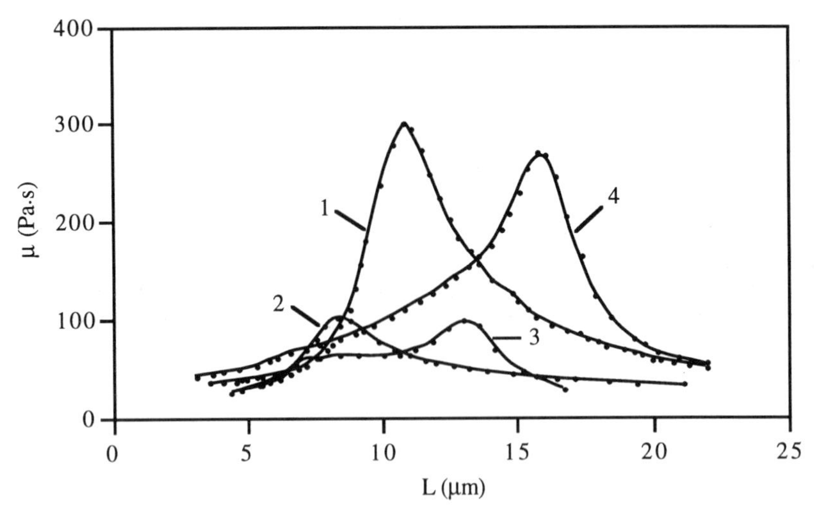

Fig. 6. Calculated instantaneous apparent viscosities (from the simplified model) for four neutrophils plotted as a function of the projection length L inside the pipet. Curves 1, 2 and 3 are for a pipet with a radius of 2.0 μm and curve 4 is for a pipet with a radius of 1.7 μm. The suction pressures for the different curves are 800 Pa (1), 1400 Pa (2), 2100 Pa (3) and 2000 Pa (4).

It is seen from Fig. 6 that the maximum values for the apparent viscosity are different for different suction pressures and for different pipet sizes. The maximum apparent viscosity is larger for smaller suction pressures with a given pipet and larger for smaller pipets at a given suction pressure. This observation is consistent with the results reported earlier (Needham and Hochmuth 1990, Hochmuth et al. 1993a), where the measured apparent viscosity in their experiments corresponds to the maximum instantaneous viscosity in Fig. 6. The initial apparent viscosity is of the same order of magnitude for all experiments and is from 25 to 40 Pa·s. These results cannot be explained by the models discussed above. Thus, some of the assumptions made by Evans and Yeung (1989) may have to be revised, namely that: 1) the neutrophil cytoplasm may not behave as a Newtonian liquid, 2) the dissipation in the cortex may not be negligible compared to the dissipation in the cytoplasm, and 3) there may be friction between the cell surface and the pipet wall.

If the apparent viscosity is considered as a function of the projection length inside the pipet (Fig. 6), its maximum value is found when about half of the nucleus is observed to be in the pipet. This observation suggests that the viscosity of the nucleus is larger than that of the rest of the cytoplasm. For small suction pressures the apparent viscosity begins to change at the very beginning of cell deformation when the nucleus is far from the pipet orifice. For a given pipet the maximum apparent viscosities for large suction pressures are similar, but they are larger than that for small suction pressures (compare curves 2 and 3 with curve 1 in Fig. 6). This result can be explained if there is friction between the cell surface and the pipet (at the pipet orifice or along its wall). In the case of friction there are two regions of frictional resistance: one at the pipet orifice and another along the pipet wall. The friction at the orifice will depend on the suction pressure and a larger frictional resistance will correspond to a larger suction pressure. The friction with the pipet wall will depend on the rate of "thinning" of the lubricating film between the pipet wall and the cell surface. This friction term will be larger for a slowly deforming cell (e.g. at low suction pressure) where the corresponding apparent viscosity will be larger. The frictional dissipation at the beginning of the deformation will be minimum, therefore, the initial apparent viscosity should not depend on the suction pressure and the pipet size as is seen in Fig. 6.

The apparent viscosity during the initial phase of deformation is studied for different suction pressures using very small pipets. Fig. 7 illustrates the entry of neutrophils into a 0.35 μm radius pipet at three different suction pressures. The cells do not flow completely into the pipet and the rate of deformation gradually decreases with time. The flow of the cell ceased before the cell membrane could be pulled smooth. The calculated apparent viscosity as a function of the projection length is shown in Fig. 8. The initial apparent viscosity is from 35 to 70 Pa·s. This is the same order of magnitude as the measurements made with large pipets (Fig. 6). Thus an estimation of the cytoplasm viscosity can be found from the initial apparent viscosity and is on the order of 30 to 60 Pa·s.

In the case of a significant friction at the pipet orifice and/or along the pipet wall, the resultant stress in the cell surface is not constant. There are three regions with different membrane tensions: 1) the hemispherical region inside the pipet, 2) the cylindrical portion along the pipet wall and 3) the cell body outside the pipet. The tension in the hemispherical cap inside the pipet is maximum, while that in the outside region is equal to the cortical tension of the resting neutrophil. Thus there is a gradient in the surface tension along the pipet wall. The membrane-cortex complex in the regions subjected to high tension undergoes some changes that result in the cessation of flow at much shorter projection lengths than the expected maximum found from the maximum available membrane surface area. The recovery of the deformed region shows that there is a significant formation of polymer in this region. (After the cell is released from the pipet, its surface in the deformed region is irregular. In many cases pseudopods grow spontaneously in this region.) Possibly because of the friction at the pipet orifice and/or along the pipet wall, the neutrophil surface inside the pipet will be subjected to an additional stress compared to the isotropic stress in the surface of the resting cell (the cortical tension). This additional stress induces an area dilation, which can be significant because of the low apparent area expansion elastic modulus (on the order of 0.04 mN/m, Needham and Hochmuth 1992). The local surface stress (or the corresponding area dilation) induces polymerization in the region inside the pipet that stiffens the cell and leads to an increase in its resistance to deformation and a subsequent cessation of flow.

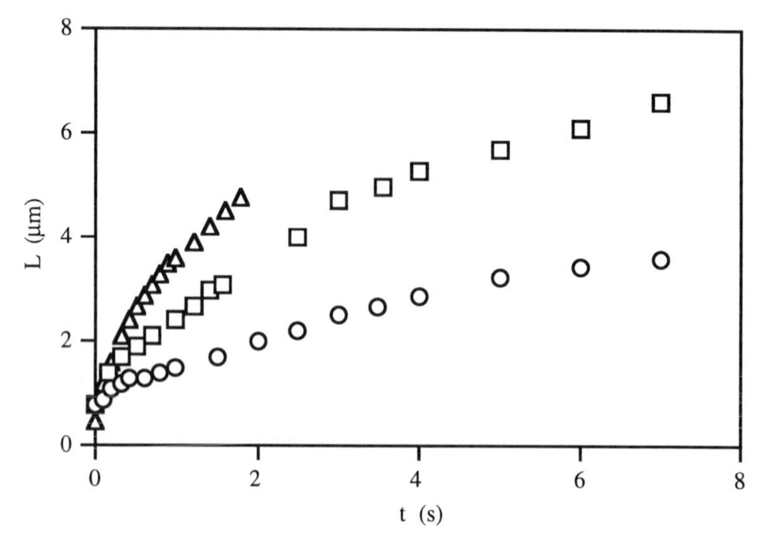

Fig. 7. Dependence of the projection length on time for entry of neutrophils flowing into a 0.35 μm radius pipet at three different suction pressures. The suction pressures are: circles - 600 Pa, squares - 1200 Pa and triangles - 2000 Pa.

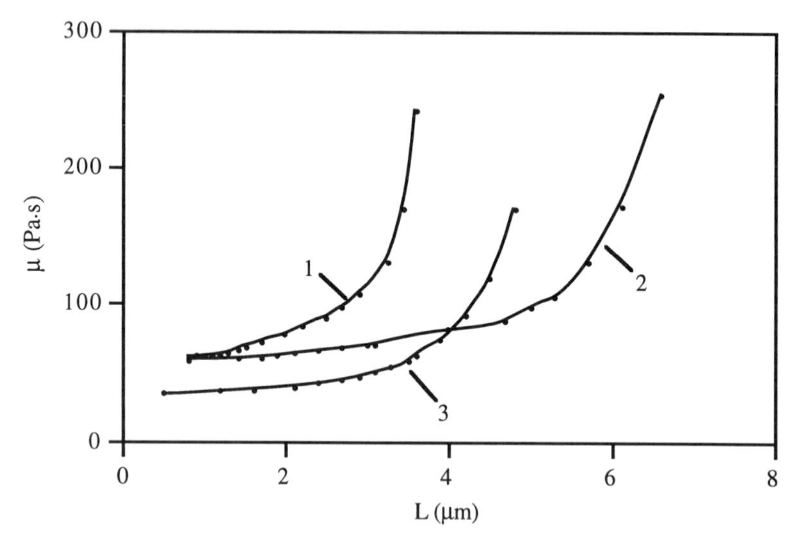

Fig. 8. Apparent viscosities for the data shown in Fig. 7. Curve 1 is for the suction pressure of 600 Pa, curve 2 for 1200 Pa and curve 3 for 2000 Pa.

Elastic Behavior of Stressed Cells

As shown in the preceding section, mechanical stresses (or strains) can induce polymerization (most probably the formation of F-actin) and cell activation. An increase of the amount of polymer filaments in the cell (especially in the stressed region), which can be coupled with an increase of the polymer density and/or crosslinking of the F-actin filaments, can cause the cell to behave as an elastic solid material. However, it will be difficult to derive some quantitative characteristic (such as tensile strength of the stressed material) or to determine the "critical" stress or strain for inducing polymerization, because neither the local stress nor the local strain are measurable during the flow.

The mechanically-induced elasticity in neutrophils is demonstrated with a 1.4 μm radius pipet (Fig. 9). Initially the cell is aspirated with a suction pressure of 180 Pa (which is larger than the critical suction pressure). During flow, the suction pressure is increased to 480 Pa, which leads to an increase in the rate of entry. During this initial deformation the cell behaves as a liquid. When the flow begins to slow down significantly, the pressure difference is reduced to its initial value. Then the projection length inside the pipet decreases as well. The suction pressure is increased again and is kept constant until the flow ceases. After this point, every change in the suction pressure is accompanied by a corresponding change in the projection length. The maximum relative area increase (the ratio of the maximum area of the aspirated cell to the area of the resting, spherical cell) in this experiment is 1.5. This value is less than the value for the maximum relative area change of 2.1 to 2.6 (Evans and Yeung 1989, Ting-Beall et al. 1993). Thus this elastic behavior is not a result of an exessive area dilation of the cell membrane.

It is seen from Fig. 9 that when the suction pressure is increased the rate of entry increases proportionally. The neutrophil behaves as a liquid in its initial entry. A liquid-like neutrophil will flow continously into a pipet at any suction pressure above the critical pressure until its membrane is pulled smooth. This is observed in the experiments with large pipets (Evans and Yeung 1989). In the experiment illustrated in Fig. 9 the neutrophil projection starts to move backwards when the suction pressure is decreased from one value above the critical pressure to another value above the critical pressure. At the same time the area of the cell surface is smaller that the maximum membrane area. This result shows that the cell begins to behave as an elastic body. During the step change in the suction pressure (when all the applied suction pressures are larger than the critical pressure), the projection length changes correspondingly and always reaches a stationary value. The stationary values of the projection lengths for different suction pressures are shown in Fig. 10. It is seen that the dependence is not linear. It is not possible to estimate the amount and/or distribution of the polymer filaments, because the cell is not in thermodynamic equilibrium. The distribution, density and crosslinking of the network providing the elastic resistance in this case may be a function of the applied stress (or the corresponding strain). The elastic resistance most probably is a result of a contraction similar to the one measured in phagocytosis (Evans et al. 1993). Then the result shown in Fig. 10 will also be a function of time.

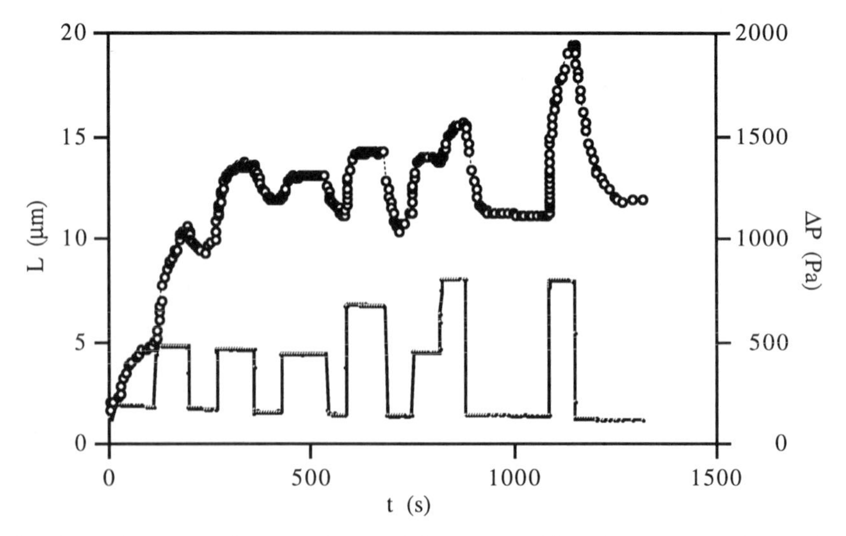

Fig. 9. Projection length verses time (top) for a neutrophil aspirated into a 1.4 µm radius pipet. The aspiration pressure (bottom) is changed in steps. After the initial flow stops, the increase or decrease in projection length follows the changes in the suction pressure. The minimum suction pressure is always above the critical one. The spherical cell has a radius of 4.2 µm.

It is possible to induce large scale polymerization in neutrophils (Zhelev and Hochmuth 1993) using small pipets (with radii in the order of 0.5 µm or less). The stressed region initially is stiffened and is more transparent than the rest of the cell. This is because the formed polymer filaments increase the rigidity of this zone while the filament network excludes the cytoplasm granules and leads to a decrease of the refractive index. The stiffening of the stressed region is followed by mass transfer from the main cell body. After tens of seconds the whole cell is activated and this is monitored as an increase in the apparent tension at the opposite side of the cell. The increase of the apparent surface tension probably results from the same mechanism as the contraction during phagocytosis (Evans et al. 1993) because the measured forces are on the same order of magnitude. A polymerization is also induced by pulling tethers from the cell surface using tiny pipets (with radii less than 0.4 µm). The polymerized region spreads inside the tether and forms a structure similar to a pseudopod. Neutrophils, fixed and stained with phallacidin, show high fluorescence intensity in the stressed region, suggesting that the polymerization is mainly a formation of F-actin filaments.

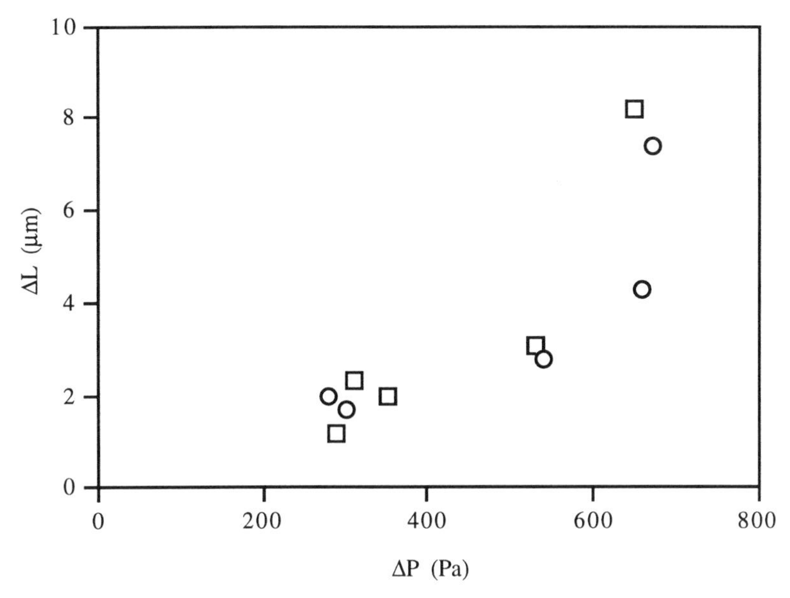

Fig. 10. Difference of the stationary projection length verses the step change in the suction pressure for the data in Fig. 9. Squares are for an increase in the suction pressure and circles are for a decrease.

Summary

The shape and the rheological properties of passive neutrophils are determined by the membrane-cortex complex and the viscous behavior of the cytoplasm. The cortex is always under an isotropic stress characterized by the cortical tension. It has a finite thickness with a resistance to bending that is important for high curvature deformations (with radii of curvatures on the order of 1 μm or less). When a neutrophil is under stress it flows like a liquid body, but during the flow its mechanical properties may change (depending on the applied stress or the corresponding strain). The apparent viscosity may vary from 30 Pa·s to infinity and the cell can behave as a viscoelastic material. These results clearly suggest that the neutrophil (even in its resting state) is far from thermodynamic equilibrium and a polymerization can be induced by mechanical stimulation. It is possible to mechanically induce large scale polymerization in a local region of the cell body that results in cell contraction. The increase in the apparent tension of the contracting cell surface is on the same order of magnitude as the contraction force measured in phagocytosis experiments (Evans et al. 1993).

Acknowledgments

This work is supported by grant 2 RO1 HL23728 from the National Institutes of Health.

References

Bagge, U.; Amundson, B.; Lauritzen, C. White blood cell deformability and Plugging of skeletal muscle capillaries in hemorrhagic shock. Acta Physiol. Scand. 108:159-163; 1980.

Bagge, U.; Skalak, R.; Attefors, R. Granulocyte Rheology. Adv. Microcirc. 7:29-48; 1977.

Bray, D.; Heath, J.; Moss, D. The membrane-associated 'cortex' of animal cells: its structure and mechanical properties. J. Cell Sci. Suppl. 4:71-88; 1986.

Chien, S.; Schmid-Schönbein, G. W.; Sung, K.-L. P.; Schmalzer, E. A.; Skalak, R. Viscoelastic properties of leukocytes. In: White cell mechanics: basic science and clinical aspects. New York: Alan R. Liss, Inc.; 1984: p. 19-51.

Esaguy, N.; Aguas, A. P.; Silva, M. T. High-resolution localization of lactoferrin in human neutrophils: Labeling of secondary granules and cell heterogeneity. J. Leukocyte Biology 46 :51-62; 1989.

Evans, E.; Kukan, B. Passive material behavior of granulocytes based on large deformation and recovery after deformation tests. Blood 64:1028-1035; 1984.

Evans, E. ; Leung, A.; Zhelev, D. Synchrony of cell spreading and contraction force as phagocytes engulf large pathogens. J. Cell Biol. 122:1295-1300; 1993.

Evans, E.; Yeung, A. Apparent viscosity and cortical tension of blood granulocytes determined by micropipet aspiration. Biophys. J. 56:151-160; 1989.

Gittes, F.; Mickey, B.; Nettleton, J.; Howard, J. Flexural rigidity of microtubules and actin filaments measured from thermal fluctuations in shape. J. Cell Biol. 120:923-934; 1993.

Hartwig, J. H.; Kwiatkowski, D. J. Actin-binding proteins. Current Opinion in Cell Biol. 3:87-97; 1991.

Hochmuth, R. M.; Ting-Beall, H. P.; Beaty, B. B.; Needham, D.; Tran-Son-Tay, R. Viscosity of passive human neutrophils undergoing small deformation. Biophys. J. 64:1596-1601; 1993a.

Hochmuth, R. M.; Ting-Beall, H. P.; Zhelev, D. V. The mechanical properties of individual passive neutrophils in vitro. In: Granger, D. N.; Schmid-Schonbain, G. W. eds. Physiology and pathophysiology of leukocyte adhesion. 1993b; in press.

Needham, D.; Hochmuth, R. Rapid flow of passive neutrophils into a 4 μm pipet and measurement of cytoplasmic viscosity. J. Biomech. Eng. 112:269-276; 1990.

Needham, D.; Hochmuth R. M. A sensitive measure of surface stress in the resting neutrophil. Biophys. J. 61 :1664-1670; 1992.

Schmid-Schönbein, G. W.; Sung, K.-L. P.; Tozeren, H.; Skalak, R.; Chien, S. Passive mechanical properties of human leukocytes. Biophys. J. 36:243-256; 1981.

Schmid-Schönbein, G. W.; Shih, Y. Y.; Chien, S. Morphometry of human leukocytes. Blood 56:866-875; 1980.

Sheterline, P.; Rickard J. E. The cortical actin filament network of neutrophil leukocytes during phagocytosis and chemotaxis. In: Hallett, M. B. ed. The neutrophil: cellular biochemistry and physiology. Boca Raton, FL: CRC Press.; 1989: p. 141-165.

Ting-Beall, H. P.; Needham, D.; Hochmuth, R. M. Volume and osmotic properties of human neutrophils. Blood 81:2774-2780; 1993.

Tran-Son-Tay, R.; Needham, D.; Yeung, A.; Hochmuth, R. Time-dependent recovery of passive neutrophils after large deformations. Biophys. J. 60:856-866; 1991.

Zhelev, D. V.; Hochmuth, R. M. Mechanically stimulated polymerization and contraction in human neutrophils. in preparation.

Zhelev, D. V.; Needham, D.; Hochmuth, R. Role of the membrane cortex in neutrophil deformation in small pipets. Biophys. J. 1993; submitted.

2
Viscous Behavior of Leukocytes

R. Tran-Son-Tay, H. P. Ting-Beall, D. V. Zhelev, and R. M. Hochmuth

Introduction

Over the past few years, enormous progress has been made toward understanding the rheological behavior of leukocytes or white blood cells, and more specifically, the neutrophils. White blood cells have been found to be more resistant to flow and to have more complex functions and structures than red blood cells. In addition to the complexities of the internal structure, the white blood cells can activate under various conditions. During activation, pseudopods (solid like protrusions) are formed on the surface of the leukocyte's membrane. In the passive state, leukocytes are spherical in outline with a ruffled membrane and are modeled as deformable spheres. In this chapter, a short overview of the viscous behavior of leukocytes is given with an emphasis on the mechanical modeling of passive neutrophils and lymphocytes, and in particular, the modeling of the recovery of these cells after large deformations. Most of the work on the rheological properties of leukocytes has been focused on the neutrophils and has used the micropipet technique. The main thrust of the studies on white blood cells is to understand their roles and functions in health and disease. Rheological studies of leukocytes other than neutrophils and lymphocytes are sparse (Sung et al., 1988), and will not be discussed in this chapter.

Neutrophils

Neutrophils are the most common granulocytes and the most studied white blood cells. The mechanical properties of neutrophils have been extensively investigated with the micropipet technique. The mechanical description of the neutrophil has been a focus of research in recent years.

Early studies involved small deformations of small portions of neutrophils as they were aspirated into a micropipet (Bagge et al., 1977; Schmid-Schönbein et al., 1981; Sung et al., 1988). The rheological behavior of the neutrophils were modeled by viscoelastic solid models. However, by studying relatively large deformations of neutrophils as they were aspirated into micropipets, Evans

(1984), Evans and Kukan (1984) and Evans and Yeung (1989) found that the neutrophil behaves more like a fluid than a solid. Therefore, by modeling the neutrophil as a spherical shell under a prestressed tension with a Newtonian fluid inside (Evans and Yeung, 1989; Needham and Hochmuth, 1990; Tran-Son-Tay et al., 1991) values between 1300 and 2000 poises were estimated for the viscosity of the cytoplasm. In the studies of Needham and Hochmuth (1990) and Tran-Son-Tay et al. (1991), a non-Newtonian behavior was reported in the deformation and recovery of neutrophils. In the investigation of the recovery of neutrophils after small deformations, Hochmuth et al. (1993) show that indeed the value for the cytoplasm viscosity is strongly dependent on the degree of deformation. This finding is important since it helps to understand the difference between the values reported for the cytoplasmic viscosity of cells undergoing large deformations and those of cells undergoing small deformations and analyzed with a viscoelastic solid model (Schmid-Schönbein et al., 1981; Sung et al., 1988) or a Maxwell liquid model (Dong et al., 1988).

The advantages of the recovery experiments over the experiments involving the flow of a cell into a pipet are that they are easier to perform and analyze, and that the potential adhesion of the cell to the pipet is not an issue. On the other hand, from the recovery experiments only, it is not possible to separate the effects of the cytoplasmic viscosity from the surface tension of the membrane. It is clear that by performing both experiments on the same cell, one will obtain information necessary for determining the constitutive properties of the cell. The present chapter focuses mainly on the recovery experiments since the pipet experiments are presented in another chapter (Zhelev and Hochmuth, 1994).

In the recovery studies, experiments are performed in which a passive human neutrophil is deformed into an elongated "sausage" shape by aspirating it into a small glass pipet. When expelled from the pipet the neutrophil recovers its natural spherical shape in about 1 min (Figure 1). This recovery process is analyzed according to a Newtonian, liquid-drop model in which a variational method is used to simultaneously solve the hydrodynamic equations for low Reynolds-number flow and the equations for membrane equilibrium with a constant membrane tension (Tran-Son-Tay et al., 1991). A typical recovery is shown in Figure 2, where the computed shapes are plotted as a function of a dimensionless time. For computational purposes, this time is scaled by the length of the major semi-axis of the initial deformed sausage shape, R_i ($R_i = D_i/2$). The index of deformation, D_i/D_o, for this particular shell is 2. The theory generates a series of recovery profiles starting from the deformed sausage shape and ending up with a sphere. The corresponding dimensionless length and width are shown below the recovery profiles as a function of the same dimensionless time.

Data reduction involves measuring the horizontal (D) and vertical (W) dimensions of the cell as it recovers from the deformed sausage shape to the resting spherical shape of diameter Do. Plots of the progress of shape recovery (D and W) versus time are shown in Figure 3 for a cell that is held in the deformed sausage shape for 8 s. The theoretical model fits the experimental data for a ratio of membrane cortical tension to cytoplasmic viscosity of approximately $1.7 \pm 0.5 \times 10^{-5}$ cm/s (37 cells). However, when the cell is held in the pipet for only a short time period of 5 s or less, and then expelled, the cell

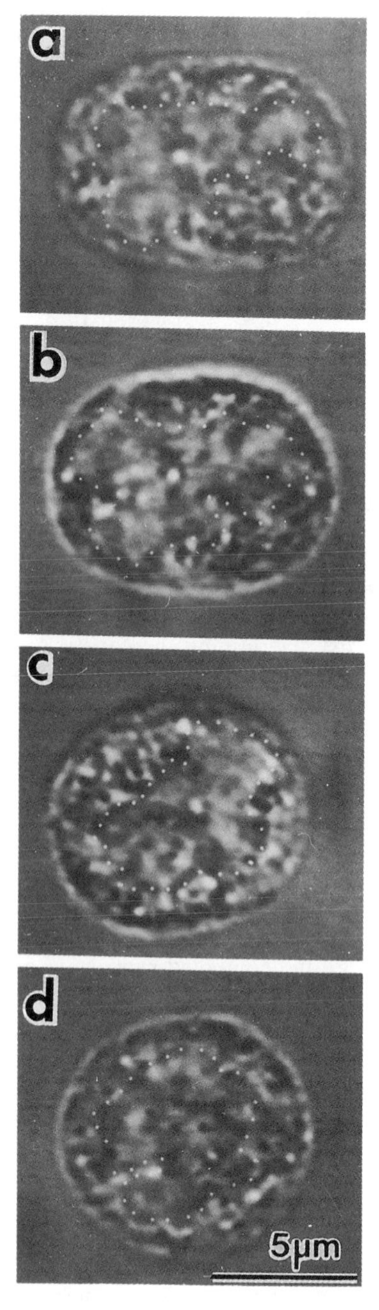

Figure 1. Recovery of a Neutrophil. (Taken from Hochmuth et al., 1994). Videomicrographs a, b, c, and d show the shape of an initially elongated neutrophil after a recovery time of 3s, 10s, 34s, and 57s, respectively.

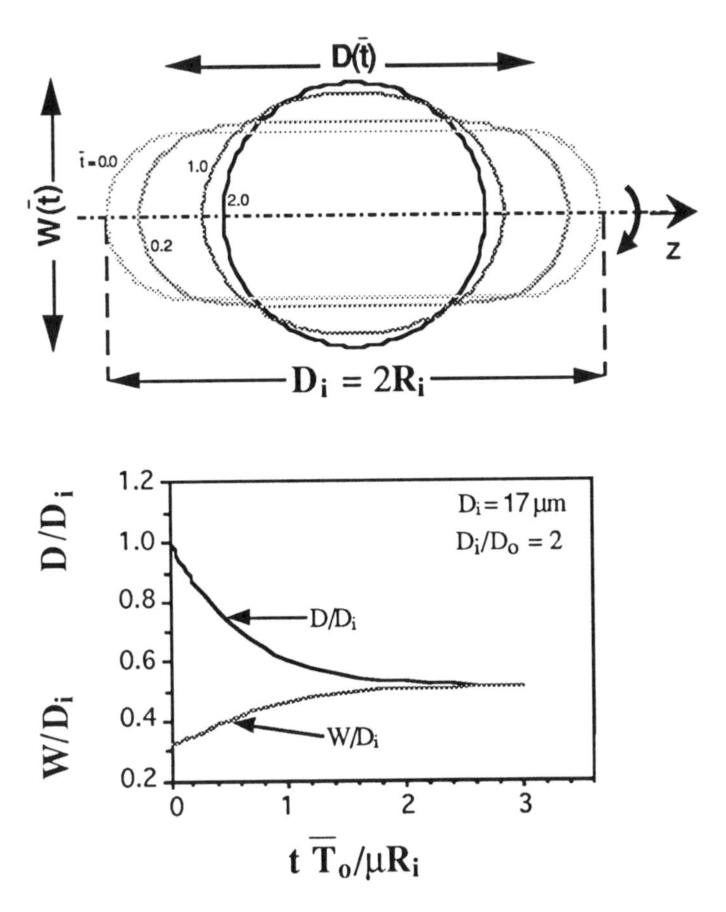

Figure 2. Time-Dependent Recovery Shapes and Dimensions. The above figure shows a series of recovery profiles starting from the deformed sausage shape and ending with a sphere generated from the theory. The dimensionless plot of the corresponding length and width of the cell is shown below the recovery profiles

undergoes an initial, rapid elastic rebound suggesting that the cell behaves in this instance as a viscoelastic liquid rather than a Newtonian liquid with constant cortical tension (Tran-Son-Tay et al., 1991).

Dong et al. (1988) have developed a theory for the recovery of a cortical shell containing a Maxwell fluid, but their theory is only valid for very small deformations. They have also presented a finite element method for the large deformation of a leukocyte aspirated into a micropipet, but the problem of recovery after large deformation has not been discussed.

Even though it is now recognized that the neutrophil behaves as a fluid, it is still not clear if the best model for this type of cell is a viscoelastic fluid model based on separate layers or one based on two intermixed Newtonian fluids. It is important to note that, to a first order of approximation, the neutrophil's nucleus is not a major contributor to the mechanical behavior of the cell for the types of deformation studied, and consequently has been neglected in the theoretical analyses.

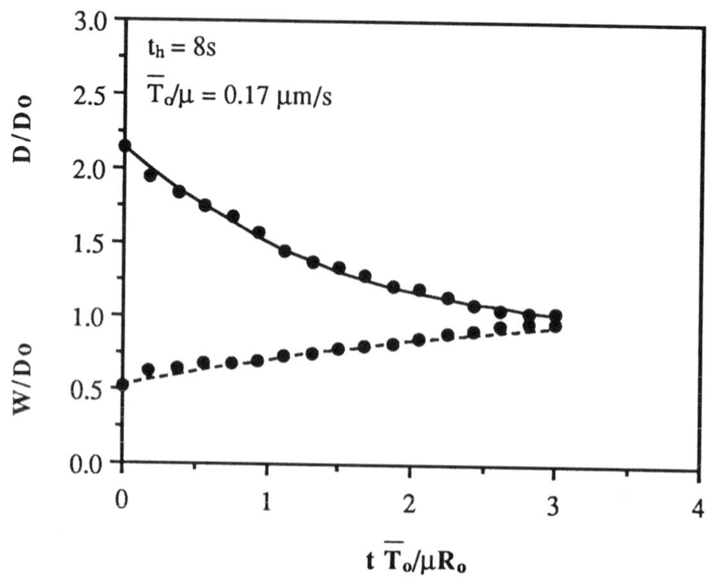

Figure 3. Recovery of a Neutrophil. This particular cell has been held inside the pipet for 8s before being released from the pipet and allowed to recover. Neutrophils that are being held inside the pipet for less than 5s do not recover as Newtonian liquid drops.

Lymphocytes

Recently, some of the studies of leukocytes have been geared towards understanding the rheological behavior of lymphocytes (Zhelev et al., 1994). The lymphocyte has a relatively larger nuclear volume than the neutrophil. Its nucleus occupies about 44% of the cell volume as compared to 21% for the neutrophil (Schmidt-Schönbein et al., 1980). When a neutrophil is aspirated into a pipet, it flows like a liquid once the suction pressure exceeds the critical suction pressure (Evans and Kukan, 1984; Evans and Yeung, 1989). (The pressure required to form a hemispherical projection length inside the pipet is called a "critical" suction pressure.) On the other hand, when a lymphocyte is aspirated into a pipet, it does not flow like a liquid. Its projection length is a function of the applied pressure (see Figure 4), even for a suction pressure above the critical suction pressure. Thus, the lymphocyte behaves as an elastic body. Zhelev et al. (1994) have found that the lymphocyte plasma membrane is under an almost constant tension of 0.035 mN/m, just as is the neutrophil membrane. The measured elastic behavior is due to the elasticity of the nucleus, which has an apparent area dilation elastic modulus on the order of 2 mN/m (Zhelev et al., 1994). This value is much smaller than the area expansion modulus for bilayer lipid membranes (on the order of 140 mN/m, Kwok and Evans, 1981) suggesting that the apparent elasticity in this case is a structural parameter. The nucleus of the neutrophil may affect the cell's apparent viscosity (Zhelev and Hochmuth, 1994), but its elasticity (if it exists) is not measurable (Evans and Yeung, 1989).

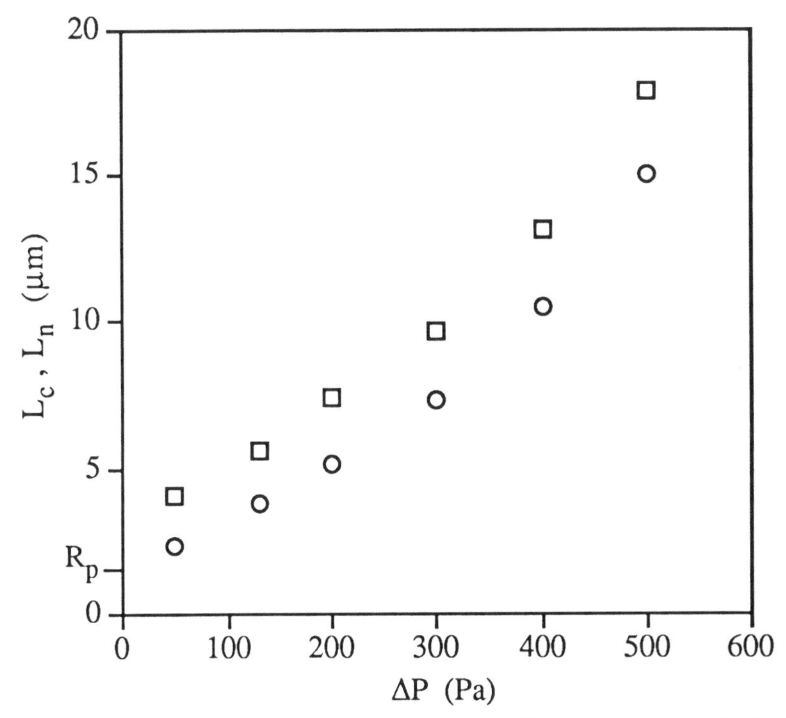

Figure 4. Projection Cell Length Versus the Applied Suction Pressure for Deforming a Lymphocyte into a Pipet. The squares represent the distance of the plasma membrane (L_c) from the pipet orifice and the circles give the distance of the nucleus (L_n) from the orifice. R_p is the pipet radius.

Videomicrographs of the time-dependent recovery of a lymphocyte are shown in Figure 5. The relatively large size of the nucleus can be seen. We are presently trying different means to enhance the quality and definition of the videomicrographs. This will allow us to better follow the recovery of the nucleus and to characterize its rheological properties. A preliminary study of the recovery has been performed and analyzed according to the method of Tran-Son-Tay et al., 1991 (Figure 6). It is found that one population of the lymphocytes

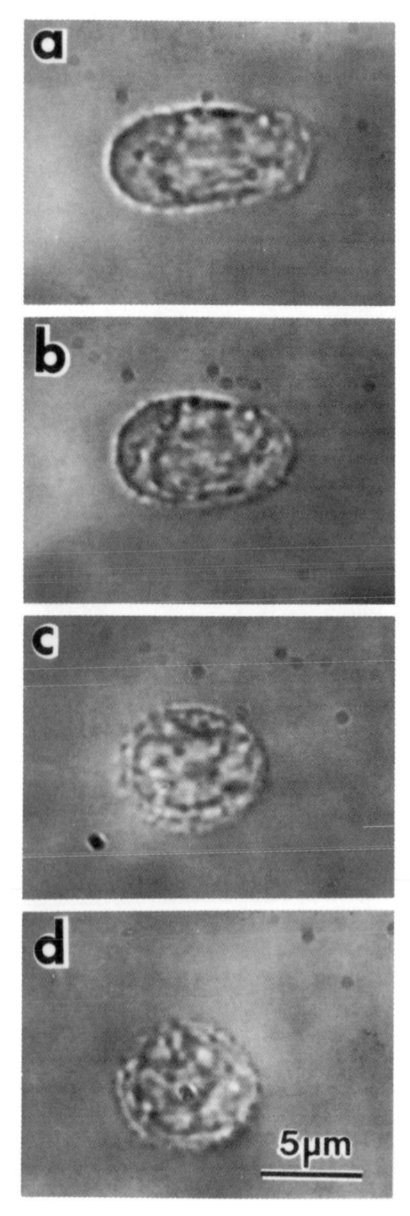

Figure 5. Recovery of a Lymphocyte. Videomicrographs a, b, c, and d show the shape of an initially elongated lymphocyte after a recovery time of 3s, 10s, 33s, and 61s, respectively.

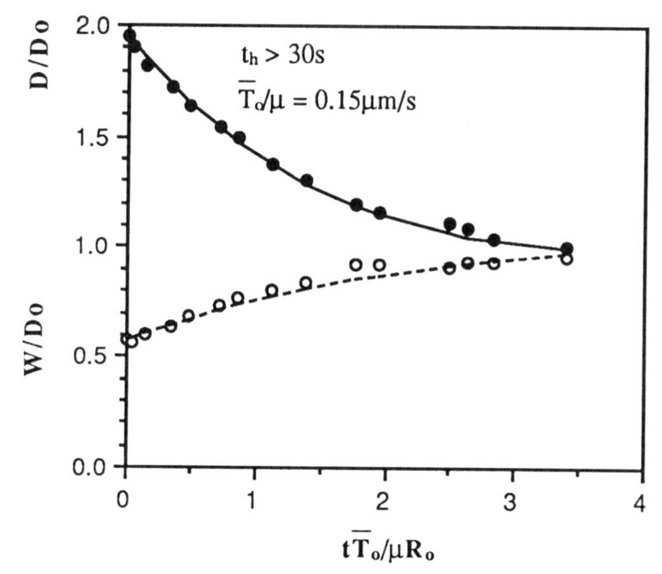

Figure 6. Recovery of a Lymphocyte. This particular cell has been held inside the pipet for more than 30s before being release from the pipet and allowed to recover. This lymphocyte recovers like a Newtonian liquid drop.

recovers as a Newtonian drop whereas the other does not. Among the population that recovers as a Newtonian drop, two groups are noted (Figure 7). The first one has an average value for the ratio of the cortical tension to cytoplasmic viscosity, $\overline{T}_o/\mu$, of $1.3 \pm 0.3 \times 10^{-5}$ cm/s (15 cells), and the second has an average value of $2.4 \pm 0.2 \times 10^{-5}$ cm/s (3 cells). The first ratio is about 25% smaller than the average value for neutrophils ($1.7 \pm 0.5 \times 10^{-5}$ cm/s, 40 cells). That is, for an identical value for the cortical tension, most of the lymphocytes that do recover as Newtonian liquid drops are about 25% more viscous than the neutrophils. Of the group of lymphocytes that recovers as Newtonian drops, only three have a value for the ratio of the cortical tension to cytoplasmic viscosity larger than the average value for the neutrophils. That is, these cells are about 42% less viscous than the average neutrophil. To complicate the issue, it is found that about half of the cells studied do not recover as Newtonian drops. The values for the apparent viscosity of these 16 cells calculated using the Newtonian model are shown for comparison in Figure 8. In all of the recovery experiments performed with lymphocytes, the time the cells are held inside the pipet before being expelled, is greater than 5 s. Again for comparison, the average value for the ratio of the cortical tension to cytoplasmic viscosity is $2.6 \pm 0.8 \times 10^{-5}$ cm/s (37 cells) for neutrophils that do not recover as Newtonian drops (Tran-Son-Tay et al., 1991). These neutrophils are held inside

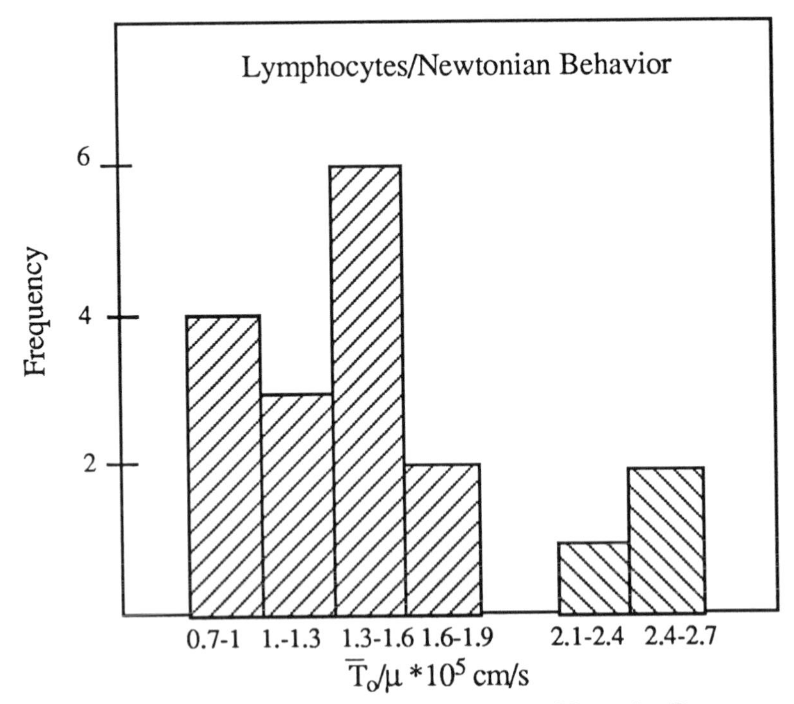

Figure 7. Distribution of the Lymphocytes that Behave as Newtonian Drops

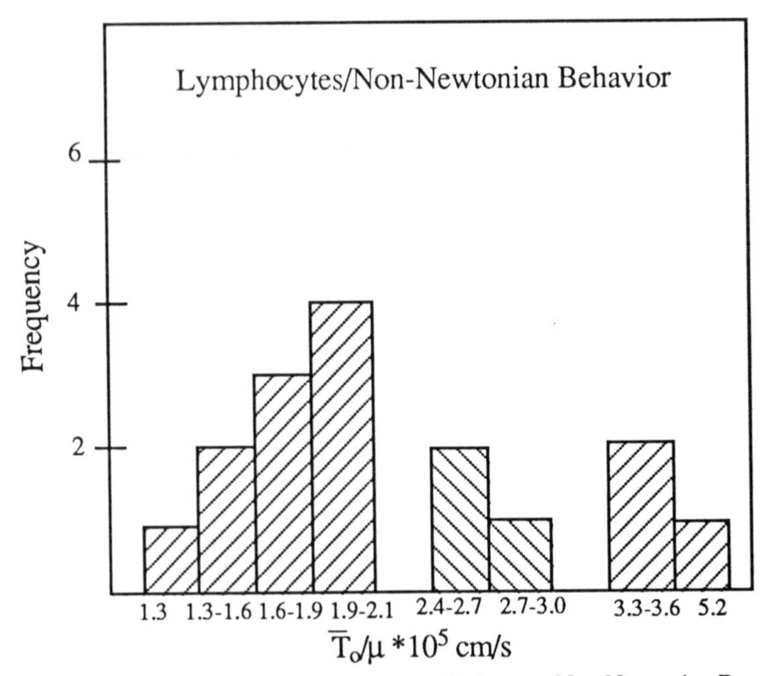

Figure 8. Distribution of the Lymphocytes that Behave as Non-Newtonian Drops

the pipet for less than 5 s. Work is in progress to develop more appropriate models to analyze and to help understanding the non-Newtonian behaviors of the neutrophils and lymphocytes.

Conclusions

It is clear that the lymphocytes, although they do not activate as neutrophils do, are also very complex cells. It is found, from the preliminary recovery experiments reported here, that the lymphocytes possess two different kinds of mechanical behavior. It is possible that these two mechanical responses correspond to the two sub-populations of lymphocytes; namely, the T and B lymphocytes. At the present time, we are unable to differentiate these two cell types in our experiments. The different values for the ratio of the cortical tension to cytoplasmic viscosity seen within a sub-population may just be a consequence of the natural variation that exists between the same type of cell, or a result of the influence of the relative size of the cell's nucleus, or a combination of both. Additional studies are definitively required in order to confirm or deny these speculations.

To conclude, the observed non-Newtonian behavior is not too surprising in light of the observations of Zhelev et al. (1994). An analysis of the recovery of lymphocytes with their relatively large nucleus modeled as a cortical shell containing a relatively large, spherical, viscoelastic solid suspended in a viscous Newtonian cytoplasm is underway and will be presented in a separate article (Tran-Son-Tay et al., 1994).

Acknowledgements

This work was supported by the National Institutes of Health through grants 2R01-HL-23728 and 7R01-HL-49060, and the National Science Foundation through grant BCS-9106452.

References

Bagge, U., Skalak, R., and Attefors, R. (1977). Granulocyte Rheology: Experimental Studies in an *in Vitro* Microflow System. Adv. Microcirc., vol. 7, pp. 29-48.

Dong, C., Skalak, R., Sung, K.-L.P., Schmid-Schönbein, G. W., and Chien, S. (1988). Passive Deformation Analysis of Human Leukocytes. J. Biomech. Eng., vol. 110, pp. 27-36.

Evans, E. A. (1984). Structural Model for Passive Granulocyte Behaviour Based on Mechanical Deformation and Recovery After Deformation Tests. *In* White Cell Mechanics: Basic Science and Clinical Aspects (Eds. Meiselman, H. J., Lichtman, M. A., and LaCelle, P. L.), Alan Liss, Inc., New York, pp. 53-74.

Evans, E. A , and Kukan, B. (1984). Passive Behavior of Granulocytes Based on Large Deformation and Recovery After Deformation Tests. Blood, vol. 64, pp. 1028-1035.

Evans, E. A., and Yeung, A. (1989). Apparent Viscosity and Cortical Tension of Blood Granulocytes Determined by Micropipet Aspiration. Biophys. J., vol. 56, pp. 151-160.

Hochmuth, R. M., Ting-Beall, H. P., Beaty, B. B., Needham, D., and Tran-Son-Tay, R. (1993). The Viscosity of Passive Neutrophils Undergoing Small Deformations. Biophys. J., vol. 64, pp. 1596-1601.

Hochmuth, R. M., Ting-Beall, H. P., and Zhelev, D. V. (1994). The Mechanical Properties of Individual Passive Neutrophils *in Vitro*. In: Physiology and Pathophysiology of Leukocyte Adhesion. Chapter 3. D.N. Granger and Schmid-Schönbein, G. W., Eds. In press.

Kwok, R., and Evans, E. (1981). Thermoelasticity of Large Lecithin Bilayer Membranes. Biophys. J. , vol. 35, pp. 637-652.

Needham, D., and Hochmuth, R. M. (1990). Rapid Flow of Passive Neutrophils into a 4 μm Pipet and Measurement of Cytoplasmic Viscosity. J. Biomech. Eng., vol. 112, pp. 269-276.

Schmid-Schönbein, G. W., Sung, K.-L.P., Tözeren, H., Skalak, R., and Chien, S. (1981) Passive Mechanical Properties of Human Leukocytes. Biophys. J., vol. 36, pp. 243-256.

Schmid-Schönbein, G. W., Shih, Y. Y, and Chien, S. (1980). Morphometry of Human Leukocytes. Blood, vol. 56, pp. 866-875.

Sung, K.-L.P., Dong, C., Schmid-Schönbein, G. W., Chien, S., Skalak, R. (1988). Leukocyte Relaxation Properties. Biophys. J., vol. 54, pp. 331-336.

Tran-Son-Tay, R., Needham, D., Yeung, A., and Hochmuth, R. M. (1991). Time Dependent Recovery of Passive Neutrophils After Large Deformation. Biophys. J., vol. 60, pp. 856-866.

Tran-Son-Tay, R., Zhelev, D. V., Ting-Beall, H. P., and Hochmuth, R. M. (1994). Mechanical Modeling of the Recovery of Lymphocytes. In preparation.

Zhelev, D. V., Ting-Beall, H.P., and Hochmuth, R. (1994). Mechanical Properties of Human Lymphocytes. In preparation.

Zhelev, D. V., and Hochmuth, R. M. (1994). Human Neutrophils Under Mechanical Stress. Published in this book.

3

Shear Rate-Dependence of Leukocyte Cytoplasmic Viscosity

R. E. Waugh and M. A. Tsai

Introduction

Although leukocytes are few in number compared to red blood cells, they have a large volume and low deformability compared to erythrocytes, and so their influence on blood flow and oxygen delivery is disproportionate to their numbers. Leukocyte mechanical deformability has been shown to play an important role in cell retention in the capillaries during ischemia or prior to tissue injury (Worthen *et al.*, 1988; Erzurum *et al.*, 1991; Harris and Skalak, 1993). In some cases of human leukemia, the number of leukocytes is elevated and a large fraction of immature leukocytes is found in the blood. Clinically, these cells are believed to be responsible for impairment of circulation to the eye, the central nervous system or the lungs, thus causing deleterious effects to the organs involved (Lichtman, 1984). Thus, it is of interest to have a more accurate picture of white cell deformability to more thoroughly understand their behavior and potential influence in the microvasculature.

The earliest mechanical model of white cell behavior was based on observations of cells undergoing small deformations as they were aspirated into a micropipette (Schmid-Schönbein *et al.*, 1981). This model described the cell as a viscoelastic solid. Subsequent observations of cells undergoing large deformations as they entered pipettes revealed that the fundamental behavior of the white cell is that of a highly viscous fluid droplet (Evans and Kukan, 1984). Numerical analysis of the entry of a whole cell into a cylindrical pipette was performed treating the cell as a liquid drop with uniform cortical tension and a

Newtonian viscous core. Subsequently, the analysis was extended to include different contributions to viscous dissipation during cell entry from cortical and more central regions of the cell (Evans and Yeung, 1989). The results of the analysis were found to be consistent with observations of whole cell aspiration into micropipettes, and it was concluded that the cortical region did not contribute disproportionately to the resistance of the cell to deformation for pipettes greater than 3.5 μm in diameter. More recent measurements of cells entering micropipettes performed in a different laboratory confirmed this general picture (Needham and Hochmuth, 1990). However, some discrepancies between theory and experiment began to emerge. The detailed prediction of the time course of cell entry was not in exact agreement with observation, and the values of the apparent viscosity obtained by the two laboratories were slightly different.

In this report, we summarize recent work performed in our laboratory to investigate the rheological properties of leukocytes. In the first section, the rheological properties of human neutrophils are examined with particular focus on the dependence of the cytoplasmic viscosity on shear rate. In the second section, an investigation of the structural basis for the cellular properties is described. In particular, the role of the actin-based cytoskeleton was assessed by treating the cells with cytochalasin B, a drug known to cause disruption of actin filaments. In the last section, preliminary observations of the properties of a human promyelocytic leukemic cell line (HL60) are presented. These data represent one of the first detailed characterizations of the properties of leukemic cells and establish a basis for using HL60 cells as a model system for investigating the properties of maturing leukocytes. These three sections are prefaced by a brief description of our experimental approach.

Approach for Testing Cell Properties

General Procedures. The experimental protocols are described in detail in a previous report (Tsai *et al.*, 1993). Briefly, the cell suspension was placed in a chamber formed by cementing two cover glasses to a plastic spacer with a U-shaped cavity 10 $mm \times 20$ $mm \times 1$ mm thick. The chamber was placed on the microscope stage and a micropipette was inserted through the open side of the "U". The tip of the pipette was positioned in the field of view and a single cell was aspirated at a fixed pressure (Fig. 1). The aspiration pressure was controlled by displacing a hydrostatic reservoir and was monitored with a pressure transducer. The applied pressures ranged from 98 Pa to 882 Pa and could be determined to an accuracy of ~ 1 Pa. The aspiration process was observed via a television camera and recorded on video tape. The time course and total time for cell entry were determined from the recordings with the use of a digital image processor.

Cortical Tension Measurements. The neutrophil cortical tension was determined by measuring the minimum (threshold) pressure Δp_t required to

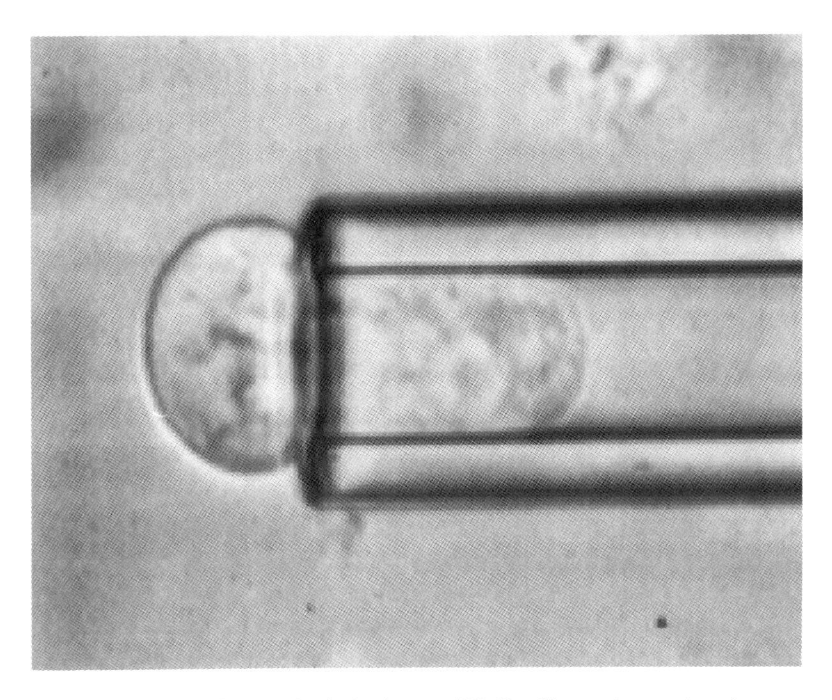

Fig. 1. Photomicrograph of a leukocyte (HL60 cell) entering a micropipette.

form a hemispherical projection in the micropipette. From simple balance of forces, it can be shown that the cortical tension T is related to the threshold pressure by:

$$T = \frac{\Delta p_t \bullet R_p R_c}{2(R_c - R_p)} \tag{1}$$

where R_p is the pipette radius and R_c is the radius of the spherical portion of the cell outside the pipette. To determine the threshold pressure, the aspiration pressure was maintained at a small value for 2 min to allow the cell to fully respond. The pressure was then increased in steps of 2.45 Pa (0.025 cm H_2O) and maintained at each step for ~ 2 min, until a stationary hemispherical bulge was formed. The pressure at this point was taken as the threshold pressure, Δp_t. The pressure was further increased by one or two steps of 2.45 Pa to ensure that the cell did not adhere to the pipette wall and that it would keep flowing into the pipette when the threshold pressure was exceeded.

Analysis of Cell Entry. Interpretation of the experimental observations was based on a numerical analysis of the entry of a highly viscous, spherical liquid drop into a smaller cylinder (Tsai *et al.*, 1993). In the analysis, the interior of the cell was treated as a fluid with a uniform viscosity, and the tension at the cell surface was assumed to be uniform and constant. At each instant in time,

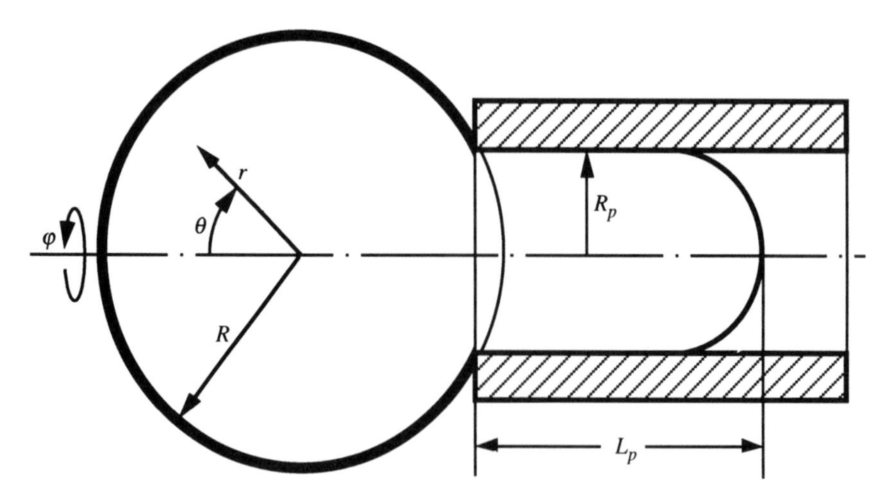

Fig. 2. Schematic diagram of a cell entering a pipette showing the cell dimensions and coordinate system.

the incremental displacement of the cell was calculated, as was the mean shear rate averaged over the cell volume. The shear rate calculation makes use of the quantity ϕ:

$$\phi = \frac{1}{2}\varepsilon_{ij}\varepsilon_{ij} \tag{2}$$

where ε_{ij} are the components of the deformation rate tensor (repeated indices are summed). At a given point and a given instant in time the shear rate is defined as:

$$\gamma = \sqrt{\phi} \tag{3}$$

The volume-averaged instantaneous mean shear rate is defined as:

$$\gamma_a = \left(\frac{3}{2} \int_0^{R(t)} \int_0^{\pi} \frac{r^2}{R^3} \phi \; sin\theta \; d\theta \; dr \right)^{1/2} \tag{4}$$

where r, θ, and φ are spherical coordinates and R is the instantaneous radius of the spherical portion of the cell outside the pipette (Fig. 2). The "total" mean shear rate includes integration over the time course of entry:

$$\gamma_m = \left(\frac{3}{2} \frac{1}{t_e} \int_0^{t_e} \int_0^{R(t)} \int_0^{\pi} \frac{r^2}{R^3} \phi \; sin\theta \; d\theta \; dr \; dt \right)^{1/2} \tag{5}$$

where t_e is the total time required for cell entry. Two schemes were used to obtain predictions for the time course of cell entry into the pipette. In the simplest case, the cytoplasmic viscosity (μ) was assumed to be constant throughout the aspiration process. In the second case, the viscosity was assumed to be spatially uniform, but to vary in time with a power-law dependence on the instantaneous mean shear rate:

$$\mu = \mu_o \left(\frac{\gamma_a}{\gamma_o} \right)^{-b} \tag{6}$$

where μ_o is the viscosity at the reference shear rate, $\gamma_o = \Delta p_h / 4\mu_o$, and Δp_h is the aspiration pressure. The values of the viscosity reported in the following sections are based on the simplified analysis in which the viscosity was assumed to be constant during the aspiration process. These values of "apparent" viscosity can be obtained simply from measurements of the initial cell radius R_o, the pipette radius R_p, and the total time required for cell entry t_e:

$$\mu = \frac{\Delta p_h \cdot t_e}{4\left[a_0 + a_1\left(\frac{R_o}{R_p}\right) + a_2\left(\frac{R_o}{R_p}\right)^2 + a_3\left(\frac{R_o}{R_p}\right)^3 \right]\sqrt{\left(\frac{R_o}{R_p}\right)^2 - 1}} \tag{7}$$

The coefficients a_i were determined from the numerical integration. They are functions of the ratio of the cortical tension T to the aspiration pressure Δp_h:

$$\beta = \frac{2T}{R_p \cdot \Delta p_h} \tag{8}$$

and can be approximated as:

$$a_0 = -1.3079 - 0.01998\beta + 0.1387\beta^2 - 0.87898\beta^3 \tag{9}$$

$$a_1 = 2.0756 - 0.21461\beta + 0.065414\beta^2 + 2.0469\beta^3 \tag{10}$$

$$a_2 = -0.37513 - 0.10393\beta - 0.4992\beta^2 - 1.6295\beta^3 \tag{11}$$

$$a_3 = 0.38012 + 0.29926\beta + 0.28995\beta^2 + 0.45455\beta^3 \tag{12}$$

Neutrophils: Shear Rate-Dependence

These studies were motivated by the observation that the apparent neutrophil viscosity was smaller when cells were aspirated rapidly into the pipette at high pressure than when aspirated slowly at lower pressure. To evaluate this rate-dependent effect, neutrophils were first isolated from whole blood by density fractionation on a Ficoll-Hypaque density gradient (Mono-Poly Resolving

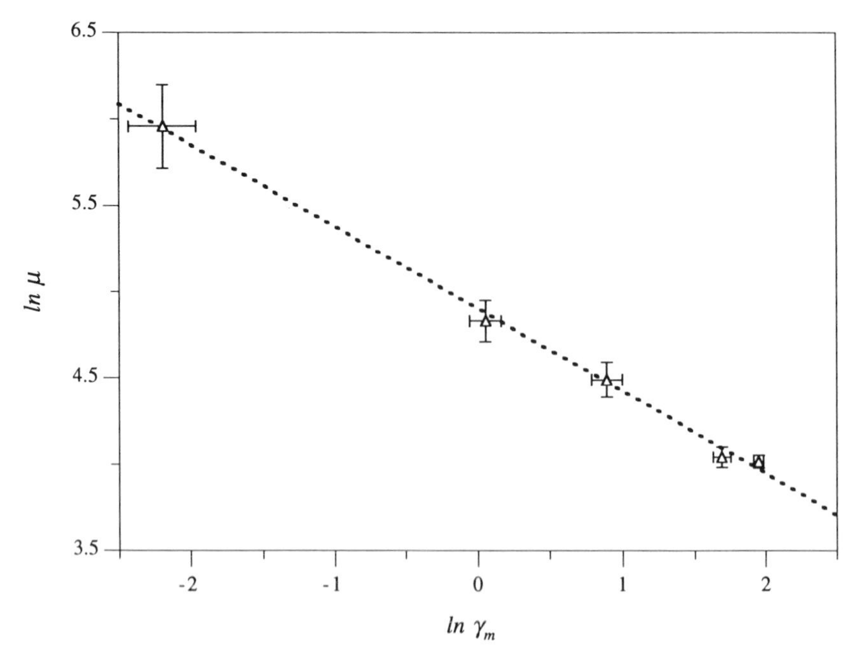

Fig. 3. Results of a typical experiment testing the shear rate dependence of the neutrophil cytoplasm. The logarithm of the apparent viscosity varies linearly with the log of the mean shear rate γ_m during cell entry. (Replotted from Tsai *et al.*, 1993.)

Medium, Flow Laboratories, Inc., McLean, VA). For mechanical measurements, cells were suspended at low concentration in HEPES-buffered saline (135.8 *mM* NaCl, 4.8 *mM* KCl, 11.3 *mM* HEPES (N-2-Hydroxyethyl-piperazine-N'-2-ethanesulfonic acid) and 9.3 *mM* HEPES sodium salt. Glucose (1 *g/l*), bovine serum albumin (2 *g/l*), streptomycin sulfate (105 units/*ml*) and penicillin-G potassium salt (95.4 units/*ml*) were also added to the buffer. The osmolality and pH were adjusted to 290±5 *mOsm/kg* and 7.40±0.05 respectively. All chemicals were purchased from Sigma Chemical Company (St. Louis, MO). The buffer was degassed and filtered through a 0.22 *μm* Millex-GS filter (Millipore Corporation, Bedford, MA). Autologous plasma 10% (v/v) was mixed into the cell suspension buffer to reduce cell adhesion to the pipette wall. Cells were then tested to measure cortical-membrane tension and cytoplasmic viscosity by the methods outlined above.

Neutrophils from a single individual on a single day were aspirated into the same pipette at different aspiration pressures. In a total of seven experiments, more than 800 cells were tested at aspiration pressures ranging from 98 to 882 *Pa*. It was found that the apparent viscosity was a strong function of the aspiration pressure. When the apparent viscosity was plotted against the total mean shear rate (γ_m) in logarithmic scale, a power-law dependence was revealed (Fig. 3). This dependence takes the form:

$$\mu = \mu_c \left(\frac{\gamma_m}{\gamma_c} \right)^{-b} \tag{13}$$

where μ_c is the characteristic viscosity at the characteristic shear rate γ_c, and b is a material constant. Choosing a convenient value for γ_c of 1 s^{-1}, $\mu_c = 130 \pm 23$ $Pa{\cdot}s$ and $b = 0.52 \pm 0.09$ for normal neutrophils.

These studies provided the first clear evidence that the neutrophil cytoplasmic viscosity depends on rate of deformation (Tsai *et al.*, 1993). This shear rate-dependence appears to account for differences in values of the viscosity reported by different laboratories. Lower values are reported in studies in which aspiration pressures and the corresponding rate of cell entry are high. The fact that the cytoplasmic viscosity is shear rate dependent points to the need for a more detailed analysis in which the viscosity varies with time and position within the cell according to the local instantaneous shear rate. Such an analysis has not yet been performed, although a preliminary analysis has been completed in which the viscosity is assumed to be spatially uniform but to vary in time according to the instantaneous shear rate averaged over the spherical cell volume. The prediction for the time course of cell entry that comes from this analysis shows good agreement with experimental observation, especially during the final stages of cell entry (Fig. 4). However the rapid initial entry of the cell into the pipette is not predicted from this approach. Thus, although some aspects of the cell behavior are accounted for by the shear rate dependence of the viscosity, the need for further refinements in the constitutive modeling of the cell is apparent.

The shear-thinning behavior of the cytosol is consistent with our knowledge of the structural character of the cell interior. The cytosol contains numerous small organelles (particles) and consists of a suspension of filamentous protein structures collectively known as the cytoskeleton. These filamentous structures fall into three main categories: actin filaments, microtubules and so-called intermediate filaments. While all of these are present in neutrophils, it has long been thought that the actin-based portion of cytoskeleton plays the dominant role in determining the rheological characteristics of neutrophils. Efforts to investigate this hypothesis and more clearly identify the role that actin filaments play in determining cell rheological behavior are described in the next section.

Actin Filaments and Cell Rheology

To assess the contribution of actin filaments to the cellular properties, cells were treated with cytochalasin B (CTB) to disrupt the filamentous actin network that forms the major structural component of the cortical cytoskeleton. Previously, Wallace *et al.* (1987) had observed that treatment of neutrophils with cytochalasin B reduced the amount of filamentous (F-) actin that remained associated with the cytoskeleton after treatment with detergent (so-called "cytoskeletal actin"). Treatment with 3.0 μM cytochalasin resulted in a 10%

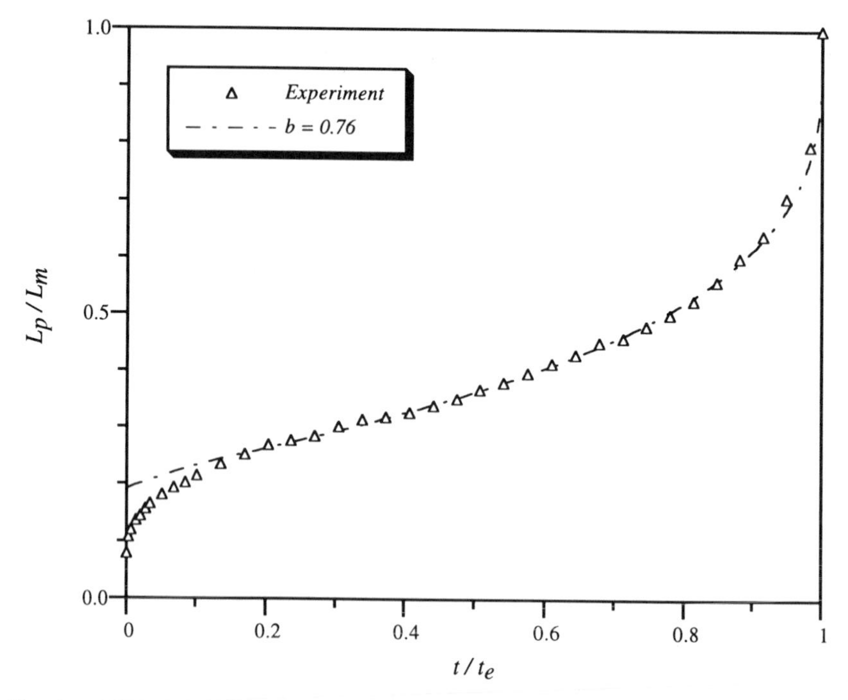

Fig. 4. Time course of cell entry. Time t and projection length L_p are normalized by t_e, the total time for cell entry, and L_m, the projection length when the cell is completely aspirated into the pipette. Dashed line shows the prediction from numerical calculations in which the instantaneous cytoplasmic viscosity was uniform within the cell, but which varied in power-law dependence on the instantaneous mean shear rate, $\dot{\gamma}_a$. Points represent measurements taken on a normal neutrophil. An arbitrary initial projection length was added to the numerical prediction to compensate for the initial rapid entry of the cell into the pipette. The initial rapid entry was not predicted by the numerical analysis. Values for the parameters to obtain this fit were: $\mu_c = 55$ *Pa*·s, and $b = 0.76$. (Data replotted from Tsai *et al.*, 1993.)

reduction in cytoskeletal actin, and treatment with 30 μM cytochalasin resulted in a 30% reduction. Interestingly, the total F-actin content in the cell prior to detergent treatment did not change appreciably as a result of cytochalasin treatment. Following the procedures of those investigators, we tested the effects of cytochalasin treatment on the viscosity and cortical tension of the neutrophil. Two concentrations (3 μM and 30 μM) were used. Cells were placed at room temperature in HEPES-buffered saline containing the appropriate concentration of CTB. Measurements were begun 5 min after the cells were exposed to the reagent and were completed within one hour.

Both the cortical tension of the cells and the cytoplasmic viscosity were reduced by cytochalasin treatment in a dose-dependent fashion. Interestingly, the dependence of the apparent viscosity on shear rate was also reduced. These

Table I. Rheological Properties of Cytochalasin-Treated Neutrophils[a]

Cell Group	*T (mN/m)*	μ_c *(Pa·s)*	*b*
Control	0.0268±0.0012[b]	130±23[c]	0.52±0.09[c]
3 *μM* CTB	0.0217±0.0007[b]	128±21[c]	0.44±0.09[c]
30 *μM* CTB	0.0135±0.0007[b]	54±15[c]	0.26±0.05[c]

a. Data taken from Tsai *et al.*, 1994*a*.
b. Mean ± SD, n = 16.
c. Mean ± SD, n = 120-130

results are summarized in Table I, the data for which were taken from Tsai *et al.* (1994*a*).

These results provide the first direct evidence of the importance of actin structures in regulating the passive mechanical properties of neutrophils. The cortical tension appears particularly sensitive to the organization of actin filaments, as under conditions in which the detergent insoluble actin is reduced by only 10%, the cortical tension was reduced by 20%. Further reductions in cytoskeleton-associated actin produced even more dramatic reductions in the tension. Thus, as expected, actin filaments play a major role in determining cortical tension in neutrophils. The reduction in cytoplasmic viscosity and the shear rate-dependence of the viscosity are also consistent with the known effects of cytochalasins on actin filament structure. *In vitro*, CTB treatment acts to shorten the length of actin filaments. Reduction of filament length in polymeric suspensions is known to reduce both the suspension viscosity and its shear rate-dependence. Thus, changes in actin filament structure can have dramatic effects on the rheological properties of the cell.

Properties of HL60 Cells

In some forms of leukemia, significant numbers of immature leukocytes may appear in the circulation. Thus, it is of interest to investigate the deformability of such cells as a basis for assessing their potential impact on flow in the microvasculature. The HL60 human promyelocytic leukemic cell line provides a convenient system in which to investigate such cells. HL60 cells were grown under standard conditions (RPMI 1640 with 10% fetal bovine serum at 37°C under 5% CO_2) and passaged twice weekly. In preliminary experiments, it was found that a growing population of HL60 cells exhibited a great deal of heterogeneity in both size and apparent viscosity. Initially, it was not clear

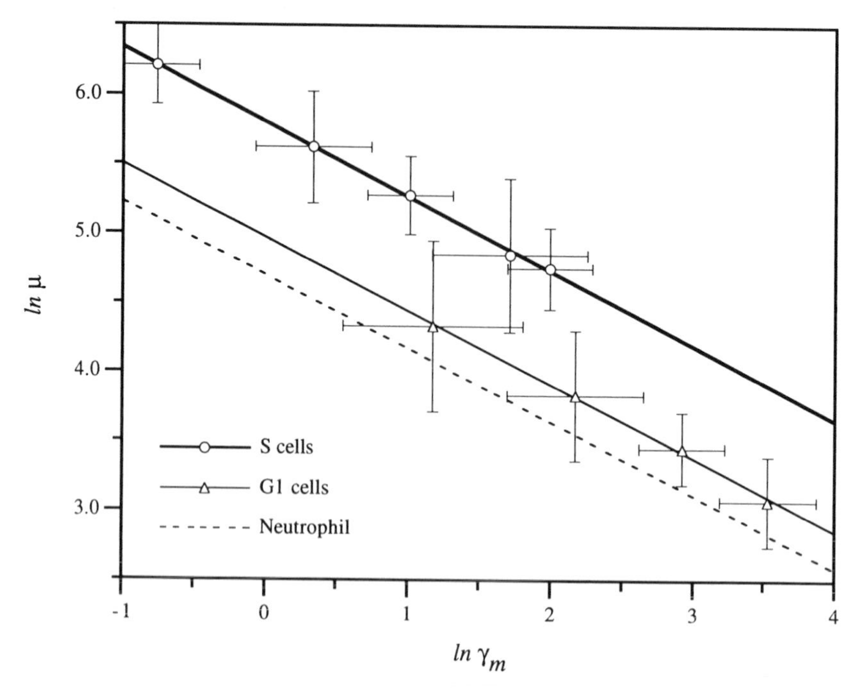

Fig. 5. Shear rate-dependence of HL60 cells in S and G1 phases. For all of the cell types tested, the log of the apparent viscosity varied in proportion with the log of the mean shear rate. Top curve (○) indicates data for S cells, middle curve (Δ) indicates data for G1 cells, and lower curve (---) indicates data for neutrophils (replotted from Tsai *et al.*, 1993). Note that although all cell types exhibit shear thinning, the viscosity at a given shear rate was different for the different leukocyte types: S cells > G1 cells > neutrophils.

whether these apparent differences in properties were due simply to differences in size, which resulted in different magnitudes of deformation for large and small cells being aspirated into the same pipette, or if the intrinsic properties of the cells were also heterogeneous. To clarify the interpretation of these observations, we used centrifugal elutriation to separate cells into different sub-populations according to their size. The elutriation method was adopted from Palis *et al.* (1988). Details of the elutriation protocol and the culturing conditions can be found in a forthcoming manuscript (Tsai *et al.*, 1994*b*).

In addition to providing information about the relationship between cell size and intrinsic properties, these measurements also inform us about the dependence of cellular properties on cell cycle. Previously, Palis *et al.* (1988) had shown that the different sized cells obtained by elutriation correspond to cells in different stages of the cell cycle. In our experiments, fractions of cells were chosen that represented the G1 and S phases of the cell cycle. To eliminate possible ambiguities in interpretation due to differences in the geometry of deformation of large and small cells, different sized cells were measured with correspondingly different-sized pipettes, so that the magnitude of deformation for the different cells was similar. The results of preliminary measurements of the rheological properties of these cells are summarized in Fig. 5. All types of cells tested exhibited more or less the same dependence of apparent viscosity on mean shear rate. However, it is clear that at the same shear rate, larger cells (corresponding to the S phase of the cell cycle) are intrinsically more viscous (less deformable) than smaller cells (corresponding to the G1 phase of the cell cycle). Interestingly, even the smallest of the immature cells (G1) were significantly more viscous than their mature counterparts, the neutrophil. Thus, consistent with clinical experience, the presence of immature cells in the circulation is likely to have serious effects on perfusion and oxygen delivery. Our preliminary results indicate that these deleterious effects are due not only to the large size of the immature cells but also to the fact that they are intrinsically more rigid.

Summary and Conclusions

The rheological behavior of leukocytes is complex, but to a large extent, the behavior of these cells can be characterized as that of a viscous liquid drop. A novel feature of the cell behavior demonstrated in the work summarized here is that the cytosolic viscosity depends strongly on rate of deformation. Such behavior is consistent with expectations for a polymeric fluid, and alterations in the polymeric structure of the cell to reduce the mean length and/or associations of actin filaments result in a decrease in the characteristic viscosity of the cell and the dependence of the viscosity on rate of deformation. Characterization of the rheological properties of the human promyelocytic leukemic cell line HL60 indicates that a shear rate-dependent viscosity is a common feature of leukocyte rheology evident in both mature and immature cell populations. In addition, the rheological characteristics of growing cell populations were shown to depend strongly on cell cycle during leukocyte proliferation.

The conclusions reached in these studies were the result of combining precise, single-cell micromechanical measurements with detailed numerical analysis of the cell deformation. While the power-law fluid model described here does not provide a perfect description of the cell behavior, it does serve to identify an essential characteristic of the cell rheology which should be included in future attempts to model the constitutive behavior of the cell. Rigorous analysis of single cell micromechanical experiments will be an essential tool for further development of our understanding of leukocyte rheology and resolution of discrepancies between existing theories and experimental observation.

Acknowledgments

This work was supported by the US Public Health Service under a grant from the National Institutes of Health, grant # HL 18208. One of the authors (MAT) is the recipient of an American Heart Association Research Fellowship.

References

Erzurum, S. C.; Kus, M. L.; Bohse, C.; Worthen, G. S. Mechanical properties of HL60 cells: Role of stimulation and differentiation in retention in capillary-sized pores. Am. J. Respir. Cell Mol. Biol. 5:230-241; 1991.

Evans, E.; Kukan, B. Passive material behavior of granulocytes based on large deformation and recovery after deformation tests. Blood 64:1028-1035; 1984.

Evans, E.; Yeung, A. Apparent viscosity and cortical tension of blood granulocytes determined by micropipet aspiration. Biophys. J. 56:151-160; 1989.

Harris, A. G.; Skalak, T. C. Leukocyte cytoskeletal structure determines capillary plugging and network resistance. Am. J. Physiol. 265:H1670-1678; 1993.

Lichtman, M. A. The relationship of excessive white cell accumulation to vascular insufficiency in patients with leukemia. In: H. J. Meiselman, M. A. Lichtman and P. L. LaCelle eds. White Cell Mechanics: Basic Science and Clinical Aspects. New York: Alan R. Liss, Inc.; 1984:p.295-306.

Needham, D.; Hochmuth, R. M. Rapid flow of passive neutrophils into a 4 micron pipet and measurement of cytoplasmic viscosity. J. Biomech. Eng. 112:269-276; 1990.

Palis, J.; King, B.; Keng, P. Separation of spontaneously differentiating and cell cycle-specific populations of HL60 cells. Leuk. Res. 12:339-344; 1988.

Schmid-Schönbein, G. W.; Sung, K.-L. P.; Tozeren, H.; Skalak, R.; Chien, S. Passive mechanical properties of human leukocytes. Biophys. J. 36:243-256; 1981.

Tsai, M. A.; Frank, R. S.; Waugh, R. E. Passive mechanical behavior of human neutrophils: Power- law fluid. Biophys. J. 65:2078-2088; 1993.

Tsai, M. A.; Frank, R. S.; Waugh, R. E. Passive mechanical behavior of human neutrophils: Effect of cytochalasin B. Biophys. J. (accepted for publication); 1994*a*.

Tsai, M. A.; Waugh, R. E.; Keng, P. C. Cell cycle dependence of HL60 cell deformability. Blood (submitted); 1994*b*.

Wallace, P. J.; Packman, C. H.; Wersto, R. P.; Lichtman, M. A. The effects of sulfhydryl inhibitors and cytochalasin on the cytoplasmic and cytoskeletal actin of human neutrophils. J. Cell. Physiol. 132:325-330; 1987.

Worthen, G. S.; Schwab, B., III; Elson, E. L.; Downey, G. P. Mechanics of stimulated neutrophils: Cell stiffening induces retention in capillaries. Science 245:183-186; 1989.

4
Cell Tumbling in Laminar Flow: Cell Velocity is a Unique Function of the Shear Rate

A. Tözeren

Introduction

Leukocyte rolling in blood venules attracted the attention of scientists for more than a century (Cohnheim, 1877; Atherton and Born 1973). Recent studies with function-blocking reagents (Abbasi *et al.* 1991) and with planar membranes containing selectin molecules (Lawrence and Springer 1991) showed that this interesting mode of cell-substrate adhesion is mediated by cell surface adhesion molecules called selectins.

Micro-mechanics of rolling has also been under investigation. Dembo *et al.* (1988) introduced a peeling analysis to model the receptor-ligand interactions that occur at the contact line between a rolling cell and a planar substrate. In this approach, the effects of the fluid force and the fluid moment acting on the cell are taken into account by introducing the concept of cell membrane tension at the edge of conjugation. However, the assumed linear correspondence between the mean velocity of blood in a vessel and the membrane tension at the contact line is phenomenological.

Hammer and Apte (1992) introduced a stochastic approach to the biophysical investigation of rolling. These authors allow for the existence of a sparse number of bonds in the contact area in their formulation of kinetics of bond formation. They compute the fluid forces and moments acting on rolling cells by using the Stokes solution valid for rigid spheres near a planar boundary in shear flow (Goldman *et al.* 1967). The numerical calculations produced by Hammer and Apte (1992) show fluctuations in rolling velocity, typically observed with leukocytes rolling in blood venules or in a flow channel. Furthermore, the model was shown to mimic experimental data on rolling velocity at one value of shear rate.

Tözeren and Ley (1992) modeled micro-mechanics of rolling by using rate equations of the sort employed in the sliding filament model of muscle contraction (Tözeren 1985). Similarly to Hammer and Apte, they computed the fluid forces and moments acting on the cell by using the solution of Goldman *et al.* (1967). The rolling cell model of Tözeren and Ley (1992) incorporates the existing biochemical data on selectins into equations governing the rate of bond formation and bond detachment. Model predictions reflect accurately the dependence of rolling velocity on the shear rate and on the ligand density. Implicit in the model is the assumption that a large number of surface receptors interact with ligands on the substrate at any one instant. The justification for this approach for cases where the adhesion bonds in the contact area are numbered in tens but not in hundreds comes from Chen and Hill (1988) which compared predictions of the deterministic approach for kinesin mediated fast axonal transport with corresponding stochastic solutions obtained using the Monte Carlo method.

Rolling interactions occur not only between leukocytes and post capillary blood venules but also between some epithelial cells and laminin, an adhesion glycoprotein which anchors epithelial and endothelial cells to the basement membrane (Tözeren *et al.* 1994). The cultured epithelial cells (MCF-10A) accumulate on laminin in the presence of fluid shear stress τ ($\tau \leq 2$ dyn/cm^2). An increase in the fluid shear stress (3.5 dyn/cm^2 $\leq \tau \leq$ 100 dyn/cm^2) initiates rolling of these cells at velocities less than the speed of flow (Tözeren *et al.* 1994). Rolling velocity averaged over (1/2) seconds is more uniform for large interphase cells with small amplitude cell projections than for cells with few cell surface projections. Thus, the kinetics of rapidly formed transient interactions appear to depend strongly on the cell geometry.

In the following, correlation between the cell velocity and the cell geometry will be investigated by using methods of mechanics. Using high magnification fluorescence light microscopy, it is possible to determine the morphology of rolling cells with some detail. This type of data can then be used to critically assess the mathematical predictions of the time course of cell paths for spiked cells rolling in simple shear flow.

Mechanics of Tumbling of Rigid Spherical Cells Decorated with Spikes

In the following, cases in which spherical cells decorated with rod shaped spikes will be considered. It is assumed that both the cell and the surface spikes are rigid so that a tumbling cell interacts with the substrate through one spike at a time. Furthermore, we will focus only on planar motion, the three dimensional motion being a straightforward extension of the formulation presented here.

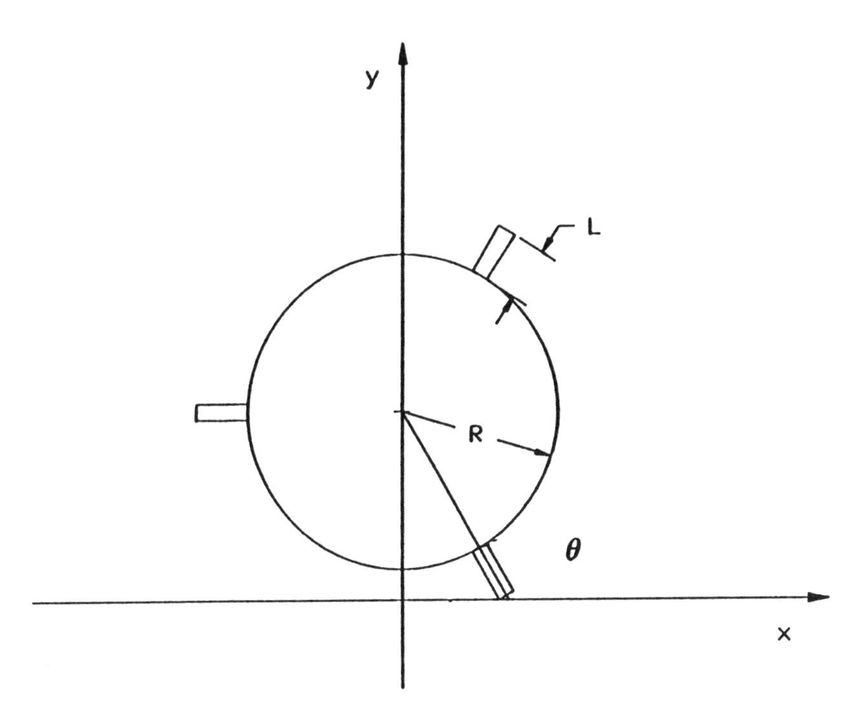

Figure 1. Schematic drawing of a spiked cell rolling on a planar substrate.

Let R be the radius of a spherical cell and L be the length of it's cell surface projections, assumed uniform here. Let (x,y,z) be a cartesian coordinate system fixed in the planar substrate such that x denotes the direction of flow and y is the distance in the direction perpendicular to the planar substrate (Figure 1). The angle θ is defined as the angle between the x axis and the line connecting the cell center to the tip of the spike in contact with the substrate (Figure 1). The parameters θ_b and θ_a denote, respectively, the minimum and maximum values of the angle θ for which the cell forms contact with the substrate through a spike. The distance between the center of the sphere and the planar substrate is then given by the following equation:

$$H = (R+L)\sin(\theta) \tag{1}$$

Assuming that the spike does not slip along the substrate, the velocity of the spherical cell decorated with rigid spikes can be written as:

$$U = -(R+L)(d\theta/dt)(\sin\theta) \tag{2a}$$

$$V = +(R+L)(d\theta/dt)(\cos\theta) \tag{2b}$$

where U and V denote the velocity components of the center of the sphere in the x and y directions, respectively, and $(d\theta/dt)$, the angular speed, is the derivative of the angle θ (Beer and Johnston 1984). Equation (2) shows that the velocity of the tumbling cell can be computed as a function of time once the time course of angular speed $(d\theta/dt)$ is evaluated.

Light microscopy studies indicate that adhesive contacts between the cell and the substrate break at the rear edge of the contact line (Tözeren and Ley 1992), providing evidence for the no-slip condition that was assumed in Equation (2). Furthermore, biophysical models developed by Hammer and Apte (1992) and Tözeren and Ley (1992) both predict that in laminar flow the slip velocity of spherical cells interacting with a planar wall is negligible compared to the cell velocity.

In order to compute the time course of the angular velocity of the tumbling cell the conditions of force balance must be utilized. Inertial forces are negligible in standard laminar flow assays because of the low Reynolds numbers involved and the small size of cells interacting with the bottom plate of the flow channel (Lawrence and Springer 1991). The conditions of force balance for cells dynamically interacting with a planar substrate in the presence of shear flow can be written as:

$$F_{xc} - 6\pi\mu(R)[U\,F_t + R\,(d\theta/dt)\,F_r - H\,S\,F_s] = 0 \tag{3a}$$

$$F_{yc} + 6\pi\mu(R)(V)(F_y) = 0 \tag{3b}$$

$$M_{zc} + H(F_{xc}) - 4\pi\mu(R^2)[2U\ M_t + 2R(d\theta/dt)M_r + R\ S\ M_s] = 0 \tag{3c}$$

where S (1/s) denotes the shear rate at the planar boundary, F_{xc}, F_{yc}, and M_{zc} denote the force and the moment exerted on the tumbling cell by the adhesion bonds, and F_t, F_r, F_s, F_y, M_r, M_t, and M_s are single valued functions of (H/R) (Hsu an Ganatos 1989). The values of these dimensionless force functions have been tabulated by Goldman *et al.* (1967) and Hsu and Ganatos (1989) at discrete values of (H/R).

The four unknown variables in equation (3) are the force and moment exerted on the cell by the adhesion bonds (F_{xc}, F_{yc}, and M_{zc}) and the angular speed $d\theta/dt$. An additional equation is needed to compute these variables as a function of time. Thus, it is assumed that the contact site acts as a pin joint, namely it can resist forces but not a torque. This is equivalent to setting $M_{zc} = 0$, yielding the following differential equation for θ:

$$(d\theta/dt) = [1.5H^2F_s + R^2M_s](S)/E \tag{4a}$$

$$E = -1.5H^2F_t + 1.5HRF_r + 2RHM_t - 2R^2M_r \tag{4b}$$

where $\theta = \theta_b$ at $t = 0$.

Equation (4) can be integrated to determine the time course of θ as the cell tumbles in shear flow. The tumbling cell velocity can then be computed by using Equation (2). The adhesion force on the tumbling cell can be evaluated as a function of time by inserting the known values of the angular and translational velocities of the tumbling cell in to Equations (3a) and (3b). The following equations can then be used to determine the magnitude of the force F_c exerted by the substrate on the tumbling cell and the orientation of this adhesive force:

$$F_c = (F_{xc}^2 + F_{yc}^2)^{1/2} \tag{5a}$$

$$\cos\phi = F_{xc}/(F_{xc}^2 + F_{yc}^2)^{1/2} \tag{5b}$$

where ϕ is the angle the contact force makes with the x axis. For cells with uniformly distributed spikes of equal size, Equation (4) needs to be integrated only in the range $\pi(1/2+1/n) \geq \theta \geq \pi(1/2-1/n)$ where n is the number of spikes per circumference at $z = 0$ plane.

In the limit as the number of spikes (n) becomes large, the cell center will remain approximately at the same distance from the substrate during tumbling interactions ($H = R + L$). Equation (4) then predicts a constant angular velocity during dynamic interactions with the planar substrate. For example, for a cell in which the spike length is half the cell radius ($L/R = 0.5$), $d\theta/dt = 0.778S$ and the translational cell velocity $U = (0.778S)(R)$. The angular speed of a neutrally buoyant cell of the same shape at the same spatial position is, on the other hand, equal to (1.377)(S)(R), indicating that spike mediated interactions with a substrate reduce the cell speed by approximately 40%.

The Role of Spike Flexure on Rolling Velocity

It is clear from light microscopic observations that spikes on the surfaces of rolling cells are not rigid but are flexible. In this section, the cell surface spikes will be treated as flexural members. As in a beam of uniform cross section, the bending moment at any cross section of the spike is assumed to be linearly proportional to the meridional curvature:

$$M_Z = E_Z I(d^2v/dx^2) \tag{6}$$

where (x,y,z) is a cartesian coordinate system in which x denotes the distance along the spike and z is perpendicular to the plane of motion. The parameter I_Z is the moment of inertia of the spike section about the z axis, E (dyn/cm^2) is the physical stiffness of the spike, and v denotes the displacement perpendicular to the spike axis in the (x,y) plane (Gere and Timoshenko 1984).

Consider a spike that is unstrained when it's axis is along the y direction, normal to the direction of the flow (x axis). Let Δx, and $\Delta\theta$ be the displacement and rotation of the end of the spike in contact with the spherical cell surface. These parameters can be expressed in terms of the fluid force and the fluid moment acting on the cell if one assumes that the contact angle between the spike and the substrate remains constant during the course of spike-substrate interaction:

$$\Delta x = (F_f L^3)/(EI_Z)+(M_f L^2)/(2EI_Z) \tag{7a}$$

$$\Delta\theta = (F_f L^2)/(2EI_Z)+(M_f L)/(2EI_Z) \tag{7b}$$

$$F_f = 6\pi\mu(R)[U\,F_t + R\,(d\theta/dt)\,F_r - H\,S\,F_s\,] \tag{7c}$$

$$M_f = R\,(F_{xc}) - 4\pi\mu(R^2)[2U\,M_t + 2R(d\theta/dt)M_r + R\,S\,M_s] \quad (7d)$$

where F_f is the fluid force acting on the tumbling cell and M_f is the fluid moment with respect to the end of the spike in contact with the cell surface. Note also that $(d\Delta x/dt) = U$, and $(d\theta/dt) = d(\Delta\theta)/dt$. Equation (7) comprises a set of first order ordinary differential equations that can be solved by standard procedures for the cell speed U and the cell angular velocity $d\theta/dt$ as a function of the fluid shear stress imposed on the bottom plate of the flow channel. The predicted path of the rolling cells may then be compared with the light microscopy data to assess the bending stiffness of spikes protruding from the surfaces of rolling cells.

Conclusions

The velocity of leukocytes rolling on post capillary blood venules shows only a weak correlation with the wall shear stress (Atherton and Born 1973). It has been suggested that the principle determinant of the rolling velocity may be the biochemical parameters of receptor counter-receptor interactions during rolling adhesion (Ley and Gaehtgens 1991). In order to test this possibility, the effect of cell surface topology on rolling velocity must be investigated. In this chapter, expressions were derived to correlate the velocity of a tumbling cell with the cell surface topology. The results suggest that the size and density of cell surface spikes strongly affects the kinetics of motion and may be responsible for the large variations commonly observed in experimental data relating rolling velocity to the fluid shear stress at the planar boundary.

References

Abbasi, O; Lane, C; Krater, S.S.; Kishimoto, T.K.; Anderson, D.C.; McIntire, L.V.; Smith, C.V. J. Immunol. 147:2107-2115; 1991.

Atherton, A; Born, G.V.R. Relationship between the velocity of rolling granulocytes and that of blood flow in venules. J. Physiol. (Lond.). 233:157-165; 1973.

Beer, F.P.; Johnston, Jr. E.R. Vector mechanics for engineers. New York: McGraw-Hill Book Co.; 1984.

Chen, Y.D.; Hill, T.L. Theoretical calculation methods for kinesin in fast axonal transport. Proc. Natl. Acad. Sci. (Wash. DC). 85:431-435; 1988.

Cohnheim, J. Vorlesungen uber allgemeine Pathologie. August Hirschwald Verlag Berlin, 1877.

Dembo, M.; Torney, D.C.; Saxman, K.; Hammer, D. The reaction limited kinetics of membrane to surface adhesion and detachment. Proc. R. Soc. Lond. B. Biol. Sci. 234:55-83; 1988.

Gere, J.M.; Timoshenko, S.P. Mechanics of materials. New York: PWS Publishers (Engineering); 1984.

Coldman, A.J.; Cox, R. G.; Brenner, H. Slow viscous motion of a sphere parallel to a plane wall. II Couette flow. Chem. Eng. Sci. 22:637-651; 1967.

Hammer, D.A.; Apte, S.M. Simulation of cell rolling and adhesion on surfaces in shear flow: General results and analysis of selectin mediated neutrophil adhesion. Biophys. J. 62:35-57; 1992.

Hsu, R.; Ganatos, P. The motion of a rigid body in viscous fluid bounded by a plane wall. J. Fluid Mech. 22:275-316; 1989.

Lawrence, M.B.; Springer, T.A. Leukocytes roll on selectin at physiologic flow rates: distinction from and prerequisite for adhesion through integrins. Cell 65:859-873; 1991.

Ley, K.; Gaehtgens, P. Endothelial, not hemodynamic differences are responsible for preferential leukocyte rolling in venules. Circ. Res. 69:1034-1041; 1991

Tözeren, A. Constitutive equations of skeletal muscle based on crossbridge mechanism. Biophys. J. 47:225-236; 1985.

Tözeren, A.; Ley, K. How do selectins mediate leukocyte rolling in blood venules. Biophys. J. 63:1-11; 1992.

Tözeren, A.; Kleinman, H.K.; Mercurio, A.M.; Byers, S.W. Integrin $\alpha 6 \beta 4$ mediates initial interactions with laminin in some epithelial and carcinoma cell lines (in review) 1994.

Part II

Flow-Induced Effects on Cell Morphology and Function

5

The Regulation of Vascular Endothelial Biology by Flow

R.M. Nerem, P.R. Girard, G. Helmlinger, O. Thoumine, T.F. Wiesner, T. Ziegler

Introduction

The inner surface of a blood vessel is composed of a monolayer of vascular endothelial cells which serves as the interface with flowing blood. Once thought to be simply a passive, non-thrombogenic barrier, the vascular endothelium is now recognized as being a dynamic participant in the biology of a blood vessel (Kaiser and Sparks 1987). Part of its dynamic character is due to the regulation of the biology of the endothelium by the mechanical environment associated with the hemodynamics of the vascular system.

As blood pulses through an artery, it imposes a physical force on the vessel wall. This force has two components to it. One component is pressure which acts normal to the wall and the other is a viscous force which acts in a direction tangential to the wall. Pressure not only acts directly on the endothelium, but also distends the vessel wall and in so doing stretches the basement membrane. The tangential viscous force per unit area is called a shear stress, and although much smaller in magnitude than the pressure, shear stress is believed to be an important regulator of vascular endothelial biology. Because of the pulsatile nature of both pressure and flow in an artery, the mechanical environment of the vascular endothelium is quite complex, with an endothelial cell being exposed to time-varying pressure and shear stress while "riding" on a basement membrane which is undergoing cyclic stretch.

It should be noted that much of the interest in flow as a regulator of vascular biology has eminated from the possible role of hemodynamics as a localizing factor in atherosclerosis, in particular in the early stages of this disease (Nerem 1992). The vascular endothelium, as the interface with flowing blood, is believed to be a prime mediator of any hemodynamic effects on the normal biology and also the pathobiology of the vessel wall. In studies of the pattern of intimal thickening and fatty streak localization, a correlation with regions of low shear stress has been found (Caro et al 1971; Grottum et al 1983; Asakura and Karino 1990; Friedman et al 1981, 1986; Ku et al 1985). This has tended to focus studies relating to hemodynamics and atherosclerosis on the role of shear stress in the disease process. This focus is not meant to in any way deny the importance of pressure in regulating vascular biology. Furthermore, regions of low shear stress are also regions of oscillatory shear stress and ones characterized by long particle residence times. Thus, whether or not it is actually low shear stress or some other hemodynamic characteristic of these regions still remains to be determined.

The endothelium also is believed to have an important role in the regulation of blood vessel diameter (Kamiya and Togawa 1980; Zarins et al 1987). Langille and O'Donnell (1986) demonstrated that reductions in arterial diameter associated with chronic decreases in blood flow were dependent on the presence of the endothelium. This suggests that the endothelium may serve as a flow or shear stress sensor.

What is clear is that an enhanced understanding of the role of shear stress in the regulation of vascular endothelial biology is required. This thus is the focus of this discussion, and in the next section a brief summary of alterations in endothelial cell structure and function due to flow will be presented. In this the emphasis will be placed on results obtained through *in vitro* studies where cultured endothelial monolayers have been exposed to a flow environment. Although cell culture does not fully simulate the *in vivo* physiologic conditions, it does provide a well-defined model, particularly in terms of the flow environment to be studied. This is in contrast to animal studies where, over the time duration required for the desired biological end points, the hemodynamic environment is time-varying and can only be defined qualitatively.

Alterations in Endothelial Cell Structure and Function

Cell culture studies on the effects of flow on endothelial biology are carried out by exposing a monolayer of endothelial cells to the sudden onset of a laminar steady flow (Dewey et al 1981; Eskin et al 1984; Levesque and Nerem 1985). Using a bovine aortic endothelial cell (BAEC) monolayer as an example, the response to this flow environment involves an adaptation in cell shape, cytoskeletal structure, and extracellular matrix as summarized in Table I.

Most dramatic is the change in cell shape which takes place over a period of 24 hours. Whereas a confluent BAEC monolayer on plastic or glass in static culture has a cobblestone appearance, once exposed to flow

Table I. Morphologic Alterations in Bovine Aortic Endothelial Cells in Response to Steady Flow

Property	Effect	References
Cell Shape	Cells elongate and align with the direction of flow	Dewey et al 1981 Eskin et al 1984 Levesque & Nerem 1985
Actin Micro-filaments	Disappearance of dense peripheral bands and formation of stress fiber bundles aligned with direction of flow	White et al 1982 Wechezak et al 1985 Sato et al 1987
Extracellular Matrix	Alteration in composition with fibronectin and laminin increasing over 48 hours; ECM structure altered into one with thicker fibrils, in the case of fibronectin aligned with the direction of flow	Girard et al 1993 Thoumine et al 1993
Focal Contact Proteins	Vinculin and $\alpha_v \beta_3$ vitronectin receptor localized preferentially upstream	Girard & Nerem 1993

and the associated shear stress, the endothelial cells change from a polygonal shape to a more elongated one, with the extent of elongation depending on the level of shear stress (Levesque and Nerem 1985). Concomitant with this elongation is an orientation of the cell's major axis with the direction of flow. These cell culture results are qualitatively consistent with observations from *in vivo* studies where endothelial cells are found in regions of high shear stress to be highly elongated and in regions of low shear stress rather polygonal in shape (Flaherty et al 1972; Silkworth and Stehbens 1975; Nerem et al 1981;Levesque et al 1986; Kim et al 1989). This suggests that the addition of flow to cell culture studies provides an additional variable which is important to the simulation of physiologic conditions.

As part of the response of a BAEC monolayer to the sudden onset of a laminar steady flow, there is a reorganization of the endothelial cell's cytoskeletal structure (White et al 1982; Wechezak et al 1985; Sato et al 1987). Most noticeable is the alteration in the localization of F-actin which occurs. Whereas in static culture the distribution of F-actin is characterized by dense peripheral bands, in response to flow these disappear, and bundles of F-actin microfilaments, i.e. stress fibers, are formed, which are oriented parallel to the cell's major axis and thus aligned with the direction of flow. If a cytoskeletal inhibitor such as cytochalasin B, which blocks actin assembly, is used to pretreat the BAEC monolayer, then only a very slight elongation is observed to occur (Levesque et al 1989). Thus, the existence of an intact microfilament structure and/or the ability to reorganize its F-actin structure appears to be critical to the cell elongation which takes place in response to flow. It should be noted that, just as with cell shape, *in vivo* observations of F-actin localization are consistent with the above noted results from cell culture studies, i.e. endothelial cells are found in regions of high shear stress to have F-actin stress fibers oriented with the cell's major axis and with the direction of flow and in regions of low shear stress to be characterized by dense peripheral bands (Kim et al 1989).

A characteristic of an endothelial cell's response to flow is that it mechanically becomes much stiffer (Sato et al 1987; Theret et al 1988). This has been demonstrated through stress-strain measurements performed using the micropipette technique and an analysis of the data based on a mathematical model of the mechanical deformation of a cell. It has been shown that the major resistance to deformation is provided by the cell's cytoskeletal structure. Furthermore, whereas endothelial cells in static culture can be characterized by a Young's modulus on the order of 1000 dynes/cm^2, after exposure to flow for 24 hours this value increases by a factor of ten. Thus, the alteration in cytoskeletal structure and cell shape, which occurs as part of an endothelial cell adapting to the onset of flow, is reflected in changes in the cell's mechanical properties.

Equally important are the changes in the extracellular matrix which accompany the response to the onset of a laminar, steady flow. This includes alterations in both the structure and composition of the extracellular matrix and changes in the distribution of certain focal contact proteins (Girard and Nerem 1993, Thoumine et al 1993). With regard to the former, over a period of 48 hours of exposure to flow, there are some important changes which have been observed. Fibronectin not only changes its structure into one with thicker fibrils which are oriented with the direction of flow, but there are also changes in the fibronectin composition of the extracellular matrix, first decreasing, but ultimately increasing such that by 48 hours it is almost a factor of two higher than that in a control, static culture monolayer. The amount of laminin also increases by a factor of two, in this case steadily over the 48 hour period; vitronectin, on the other hand, undergoes no change in the amount in the ECM. Although we have no data on any alteration in the ECM composition of collagen IV, we have shown that it changes in a structure characterized by thicker fibrils, with no orientation apparent. As noted earlier, there also are changes in the distribution of certain focal contact proteins, with vinculin and the $\alpha_v\beta_3$ vitronectin receptor relocating to the upstream end of the cell. This is in contrast to the protein, talin, which exhibits no real change in distribution, except for the alignment of the talin label in focal streaks with the direction of flow, and the $\alpha_v\beta_1$ fibronectin receptor which, though exhibiting a distribution that includes both the periphery and more central areas of the cell, shows no upstream or downstream preference. Since it is now recognized that the extracellular matrix is also involved in cell signaling, it must be presumed that these changes in various facets of the extracellular matrix are part of what might be called a carefully orchestrated response of an endothelial cell to flow.

With these alterations in cell structure, including cell shape and the extracellular matrix, there are important changes in cell function. For sub-confluent monolayers, not only is there a decrease in the rate of proliferation with flow (Levesque et al 1990), this being due to an inhibition of entry into S-phase (Mitsumata et al 1991), but the process of cell division takes place in a markedly different manner (Ziegler and Nerem 1994). Whereas polygonal-shaped bovine aortic endothelial cells in static culture, or even shortly after the onset of flow, round up as part of the cytokinesis process, once elongated and aligned with the direction of flow the morphology of cell division is different. In this case the cell "hugs" the surface as it goes through cytokinesis, with a contractile ring forming perpendicular to the direction of flow, "pinching" off the cell and producing an upstream and a downstream daughter cell. After completion of cytokinesis, there is a highly active pseudopod activity, one which initially is preferentially located at the downstream edge of each of the two daughter cells, but ultimately is extended to include the cell's entire periphery.

Cell culture studies also demonstrate that there is an important influence of flow on endocytotic processes. This includes smaller molecules such as horseradish peroxidase (Davies et al 1986) and large molecules such as low density lipoproteins, the latter being through an effect on receptor expression (Sprague et al 1987).

The effect of flow and the associated shear stress on cell function also includes the regulation of the synthesis and secretion of bioactive substances (Frangos et al 1985; Grabowski et al 1985; Diamond et al 1989). This influence extends to the level of gene expression. As noted in Table II, measurements of messenger RNA in endothelial cells exposed to moderate and high levels of shear stress indicate a differential regulation, in some cases there is an up regulation, in some cases a down regulation, and in some cases no effect at all.

The effect of flow also includes an influence on the ability of an endothelial cell to respond to chemical agonists. For example, the induction by α-thrombin of the proto-oncogene, c-myc, is dramatically inhibited in BAEC after exposure to flow (Berk et al 1990). Another example is the induction of the cell adhesion molecule, VCAM-1, by the cytokine, IL-1ß, which is markedly inhibited by flow. This is in contrast to such cell adhesion molecules as ICAM-1 and ELAM-1, which if anything exhibits an increase, and is again evidence of a differential regulation by flow at the gene expression level.

Table II. Some Alterations in Gene Expression in Endothelial Cells in Response to Flow and the Associated Shear Stress

Messenger RNA	Effect	References
Basic Fibroblast Growth Factor	No effect in HUVEC	Diamond et al 1990
Cell Adhesion Molecule VCAM-1	Downregulated in BAEC and Mouse EC	Sprague et al 1992 Ohtsuka et al 1993
c-fos	No effect in BAEC	Mitsumata et al 1993
Monocyte Chemotatic Protein-1	Downregulated in BAEC	Sprague et al 1992
Nitric Oxide Synthase	Upregulated in BAEC	Nishida et al 1993
Platelet-Derived Growth Factor	Upregulated in BAEC and HUVEC	Hsieh et al 1991 Mitsumata et al 1993 Resnick et al 1993
Tissue Plasminogen Activator	Upregulated in HUVEC	Diamond et al 1990
c-myc	Response to thrombin is downregulated in BAEC	Berk et al 1990
Preproendothelin	Downregulated in HUVEC	Sharefkin et al 1991

Cell Signaling Mechanisms

From the previous section there clearly are major alterations in endothelial cell structure and function due to flow and the associated shear stress. An important question is how does a vascular endothelial cell recognize the flow environment in which it resides, and having done so, what are the second messengers involved in the transduction of this signal into the structure and functional changes already noted?

The easier question seems to be the second one, for there are second messengers involved in this transduction process which are already known due to their role in transducing chemical stimuli. Several studies have indicated that shear stress stimulates the phosphoinositide system (Bhagyalakshmi et al 1992; Nollert et al 1990; Prasad et al 1993). There is not only an early peak, but also a bimodal character to the time history of inositol trisphosphate, IP_3, with there being a second peak prior to IP_3 decreasing to a level below baseline. There also is evidence of a translocation in protein kinase C from cytosol to membrane (Girard and Nerem 1993).

Of particular interest has been the elevation in intracellular calcium caused by the sudden onset of a laminar, steady flow. A number of laboratories have investigated this (Ando et al 1988; Mo et al 1991; Dull et al 1991; Shen et al 1992; Geiger et al 1992), and although it is clear that such an elevation occurs, there is considerable disagreement as to whether this is due to a direct effect of shear stress or a shear effect associated with the control of the transport of a small molecule by flow, e.g. ATP or some other molecule acting as an agonist. Based on results from our laboratory, we believe that there is a direct effect of shear stress, but one which produces a slowly rising, small amplitude elevation in intracellular calcium. What type of recognition event could lead to such a response? To this question we do not have an answer; however, a possible candidate is a mechanically activated ion channel (Lansman et al 1987). In fact, there is evidence supporting the existence of both calcium-permeable and potassium ion channels which are shear stress sensitive (Olesen et al 1988; Schwarz et al 1992).

Separate from the possibility of a direct effect of shear stress, there is general agreement that an agonist, either ATP or some other biologically active molecule, could be the vehicle of a shear effect through a convection-diffusion coupling mechanism (Nollert et al 1992; Shen et al 1993). In this case we believe that the elevation in intracellular calcium could be considerably enhanced as compared to that associated with a direct shear stress effect.

The issue of the recognition and transduction of a mechanical signal by an endothelial cell is not limited to the application of flow and the associated shear stress. As noted in the introduction, pressure acts directly on an endothelium which rides on a basement membrane undergoing cyclic stretch while being exposed to flow and the associated tangentially-acting shear stress. The effects of cyclic stretch have been studied, and these indicate that this mechanical stimulus also produces changes in EC structure and function (Birukov et al 1993; Dartsch et al 1989; Gorfien et al 1989; Shirinsky et al 1989; Sumpio and Banes 1988; Sumpio et al 1987, 1988; Winston et al 1993). Thus, it is this rather complex mechanical environment which must be recognized, and it is not unreasonable to hypothesize that recognition takes place through multiple and parallel pathways. We must even wonder whether the recognition by an endothelial cell of its mechanical environment takes place on its lumenal surface, its ablumenal surface, or even throughout the cell. In addition to mechanisms already mentioned, other possibilities include a biologic "strain gauge" which senses deformation of a cell's cytoskeletal structure and a membrane protein acting as a "shear" receptor, perhaps even one of the focal contact proteins involved in cell adherence.

Pulsatile Flow Effects

Over the past decade most of the studies carried out have been using a laminar, steady flow. A few studies have employed a pulsatile flow, and there also has been a report on the influence of a turbulent flow (Davies et al 1986). A few years ago we initiated a study in which the goal has been to investigate in a systematic way the influence of a laminar, unsteady flow (Helmlinger et al 1991). Three distinct types of flow are being studied, all with a sinusoidal waveform and a frequency of 1 Hz. One is a non-reversing pulsatile flow, where although the flow rate and shear stress are time varying, the direction is always the same. A second is a reversing pulsatile flow, where during a part of the cycle the flow and the associated shear stress reverse in direction. However, the most interesting flow has turned out to be the third type of pulsatile flow. This is, for all practical purposes, a purely oscillatory flow, one with a negligible mean component such that the flow simply "sloshes" back and forth over the monolayer in a sinusoidal manner.

In comparing the response of an endothelial monolayer to either a non-reversing or reversing pulsatile flow with that to a steady flow, there were some differences quantitatively, but not qualitatively. In both of these pulsatile flow cases and just as with steady flow, BAECs elongated in shape, aligned themselves with the direction of flow, and developed F-actin stress fibers, also oriented with the direction of flow. In the case of a purely oscillatory flow, in which there also are pressure oscillations, there was no change in cell shape. This implies that the elongation response to flow by endothelial cells requires that there exist a mean flow component; however, it does not mean that there was no influence of this particular flow environment. For this purely oscillatory flow case, the F-actin microfilaments and the fibronectin network became denser, although still randomly oriented, and there were other noticeable changes in the extracellular matrix, with there being a restructuring into thicker fibers of laminin and collagen IV and an increase in the amount of fibronectin and laminin (Thoumine et al 1993). There also was a clustering of the fibronectin receptor into streaks, often co-localized with vinculin.

If, in fact, there is an endothelial alteration due to a purely oscillatory flow environment, then there must be signaling events which cause such changes. We have investigated the elevation in intracellular calcium for all three types of pulsatile flow studied (Helmlinger et al 1993). As with our previous steady flow results, both non-reversing and reversing pulsatile flows elicited an overall elevation in cytosolic calcium, the amplitude and rise time of which were dependent on the exact flow and media conditions. Individual cell analyses revealed asynchronous calcium oscillations. The purely oscillatory flow environment was again found to be the most interesting. Here, no matter the medium employed, the monolayer exhibited no average response at all in terms of an increase in calcium. Even so, for all medium conditions used, there were asynchronous calcium oscillations at the single cell level. It should be noted that, for all three types of pulsatile flow studied, the percentage of cells in the monolayer and the frequency of the calcium oscillations were dramatically higher than that observed for steady flow.

From these results, it is clear that endothelial cells exhibit differences due to these different pulsatile flow environments. Most notably, an endothelial monolayer in the presence of a purely oscillatory flow, with no mean flow component is not the same as a similar monolayer for a static, no-flow condition, this in spite of the fact that both monolayers have a similar cobblestone appearance. This suggests that *in vivo* the vascular endothelium in oscillatory flow regions may possess different characteristics from those exhibited in regions of stasis.

A Mathematical Model of Signal Recognition and Transduction

Recently, we have embarked on the development of a mathematical model of the signaling mechanisms involved in the response of a vascular endothelial cell to flow. To be included in this model are both the recognition of the flow environment in which the cell resides and the transduction of that signal by second messengers. Although perhaps an ambitious undertaking, the development of such a model is believed to be a necessary complement to the experimental work in our laboratory. It not only helps structure one's thinking and the information already available, but it logically raises certain questions and thus assists in the design of future experiments.

As a first step in the development of such a model, we have addressed the elevation in intracellular calcium caused by an external chemical agonist (Wiesner et al 1993). The specific application is the response of human umbilical vein endothelial cells (HUVECs) to a thrombin-containing medium in a square capillary tube flow chamber. The species continuity equation for thrombin is coupled to a kinetic equation for the reversible binding of thrombin to a membrane receptor, the degradation of which then stimulates the mobilization of calcium from intracellular stores. This calcium oscillation model is based on the work of Meyer and Stryer (1988) and includes the time rate-of-change in cytosolic calcium, its relationship to the endoplasmic reticulum concentration as well as to the intracellular concentration of IP_3, and the sequestration of calcium in the mitochondria. Here the influence of flow enters in through the convection-diffusion coupling represented by the species continuity equation for thrombin.

The choice of both thrombin as the agonist and HUVECs as the cell was based on the availability in the literature of a reasonable amount of the quantitative data necessary to describe the kinetics of this model system. It is thus not surprising, perhaps, that calculations based on this agonist pathway model describe reasonably well experimental results on the intracellular calcium time course as obtained from cell culture flow studies. What these calculations do demonstrate is that mass transfer, receptor binding, and intracellular mobilization all contribute to the calcium dynamics. This is particularly true at very low shear stresses, as the response moves from being transport limited to one which is kinetically limited over a flow rate range corresponding to shear stresses between 1 and 5 dynes/cm^2.

This model is now being extended in a variety of ways. First, mitochondrial sequestration has been deleted since it is believed to be relatively unimportant in endothelial cells. Secondly, calcium buffering and the extrusion of calcium across the plasma membrane are being incorporated into the model. Finally, a model of the shear stress regulation of a membrane ion channel is being developed. Critical to this is the tension in the plasma membrane (Fung and Lin 1993), i.e. how much of the load is borne by the membrane and how much by the internal cytoskeletal structure. Taken together these modifications should result in a model of the response of an endothelial cell to flow which represents a first step in the quantitative description of the underlying mechanisms. It is our long-term goal to relate these signaling events to the relocalization in F-actin which occurs in response to flow and the associated elongation in cell shape.

Summary

The response of an endothelial cell to the sudden onset of a laminar, steady flow is an active process, one involving a localization of the F-actin microfilaments into stress fiber bundles, an alteration in the extracellular matrix including focal contact proteins, and an elongation in cell shape. This response is modified in the presence of a pulsatile flow environment, with the purely oscillatory flow representing a particularly fascinating type of flow in that, even with no mean flow component, there are alterations which take place. Concomitant with this adaptation to its mechanical environment, the endothelial cell exhibits a number of other changes. This includes the inhibition of proliferative activity, differential regulation in the synthesis and secretion of bioactive substances, and a selective alteration in the expression of cell adhesion molecules. These functional changes extend to the gene expression level. Critical to the endothelial cell's response to flow and the associated shear stress is its recognition of the mechanical environment in which it resides and the transduction of this mechanical signal into changes in structure and function. A key second messenger in this transduction process is the elevation in intracellular calcium which occurs as part of the response mechanism. Initial efforts to mathematically model the calcium dynamics which occur in response to flow have focused on the presence of a chemical agonist, e.g. thrombin, in the flowing medium. The results obtained indicate that, depending on the level of shear, mass transfer, receptor binding, and intracellular mobilization all contribute to the shape of the calcium transient.

Acknowledgments

This chapter was written while the senior author (RMN) was the Sackler Visiting Scholar in the Institute of Advanced Studies at Tel Aviv University in Israel. The authors acknowledge with thanks the support of the research discussed here by the National Institutes of Health (HL-

26890 and HL-48667) and the National Science Foundation (ECS-8815656 and BCS-9111761). The authors thank R.W. Alexander, B.C. Berk, R.M. Medford, C.J. Schwartz, and E.A. Sprague for their contributions to the work and ideas reflected in this chapter.

References

Ando, J.; Komatsuda, T.; Kamiya, A. Cytoplasmic calcium responses to fluid shear stress in cultured vascular endothelial cells. *In Vitro* Cell Dev. Biol. 24:871-877, 1988.

Asakura, T.; Karino, T. Flow patterns and spatial distribution of atherosclerotic lesions in human coronary arteries. Circ. Res. 66:1045-1066, 1990.

Berk, B.C.; Girard, P.R.; Mitsumata, M.; Alexander, R.W.; Nerem, R.M.; Shear stress alters the genetic growth program of cultured endothelial cells. Proc. World Congress Biomechs., Vol. II, 315, 1990 (Abstract).

Bhagyalakshmi, A.; Berthiaume, F.; Reich, K.M.; Frangos, J.A. Fluid shear stress stimulates membrane phospholipid metabolism in cultured human endothelial cells. J. Vasc. Res. 29:443-449, 1992.

Birukov, K.G.; Shirinsky, V.P.; Stepanova, O.V.; Tkachuk, V.A.; Resink, T.J. Cyclic stretch contributes to smooth muscle cell differentiation in culture. Ann. Biomed. Engr. 21 (Supp. 1):29, 1993 (Abstract).

Caro, C.G.; Fitz-Gerald, J.M.; Schroter, R.C. Atheroma and arterial wall shear. Observation, correlation and proposal of a shear-dependent mass transfer mechanism for atherogenesis. Proc. Roy. Soc., London, B 177:109-159, 1971.

Dartsch, P.C.; Betz, E. Response of cultured endothelial cells to mechanical stimulation. Basic Res. Cardiol. 84:268-281, 1989.

Davies, P.F.; Dewey, C.F. Jr.; Bussolari, S.R.; Gordon, E.J.; Gimbrone, M.A. Jr. Influence of hemodynamic forces on vascular endothelial function. *In vitro* studies of shear stress and pinocytosis in bovine aortic endothelial cells. J. Clin. Invest. 73:1121-1129, 1983.

Davies, P.F.; Remuzzi, A.; Gordon, E.J.; Dewey, C.F. Jr.; Gimbrone, M.A. Jr. Turbulent shear stress induces vascular endothelial cell turnover *in vitro*. Proc. Natl. Acad. Sci. USA 83:2114-2117, 1986.

Dewey, C.F.; Bussolari, S.R.; Gimbrone, M.A. Jr.; Davies, P.F. The dynamic response of vascular endothelial cells to fluid shear stress. ASME J. Biomech. Engr. 103:177-181, 1981.

Diamond, S.L.; Eskin, S.G.; McIntire, L.V. Fluid flow stimulates tissue plasminogen activator secretion by cultured human endothelial cells. Science 243:1483-1485, 1989.

Diamond, S.L.; Sharefkin, J.B.; Dieffenbach, C.; Frazier-Scott, K.; McIntire, L.V.; Eskin, S.G. Tissue plasminogen activator messenger RNA levels increase in cultured human endothelial cells exposed to laminar shear stress. J. Cell Physiol. 143:364-371, 1990.

Dull, R.O.; Davies, P.F. Flow modulation of agonist (ATP)-response (Ca^{++}) coupling in vascular endothelial cells. Am. J. Physiol. 261:H149-154, 1991.

Eskin, S.G.; Ives, C.L.; McIntire, L.V.; Navarro, L.T. Response of cultured endothelial cells to steady flow. Microvasc. Res. 28:87-94, 1984.

Flaherty, J.R.; Pierce, J.R.; Ferrans, V.J.; Patel, D.J.; Tucker, W.K.; Fry, D.L. Endothelial nuclear patterns in the canine arterial tree with particular reference to hemodynamic events. Circ. Res. 30:23-33, 1972.

Frangos, J.A.; Eskin, S.G.; McIntire, L.V.; Ives, C.L. Flow effects on prostacyclin production by cultured human endothelial cells. Science 227:1477-1479, 1985.

Friedman, M.H.; Hutchins, G.M.; Bargeron, C.B.; Deters, O.J.; Mark, F.F. Correlation between intimal thickness and fluid shear in human arteries. Atherosclerosis 39:425-436, 1981.

Friedman, M.H.; Peters, O.J.; Bargeron, C.B.; Hutchins, G.M.; Mark, F.F. Shear-dependent thickening of the human arterial intima. Atherosclerosis 60:161-171, 1986.

Fung, Y.C.; Lin, S.Q. Elementary mechanics of the endothelium of blood vessels. ASME J. Biomech. Engr. 115:1-12, 1993.

Geiger, R.V.; Berk, B.C.; Alexander, R.W.; Nerem, R.M. Flow-induced calcium transients on single endothelial cells: spatial and temporal analysis. Am. J. Physiol.: Cell Physiol. 262: C1411-1417, 1992.

Girard, P.R.; Nerem, R.M. Endothelial cell signaling and cytoskeletal changes in response to shear stress. Frontiers Med. Biol. Eng. 5:31-36, 1993.

Girard, P.R.; Helmlinger, G.; Nerem, R.M. Shear stress effects on the morphology and cytomatrix of cultured vascular endothelial cells. In: Physical Forces and the Mammalian Cell. J.A. Frangos (Ed.). Academic Press, NY, pp. 193-222, 1993.

Gorfien, S.F.; Winston, S.K.; Thibault, L.E.; Macarak, E.J. Effects of biaxial deformation on pulmonary artery endothelial cells. J. Cell Physiology 139:492-500, 1989.

Grabowski, E.F.; Jaffe, E.A.; Weksler, B.B. Prostacyclin production by cultured human endothelial cells exposed to step increases in shear stress. J. Lab. Clin. Med. 105:36-43, 1985.

Grottum, P.; Svindland, A.; Walloe, L. Localization of atherosclerotic lesions in the bifurcation of the left main coronary artery. Atherosclerosis, 47:55-62, 1983.

Helmlinger, G.; Geiger, R.V.; Schreck, S.; Nerem, R.M. Effects of pulsatile flow on cultured vascular endothelial cell morphology. ASME J. Biomech. Engr. 113:123-131, 1991.

Helmlinger, G.; Nerem, R.M. The intracellular free calcium response in endothelial cells subjected to steady and pulsatile laminar flow. Ann. Biomed. Engr. 21 (Supp. 1):39, 1993 (Abstract).

Hsieh, H.J.; Li, N.Q.; Frangos, J.A. Shear stress increases endothelial platelet-derived growth factor messenger RNA levels. Am. J. Physiol. 260 (Heart Circ. Physiol. 29):H642-646, 1991.

Kaiser, L.; Sparks, H.V. Endothelial cells: not just a cellophane wrapper. Arch. Intern. Med. 147:569-573, 1987.

Kamiya, A.; Togawa, T. Adaptive regulation of wall shear stress to flow and change in the canine carotid artery. Am. J. Physiol. 239:H14-21, 1980.

Kim, D.W.; Gotlieb, A.I.; Langille, B.L. *In vivo* modulation of endothelial F-actin microfilaments by experimental alterations in shear stress. Arteriosclerosis 9:439-445, 1989.

Ku, D.N.; Giddens, D.P.; Zarins, C.K.; Glagov, S. Pulsatile flow and atherosclerosis in the human carotid bifurcation. Arteriosclerosis 5:293-302, 1985.

Langille, B.L.; O'Donnell, F. Reductions in arterial diameter produced by chronic decreases in blood flow are endothelium-dependent. Science 231:405-407, 1986.

Lansman, J.B.; Hallam, T.J.; Rink, T.J. Single stretch-activated ion channels in vascular endothelial cells as mechanotransducers? Nature 325:811-813, 1987.

Levesque, M.J.; Liepsch, D.; Moravec, S.; Nerem, R.M. Correlation of endothelial cell shape and wall shear stress in a stenosed dog aorta. Arteriosclerosis 6:220-229, 1986.

Levesque, M.J.; Nerem, R.M. The elongation and orientation of cultured endothelial cells in response to shear stress. ASME J. Biomech. Engr. 107:341-347, 1985.

Levesque, M.J.; Sprague, E.A.; Nerem, R.M. Vascular endothelial cell proliferation in culture and the influence of flow. Biomaterials 11:702-707, 1990.

Meyer, T.; Stryer, L. Molecular model for receptor-stimulated calcium spiking. Proc. Nat. Aca. Sci. USA, 85:5051-5055, 1988.

Mitsumata, M.; Fishel, R.S.; Nerem, R.M.; Alexander, R.W.; Berk, B.C. Fluid shear stress stimulates platelet-derived growth factor expression in endothelial cells. Am. J. Physiology 265:H3-8, 1993.

Mitsumata, M.; Nerem, R.M.; Alexander, R.W.; Berk, B.C. Shear stress inhibits endothelial cell proliferation by growth arrest in the G_0/G_1 phase of the cell cycle. FASEB J. 5(4):A527 (Abstract), 1991.

Mo, M.; Eskin, S.G.; Schilling, W.P. Flow-induced changes in Ca^{2+} signaling of vascular endothelial cells: effect of shear stress and ATP. Am. J. Physiol. 260:H1698-1707, 1991.

Nerem, R.M. Vascular fluid mechanics, the arterial wall, and atherosclerosis. ASME J. Biomech. Engr. 114:274-282, 1992.

Nerem, R.M.; Levesque, M.J.; Cornhill, J.F. Vascular endothelial morphology as an indicator of blood flow. ASME J. Biomech. Engr. 103:172-176, 1981.

Nishida, K.; Harrison, D.G.; Navas, J.P.; Fisher, A.A.; Dockery, S.P.; Uematsu, M.; Nerem, R.M.; Alexander, R.W.; Murphy, T.J. Molecular cloning and characterization of the constitutive bovine aortic endothelial cell nitric oxide synthase. J. Clinical Invest. 90:2092-2096, 1992.

Nollert, M.U.; Eskin, S.G.; McIntire, L.V. Shear stress increases inositol trisphosphate levels in human endothelial cells. Biochem. Biophys. Commun. 170:281-287, 1990.

Nollert, M.V.; McIntire, L.V. Convective mass transfer effects on the intracellular calcium response of endothelial cells. ASME J. Biomech. Engr. 114:321-326, 1992.

Ohtsuka, A.; Ando, J.; Korenaga, R.; Kamiya, A.; Toyama-Sorimachi, N.; Miyasaka, M. The effect of flow on the expression of vascular adhesion molecule-1 by cultured mouse endothelial cells. Biochem. Biophys. Res. Comm. 193:303-310, 1993.

Olesen, S.P.; Clapham, D.E.; Davies, P.F. Haemodynamic shear stress activates a K+ current in vascular endothelial cells. Nature 331:168-170, 1988.

Prasad, A.R.S.; Logan, S.A.; Nerem, R.M.; Schwartz, C.J.; Sprague, E.A. Flow-related responses of intracellular inosital phosphate levels in cultured aortic endothelial cells. Circ. Res. 72/4:827-836, 1993.

Resnick, N., Dewey, C.F.; Atkinson, B.; Collins, T.; Gimbrone, M.A. Jr. Platelet-derived growth factor B chain promoter contains a cis-acting fluid shear stress responsive element. Proc. Natl. Acad. Sci. USA 90:4591-4595, 1993.

Sato, M.; Levesque, M.J.; Nerem, R.M. Micropipette aspiration of cultured bovine aortic endothelial cells exposed to shear stress. Arteriosclerosis 7:276-286, 1987.

Sato, M.; Theret, D.P.; Wheeler, L.T.; Ohshima, N.; Nerem, R.M. Application of the micropipette technique to the measurement of cultured porcine aortic endothelial cell viscoelastic properties. ASME J. Biomech. Eng. 112:263-268, 1990.

Schwarz, G.; Droogmans, G.; Nilius, B. Shear stress induced membrane currents and calcium transients in human vascular endothelial cells. Pflügers Arch. 421:394-396, 1992.

Sharefkin, J.B.; Diamond, S.L.; Eskin, S.G.; McIntire, L.V.; Dieffenbach, C.W. Fluid flow decreases preproendothelin mRNA levels and suppresses endothelin-1 peptide release in cultured human endothelial cells. J. Vasc. Surg. 14:1-9, 1991.

Shen, J.; Luscinskas, F.W.; Connolly, A.; Dewey, C.F. Jr., Gimbrone, M.A. Jr. Fluid shear stress modulates cytosolic free calcium in vascular endothelial cells. Am. J. Physiol. 262 (Cell Physiol. 31): C384-390, 1992.

Shen, J.; Gimbrone, M.A. Jr.; Luscinskas, F.W.; Dewey, C.F. Jr. Regulation of adenine nucleotide concentration at endothelium-fluid interface by viscous shear flow. Biophys. J. 64:1323-1330, 1993.

Shirinsky, V.P.; Antonov, A.S.; Birukov, K.G.; Sobolevsky, A.V.; Romanov, Y.A.; Kabaeva, N.V.; Anonova, G.N.; Smirnov, V.N. Mechano-chemical control of human endothelium orientation and size. J. Cell Biol. 109:331-339, 1989.

Silkworth, J.B.; Stehbens, W.E. The shape of endothelial cells in *en face* preparations of rabbit blood vessels. Angiology 26:474-487, 1975.

Sprague, E.A.; Cayatte, A.J.; Nerem, R.M.; Schwartz, C.J. Cultured endothelial cells conditioned to prolonged low shear stress exhibit enhanced monocyte adherence and expression of related genes, MCP-1 and VCAM-1. In: Proceedings of the Cardiovascular Science and Technology Conference. LH Edmunds, Jr. (ed)., Association for the Advancement of Medical Instrumentation, Washington, D.C. 100, 1992.

Sprague, E.A.; Steinbach, B.L.; Nerem, R.M.; Schwartz, C.J. Influence of a laminar steady-state fluid-imposed wall shear stress on the binding, internalization, and degradation of low density lipoproteins by cultured arterial endothelium. Circulation 76:648-656, 1987.

Sumpio, B.E.; Banes, A.J. Prostacyclin synthetic activity in cultured aortic endothelial cells undergoing cyclic stretch. Surgery 104:383-389, 1988.

Sumpio, B.E.; Banes, A.J.; Levin, L.G.; Johnson, G. Jr. Mechanical stress stimulates aortic endothelial cells to proliferate. J. Vasc. Surg. 6:252-256, 1987.

Sumpio, B.E.; Banes, A.J.; Levin, L.G.; Johnson, G. Jr. Alternations in aortic endothelial cell morphology and cytoskeleton protein synthesis during cyclic tensional deformation. J. Vasc. Surg. 7:130-138, 1988.

Theret, D.P.; Levesque, M.J.; Sato, M.; Nerem, R.M.; Wheeler, L.T. The application of a homogeneous half-space model in the analysis of endothelial cell micropipette measurement. ASME J. Biomech. Eng. 110:190-199, 1988.

Thoumine, O.; Girard, P.R.; Nerem, R.M. The effects of shear stress on the extracellular matrix of cultured bovine aortic endothelial cells. J. Cell. Biochem. (Supplement) 17E:157, 1993 (Abstract).

Watson, P.A. Function follows form: generation of intracellular signals by cell deformation. FASEB J. 5:2013-2019, 1991.

Wechezak, A.R.; Viggers, R.F.; Sauvage, L.R. Fibronectin and F-actin redistribution in cultured endothelial cells exposed to shear stress. Lab. Invest. 53:639-647, 1985.

White, G.E.; Fujiwara, K.; Shelton, E.J.; Dewey, C.F. Jr.; Gimbrone, M.A. Jr. Fluid shear stress influences cell shape and cytoskeletal organization in cultured vascular endothelium. Fed. Proc. 41:321 (Abstract), 1982.

White, G.E.; Gimbrone, M.A. Jr.; Fujiwara, K. Factors influencing the expression of stress fibers in vascular endothelial cells *in situ*. J. Cell Biol. 97:416-424, 1983.

Wiesner, T.F.; Helmlinger, G.; Nerem, R.M. A model of thrombin-mediated cytosolic calcium mobilization in HUVECs responding to flow. Ann. Biomed. Engr. 21 (Supp. 1):38, 1993 (Abstract).

Winston, F.K.; Thibault, L.E.; Macarak, E.J. An analysis of the time-dependent changes in intracellular calcium concentration in endothelial cells in culture induced by mechanical stimulation. ASME J. Biomech. Engr. 115:160-168, 1993.

Zarins, C.K.; Zatina, M.A.; Ku, D.N.; Glagov, S.; Giddens, D.P. Shear stress regulation of artery lumen diameter in experimental atherogenesis. J. Vasc. Surg. 5:413-420, 1987.

Ziegler, T.; Nerem, R.M. The effect of flow on the process of endothelial cell division. Arteriosclerosis and Thrombosis (in press).

6
Flow Modulation of Receptor Function in Leukocyte Adhesion to Endothelial Cells

D. A. Jones, C. W. Smith, L. V. McIntire

Introduction

Proper function of the immune system depends on the accurate localization and recirculation of leukocytes throughout the body. This leukocyte homing is important in physiological processes such as neutrophil accumulation at sites of infection and lymphocyte recirculation through the blood and lymphatics. It is also important in pathological processes such as neutrophil-mediated myocardial reperfusion injury and the accumulation of monocytes at atherosclerotic lesions. All of the various localization processes have in common the specific, receptor-mediated adhesion of a particular population of leukocytes to a particular area of vascular endothelium, and a thorough understanding of this adhesion has great potential to allow development of useful new medications. Research in this area is complicated by the large number of receptors involved and the differential regulation of their expression on particular cell subpopulations. An extremely important additional complication arises from the fact that these interactions occur within the flowing bloodstream. Research on the effect of flow on leukocyte adhesion to vascular endothelium has revealed that different types of receptors are capable of mediating distinct types of adhesive events, such as initial rolling of leukocytes along the endothelium or later firm adhesion of leukocytes. This chapter will discuss leukocyte/endothelial cell interactions under flow conditions focusing on the receptors involved, receptor function in various physiological settings, experimental methods in common use, and properties of the bonds formed by different receptor types.

Properties of Receptors Involved in Leukocyte-Endothelial Cell Adhesion

Integrins

Integrins are heterodimeric glycoproteins consisting of one of several α subunits plus one of several β subunits bound noncovalently. Combinations of the thirteen known α subunits and the seven known β subunits produce at least nineteen receptors which have been extensively reviewed (Bevilacqua 1993; Sanchez-Madrid and Corbi 1992). The four integrin receptors which to date are most heavily implicated in specific leukocyte/endothelial cell interactions are $\alpha_L\beta_2$ (CD11a/CD18, LFA-1), $\alpha_M\beta_2$ (CD11b/CD18, Mac-1), $\alpha_4\beta_1$ (CD49d/CD29, VLA-4), and $\alpha_4\beta_7$. A schematic diagram of the structures of Mac-1 and VLA-4 is given in Fig. 1. Other integrins are

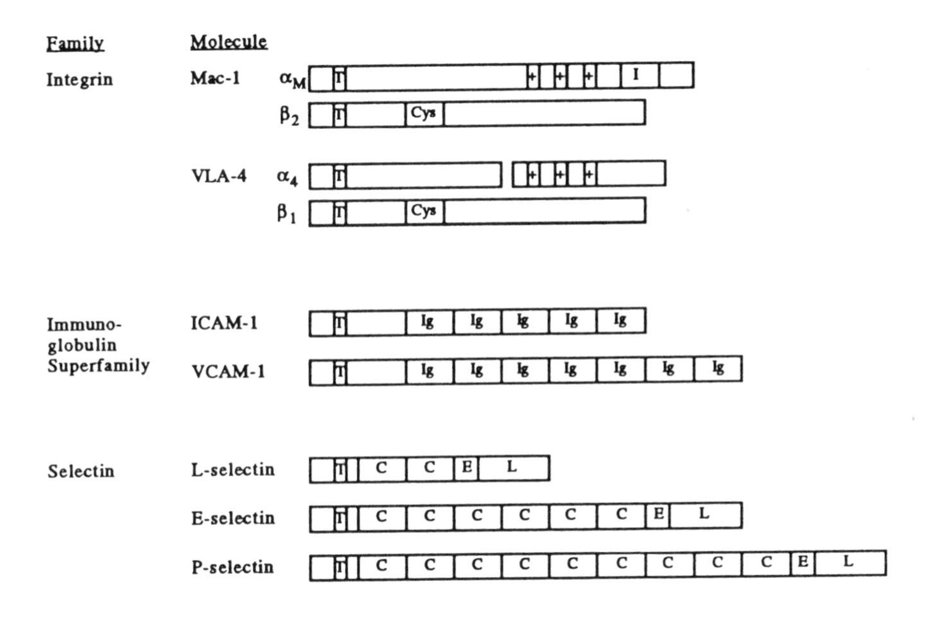

Fig. 1: Schematic diagrams of the primary structure of representative adhesion molecules. Nomenclature: T, hydrophobic transmembrane segment; +, divalent cation binding site; I, integrin I domain; Cys, cysteine-rich region; Ig, immunoglobulin domain; C, complement binding protein-like domain; E, epidermal growth factor-like domain; L, lectin-like carbohydrate binding domain.

very similar in structure, with α subunits averaging approximately 1100 amino acids and β subunits averaging approximately 750 amino acids. The integrins have a very broad distribution and are important in a wide variety of cell-cell and cell-matrix interactions. With regard to leukocytes, the integrins vary in expression among the various leukocyte groups, and this is likely an important means of providing specificity in leukocyte targeting. For instance, LFA-1 is expressed on all circulating leukocytes while Mac-1 is expressed on monocytes and granulocytes but not on lymphocytes. VLA-4 is expressed on monocytes and lymphocytes, but not on neutrophils. $\alpha_4\beta_7$ expression varies even among subpopulations of memory T lymphocytes.

Regulation of integrin-mediated binding occurs at three levels: modulation of receptor synthesis, mobilization of an intracellular pool of receptors, and modulation of the binding affinity of individual receptors. Modulation of integrin synthesis is generally associated with cell differentiation rather than with the rapid response of terminally differentiated cells to particular challenges, and refers to the differences in expression noted above. Cell adhesion can be modulated on a time scale of minutes by mobilizing intracellular pools of integrin receptors. For example, in neutrophils and monocytes chemotactic factors cause up to a 10-fold increase in surface expression of Mac-1 and p150,95 (reviewed in Smith 1993). The mechanism of this form of upregulation involves fusion of leukocyte granules containing the receptors with the cell surface. An even more rapid mechanism of cell adhesion modulation is the ability of some integrins to increase their binding affinity through conformational changes upon stimulation of the cell (reviewed in Hogg and Landis 1993). This effect has been demonstrated for all of the four integrins mentioned above, and mechanistic details of this integrin "activation" are still an area of active research. It is also important to note that in addition to serving as effector molecules of leukocyte activation, they can also serve as sensors, triggering a variety of leukocyte activation responses upon ligand binding.

Immunoglobulin Superfamily

The immunoglobulin superfamily of receptors is defined by the presence of the immunoglobulin domain, which is composed of 70-110 amino acids arranged in a well-characterized structure (reviewed in Bevilacqua 1993; Buck 1992). This group of receptors is much more diverse than the integrins, ranging from soluble and membrane-bound immunoglobulins to the multi-receptor T-cell antigen receptor complex to single-chain cellular adhesion molecules. This review will focus on the cellular adhesion molecules ICAM-1 (CD54), VCAM-1, and MAdCAM-1, as these have been most strongly implicated in specific leukocyte/endothelial cell recognition. ICAM-2, PECAM-1 (CD31), and LFA-3 (CD58) play important adhesive roles as well. ICAM-1, VCAM-1, and MAdCAM-1 serve as protein ligands for the integrins LFA-1 and Mac-1, VLA-4, and $\alpha_4\beta_7$, respectively. Schematic diagrams of ICAM-1 and VCAM-1 are shown in Figure 1. The others mentioned above are similar in structure, with various numbers of immunoglobulin homology domains.

ICAM-1 is constitutively expressed on only a few cell types, but at the site of an inflammatory response is induced on a wide variety of cells. Endothelial cells upregulate ICAM-1 within several hours of cytokine stimulation, and peak expression occurs after 24-48 hours. Since it is a counter-receptor for the leukocyte integrins LFA-1 and Mac-1, ICAM-1 induction provides an important means of recruiting leukocytes specifically to inflammatory sites. VCAM-1 expression is induced on endothelial cells by many of the same stimuli as ICAM-1, and with a similar time course. It serves as a counter-receptor for leukocyte VLA-4 in the recruitment of leukocytes to inflammatory sites. Recent *in vivo* studies have demonstrated that PECAM-1 is essential for proper recruitment of neutrophils in a peritonitis model, although its exact role is not yet understood (Vaporciyan et al. 1993). It was also recently reported that this molecule is expressed on unique T cell subsets and that CD31 binding has the capacity to activate integrin-mediated adhesion, indicating that CD31 signalling may provide another means of selective T cell recruitment (Tanaka et al. 1992).

Selectins

The selectin family of adhesion receptors has three known members: L-selectin (LECCAM-1, LAM-1, murine gp90^{Mel14}, peripheral lymph node homing receptor), E-selectin (ELAM-1), and P-Selectin (GMP-140, PADGEM, CD62) which have been extensively reviewed (Bevilacqua 1993; Vestweber 1992). They are grouped as selectins based on structural similarities diagrammed in Fig. 1. The processed proteins have an NH_2-terminal domain homologous to the type C (calcium-dependent) lectins, which is the ligand binding site. As with other lectins, ligands are carbohydrate structures.

Selectin expression is regulated by a variety of mechanisms. L-selectin is constitutively expressed on unactivated neutrophils, eosinophils, and monocytes, as well as on naive T cells and one subset of memory T cells. On neutrophils, expression of L-selectin has been shown to be localized to the tips of microvillus-like projections on the cell surface (Picker et al. 1991b). This striking localization should facilitate the proposed role of neutrophil L-selectin as an adhesion receptor for endothelial cell carbohydrate structures. Neutrophils and lymphocytes also have the remarkable property of proteolytically cleaving L-selectin from the cell surface upon activation (Kishimoto et al. 1989). This shedding is quite rapid, with 50% of the protein shed within five minutes of stimulation with chemotactic factors such as formyl-Met-Leu-Phe (fMLP) (Smith et al. 1991, reviewed in Smith 1993).

E-selectin has been detected *in vivo* only in inflammed tissues. It is induced on endothelial cells upon stimulation by the cytokines IL-1, TNFα, or LPS, and this induction appears to be predominantly localized to postcapillary venules (Rohde et al. 1992), the principal site of leukocyte emigration in inflammation. Peak levels of E-selectin are reached after approximately four hours of cytokine stimulation and return to basal levels after 16-24 hours. However, much longer-lasting expression has been observed in the setting of chronic inflammation in the skin (Picker et al. 1991a).

P-selectin is found constitutively within α granules of platelets and Weibel-Palade bodies of endothelial cells. In endothelial cells, histamine, thrombin, bradykinin, and substance P cause a rapid mobilization of Weibel-Palade bodies to the cell periphery, where they fuse with the cell membrane resulting in expression of the receptors on the cell surface. Under these conditions, expression begins within seconds of stimulation, followed by peak expression at approximately 8-10 min, and endocytosis of most receptors by 45-60 min.

Oligosaccharide ligands for selectins are the focus of considerable current research. An important concept in this area is that differential glycosylation features of surface glycoproteins appear to be a mechanism of determining specificity in leukocyte interactions. The exact nature of the glycosylation features is not clear, although a large body of evidence suggests an important role for the oligosaccharides sialyl Lewis x, sialyl Lewis a, and related structures. These are widespread glycocalyx components, but nonetheless can contribute to specificity. For example, neutrophils but not lymphocytes express sLe^x and are bound by endothelial E- and P-selectins (Picker et al. 1991b). One subset of lymphocytes, on the other hand, expresses the closely related cutaneous lymphocyte antigen (CLA) which is a sialylated oligosaccharide that can serve as an alternate ligand for E-selectin (Berg et al. 1991). Several of the peptide cores which carry the oligosaccharide binding sites have recently been characterized. Three of these contain mucin-like domains which likely bear the glycosylations bound by L-selectin. These are GlyCAM-1 (Sgp50, the 50kd component of peripheral node addressin)(Dowbenko et al. 1993), CD34 (Sgp90, the 90kd component of peripheral node addressin)(Baumhueter et al. 1993), and MAdCAM-1 (Berlin et al. 1993). MAdCAM-1 is thus capable of serving as a ligand for both L-selectin and $\alpha_4\beta_7$ integrin at sites such as mucosal lymph nodes where the necessary glycosylation takes place (Berlin et al. 1993).

Other Adhesion Molecules

Two other adhesion molecules of special interest are CD44 and VAP-1. CD44 (H-CAM, the Hermes antigen, pgp1, ECMRIII) exists in many forms, and has had many potential functions ascribed to it (Lesley et al. 1993). Like the selectins, its activity in leukocyte adhesion appears to be mediated by carbohydrate binding, but in this case hyaluronic acid is a principle component of the ligand. The wide distribution of CD44 suggests that it may play an accessory role, augmenting specific interactions initiated by other adhesion receptors (Picker and Butcher 1992) or may function in activation steps. A recently characterized endothelial cell adhesion molecule designated VAP-1 (Salmi and Jalkanen 1992) has also been shown to mediate adhesion of lymphocytes to high endothelial venules. The complete primary structure of this molecule is not yet known, although it does appear to be a novel receptor.

Leukocyte Adhesion Under Flow Conditions

The Multistep Process

One of the most important aspects of leukocyte extravasation is that it is a multistep process. Fig. 2 shows the current model of this process, divided for the sake of discussion into five steps: initial contact, primary adhesion (often rolling), activation, secondary adhesion, and transmigration. Initial contact with endothelium is aided by the size of postcapillary

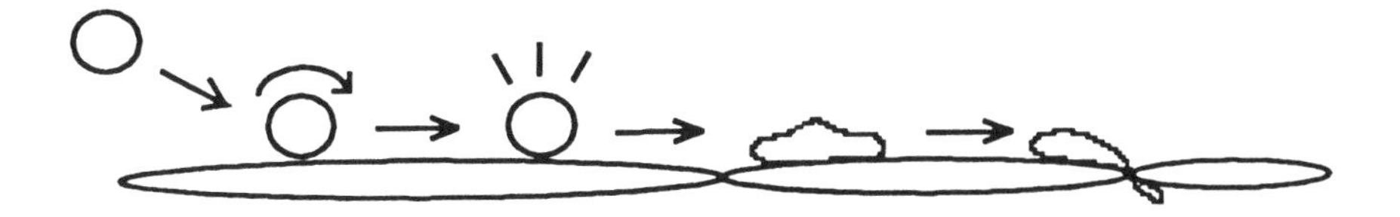

Fig. 2: The multistep process in which leukocytes adhere to vascular endothelium and extravasate. The steps illustrated are initial contact, primary adhesion, activation, secondary adhesion, and transmigration.

venules, which are the principle sites of selective leukocyte extravasation (Fiebig et al. 1991), and by increased vascular permeability. The size of postcapillary venules (approximately 20-60 μm) is small enough that contact between leukocytes and endothelium is frequent, but large enough that initial contact is brief, necessitating specific binding for further interactions (Ley and Gaehtgens 1991). In an inflammatory setting, increased vascular permeability leads to plasma leakage and an increase in local hematocrit which changes the characteristics of the flow profile through the vessel and allows more frequent contact between leukocytes and the vessel wall (Chien 1982; Tangelder et al. 1986). Following this initial contact, leukocytes often roll slowly along the endothelium for some distance before either reentering the free stream or establishing firm adhesion. This rolling is believed to keep the cells in close enough contact with the endothelium that they can be stimulated by substances in the local environment and engage additional binding mechanisms. Rolling requires a definite adhesive interaction, as was first demonstrated by the observation that neutrophils roll with a much slower velocity than that predicted for cells tumbling in the fluid stream adjacent to the vessel wall without any adhesion (Atherton and Born 1973; Goldman et al. 1967). Firm adhesion can follow the rolling interaction and is often dependent upon the leukocyte receiving an activating stimulus while in contact with the endothelium. There is evidence that these activating signals can be soluble substances derived from the endothelium, tissues, or pathogens, factors bound to the endothelium, or direct transduction of activating singals by adhesion receptors themselves (Palecanda et al. 1992, reviewed in Hogg and Landis 1993). Activation of the leukocytes can bring about several changes which alter adhesive characteristics and allow the establishment of firm adhesion, as will be discussed later. Once firm adhesion has been established, the leukocytes migrate to interendothelial junctions and diapedese (reviewed in Smith 1993).

It is important to note that in this multistep process of leukocyte extravasation, specificity in leukocyte recruitment can be generated at any of the steps (Butcher 1991). For instance, a leukocyte which contains primary adhesion machinery to bind a particular endothelial surface might not emigrate there due to a lack of responsiveness to local activating substances or a lack of the proper activation-dependent adhesion machinery. It is also important to clarify our particular use of the terms primary and secondary adhesion. Primary adhesion refers here to adhesive mechanisms which are capable of binding leukocytes which are moving with the bloodstream. These mechanisms would normally be activation-independent, although one should not preclude the possibility of, for example, constitutively activated integrin receptors on a leukocyte subpopulation mediating primary adhesion. Secondary adhesion refers to adhesive mechanisms which are capable of producing long-lasting, stable adhesion of leukocytes to endothelial cells. Some of these mechanisms, such as Mac-1/ICAM-1 binding, are activation-dependent, although others might simply require the close apposition provided by previous primary adhesion.

Neutrophil Adhesion Under Flow

The first evidence for a distinction between primary and secondary adhesion came with studies of neutrophil adhesion to endothelial cells *in vitro* (Lawrence et al. 1987; Lawrence et al. 1990) and *in vivo* (Arfors et al. 1987) demonstrating CD18-independent and CD18-dependent adhesion mechanisms under flow conditions. These studies were followed by many further studies (reviewed in Smith 1993) which clearly distinguish separate mechanisms for initial adhesion and rolling *versus* firm adhesion and migration. This research has further shown that selectin/carbohydrate interactions are primarily responsible for initial adhesion and rolling, while firm adhesion and leukocyte migration are mediated primarily by integrin/polypeptide interactions. In the neutrophil system, all three selectins have been implicated in primary adhesion. E-selectin appears to be the predominant endothelial receptor for primary adhesion when the endothelium is stimulated with cytokines such as IL-1, TNFα, or lipopolysaccharide on a time scale of hours (Abbassi et al. 1993). P-selectin, on the other hand, mediates neutrophil rolling on a time scale of minutes following endothelial stimulation with rapid-acting inflammatory agonists such as histamine, thrombin, or bradykinin (Jones et al. 1993a). L-selectin appears to contribute to primary adhesion via several endothelial ligands not yet fully characterized (Abbassi et al. 1991, reviewed in Smith 1993).

The transition from selectin-mediated initial adhesion and rolling to integrin-mediated firm adhesion and migration in the neutrophil is a remarkable process. Activation of the neutrophil appears to simultaneously upregulate CD18 integrin binding mechanisms and downregulate L-selectin binding mechanisms. Upregulation of integrin binding is accomplished primarily through conformational changes in the integrins which increases their affinity, as discussed earlier. Also, surface expression of Mac-1 is increased up to ten-fold by rapid mobilization of an intracellular pool of receptors (Smith et al. 1991; Hughes et al. 1992), although additional activation of this receptor pool appears to be required for their use (Hughes et al. 1992; Diamond and Springer 1993). Downregulation of L-selectin-mediated adhesion is accomplished through rapid proteolytic shedding of surface L-selectin (Kishimoto et al. 1989). This is very rapid in the case of the neutrophil, with 50% of the protein shed within 5 minutes of stimulation (Smith et al. 1991). This may be a way for the neutrophil to disengage L-selectin-mediated binding to allow free migration along the endothelial surface.

The basic idea that selectin/carbohydrate bonds mediate initial adhesion and leukocyte rolling while integrin/polypeptide bonds mediate firm adhesion and migration has recently received dramatic confirmation through the study of two immunodeficiency syndromes called leukocyte adhesion deficiency types 1 and 2 (LAD-1 and LAD-2). The molecular defects underlying LAD-1 are known to be defects in the β_2 subunit of the leukocyte integrins (Anderson and Springer 1987; Kishimoto et al. 1987) and those underlying LAD-2 are likely to be fucose metabolism defects (Etzioni et al. 1992). In particular, neutrophils from LAD-2 patients lack the fucose-containing oligosaccharide moiety sialyl Lewis x which serves as a ligand for vascular selectins. Experiments performed (von Andrian et al. 1993) using isolated neutrophils from both LAD-1 and LAD-2 patients in a mesenteric venule experimental model of inflammation in rabbits gave the following results: integrin-deficient LAD-1 neutrophils rolled along inflammed venules normally, but were unable to bind firmly and transmigrate even under conditions of reduced flow. Sialyl Lewis x-deficient LAD-2 neutrophils on the other hand were markedly deficient in their ability to bind to the endothelium and roll under flow conditions, but bound firmly and transmigrated when flow was reduced to low levels.

Lymphocyte Adhesion Under Flow Conditions

Lymphocyte adhesion to endothelium is much less well understood than neutrophil adhesion. This is because the physiology of lymphocyte trafficking is much more complex and because many more adhesion receptors have been implicated in lymphocyte/endothelial cell interactions with many more still likely to be discovered (Picker 1992). As discussed earlier, L-selectin is an important adhesion molecule in lymphocyte binding to high endothelial venules with evidence for interactions with the endothelial ligands GlyCAM-1, CD34, and MAdCAM-1. E-selectin has also been demonstrated to be important in recruitment of a skin-homing subset of T lymphocytes which bear the cutaneous lymphocyte antigen (CLA) (Picker et al. 1991a; Berg et al. 1991). Flow studies in our lab indicate that CLA^+ T cells isolated by fluorescence activated cell sorting bind at very high levels to E-selectin under flow conditions, whereas CLA^- T cells do not show significant adhesion (Jones et al. 1993b). Other studies, including work in our own laboratories point to VLA-4 ($\alpha_4\beta_1$ integrin)/VCAM-1 and β_2 integrin/ICAM-1 interactions as also being very important in lymphocyte binding to endothelial cells.

Given the evidence for both selectin/carbohydrate interactions and integrin/polypeptide interactions in lymphocyte adhesion, it is appealing to think that the adhesion sequence elucidated for neutrophils may hold for lymphocytes as well. It is just as likely, however, that a variety of adhesion sequences are possible which allow for greater specificity in the localization of particular lymphocyte subsets to particular endothelial environments. For instance, in one *in vivo* study of lymphocyte binding, rolling of lymphocytes along high endothelial venules (HEV) was not observed (Bjerknes et al. 1986). Another *in vivo* study did observe rolling, however, and found that a G-protein mediated activation event is associated with the establishment of firm adhesion (Bargatze and Butcher 1993).

Detailed studies in our laboratories have addressed this issue (Jones et al. 1993b), and give evidence for well-defined primary and secondary adhesion mechanisms in the binding of T lymphocytes to endothelial cells *in vitro*. We find that with IL-1 stimulated HUVEC and postcapillary venular flow rates, some lymphocytes roll for extended periods, others roll briefly then adhere firmly, and others adhere firmly without rolling at all. Static and flow adhesion assays using murine L cells transfected with E-selectin, ICAM-1, or VCAM-1 show that E-selectin is capable of mediating only primary adhesion (pure rolling under flow conditions), ICAM-1 is capable of mediating only secondary adhesion (firm adhesion under static conditions only), and VCAM-1 is capable of mediating both primary and secondary adhesion (firm adhesion under static or flow conditions). Studies with IL-1-stimulated HUVEC monolayers and receptor-blocking monoclonal antibodies show that secondary adhesion is mediated by the integrin pathways VCAM-1/VLA-4 and ICAM-1/LFA-1, but that a primary adhesion mechanism exists which does not appear to utilize any of the known adhesion molecules. This mechanism by itself mediates T cell rolling and is a necessary prerequisite to adhesion by the two integrin pathways.

Monocyte Adhesion Under Flow Conditions

Monocyte adhesion to endothelial cells may be similar in many respects to neutrophil adhesion. The special case of monocyte recruitment to atherosclerotic lesions, however, has several special characteristics which are influenced greatly by arterial blood flow. The specific localization of monocytes at atherosclerotic lesions is likely to be determined by several factors including receptor expression at the lesion site, activating substances released from the lesion, the low fluid shear and longer blood cell residence times noted at lesion prone sites, and flow-dependent alterations to the endothelial cells at these sites. Receptors involved in the various adhesive events which lead to monocyte extravasation into atherosclerotic lesion-prone sites have not yet been thoroughly determined. It is appealing to think that, as with neutrophils and lymphocytes, there are primary and secondary adhesion mechanisms involved, but this need not be the case if the flow field is such that integrin binding can occur without a primary adhesion mechanism. E-selectin (Carlos et al. 1990; Carlos et al. 1991; Leeuwenberg et al. 1992; Hakkert et al. 1991) and L-selectin (Spertini et al. 1992) mediated adhesion of mononuclear cells to endothelial cells stimulated with cytokines for four hours has been shown, although these mechanisms may be more important in monocyte recruitment to inflammatory sites rather than atherosclerotic plaques. There is no evidence that E- or P-selectins are present on plaques. Firm adherence could potentially involve β_2 integrins, which are present on monocytes, and there is evidence that the ligand ICAM-1 is present on the plaques (van der Wal et al. 1992; Poston et al. 1992). There is also evidence that VCAM-1 is present on the surface of atherosclerotic plaques (Cybulsky and Gimbrone 1991), which mediates adhesion of monocytes and lymphocytes via the integrin VLA-4, as discussed earlier.

The flow field over lesion prone arterial sites is much different than the steady flow through postcapillary venules or high endothelial venules (reviewed in Schwartz et al. 1991), and this could potentially allow monocyte adhesion to atherosclerotic plaques directly via integrin mechanisms. Lesion-prone sites are known to be at low-shear locations within arteries, often just downstream from bifurcations. Flow can reverse at these locations, allowing for periodically alternating flow. Oscillatory wall shear stresses in this case have been shown in one study to range from -13 to 9 dyn/cm^2 with a time averaged mean of only -0.5 dyn/cm^2 (Ku and Giddens 1987). The effect of such oscillatory flow conditions on leukocyte adhesion is likely to be very important (Satcher et al. 1992; DePaola et al. 1992). For comparison, at non-lesion prone arterial sites flow is continuous and wall shear stresses range from approximately 10 to 50 dyn/cm^2. These shear stresses are most likely too high for even selectins to mediate adhesion (Jones et al. 1993a).

In addition to affecting binding through direct forces on leukocytes, blood flow has dramatic effects on endothelial cell function which could influence leukocyte adhesion. The most readily apparent effect is the elongation and orientation in the direction of flow characteristic of endothelial cells exposed to arterial shear stresses. This has been observed *in vivo* at high shear arterial locations (Nerem et al. 1981) and with cultured endothelial cells exposed to steady arterial wall shear stresses (Ives et al. 1986; Eskin et al. 1984). Endothelial cells exposed to low shear stresses, on the other hand, take on a cobblestone appearance, which correlates with the more cobblestone endothelial morphology found at lesion-prone sites. Many other differences between endothelial cells subjected to high and low shear stress have been observed (Schwartz et al. 1991; Nollert et al. 1991; Diamond et al. 1990; Nollert et al. 1992; Panaro and McIntire 1993) which could contribute to the increased susceptibility of low shear arterial surfaces to atherosclerotic lesions. In particular, there is evidence for a direct effect of shear stress on adhesion receptor expression (Sampath et al. 1992). Finally, arterial endothelial cells are exposed to higher mean hydrostatic pressures and cyclic stretching, which can affect function and possibly adhesion receptor expression as well (Carosi et al. 1992; Carosi and McIntire 1993).

Model Systems for Studying Leukocyte-Endothelial Cell Adhesion

Static Assays

Leukocyte-endothelial cell adhesion is generally studied using three types of assays: static assays, *in vitro* flow assays, and *in vivo* flow assays. Static assays are the most straightforward and inexpensive systems available for basic adhesion studies. In these systems, the substrate, which can be, for example, a frozen section, a cultured cell monolayer, or purified ligand, is overlaid with a leukocyte suspension for some period of time. Non-adherent leukocytes are then rinsed or centrifuged away, and adherent leukocytes are quantified. These systems allow controlled stimulation of both the leukocytes and the endothelial cells, allow quantitation of leukocytes which transmigrate through the monolayer onto the substrate (or through it, in the case of a porous substrate), and with centrifugation can give a measure of the strength of adhesion (Charo et al. 1985). Variants utilize a multilayered artificial blood vessel wall construct into which leukocytes can migrate following diapedesis (Huber et al. 1991; Hakkert et al. 1990) to examine transmigration and chemotactic/chemokinetic stimuli. The common disadvantage of all static assay systems is the impossibility of distinguishing primary and secondary adhesion events.

In vitro Flow Assays

Flow assays provide more realistic information than static assays by allowing descrimination of primary adhesion events such as rolling from secondary adhesion events such as firm adhesion and transmigration. They are also useful in examining the activation of leukocytes by soluble or cell-associated factors. However, these systems require substantially more sophistication in design, construction, and use than static assays. The parallel-plate flow chamber is a commonly used configuration (Lawrence et al. 1987; Hochmuth et al. 1973; Gallik et al. 1989; Jones et al. 1993a). The parallel-plate geometry produces flow for which the mathematical analysis is relatively straightforward. The flow field is typically characterized by either its shear rate (defined as the change of downstream velocity with change in distance from the wall, in units of reciprocal time) or shear stress (defined as viscosity × shear rate, in units of force per unit area) at the wall, and these parameters can be set equal to the wall shear rate or shear stress characteristic of flow through a blood vessel. Flows corresponding to capillaries or large arteries are easily attained by adjusting either flow rate or chamber geometry according to the relation

$$\tau_w = \frac{6Q\mu}{wh^2}$$

where τ_w is the wall shear stress, Q is volumetric flow rate, μ is viscosity, w is channel width and h is channel height.

Our own current implementation of the parallel plate flow chamber is shown in Fig. 3. The substrate is generally cultured human umbilical vein endothelial cells (HUVECs) grown in the 35 mm plastic tissue culture dish, although monolayers of transfected cell lines and dishes coated with purified ligands have also been used successfully. Two flow channels approximately 0.6 by 1.2 cm are available, which allows an experiment with, for example, a receptor-blocking antibody to be run next to a control experiment using the same endothelial monolayer. The entire flow system is maintained at 37°C in a warm air box surrounding the microscope.

Fig. 4 shows some example images acquired using our flow system. 4A is a monolayer of 24 hr IL-1 stimulated HUVECs before flow and 4B shows the same monolayer after 10 min of continuous flow of T cells through the chamber at the postcapillary venular wall shear stress of 2 dyn/cm^2. 4C is a maximization image (which is analogous to a 4 sec photographic exposure) of T cells adhering to 24 hr IL-1 stimulated HUVECs. The long blurs are paths of T cells which roll along the monolayer during the 4 sec of image acquisition. Images such as these are used to calculate rolling velocities.

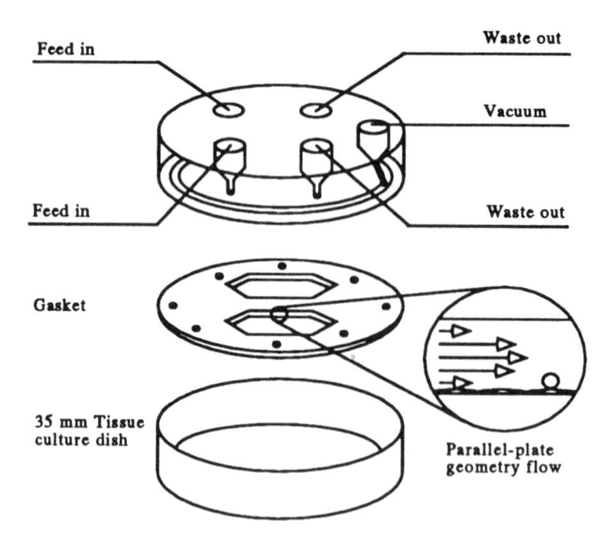

Fig. 3: Our current implementation of the parallel-plate flow chamber. Two flow channels are formed by cutouts in the gasket, which is made from silicone rubber sheeting. The top plate of each flow channel is the surface of the polycarbonate disc and the bottom plate is the tissue culture dish. Adhesion events are observed using either phase-contrast or fluorescence videomicroscopy.

4D is similar to 4C, except that in this case MAbs which block the VCAM-1/VLA-4 and ICAM-1/CD18 integrin adhesion pathways are blocked. Note that under these conditions, T cells roll along the endothelial cells but do not establish firm adhesion. They also roll approximately 80% more quickly (Jones et al. 1993b). 4E and F show the adhesion of neutrophils to HUVEC stimulated with histamine (10 min.) or IL-1 (4 hr.), respectively, after 10 min of flow. With histamine stimulation, most neutrophils roll along the endothelium and do not adhere firmly or transmigrate, whereas with IL-1 stimulation many neutrophils transmigrate beneath the monolayer and are seen as phase-dark patches (Jones et al. 1993a).

A second flow configuration that can be used is radially outward flow between parallel discs. This configuration has been used in more general studies of receptor-mediated cell adhesion (Fowler and McKay 1980; Cozens-Roberts et al. 1990), and could easily be adapted to study leukocyte-endothelial cell interactions. The flow pattern in this configuration has the characteristic of providing high shear near the center of the discs and progressively lower shear toward the outer edges. This could in some cases be an advantage over the parallel-plate geometry which provides only a single shear rate at a time. However, a serious potential problem in the study of leukocyte adhesion is that many adhesion events are dependent not only on

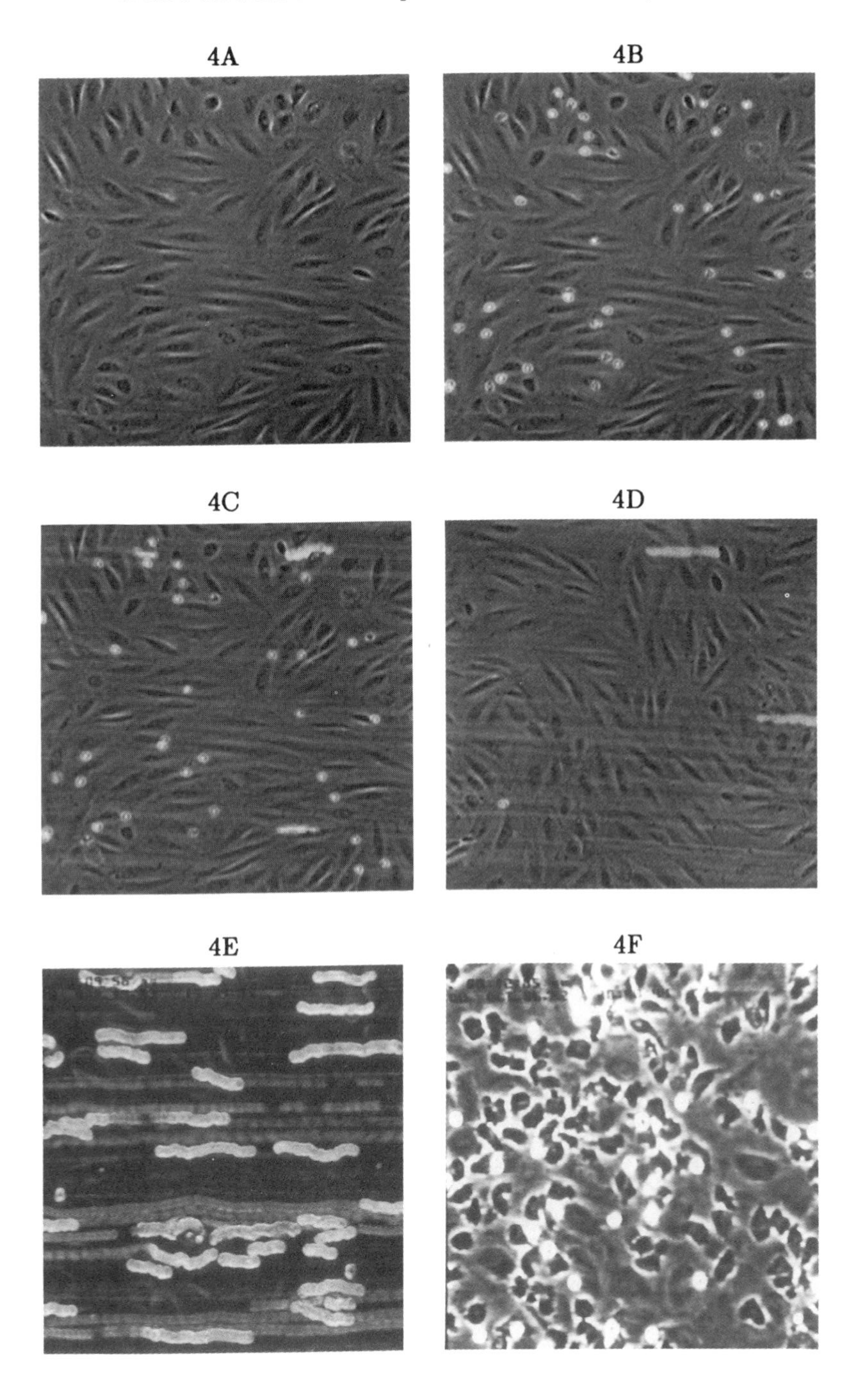

Fig. 4: Example images acquired using our flow system (see text).

the shear, but also on leukocyte activation. Thus, increased cell adhesion near the periphery could be due either to the lower shear there, or to longer exposure of the leukocytes to endothelial-derived activating substances. In addition, determination of leukocyte rolling velocities is more difficult since the local wall shear rate is continuously changing.

In vivo Flow Assays

A variety of systems have been used to study leukocyte adhesion *in vivo*. Intact translucent tissues such as rodent ears, the hamster cheek pouch, and nude mouse skin (Atherton and Born 1972; Mayrovitz 1992) are popular due to the minimal stimulation of tissues involved and relative simplicity of the preparations. Exteriorized mesenteric venules are also a very popular model (Ley et al. 1987; Zeintl et al. 1989; oude Egbrink et al. 1992). This system has considerable flexibility. For example, the mesentery can be superfused with a variety of stimulating cytokines, and neutrophils can be pretreated, labeled and reintroduced to the vessel under study at known concentrations for quantitative measurements of adhesion (Perry and Granger 1991; Yuan and Fleming 1990). The most serious potential drawback in this system is that exteriorization of the tissue can cause undesired stimulation of the endothelium. This often results in "spontaneous rolling" of leukocytes along the venules within minutes (Fiebig et al. 1991). A recently developed monoclonal antibody to canine P-selectin has been found to almost completely block this "spontaneous rolling" in canine mesenteric vessels (Dore et al. 1993).

Micromechanics of Leukocyte-Endothelial Cell Adhesion

Properties of Selectin-Mediated Adhesion

There are several possible reasons why selectin/carbohydrate bonds are more effective in the initial arrest of leukocytes from the bloodstream (see Fig. 5). One of the most striking considerations is the location of L-selectin on the tips of microvillus-like projections from resting neutrophils (Picker et al. 1991b; Erlandsen et al. 1993). This is especially important in light of the relatively small number of receptors on a neutrophil (approximately $2\text{-}4 \times 10^4$ sites per cell (Smith et al. 1991; Lawrence and Springer 1991)) and the length of L-selectin (approximately 15 nm) relative to E- and P-selectins (approximately 30 and 40 nm, respectively (Springer 1990)). Even though L-selectin is not a long molecule relative to other selectins, direct coupling to endothelial E- or P-selectins is predicted to form quite long tethering structures (on the order of 40 and 50 nm, respectively (Springer 1990)).

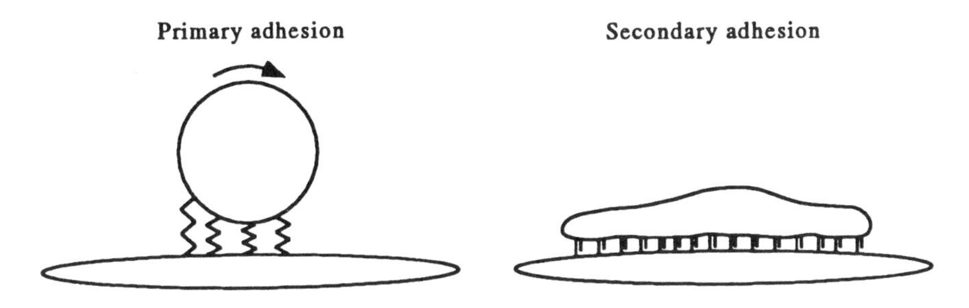

Fig. 5: Selectin-mediated rolling *versus* integrin-mediated firm adhesion. Selectin/carbohydrate interactions are well-suited to mediate initial leukocyte attachment and rolling for several possible reasons. These include the localization of L-selectin to tips of microvillus-like projections, the length and flexibility of receptor/ligand tethers, the ability of these tethers to withstand considerable strain before breaking, and a rapid rate of bond formation. On the other hand, integrin/polypeptide interactions have characteristics which make them well-suited to mediate firm adhesion and migration. Conformational changes in surface integrins can modulate binding affinity very rapidly, and on a slightly longer time-scale, mobilization of intracellular stores can provide additional surface receptors. Flattening of the leukocyte upon activation greatly increases the leukocyte/endothelial cell contact area which allows a large number of integrin bonds to form and decreases fluid forces on the cell. These characteristics combine to allow adhesion which is highly shear-resistant but can be modulated to allow migration. Further, leukocyte activation causes shedding of L-selectin, so that migration is not hindered by these bonds.

The intermembrane separation resulting from these tethers is large enough that energetically unfavorable glycocalyx interdigitation would not be necessary to form the selectin bonds (Bell et al. 1984). In addition to avoiding glycocalyx steric hinderance, selectin/carbohydrate interactions avoid the steric hinderance that glycoprotein surface decorations can produce for integrin/peptide core interactions. The flexibility of long single-chain selectins relative to two-chain integrins may also be an advantage in mediating initial interactions (Tozeren and Ley 1992; Lawrence and Springer 1991). Alternatively, some flexibility could arise in the carbohydrate chain containing the site bound by P-selectin (Fukada et al. 1984). In addition to buckling of receptors themselves, flexibility could result from deformation of the microvillus-like projections on which the receptors are located (Tozeren and Ley 1992). Any such deformations or buckling of receptors would have the effect of increasing the contact area between a spherical neutrophil and the planar endothelial cell surface allowing more bonds to form. Flexibility might also have the important consequence of enhancing the mobility of binding sites so that enhanced diffusivity could increase the rate of formation of bonds (Lawrence and Springer 1991). An extremely rapid rate of bond formation is necessary to arrest leukocytes moving quickly in the bloodstream.

The actual bonds formed between selectins and their carbohydrate ligands may also have properties which enhance their function in mediating initial adhesion and rolling. These bond properties have been addressed in detail in two recent mathematical models of leukocyte rolling (Tozeren and Ley 1992; Hammer and Apte 1992). Both model the tethers formed by receptor/ligand pairs as springs, such that the force the tethers exert is directly proportional to the length they are stretched from their equilibrium configurations. This force is also the force which tends to break the receptor/ligand bonds. In the model of Tozeren and Ley, unstrained bonds (which carry no force) form and break as a simple bimolecular reaction. Strained bonds, on the other hand, break at a uniform rate which is determined by fitting model predictions to experimental data. Their computational experiments point to an adhesion mechanism in which the rate of bond formation is high and the detachment rate low, except at the rear of the contact area where the stretched bonds detach at a high uniform rate. They find that a high value for the strain resistance of the selectin bonds (10^{-5} dyne) is consistent with rolling, which allows cell adhesion despite a relatively small number of L-selectin molecules available for bonding. Hammer and Apte reach similar conclusions using a more realistic model for the bonds based on earlier work dealing with the effect of strain on receptor binding (Dembo et al. 1988). In this case, receptor/ligand tethers are again modelled as springs, but the expressions for bond formation and breakage account for receptor/ligand separation distance. Fitting this model to experimental data leads them to the conclusion that selectin bonds break very nearly as "ideal" bonds, which means that the breakage rate is nearly independent of the amount of strain on the bond. As a result, the bonds can withstand high strain (as in the model of Tozeren and Ley), which allows the cell to be arrested from the free stream and roll.

Another important property of selectin bonds that suits them to initial adhesion and rolling is that these bonds are functional without activation of the leukocyte. Activation-independent adhesion has been demonstrated for neutrophils, lymphocytes, and monocytes, and selectins contribute to this adhesion in all cases. This makes sense physiologically since circulating leukocytes are normally unactivated, but can bind to specific endothelial sites when necessary. It is nonetheless possible that later activation can affect selectin adhesion. It has been reported, for instance, that fMLP stimulation of neutrophils induces a brief increase in L-selectin on the cell surface followed by rapid proteolytic shedding (Smith et al. 1991), and that leukocyte activation can augment the interaction between L-selectin and a yeast-derived polysaccharide (Spertini et al. 1991).

Properties of Integrin-Mediated Adhesion

An important property of integrins which makes them well-suited to mediate firm adhesion and migration is the fact that they can be modulated by cell activation. As discussed earlier, activation produces rapid conformational changes in leukocyte integrins which substantially modify their affinity for ligands. Internal cell stimuli such as phorbol esters are known to induce the high affinity state, and external Mn^{2+} binding as well as binding by monoclonal antibodies can "lock" the receptors in the high affinity conformation (reviewed in Hogg and Landis 1993). Integrin affinity modulation occurs very rapidly and has been shown to mediate increased adhesion of neutrophils to surfaces within 15-30 seconds (Hughes et al. 1992). Neutrophil activation also leads to mobilization of additional Mac-1 receptors from an intracellular pool. However, this newly mobilized Mac-1 does not appear to be important in mediating adhesion unless the cell is restimulated following expression of the new receptors (Hughes et al. 1992; Diamond and Springer 1993). Morphological changes in the neutrophil upon activation can influence adhesion as well. When the neutrophil is activated, it flattens out on the endothelium greatly increasing its contact area. Increased contact area allows many bonds to form, taking advantage of the large number of integrin receptors (30×10^4 Mac-1 receptors on an unstimulated neutrophil (Smith et al. 1991)) relative to selectin receptors ($2 - 4 \times 10^4$ (Smith et al. 1991; Lawrence and Springer 1991)). This is especially important because CD18 integrins are not localized on the tips of cell projections, as is L-selectin. Flattening of leukocytes also greatly decreases the fluid drag and torque leukocytes experience (Firrell and Lipowsky 1989). All of these factors combine to produce adhesion which is up to 100-fold more shear-resistant than selectin-mediated binding (Lawrence and Springer 1991).

The ability of integrins to rapidly modulate their affinity appears to be very important in neutrophil migration. Monoclonal antibodies which "lock" CD18 integrins in the high affinity state inhibit neutrophil migration induced by chemotactic stimulation. The failure of integrins to return to the low affinity state appears to tether the neutrophil to the endothelium and prevent migration (Dransfield et al. 1992; Dransfield and Hogg 1989; Robinson et al. 1992). In one study (Sung et al. 1993), the force required to break integrin bonds was found to be 2.1×10^{-7} dyne for activated GpIIb-IIIa ($\alpha_{IIb}\beta_3$ integin) and 5.7×10^{-8} dyne for nonactivated GpIIb-IIIa binding to fibrinogen. Corresponding data for selectins is not yet available, but in the model of Tozeren and Ley, a maximum selectin bond force of 10^{-5} dyne was found to be consistent with rolling adhesion. As described above, the smaller bond strength of integrins might be compensated for by the larger number of integrin receptors and morphological changes accompanying neutrophil activation. In addition, binding strength might be substantially higher for CD18 integrin bonds than for GpIIb-IIIa.

As a final point of discussion, we mention the transition from a rolling state to a stopped state. Most current experimental evidence in this area is largely indirect, since studies generally aim at examining either the rolling phenomenon or aspects of secondary adhesion separately. Several lines of evidence, however, point to integrin activation as an event of central importance in the transition. First, although integrin mechanisms generally do not appear to be primary mediators of rolling adhesion, they do serve to decrease the average rolling velocity. This has been shown in the case of neutrophils, where the speed of P-selectin-mediated rolling on histamine-stimulated HUVECs increases by approximately 25% with antibodies blocking CD18 and ICAM-1 (Jones et al. 1993a). This has also been shown in the case of lymphocytes, where blocking either LFA-1/ICAM-1 or VLA-4/VCAM-1 increases the average rolling velocity by approximately 33% and blocking both pathways increases the rolling velocity by approximately 80% (Jones et al. 1993b). This indicates that integrins are capable of interacting with endothelial cells while the leukocyte is rolling, so that complete cell stopping is not a prerequisite for at least some integrin function. Second, a direct correlation between decreasing average rolling velocity and the fraction of rolling neutrophils which stop has been clearly demonstrated (Kaplanski et al. 1993). Third, *in vivo* experiments with lymphocytes (Bargatze and Butcher 1993) show a direct correlation between a G-protein mediated activation event and cell stopping. Taken together, these findings indicate that integrins do slow rolling leukocytes, and suggest that integrin activation is important for the complete stopping of rolling leukocytes.

Acknowledgments

This work was supported by NIH grants HL-18672, NS-23327, HL-42550, Robert A. Welch Foundation grant C-938, and a grant from the Butcher Fund.

References

Abbassi, O.; Kishimoto, T. K.; McIntire, L. V.; Anderson, D. C.; Smith, C. W. E-selectin supports neutrophil rolling in vitro under conditions of flow. J. Clin. Invest. 92:2719–30; 1993.

Abbassi, O.; Lane, C. L.; Krater, S.; Kishimoto, T. K.; Anderson, D. C.; McIntire, L. V.; Smith, C. W. Canine neutrophil margination mediated by lectin adhesion molecule-1 in vitro. J. Immunol. 147:2107–15; 1991.

Anderson, D. C.; Springer, T. A. Leukocyte Adhesion Deficiency: an inherited defect in the Mac-1, LFA-1, and p150,95 glycoproteins. Annu. Rev. Med. 38:75–94; 1987.

Arfors, K. E.; Lundberg, C.; Lindbom, L.; Lundberg, K.; Beatty, P. G.; Harlan, J. M. A monoclonal antibody to the membrane glycoprotein complex CD18 inhibits polymorphonuclear leukocyte accumulation and plasma leakage in vivo. Blood 69:338–40; 1987.

Atherton, A.; Born, G. V. R. Quantitative investigation of the adhesiveness of circulating polymorphonuclear leucocytes to blood vessel walls. J. Physiol. 222:447–474; 1972.

Atherton, A.; Born, G. V. R. Relationship between the velocity of rolling granulocytes and that of the blood flow in venules. J. Physiol. 233:157–165; 1973.

Bargatze, R. F.; Butcher, E. C. Rapid G protein-regulated activation event involved in lymphocyte binding to high endothelial venules. J. Exp. Med. 178:367–72; 1993.

Baumhueter, S.; Singer, M. S.; Henzel, W.; Hemmerich, S.; Renz, M.; Rosen, S. D.; Lasky, L. A. Binding of L-selectin to the vascular sialomucin CD34. Science 262:436–8; 1993.

Bell, G. I.; Dembo, M.; Bongrand, P. Cell adhesion: competition between nonspecific repulsion and specific bonding. Biophys. J. 45:1051–1064; 1984.

Berg, E. L.; Yoshino, T.; Rott, L. S.; Robinson, M. K.; Warnock, R. A.; Kishimoto, T. K.; Picker, L. J.; Butcher, E. C. The cutaneous lymphocyte antigen is a skin lymphocyte homing receptor for the vascular lectin endothelial cell-leukocyte adhesion molecule 1. J. Exp. Med. 174:1461–6; 1991.

Berlin, C.; Berg, E. L.; Briskin, M. J.; Andrew, D. P.; Kilshaw, P. J.; Holzmann, B.; Weissman, I. L.; Hamann, A.; Butcher, E. C. Alpha 4 beta 7 integrin mediates lymphocyte binding to the mucosal vascular addressin MAdCAM-1. Cell 74:185–95; 1993.

Bevilacqua, M. P. Endothelial-leukocyte adhesion molecules. Annu. Rev. Immunol. 11:67–804; 1993.

Bjerknes, M.; Cheng, H.; Ottaway, A. Dynamics of lymphocyte-endothelial interactions in vivo. Nature 231:402–5; 1986.

Buck, C. A. Immunoglobulin superfamily: structure, function and relationship to other receptor molecules. Semin. Cell Biol. 3:179–88; 1992.

Butcher, E. C. Leukocyte-endothelial cell recognition: three (or more) steps to specificity and diversity. Cell 67:1033–6; 1991.

Carlos, T.; Kovach, N.; Schwartz, B.; Rosa, M.; Newman, B.; Wayner, E.; Benjamin, C.; Osborn, L.; Lobb, R.; Harlan, J. Human monocytes bind to two cytokine-induced adhesive ligands on cultured human endothelial cells: endothelial-leukocyte adhesion molecule-1 and vascular cell adhesion molecule-1. Blood 77:2266–71; 1991.

Carlos, T. M.; Dobrina, A.; Ross, R.; Harlan, J. M. Multiple receptors on human monocytes are involved in adhesion to cultured human endothelial cells. J. Leukoc. Biol. 48:451–6; 1990.

Carosi, J. A.; Eskin, S. G.; McIntire, L. V. Cyclical strain effects on production of vasoactive materials in cultured endothelial cells. J. Cell Physiol. 151:29–36; 1992.

Carosi, J. A.; McIntire, L. V. Effects of cyclical strain on production of vasoacive materials by cultured human and bovine endothelial cells. Eur. Respir. Rev. 3:589–608; 1993.

Charo, I. F.; Yuen, C.; Goldstein, I. M. Adherence of human polymorphonuclear leukocytes to endothelial monolayers: effects of temperature, divalent cations, and chemotactic factors on the strength of adherence measured with a new centrifugation assay. Blood 65:473–79; 1985.

Chien, S. Rheology in the microcirculation in normal and low flow states. Adv. Shock Res. 8:71–80; 1982.

Cozens-Roberts, C.; Quinn, J. A.; Lauffenburger, D. A. Receptor-mediated cell attachment and detachment kinetics. II. Experimental model studies with the radial-flow detachment assay. Biophys. J. 58:857–72; 1990.

Cybulsky, M. I.; Gimbrone, M. A. Endothelial expression of a mononuclear leukocyte adhesion molecule during atherogenesis. Science 251:788–91; 1991.

Dembo, M.; Torney, D. C.; Saxman, K.; Hammer, D. A. The reaction-limited kinetics of membrane-to-surface adhesion and detachment. Proc. R. Soc. Lond. B Biol. Sci. 234:55–83; 1988.

DePaola, N.; Gimbrone, M. A.; Davies, P. F.; Dewey, C. F. Vascular endothelium responds to fluid shear stress gradients. Arterioscler. Thromb. 12:1254–7; 1992.

Diamond, M. S.; Springer, T. A. A subpopulation of Mac-1 (CD11b/CD18) molecules mediates neutrophil adhesion to ICAM-1 and fibrinogen. J. Cell Biol. 120:545–556; 1993.

Diamond, S. L.; Sharefkin, J. B.; Dieffenbach, C.; Frasier-Scott, K.; McIntire, L. V.; Eskin, S. G. Tissue plasminogen activator messenger RNA levels increase in cultured human endothelial cells exposed to laminar shear stress. J. Cell Physiol. 143:364–71; 1990.

Dore, M.; Korthuis, R. J.; N., G. D.; Entman, M. L.; Smith, C. W. P-selectin mediates spontaneous leukocyte rolling in vivo. Blood 82:1308–16; 1993.

Dowbenko, D.; Andalibi, A.; Young, P. E.; Lusis, A. J.; Lasky, L. A. Sructure and chromosomal localization of the murine gene encoding GLYCAM 1. A mucin-like endothelial ligand for L selectin. J. Biol. Chem. 268:4525–9; 1993.

Dransfield, I.; Cabanas, C.; Craig, A.; Hogg, N. Divalent cation regulation of the function of the leukocyte integrin LFA-1. J. Cell Biol. 116:219–26; 1992.

Dransfield, I.; Hogg, N. Regulated expression of Mg2+ binding epitope on leukocyte integrin alpha subunits. EMBO J. 8:3759–65; 1989.

Erlandsen, S. L.; Hasslen, S. R.; Nelson, R. D. Detection and spatial distribution of the beta-2 integrin (Mac-1) and L-selectin (LECAM-1) adherence receptors on human neutrophils by high-resolution field emmision SEM. J. Histochem. Cytochem. 41:327–333; 1993.

Eskin, S. G.; Ives, C. L.; McIntire, L. V.; Navarro, L. T. Response of cultured endothelial cells to steady flow. Microvasc. Res. 28:87–94; 1984.

Etzioni, A.; Frydman, M.; Pollack, S.; Avidor, I.; Phillips, M. L.; Paulson, J. C.; Gershoni-Baruch, R. Recurrent severe infections caused by a novel leukocyte adhesion deficiency. New Engl. J. Med. 327:1789–92; 1992.

Fiebig, E.; Ley, K.; Arfors, K. E. Rapid leukocyte accumulation by "spontaneous" rolling and adhesion in the exteriorized rabbit mesentery. Int. J. Microcirc. Clin. Exp. 10:127–44; 1991.

Firrell, J. C.; Lipowsky, H. H. Leukocyte margination and deformation in mesenteric venules of rat. Am. J. Physiol. 256:H1667–74; 1989.

Fowler, H. W.; McKay, A. J.: 1980. in R. C. W. Berkeley et al. (eds.), *Microbial adhesion to surfaces*; pp 143–161, Ellis Horwood, Chincester

Fukada, M.; Spooncer, E. S.; Oates, J. E.; Dell, A.; Klock, J. C. Structure of sialylated fucosyl lactosaminoglycan isolated from human granulocytes. J. Biol Chem. 259:10925–35; 1984.

Gallik, S.; Usami, S.; Jan, K.-M.; Chien, S. Shear stress-induced detachment of human polymorphonuclear leukocytes from endothelial cell monolayers. Biorheology 26:823–834; 1989.

Goldman, A. J.; Cox, R. G.; Brenner, H. Slow viscous motion of a sphere parallel to a plane wall. Chem. Eng. Sci. 22:637–660; 1967.

Hakkert, B. C.; Kuijpers, T. W.; Leeuwenberg, J. F. M.; van Mourik, J. A.; Roos, D. Neutrophil and monocyte adherence to and migration across monolayers of cytokine-activated endothelial cells: the contribution of CD18, ELAM-1, and VLA-4. Blood 78:2721–6; 1991.

Hakkert, B. C.; Rentenaar, J. M.; Van Aken, W. G.; Roos, D.; Van Mourik, J. A. A three-dimensional model system to study the interactions between human leukocytes and endothelial cells. Eur. J. Immunol. 20:2775–2781; 1990.

Hammer, D. A.; Apte, S. M. Simulation of cell rolling and adhesion on surfaces in shear flow: general results and analysis of selectin-mediated neutrophil adhesion. Biophys. J. 63:35–57; 1992.

Hochmuth, R. M.; Mohandas, N.; Blackshear, P. L. Measurement of the elastic modulus for red cell membrane using a fluid mechanical technique. Biophys. J. 13:747–62; 1973.

Hogg, N.; Landis, R. C. Adhesion molecules in cell interactions. Curr. Opin. Immunol. 5:383–90; 1993.

Huber, A. R.; Kunkel, S. L.; Todd, R. F.; Weiss, S. J. Regulation of transendothelial neutrophil migration by endogenous interleukin-8. Science 254:99–102; 1991.

Hughes, B. J.; Hollers, J. C.; Crockett-Torabi, E.; Smith, C. W. Recruitment of CD11b/CD18 to the neutrophil surface and adherence-dependent cell locomotion. J. Clin. Invest. 90:1687–1696; 1992.

Ives, C. L.; Eskin, S. G.; McIntire, L. V. Mechanical effects on endothelial cell morphology: in vitro assessment. In Vitro Cellular and Developmental Biology 22:500–7; 1986.

Jones, D. A.; McIntire, L. V.; McEver, R. P.; Smith, C. W. P-selectin supports neutrophil rolling on histamine-stimulated endothelial cells. Biophys. J. 65:1560–69; 1993.

Jones, D. A.; McIntire, L. V.; Smith, C. W.; Picker, L. J. T-cell adhesion to endothelial cells under flow conditions: functional characterization of a new adhesion pathway. J. Cell Biol.; 1993. Submitted

Kaplanski, G.; Farnarier, C.; Tissot, O.; Pierres, A.; Benoliel, A. M.; Alessi, M. C.; Kaplanski, S.; Bongrand, P. Granulocyte-endothelium initial adhesion: analysis of transient binding events mediated by E-selectin in a laminar shear flow. Biophys. J. 64:1922–33; 1993.

Kishimoto, T. K.; Hollander, N.; Roberts, T. M.; Anderson, D. C.; Springer, T. A. Heterogeneous mutations in the beta subunit common to the LFA-1, Mac-1, and p150,95 glycoproteins cause leukocyte adhesion deficiency. Cell 50:193–202; 1987.

Kishimoto, T. K.; Jutila, M. A.; Berg, E. L.; Butcher, E. C. Neutrophil Mac-1 and MEL-14 adhesion proteins inversely regulated by chemotactic factors. Science 245:1238–41; 1989.

Ku, D. N.; Giddens, D. P. Laser Doppler anemometer measurements of pulsatile flow in a model carotid bifurcation. J. Biomech. 20:407–21; 1987.

Lawrence, M. B.; McIntire, L. V.; Eskin, S. G. Effect of flow on polymorphonuclear leukocyte/endothelial cell adhesion. Blood 70:1284–90; 1987.

Lawrence, M. B.; Smith, C. W.; Eskin, S. G.; McIntire, L. V. Effect of venous shear stress on CD18-mediated neutrophil adhesion to cultured endothelium. Blood 75:227–37; 1990.

Lawrence, M. B.; Springer, T. A. Leukocytes roll on a selectin at physiologic flow rates: distinction from and prerequisite for adhesion through integrins. Cell 65:859–73; 1991.

Leeuwenberg, J. F.; Jeunhomme, T. M.; Buurman, W. A. Role of ELAM-1 in adhesion of monocytes to activated human endothelial cells. Scand. J. Immunol. 35:335–41; 1992.

Lesley, J.; Hyman, R.; Kincade, P. W. CD44 and its interaction with extracellular matrix. Adv. Immunol. 54:271–335; 1993.

Ley, K.; Gaehtgens, P. Endothelial, not hemodynamic, differences are responsible for preferential leukocyte rolling in rat mesenteric venules. Circ. Res. 69:1034–41; 1991.

Ley, K.; Pries, A. R.; Gaehtgens, P. A versatile intravital microscope design. Int. J. Microcirc. Clin. Exp. 6:161–7; 1987.

Mayrovitz, H. N. Leukocyte rolling: a prominent feature of venules in intact skin of anesthetized hairless mice. Am. J. Physiol. 262:H157–61; 1992.

Nerem, R. M.; Levesque, M. J.; Cornhill, J. F. Vascular endothelial morphology as an indicator of patterns of blood flow. J. Biomech. Eng. 103:172; 1981.

Nollert, M. U.; Diamond, S. L.; McIntire, L. V. Hydrodynamic shear stress and mass transport modulation of endothelial cell metabolism. Biotechnology and Bioengineering 38:588–602; 1991.

Nollert, M. U.; Panaro, N. J.; McIntire, L. V. Regulation of genetic expression in shear stress stimulated endothelial cells. Ann. N. Y. Acad. Sci. 665:94–104; 1992.

oude Egbrink, M. G.; Tangelder, G. J.; Slaaf, D. W.; Reneman, R. S. Influence of platelet-vessel wall interactions on leukocyte rolling in vivo. Circ. Res. 70:355–63; 1992.

Palecanda, A.; Walcheck, B.; Bishop, D. K.; Jutila, M. A. Rapid activation-independent shedding of leukocyte L-selectin induced by cross-linking of the surface antigen. Eur. J. Immunol. 22:1279–86; 1992.

Panaro, N. J.; McIntire, L. V.: 1993. in B. E. Sumpio (ed.), *Hemodynamic forces in modulating vascular cell biology*; pp 47–65, R. G. Landes Company, Austin, TX

Perry, M. A.; Granger, D. N. Role of CD11/CD18 in shear rate-dependent leukocyte-endothelial cell interactions in cat mesenteric venules. J. Clin. Invest. 87:1798–804; 1991.

Picker, L. J. Mechanisms of lymphocyte homing. Curr. Opin. Immunol. 4:277–86; 1992.

Picker, L. J.; Butcher, E. C. Physiological and molecular mechanisms of lymphocyte homing. Annu. Rev. Immunol. 10:61–91; 1992.

Picker, L. J.; Kishimoto, T. K.; Smith, C. W.; Warnock, R. A.; Butcher, E. C. ELAM-1 is an adhesion molecule for skin-homing T cells. Nature 349:796–9; 1991.

Picker, L. J.; Warnock, R. A.; Burns, A. R.; Doerschuk, C. M.; Berg, E. L.; Butcher, E. C. The neutrophil selectin LECAM-1 presents carbohydrate ligands to the vascular selectins ELAM-1 and GMP-140. Cell 66:921–33; 1991.

Poston, R. N.; Haskard, D. O.; Coucher, J. R.; Gall, N. P.; Johnson-Tidey, R. R.

Expression of intercellular adhesion molecule-1 in atherosclerotic plaques. Am. J. Pathol. 140:665–73; 1992.

Robinson, M. K.; Andrew, D.; Rosen, H.; Brown, D.; Ortlepp, S.; Stephens, P.; Butcher, E. C. Antibody against the Leu-CAM beta-chain (CD18) promotes both LFA-1- and CR3-dependent adhesion events. J. Immunol. 148:1080–5; 1992.

Rohde, D.; Schluter-Wigger, W.; Mielke, V.; von den Driesch, P.; von Gaudecker, B.; Sterry, W. Infiltration of both T cells and neutrophils in the skin is accompanied by the expression of endothelial leukocyte adhesion molecule-1 (ELAM-1): an immunohistochemical and ultrastructural study. J. Invest. Dermatol. 98:794–9; 1992.

Salmi, M.; Jalkanen, S. A 90-kilodalton endothelial cell molecule mediating lymphocyte binding in humans. Science 257:1407–9; 1992.

Sampath, R.; McIntire, L. V.; Eskin, S. G.: 1992. in *Extended Abstracts, American Institute of Chemical Engineers 1992 Annual Meeting*; p. Abstract 169e

Sanchez-Madrid, F.; Corbi, A. L. Leukocyte integrins: structure, function and regulation of their activity. Semin. Cell Biol. 3:199–210; 1992.

Satcher, R. L.; Bussolari, S. R.; Gimbrone, M. A.; Dewey, C. F. The distribution of fluid forces on model arterial endothelium using computational fluid dynamics. J. Biomech. Eng. 114:309–16; 1992.

Schwartz, C. J.; Valente, A. J.; Sprague, E. A.; Kelley, J. L.; Nerem, R. M. The pathogenesis of atherosclerosis: an overview. Clin. Cardiol. 14:I1–16; 1991.

Smith, C. W. Endothelial adhesion molecules and their role in inflammation. Can. J. Physiol. Pharmacol. 71:76–87; 1993.

Smith, C. W.; Kishimoto, T. K.; Abbassi, O.; Hughes, B.; Rothlein, R.; McIntire, L. V.; Butcher, E.; Anderson, D. C. Chemotactic factors regulate lectin adhesion molecule 1 (LECAM-1)-dependent neutrophil adhesion to cytokine-stimulated endothelial cells in vitro. J. Clin. Invest. 87:609–18; 1991.

Spertini, O.; Kansas, G. S.; Munro, J. M.; Griffin, J. D.; Tedder, T. F. Regulation of leukocyte migration by activation of the leukocyte adhesion molecule-1 (LAM-1) selectin. Nature 349:691–4; 1991.

Spertini, O.; Luscinskas, F. W.; Gimbrone, M. A.; Jr.Tedder, T. F. Monocyte attachment to activated human vascular endothelium in vitro is mediated by leukocyte adhesion molecule-1 (L-selectin) under nonstatic conditions. J. Exp. Med. 175:1789–92; 1992.

Springer, T. A. Adhesion receptors of the immune system. Nature 346:425–34; 1990.

Sung, K.-L. P.; Frojmovic, M. M.; O'Toole, T. E.; Zhu, C.; Ginsberg, M. H.; Chien, S. Determination of adhesion force between single cell pairs generated by activated GpIIb-IIIa receptors. Blood 81:419–23; 1993.

Tanaka, Y.; Albelda, S. M.; Horgan, K. J.; van Seventer, G. A.; Shimizu, Y.; Newman, W.; Hallam, J.; Newman, P. J.; Buck, C. A.; Shaw, S. CD31 expressed on distinctive T cell subsets is a preferential amplifier of beta 1 integrin-mediated adhesion. J. Exp. Med. 176:245–53; 1992.

Tangelder, G. J.; Slaaf, D. W.; Muijtjens, A. M.; Arts, T.; oude Egbrink, M. G.; Reneman, R. S. Velocity profiles of blood platelets and red blood cells flowing in arterioles of the rabbit mesentery. Circ. Res. 59:505–14; 1986.

Tozeren, A.; Ley, K. How do selectins mediate leukocyte rolling in venules? Biophys. J. 63:700–709; 1992.

van der Wal, A. C.; Das, P. K.; Tigges, A. J.; Becker, A. E. Adhesion molecules on the endothelium and mononuclear cells in human atherosclerotic lesions. Am. J. Pathol. 141:1427–33; 1992.

Vaporciyan, A. A.; DeLisser, H. M.; Yan, H.-C.; Mendiguren, I. I.; Thom, S. R.; Jones, M. L.; Ward, P. A.; Albelda, S. M. Involvement of Platelet-Endothelial Cell Adhesion Molecule-1 in neutrophil recruitment in vivo. Science 262:1580–2; 1993.

Vestweber, D. Selectins: cell surface lectins which mediate the binding of leukocytes to endothelial cells. Semin. Cell Biol. 3:211–20; 1992.

von Andrian, U. H.; Berger, E. M.; Ramezani, L.; Chambers, J. D.; Ochs, H. D.; Harlan, J. M.; Paulson, J. C.; Etzioni, A.; Arfors, K. E. In vivo behavior of neutrophils from two patients with distinct inherited leukocyte adhesion deficiency syndromes. J. Clin. Invest. 91:2893–7; 1993.

Yuan, Y.; Fleming, B. P. A method for isolation and fluorescent labeling of rat neutrophils for intravital microvascular studies. Microvasc. Res. 40:218–29; 1990.

Zeintl, H.; Sack, F. U.; Intaglietta, M.; Messmer, K. Computer assisted leukocyte adhesion measurement in intravital microscopy. Int. J. Microcirc. Clin. Exp. 8:293–302; 1989.

7
Osteoblast Responses to Steady Shear Stress

H.Y. Shin, R.D. Iveson, F.A. Blumenstock, R. Bizios

Introduction

Like many tissues in the human body, bone is constantly undergoing changes in its microstructure. The homeostasis of bone can be described as a dynamic equilibrium existing between resorption and deposition of mineral. Although each bone has a genetically determined minimal mass and structure, the normal state of bone in the physiologic environment is a consequence of adaptations to various chemical and physical stimuli (Lanyon 1984). These adaptations are achieved through an active remodeling of the internal and surface characteristics (*e.g.*, structure and mineral content) of bone.

Lanyon (1984) hypothesized that under mechanical loads bone formation is primarily induced by mechanical stimuli while resorption is mostly controlled by hormonal, biochemical, and genetic factors. An equilibrium between the biochemically-controlled resorptive and the mechanically-induced formative processes of bone results in a "remodeling equilibrium. Cowin (1984) also described a "remodeling equilibrium" in which the net resorption and deposition of mineral is zero in physiologically-loaded bones.

Attempts have been made to determine the mechanical parameters which influence the adaptation of bone. Cowin (1984) proposed that, although bone experiences stresses due to imposed mechanical loads, the resultant strains developed in the mineralized matrix are the functional consequences leading to bone remodeling. Above a physiologic range of strain levels, net deposition of bone is stimulated; below such a range, net resorption occurs (Cowin 1984).

In order to predict the determinants of mechanically-induced bone remodeling, Brown *et al.* (1990) quantified observations of periosteal remodeling from *in vivo* loading of ulnar specimens from turkeys and incorporated their results into a finite element model of the turkey ulna. The finite element model identified three parameters as the most probable initiators of bone remodeling: strain energy density, principal stress/strain, and longitudinal shear stress (Brown *et al.* 1990).

Literature reports provide evidence that fluid flow in bone can occur as a result of imposed mechanical loads. Dillaman *et al.* (1991) proposed that loads applied to long bones, such as the femur, cause a hydrostatic pressure gradient between the endosteal and periosteal surfaces; such a pressure gradient can drive fluid through the interstitial fluid spaces of bone. Kelly and Bronk (1990) showed that an increase in saphenous vein pressure of the canine tibia resulted in an increase of fluid flow from the capillaries to the mineralized matrix of bone and an increase in periosteal bone deposition. Furthermore, this increase in bone formation was independent of increases in blood flow in the vasculature, alterations in oxygen saturation, and changes in the partial pressure of oxygen (Kelly and Bronk 1990). Weinbaum *et al.* (1994) developed a model for the mechanosensory transduction mechanism at the cellular level of bone. These investigators proposed that cyclic, mechanical (combined axial and bending) loads applied to whole bone induce fluid flow (8 - 30 dynes/cm^2) within canalicular-lacunar pores filled with proteoglycans (whose glycosaminoglycans were ordered by albumin). These small shear stresses act on the membranes of osteocytic responses (extending in the canaliculi) and may induce stress-related bone resorption and deposition of bone tissue (Weinbaum *et al.* 1994).

The fact that mechanical loads or stresses influence the deposition of new bone suggests that mechanical stimuli may influence/trigger bone cell function(s). The bone cells which are thought to respond to alterations in mechanical stress/strain states in the mineralized matrix are the osteocytes (differentiated bone cells found in the lacunar spaces of bone), the osteoclasts (the bone resorbing cells), bone-lining cells (resting cells residing on inactive bone surfaces which may be stimulated to become a layer of osteoblasts), and the osteoblasts (the bone-forming cells) (Cowin *et al.* 1991). Although the normal function and remodelling of bone is attributed to the various types of bone cells, the osteoblasts are the primary cells responsible for the production, secretion, and organization of the mineralized matrix (Robey *et al.* 1989). Formation of new bone requires collagenous and non-collagenous proteins which sustain the integrity of the mineralized matrix. Type I collagen constitutes ~70% of the organic constituents and ~90% of the protein in bone (Sodek *et al.* 1991); the non-collagenous proteins include proteoglycans (*e.g.*, decorin and biglycan), glycoproteins (*e.g.*, osteonectin), and sialoproteins (*e.g.*, osteopontin and bone sialoprotein) (Robey 1989).

The present study was an *in vitro* investigation of the effect(s) of steady, fluid shear stress on the synthesis and release of proteins by osteoblasts.

Materials and Methods

Osteoblast Isolation and Culture

Osteoblasts were isolated from the calvaria of one-day-old Sprague-Dawley rats via sequential enzymatic digestion using established techniques (Puleo *et al.* 1991). The osteoblasts were cultured in Dulbecco's modified Eagle medium (DMEM; Gibco BRL) containing 10 % fetal bovine serum (Hyclone) in a standard culture environment, that is, a humidified 5% CO_2/95% air incubator maintained at 37°C. When needed, osteoblasts were passaged with PET (0.05% trypsin, 0.02% ethylene glycol bis-(β-aminoethyl ether) N,N,N',N'-tetraacetic acid (EGTA), 1% polyvinylpyrrolidone, and 18 mM 4(2-hydroxylethyl)-1-piperazine-ethane-sulfonic acid (HEPES) buffer in 0.5% NaCl) (Puleo *et al.* 1991).

Characterization of the cells as osteoblasts was established by alkaline phosphatase activity, production cAMP in response to parathyroid hormone, synthesis of type I collagen, formation of calcium phosphate mineral deposits, and expression of mRNA for bone-related proteins such as osteocalcin, osteonectin, and osteopontin (Puleo *et al.* 1993).

Substrate Preparation

The substrates used in this study were 38 mm x 75 mm glass plates (Fisher Scientific). The glass plates were washed in 1% Liquinox detergent (Fisher Scientific), degreased in acetone and 70% ethanol, etched in 0.5N NaOH for 2 hours, sterilized in a steam autoclave at 270°F for 20 minutes, and oven-dried overnight.

Substrate Seeding

Osteoblasts (passage 4 to 6) were seeded onto the etched-glass plates (500,000 cells/plate) under sterile conditions and were cultured in DMEM containing 10% fetal bovine serum in a humidified 5% CO_2/95% air atmosphere at 37°C for 48 hours prior to all experiments. At that time, formation of a subconfluent monolayer of osteoblasts on the glass plates was confirmed through phase-contrast microscopy examination.

Parallel Plate Flow System

The parallel plate flow system was adapted from Frangos *et al.* (1988) and consisted of a parallel plate flow chamber (Figure 1) and a flow loop (Figure 2). When assembled, the parallel plate (5.7 cm x 2.5 cm) flow chamber contained a flow channel 0.0183 to 0.022 cm high. A slight vacuum pressure was used during the shear experiments to hold the parallel plate flow chamber assembly together.

Flow of fresh, serum-free DMEM (30 mLs) was maintained through the parallel plate flow chamber by a pressure difference generated in the flow loop. A constant pressure head was established and maintained (for the duration of the shear stress experiments) by varying the distance between two reservoirs containing serum-free DMEM; serum-free DMEM was also circulated around the flow loop with the help of a roller pump (Masterflex). All shear stress experiments were conducted in a humidified 5% CO_2/95% air atmosphere at 36°C.

Shear Stress Experiments

Prior to all shear stress experiments, the osteoblasts grown on glass plates were rinsed 3 times with isotonic PBS (to remove any trace amounts of fetal bovine serum which was a constituent of the medium used to culture the cells). The glass plates with cultured cells were then incorporated into the parallel plate flow system; steady shear stresses ranging from 2 to 15 dynes/cm^2 were applied to the osteoblasts for 6 hours. Osteoblasts cultured in 30 mLs of serum-free DMEM under standard culture conditions were the control preparations.

Experiments were also conducted to investigate the mechanism by which osteoblasts release proteins in response to shear stress. Specifically, the osteoblasts on glass plates were pretreated with either 14 μM monensin in DMEM or 30 μg/mL of cycloheximide in DMEM for 3 hours immediately prior to shear stress exposure. The osteoblasts on the glass substrates were gently rinsed 3 times with isotonic PBS before each experiment and were exposed to shear stresses (2 to 15 dynes/cm^2; 6 hours) using 30 mLs of fresh serum-free DMEM as the circulating medium. Controls were osteoblasts under static, no-flow conditions in serum-free DMEM (30 mLs; 6 hours) in the presence and absence of these chemical compounds.

At the end of each experiment, 20 mLs of supernatant fluid were collected and stored at -70°C in polypropylene tubes (Falcon) until protein analysis.

Osteoblast Morphology

Osteoblast morphology was examined via phase-contrast microscopy using a Diaphot-TMD inverted microscope (Nikon, Inc.). Briefly, the osteoblasts on the glass plates were rinsed with isotonic PBS, fixed with Buffered Formalde-

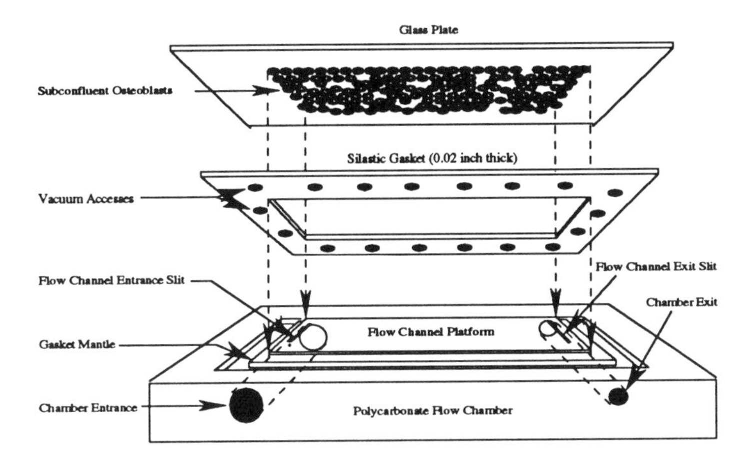

Figure 1. Schematic of the parallel-plate flow channel assembly

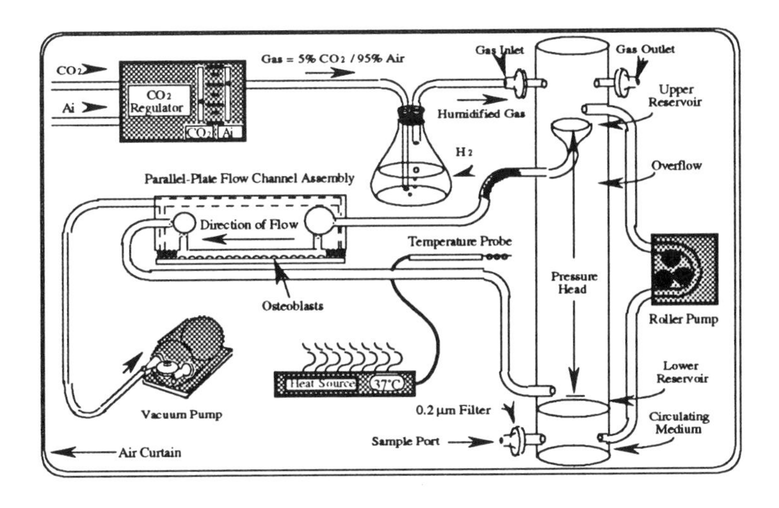

Figure 2. Schematic of the laboratory set-up used for studying the effects of shear stress on osteoblasts

Fresh (10% Formalin; Fisher Scientific), and stained with a solution containing 0.1% Coomasie Brilliant Blue R-250 (Sigma), 10% acetic acid, and 45% methanol in distilled water. The stained cell preparations were then rinsed with distilled water and allowed to airdry.

Sodium Dodecyl Sulfate-Polyacrylamide Gel Electrophoresis (SDS-PAGE)

Preparation of Gradient Gels

Gradient gels (4% - 15%) were polymerized from a stock solution of 40% acrylamide and 1.5% BIS in distilled water using 0.025% ammonium persulfate and tetramethylethylenediamine (TEMED). Solutions for 15% acrylamide gels (pH 8.8) and 4% acrylamide gels (pH 8.8) were poured into separate compartments of a gradient chamber. The gradient chamber was then utilized to pour a gradient gel which was 4% acrylamide and 15% (v/v) at the top and bottom, respectively. After polymerization, 4% acrylamide stacking gels (pH 6.8) were made from 40% acrylamide stock solutions and polymerized on top of the separation gels with 0.025% ammonium persulfate and TEMED. A 15 well column template was placed into the stacking gel to produce a number of wells or lanes.

Preparation of Protein Samples

The volume of the supernatant fluid samples (20 mLs) was reduced down to 5 mLs using dialysis (removal of excess water) against polyethylene glycol (Fisher Scientific) for 90 minutes. The proteins in the small volume samples were then precipitated by incubating with 1 mL of 100% trichloroacetic acid (Sigma) in an ice bath for 2 hours and, then, pelleted by centrifugation at 5,500 rpm at 4°C for 20 minutes. Each protein pellet was resuspended in 2 mLs of a 1:1 ethanol:ether (v/v) solution and centrifuged at 5,500 rpm at 4°C for 30 minutes. Finally, resulting protein pellets were resuspended in 100 μL of a reducing buffer containing 0.2% β-mercaptoethanol, 0.125 M tris-HCl, 4 % sodium dodecyl sulfate, and 20% glycerol.

Prestained molecular weight markers (Gibco BRL) were prepared by addition of 60 μL of non-reducing buffer (0.125 M tris-HCl, 4% sodium dodecyl sulfate, and 20% glycerol) to prestained proteins of known molecular weights (14,880 to 219,600 Da).

The protein solutions from experimental samples and the molecular weight markers were then vortexed, boiled for 5 minutes, and frozen at -70°C until SDS-PAGE electrophoresis.

Electrophoresis

Aliquots (30 μL) of the protein solutions were loaded into individual wells of the stacking gel. Molecular weight markers were loaded in a different well and in parallel with all supernatant samples. Electrophoresis was carried out at room temperature in a vertical slab gel unit (Hoeffer Scientific Instruments) containing electrode buffer which was continuously stirred. A constant electric current of 30 mA was delivered to the gel unit while the leading edge of the molecular weight markers was traveling through the stacking gel (approximately 30 - 60 minutes). The electric current was increased to 60 mA when the leading edge of the molecular weight markers reached the separation gel. The electrophoretic process was terminated when the leading edge of the molecular weight markers reached the bottom of the separation gel (approximately 1 hour).

Visualization and Identification of Protein/Peptide Bands

After electrophoresis, the SDS-PAGE gels were removed from vertical slab gel unit, stained with Coomasie blue stain (0.7 % Coomasie Brilliant Blue R-250 in a solution containing 30% methanol and 6% acetic acid) at room temperature for 30 minutes, and destained by immersing in fresh destaining solution (45% methanol, 10% acetic acid) for different time intervals (5, 10, 15, minutes and 12 hours). The stained SDS-PAGE gels were stored in plastic bags containing distilled water at room temperature until protein/peptide band analysis.

A number of distinct protein/peptide bands were visible on the stained gels and were identified with the help of molecular weight markers and their retention factors. Specifically, retention factors (Rf values) of each proteins/peptide were determined as the ratio of the distance each protein/peptide travelled over the distance travelled by the leading edge of the molecular weight markers. The molecular weight for each protein/peptide contained in the supernatant of the experimental samples was determined from a standard plot of Rf values *versus* the logarithm of the molecular weights of protein markers.

Immunoblotting (Western Blots)

Unstained SDS-PAGE gels were mounted in a transblot sandwich with a sheet of nitrocellulose paper adjacent to the gel. The transblot sandwich was then placed into a transblot unit (Hoeffer Scientific) and the protein bands on the SDS-PAGE gels were electrophoretically transferred to the nitrocellulose paper at a constant electric current of approximately 0.75 A for 2 hours. Upon completion of the electrotransfer process, the nitrocellulose paper was placed in a hydration chamber, containing a solution of 1% bovine serum albumin (BSA) in tris-buffered saline (TBS), and rocked gently at room temperature for 1 hour.

At that time, 10 μLs of primary antibody (rabbit-antirat osteonectin and rabbit anti-rat bone sialoprotein) was added to the hydration chamber and allowed to react with the appropriate protein/peptide bands under well-mixed condtions overnight. The nitrocellose paper in the hydration chamber was then allowed to react with the secondary antibody solution (5% Casein, 2% rat serum, and a 1:10,000 dilution of horse radish peroxidase-conjugated GARIgG in TBS) with stirring at room temperature for 2 hours. Finally, the protein bands were visualized by treating with a 5:5:1 solution of TMB Peroxidase Substrate: Peroxidase Solution B: TMB Membrane enhancer (TMB Membrane Peroxidase Substrate System) under constant stirring for 5 minutes. The nitrocellulose paper with protein bands was then rinsed with distilled water, dried on filter paper and sprayed with Matte fixative (Chartpak) to preserve visualization of the protein/peptide bands.

Results

Oseoblast Morphology

Before and after exposure to steady shear stresses (values in the range of 2 - 15 dynes/cm^2), osteoblast morphology was examined using phase-contrast microscopy. There was no change in the morphology of osteoblasts exposed to the steady shear stresses used in this study. The absence of morphological changes was confirmed by fluorescence microscopic examination of the F-actin filaments of the cytoskeleton (stained with rhodamine phalloidin) and of tubulin (labelled with mouse anti-β-tubulin (Tu27b) and treated with rhodamine-labelled, goat anti-mouse antibody).

Proteins/Peptides Produced and Released by Osteoblasts Under Steady Shear Stress

The protein/peptide bands that were visualized on SDS-PAGE gels of the circulating supernatants from osteoblasts exposed to shear stress for 6 hours were compared to the supernatants from controls (osteoblasts under static, no-flow conditions) and to the samples of fresh, serum-free DMEM. The retention factors (Rf values) and the approximate molecular weights of unknown proteins/peptides were calculated using the results obtained from known molecular weight markers run on the same gels in parallel with the experimental samples (Table 1).
Compared to supernatants of osteoblasts under control conditions and to serum-free DMEM, three additional proteins/peptides, with molecular weights of 150 kDa, 60 kDa, 55 kDa, and 43 kDa, were detected in the circulating supernatants of osteoblasts exposed to 2 dynes/cm^2 of shear stress for 6 hours. The supernatants of osteoblasts exposed to 15.9 dynes/cm^2 for 6 hours contained three additional proteins/peptides with estimated molecular weights

Table 1. Proteins Released by Osteoblasts After 6 Hours of Steady Shear Stress

Shear Stress (dynes/square cm)	Released Proteins	
	Rf Value	*Estimated Molecular Weight (kDa)*
2	0.19	150
	0.45	60
	0.48	55
	0.57	43
15	0.46	60
	0.53	55
	0.58	43

of 60 kDa, 55 kDa, and 43 kDa; supernatants from controls and from fresh, serum-free media did not contain these protein/peptides.

Western blotting or immunoblotting techniques were used to identify protein bands which were electrophorectically transferred to nitrocellulose paper from SDS-PAGE gels of the circulating supernatants of osteoblasts exposed to 2 dynes/cm^2 of steady shear stress, from supernatants of osteoblasts under control (no-flow) conditions, and from fresh, serum-free DMEM for 6 hours. Some protein bands reacted with the rabbit anti-rat osteonectin and with the rabbit anti-rat bone sialoprotein antibodies. Pretreatment of osteoblasts with either cycloheximide or monensin (3 hours prior to exposure to 2 dynes/cm^2 for 6 hours) resulted in the absence of the 60 kDa, 55 kDa, and 43 kDa proteins/peptides which were detected on Western blots of supernatants retrieved from shear stress experiments.

Discussion

Normal, daily function of the body involves exposure of bones to mechanical forces. In the past, experimental studies of bone focused on explants and on animal models. Many hypotheses (including those addressing bone formation, remodeling, and healing at the cellular/molecular level) could not be definitively tested in the laboratory using these macroscopic models. Advances in cell culture techniques and availability of well-characterized cell lines have provided researchers with versatile, *in vitro* models for simulating physiological events (such as mechanical loading of bone) and studying these systems at the cellular level. The importance of the microscopic approaches to physiological systems is also reflected in recent theoretical models of bone. Weinbaum *et al.* (1994) predicted that mechanical loading of bone induces fluid shear stresses (8 - 30 dynes/cm^2) past osteocytes in the proteoglycan-filled canaliculi. In conjunction with recent experimental studies, these results advance intriguing implications regarding the mechanism of stress-related resorption and deposition of bone tissue; most important, the Weinbaum *et al.* (1994) study provides the first set of specific numerical values of fluid shear stresses at the cellular level of bone.

Exposure of cells to mechanical stresses induces morphological changes and functional responses. For example, human umbilical vein (Eskin *et al.* 1985) and bovine aortic endothelial cells (Dewey *et al.* 1981; Levesque and Nerem 1985) exposed to steady shear (1 - 85 dynes/cm^2) for periods of 1 - 2 days exhibited rearrangement of the microtubules, concomitant cell elongation, and alignment parallel to the direction of flow. Furthermore, exposure of human umbilical vein endothelial cells to steady shear stresses (10 dynes/cm^2; 7 - 8 hours) resulted in increased production of prostacyclin (Frangos *et al.* 1985) while exposure to 15 and 25 dynes/cm^2 caused secretion of tissue plasminogen activator (Diamond *et al.* 1989).

The results of the present study showed that morphological changes and concomitant cytoskeletal rearrangements were absent in viable (trypan blue dye exclusion results) osteoblasts exposed to steady shear stresses (2 - 30 dynes/cm^2) for 6 hours. There is evidence in the literature, however, that fluid shear stresses induce functional responses from osteoblasts. Specifically, neonate rat calvarial osteoblasts exposed to shear rates in the range of 10 - 3,500 sec^{-1} (corresponding to shear stresses of 0.1 - 35 dynes/cm^2) for 0.5 - 15 minutes exhibited increased levels of intracellular cyclic adenosine monophosphate (Reich *et al.* 1990); exposure of these cells to 24 dynes/cm^2 for up to 2 hours resulted in elevated levels of prostaglandin E_2 and inositol triphosphate (Reich *et al.* 1991). The present study extends the range of osteoblast functions triggered by shear stress to protein synthesis and release. Compared to osteoblasts under static, no-flow conditions, the osteoblasts exposed to 2 or to 15.9 dynes/cm^2 for 6 hours produced/secreted three proteins/peptides with molecular weights of 43, 55, and 60 kDa; when exposed to 2 dynes/cm^2, osteoblasts produced/secreted an additional protein/peptide with a molecular weight of 150 kDa. The possibility that proteins/peptides "peeled off" osteoblast membranes or were released due to cellular lysis during shear stress exposure was addressed in the present study by pretreating osteoblasts with either monensin, an inhibitor of protein release at the cell membrane, or cycloheximide, an inhibitor of protein synthesis, prior to shear stress experimentation. The supernatants of these pretreated osteoblasts following exposure to shear stresses contained no proteins/peptides. These results imply that protein production by sheared osteoblasts is a *de novo* process.

Tensile stresses (imposed when the flexible substrates, on which cells were grown, were deformed) induced changes in the synthesis of non-collagenous proteins by osteoblasts and various bone-related cells. Brighton *et al.* (1990) reported decreased synthesis of non-collagenous proteins in the supernatant medium as well as in the homogenates of rat calvarial bone-cell cultures subjected to 0.04% cyclic biaxial strains for various time intervals (15 minutes -

72 hours). Buckley *et al.* (1990) analyzed the cell extracts and the supernatants of chick, osteoblast-like cells exposed to a maximum 24% elongation of the substrate in a three cycle per minute (0.05 Hz) mode and reported decreased levels of non-collagenous proteins at day 1 but elevated levels at day 3. Hasegawa *et al.* (1985) observed increased levels of a 58 kDa and a 40 kDa protein on radiolabelled SDS-PAGE gradient (5 - 15%) gels of rat calvarial bone cell extracts following exposure to mechanical stretching (4% increase in the surface area of the substrate; 2 hours). The present study is the first to document *de novo* protein production and release by osteoblasts in response to steady shear stresses. The 43, 55, and 60 kDa proteins produced by rat calvarial osteoblasts exposed to 2 and to 15.9 dynes/cm^2 for 6 hours (in the present study) were in the molecular weight range of proteins produced by bone cells subjected to mechanical stretching (Hasegawa *et al.* 1985) for time intervals of hours (6 hours of shear stress *versus* 2 hours of mechanical stretch) and analyzed by similar techniques (4 - 15% SDS PAGE gradient gels *versus* 5 - 15% SDS-PAGE gels).

In summary, exposure of osteoblasts (and of related bone-cells) to mechanical stresses induces functional responses, specifically, synthesis and/or release of collagenous (Hasegawa *et al.* 1985; Brighton *et al.* 1990; Buckley *et al.* 1990) and non-collagenous (Hasegawa *et al.* 1985; Brighton *et al.* 1990; Buckley *et al.* 1990; present study) proteins which are necessary for the formation and maintenance of bone (Robey 1989). In this respect, protein synthesis triggered by mechanical stimuli may have implications regarding the mechanism(s) of proper bone formation during remodeling and/or healing.

Acknowledgements

We gratefully acknowlede and thank: Dr. John A. Frangos, The Pennsylvania State University, for the parallel plate flow chambers used in this study; Dr. Donald R. Bertolini, Department of Cellular Biochemistry, SmithKline Beecham Pharmaceuticals, for the osteopontin antibody; Dr. Larry W. Fisher, Bone and Research Branch, National Institutes of Health, for the osteonectin and bone sialoprotein antibodies; Ms. Christine Keenan, Department of Physiology and Cell Biology, Albany Medical College, for technical assistance with SDS-PAGE; and Ms. Kay C Dee, Department of Biomedical Engineering, Rensselaer Polytechnic Institute, for isolation and characterization of the osteoblasts.

References

Brighton C. T.; Strafford, B.; Gross, S. B.; Leatherwood, D. F.; Williams, J. L.; Pollack, S. R. The proliferative and synthetic response of isolated calvarial bone cells of rats to cyclic biaxial strain. J. Bone Jt. Surg. 73-A: 320-331; 1991.

Brown, T. D.; Pedersen, D. R.; Gray, M. L.; Brand, R. A.; Rubin, C. T. Toward identification of mechanical parameters initiating periosteal remodeling. J. Biomech. 23: 893-905; 1990.

Buckley, M. J.; Banes, A. L.; Jordan, R. D. The effects of mechanical strain on osteoblasts *in vitro*. J. Oral Maxillofac. Surg. 48:276-282:1990.

Cowin, S. C. Mechanical modeling of the stress adaptation process in bone. Calc. Tiss. Int. 36: S98-S103; 1984.

Cowin, S. C.; Moss-Salentijn, L.; Moss, M. L. Candidates for the mechanosensory system in bone. 113: 191-197; 1991.

Dewey, C. F.; Bussolari, S. R.; Gimbrone, M. A.; Davies, P. F. The dynamic response of endothelial cells to fluid shear stress. J. Biomech. Eng. 103: 177-185;1981.

Diamond, S. L.; Eskin, S. G.; Mcintire, L. V. Fluid flow stimulates tissue plasminogen factor secretion by cultured human endothelial cells. Science 243: 1483-1485; 1989.

Dillaman, R. M.; Roer, R. D.; Gay, D. M. Fluid movement in bone: theoretical and empirical. J. Biomech. 23: 893-905; 1991.

Eskin, S. G.; Ives, C. L., Frangos, J. A.; McIntire, L. V. Cultured endothelium: the response to flow. ASAIO. 8: 109-112; 1985.

Frangos, J. A.; Eskin, S. G.; McIntire, L. V.; Ives, C. L. Flow effects on prostacyclin production by cultured human endothelial cells. Science 227: 1477-1479; 1985.

Frangos, J. A.; Mcintire, L. V.; Eskin, S. G. Shear stress induced stimulation of mammalian cell metabolism. Biotech. and Bioeng. 32:1053-1060; 1988.

Hasegawa, S.; Sato, S.; Saito, S.; Suzuki, Y.; Brunette, D. M. Mechanical stretching increases the number of cultured bone cells synthesizing DNA and alters their pattern of protein synthesis. Calcif. Tiss. Int. 37: 431-436; 1985.

Kelly, P. J.; Bronk, J. T. Venous pressure and bone formation. Microvas. Res. 39: 364-375; 1990.

Lanyon, L. E. Functional strain as a determinant for bone remodeling. Calc. Tiss. Int. 36: S56-S61; 1984.

Levesque, M. J.; Nerem, R. M.; The elongation and orientation of cultured endothelial cells in response to shear stress. J. Biomech. Eng. 107: 341-347; 1985.

Puleo, D. A.; Holleran, L. A.; Doremus, R. H.; Bizios, R. Osteoblast response to orthopedic implants. J. Biomed. Mat. Res. 25: 711-723; 1991.

Puleo, D. A.; Preston, K. E.; Shaffer, J. B.; Bizios, R. Examination of osteoblast-orthopedic biomaterial interactions using molecular techniques. 14: 111-114 ;1993.

Reich, K. M.; Gay, C. V.; Frangos, J. A. Fluid shear stress as a mediator of osteoblast cyclic adenosine monophosphate production. 143: 100-104; 1990.

Reich, K. M.; Frangos, J. A. Effects of flow on prostaglandin E_2 and inositol triphosphate levels in osteoblasts. Am. J. Physiol. 261 (Cell Physiol. 30): C428-C432; 1991.

Robey, P. G. The biochemistry of bone. Endocrinol. Metab. Clin. North AM. 18: 859-902; 1989.

Sodek, J.; Zhang, Q.; Goldberg, H. A.; Domenicucci, C.; Kasugai, S.; Wrana, J. L.; Shapiro, H.; Chen, J. Non-collagenous bone proteins and their role in substrate-induced bioactivity. In: Davies, J. E., ed. The bone-biomaterial interface. Toronto: University of Toronto Press; 1991: p. 97-110.

Weinbaum, S.; Cowin, S. C.; Zeng, Y. A model for the excitation of osteocytes by mechanical loading-induced bone fluid shear stresses. J. Biomech (in press; 1994).

8

Effects of Shear Stress on Cytoskeletal Structure and Physiological Functions of Cultured Endothelial Cells

N. Ohshima, K. Ookawa

Introduction

Endothelial cell monolayer, which lines the inner surface of the blood vessel, plays an important role in controlling the vascular permeability to a wide variety of solutes to be exchanged between blood and tissues. At the same time, this endothelial monolayer is an unique tissue that ceaselessly is exposed to fluid mechanical stimuli caused by the blood flow. Consequently, the physiological functions of vascular endothelial cells must be discussed taking into consideration numerous factors related to their environment as a whole. Such factors include not only the soluble humoral factors but also mechanical stimuli as fluid-imposed shear stress or cyclic stretch of the vessel wall.

In the present paper, therefore, the effects of shear stress on the structure and physiological functions of vascular endothelial cells after exposure to physiological level of shear stress were studied in terms of; a) morphometric parameters of the cell and microfilament, b) transendothelial permeability to high molecular weight tracers, and c) expression of adhesion molecules known to mediate leukocyte-endothelial cell adhesion.

Materials and Methods

Endothelial cell culture
Porcine aortic endothelial cells (PAECs) were harvested from thoracic and abdominal aortae as described previously (Ookawa et al 1992; Ookawa et al 1993). The aorta was longitudinally opened and then the lumen was gently stroked with a surgical blade to obtain endothelial cells. The adherent cells were dispersed into 25-cm^2 tissue culture flasks filled with Dulbecco's modified Eagle medium (DME; Gibco Laboratories, NY) containing 10% heat-inactivated fetal bovine serum (HyClone Laboratories, UT), penicillin, streptomycin, and amphotericin B (1% v/v, Gibco). When confluent, cells were subcultured with 0.05% trypsin-EDTA (Gibco) and passaged at 1:4 split ratios. For the experiments, PAECs (0.8-1.0 $\times 10^5$ cm^{-2}) were inoculated either on a glass slide for the purpose of morphological observations, or on the outer surface of the filter membrane unit (Falcon® Cell Culture Insert: porous polyethylene telephthalate membrane unit; pore size 0.45 μm, 4.9 cm^2; Becton Dickinson, NJ) for experiments to evaluate transendothelial permeability. Thereafter, the PAEC monolayers were subjected to shear stress exposure when confluent (7 days after seeding, passage 3-8). Endothelial cells were morphologically identified by their typical polygonal shape observed microscopically and positive fluorescence staining after incubation with acetylated low density lipoprotein labeled with tetramethylindocarbocyanine perchlorate (DiI-Ac-LDL; Biomedical Technologies, MA) (Voyta et al 1984).

Shear stress exposure
Fully confluent cultured PAEC populations were first subjected to fluid-imposed shear stress in a parallel plate flow chamber described in detail previously (Ookawa et al 1992) or in a modified and newly developed type of the similar device ("tapered-plate flow chamber") to attain an increasing level of shear stress along the fluid pass within the chamber.

A glass slide or a filter membrane unit with PAEC monolayer was positioned in the flow chamber, and the perfusion circuit consisted of this flow chamber, two reservoirs and a roller pump were filled with the same culture medium as stated above. An arterial level of wall shear stress (20 dyn cm^{-2}) was applied for up to 24 h in these flow chambers. Measurements of transendothelial permeability of the PAEC after shear stress exposure were performed using a separate type of diffusion chamber as will be described later. For the observation of cell adhesion molecules after shear stress exposure, PAECs were exposed to shear stress beforehand using the tapered-plate flow chamber.

Morphometry of cytoskeletal structure
In the first series of experiments, morphological changes of PAECs were studied using the parallel plate flow chamber. After flow exposure, the cells were fixed and photographed after staining with rhodamine-phalloidin to visualize the cytoskeletal F-actin filament as described previously (Ookawa et al 1992; Ookawa et al 1993). On the

photomicrographs of PAECs, the following morphometric parameters were determined: area (A) and perimeter (P) of each cell, angle of cell orientation (θ_{cell}) defined as the deviation of the longest diagonal of the cell from the direction of the flow, angle of F-actin filament orientation (θ_{fil}) defined as the deviation of each stress fiber within the cell from the direction of the flow, and shape index (SI) defined as $4\pi \cdot A\ P^{-2}$ (Cornhill et al 1980). According to this definition, the value of SI equals 1.0 for a complete circle and approaches zero for a highly elongated cell shape. These parameters were measured for 30-50 individual cells observed on each coverslip. In some experiments, the cells which had highly uniform orientation of microfilaments with the values of $\theta_{fil} < 15\ °$, were selectively subjected to shape index measurement in order to determine whether the change of filament orientation precedes the change of cell elongation.

Measurement of transendothelial permeability
Transendothelial permeability after shear stress exposure was evaluated in a different type of permeability testing chamber with tracer macromolecules using fully confluent PAEC populations cultured beforehand on porous filter membrane units. The effects of shear stress exposure to transendothelial permeability was compared with those of a known permeability agonist, phorbol mirystate acetate (PMA), chosen as a positive control. Tracers of different molecular weight were used for permeability determination; bovine serum albumin (BSA, molecular weight = 67000; Sigma) and fluorescently labeled dextrans, i. e., rhodamine-labeled dextran (RD-70, average molecular weight = 73100; Sigma) and FITC-labeled dextran (FD-150, average molecular weight = 148900; Sigma). These tracers were dissolved in Hanks' balanced salt solution (HBSS; Gibco) beforehand, and were loaded in the luminal compartment of the membrane unit. Time course changes of the tracer concentration in the subluminal compartment was measured after loading the tracer solution. The luminal compartment of the permeability testing chamber was continuously stirred by use of an orbital shaker (60 rpm) to minimize mass transfer resistance between the two compartments. All experiments were performed in a humidified incubator (100% room air) at 37 °C unless stated otherwise.

Permeability of the filter membrane unit with PAEC monolayer ($P_{overall}$) to each tracer was calculated based on Eq. 1 proposed by Casnocha et al (1989) who applied the Fick's law of diffusion to transendothelial permeation studies similar to those used in the present experiments:

$$\ln\left\{\frac{1}{1-(V_L+V_S)\,C_{S_{(t)}}/M_T}\right\} = P_{overall}\ A\ \frac{(V_L+V_S)}{V_L V_S}\ t \qquad (1)$$

where A, area of membrane (cm^2); V_L, volume of luminal chamber (ml); V_S, volume of subluminal chamber (ml); t, sampling time (s);

$C_{S(t)}$, tracer concentration in subluminal chamber at each sampling time (mg/ml); and M_T, total mass of tracer (mg).

Assuming the nature of additive diffusion resistance, determination of the permeability of endothelial cell monolayer *per se* ($P_{endothelial\ cell}$) was made by subtracting the mass transfer resistance of the filter membrane ($1/P_{membrane}$) from that of the overall membrane unit with PAEC monolayer ($1/P_{overall}$), as described in Eq. 2 (Cooper et al 1987).

$$\frac{1}{P_{endothelial cell}} = \frac{1}{P_{overall}} - \frac{1}{P_{membrane}} \qquad (2)$$

Observation of cell surface adhesion molecules

To elucidate the functional modification of PAECs exposed to shear stress, expression of endothelium-bound adhesion molecules were investigated using the "tapered-plate flow chamber." Adhesion molecules tested were; VCAM-1 (vascular cell adhesion molecule-1), ELAM-1 (endothelial cell-leukocyte adhesion molecule-1).

To visualize the distribution of adhesion molecules on the cell surface, the PAEC monolayers were immunostained with antibodies against VCAM-1 (M084; British Bio-technology, U. K.) and ELAM-1 (0805; Immunotech S. A., France). Briefly, immediately after exposure to shear stress for 24 h, the PAEC monolayer was first incubated with the first antibodies (1:100) at 37 °C for 1 h in order to stain exclusively the molecules that exist on cell surface. Then, the cells were fixed with acetone at -20 °C for 5 min, and subsequently incubated with PBS containing 5% normal calf serum for 1 h. The cells were then immunostained with biotinylated second antibodies (1:500; Vector Laboratories, CA). Between these steps, the cells were intensively rinsed with PBS containing 0.1% Tween-20. Finally the adhesion molecules were visualized by DAB (diaminobenzidine)-H_2O_2 solution as suggested by the manufacturer, then observed and photographed under the microscope system as stated above.

Cell morphology and cytoskeletal structure

After shear stress exposure, PAECs cultured on a glass coverslip were found to elongate and orient to the direction of the flow. These changes were predominantly observed 6 h after the onset of the flow or later. Whereas, the stress fiber-like structure of an F-actin bundle was formed in the central part of the cells at 3 h, and these filaments were also oriented to the direction of the flow. Thus, the changes in cytoskeletal F-actin filaments were considered to precede cell elongation and orientation. After 24 h, peripheral filaments as well as the central stress fibers in highly elongated PAECs were again observable.

In a no-flow condition (Fig. 1-a), F-actin filaments were mainly localized at the periphery of the cells, although some filaments were observable in more central portions. The angle of filaments were observed to be randomly distributed. After 3 h (Fig. 1-b), the stress fiber-like structure of an F-actin bundle was formed in the central part of some cells. Although the averaged values of filament angles to the flow direction did not show statistically significant changes as stated below, some filaments began to orient toward the flow direction. However, endothelial cells maintained a polygonal shape. After 6 h (Fig. 1-c), such changes in cytoskeleton became much more prominent. At this stage, cell elongation was completely established. After 24 h (Fig. 1-d), cell elongation and orientation were clear, and bright F-actin bundles were also observed as stress fibers and dense peripheral bands (DPBs).

As the exposure time to shear stress became longer, the shape index decreased (Fig. 2), that is, the cell became more elongated. The cells were significantly elongated after 6 h or later (0.83±0.01, no-flow; 0.77±0.01, 6 h; 0.70±0.06, 12 h; 0.65±0.05, 24 h), although there was almost no change within 3 h (0.84±0.01). In the case of the subpopulation of the cells that showed highly oriented filaments ($\theta_{fil} <$ 15 °) after 3 h shear stress exposure, the values of SI (0.85±0.02) showed no significant decrease. Therefore, the change of filament orientation again seemed to precede the change of cell elongation (Ookawa et al 1992).

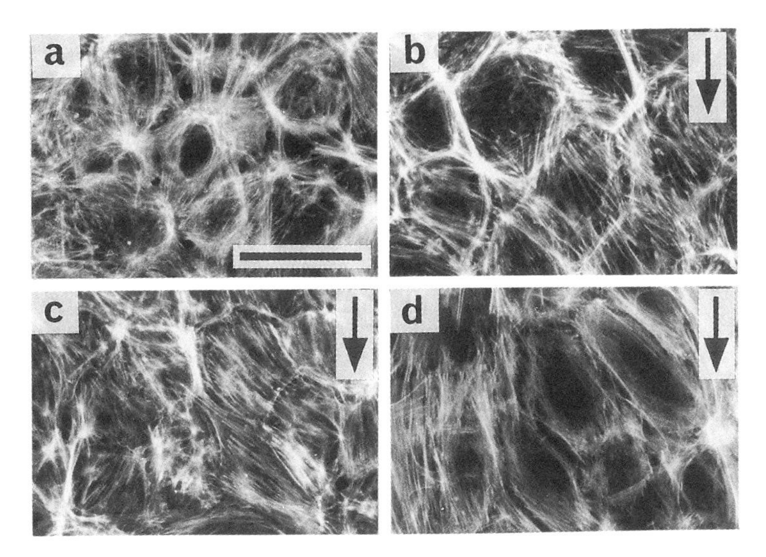

Figure 1. Typical photomicrographs of rhodamine phalloidin-stained porcine aortic endothelial cells exposed to fluid-imposed shear stress of 20 dyn cm^{-2} for (a) 0 h, (b) 3 h, (c) 6 h, and (d) 24 h. Bar = 50 μm. Arrows = direction of the flow applied.

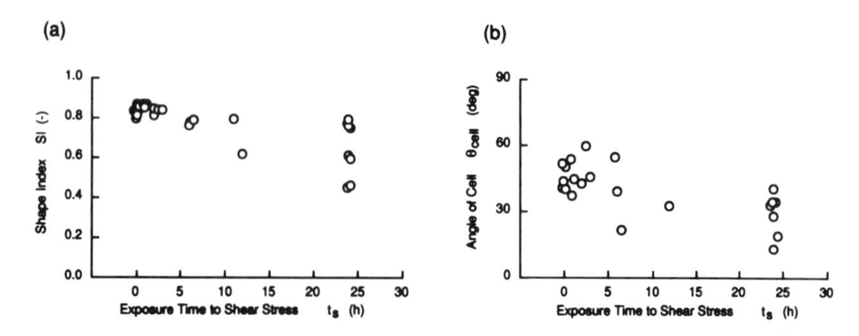

Figure 2. Effect of exposure time to shear stress on the shape index (a) and on the angle of cell orientation (b) of porcine aortic endothelial cells.

The effects of flow on cytoskeletal structures after the application of arterial or venous level of shear stress on cultured endothelial cells have thus far been analyzed only qualitatively (Franke et al 1984; Wechezak et al 1989). Wechezak et al (1985) reported that cell reorientation was dependent on both the time of exposure and the magnitude of shear stress applied. They also reported that this orientation was accompanied with a reorganization in cellular fibronectin and F-actin. Franke et al (1984) reported that human vascular endothelial stress fiber, namely reoriented F-actin bundles, was induced even by the 3-h exposure of confluent monolayer cultures exposed to a fluid shear stress of venous level (2 dyn cm^{-2}) without inducing any changes in cell shape. Therefore, the threshold level of shear stress for the induction of cytoskeletal changes is much lower than that for cell reorientation.

Our results obtained from the fluorescence microscopic observation of the cytoskeleton were consistent with that of Franke et al (1984) with respect to the time course of rearrangement of F-actin bundles. Within 3 h after the onset of the flow, numbers of peripheral filaments decreased and some filaments began to orient toward the flow direction, while cell shape remained unchanged (Fig. 2). The SI value calculated from the cells which had apparently-oriented microfilaments was not significantly different from that of the cells under the control, i. e., no flow condition. Such orientation of filaments was more prominent at 6 h after the onset of the flow, although the orientation of the cell was not significantly differed yet. These results strongly suggests that the re-distribution of microfilaments preceded the changes in cell shape and orientation. Another series of experiments of the authors' using cytochalasin B-treated endothelial cells (data not shown) revealed that these changes were retarded.

Present experiments seemed to provide a first stage clue to understand the *in vivo* endothelial responses to the blood flow. The morphometric analysis adopted was focused on the very early stage change for up to 24 h after the onset of the flow, whereas the endothelial cells *in vivo* ceaselessly experience the shear stress. The nature of the flow applied in the present study was also limited to the steady laminar flow. However, the evidence of such acute responses demonstrated by the present study is expected to hold fundamentally for *in vivo* situation and to serve to understand the mechanisms of local modulations or differentiation of the endothelial cells *in vivo*. Extensive studies under various flow conditions, such as in the turbulent or non-steady flow, are thus needed toward the thorough understanding.

Involvement of major cytoskeletal components other than F-actin filaments should not be overlooked. Ives et al (1986) demonstrated that the flow-induced orientation of microtubules in the *in vitro* cultured bovine aortic and human umbilical vein endothelial cell monolayers. As Sato et al reported (1990), microtubules can also contribute to the change in viscoelasticity of the static cultured endothelial cells. The contribution of other cytoskeletal components and their interactions among them need further considerations.

Summarizing this section, cultured porcine aortic endothelial cells elongated and oriented to the flow direction when they were exposed to shear stress, and re-distribution of F-actin filaments was one of the early cellular responses to the onset of shear stress. This re-distribution from the periphery to the central part of the cell preceded the cell elongation and orientation. Also, normalized fluorescence intensity of F-actin filaments showed rather slow increase (data not shown), and it was suggested that not only F-actin distribution but also F-actin content might be affected by the onset of shear stress.

Transendothelial permeability to macromolecules

Transendothelial permeability was studied under static culture conditions with control PAECs and those after the exposure to shear stress for 24 h. This exposure period was confirmed to be long enough for the cell elongation and stress fiber formation to develop, as summarized in the previous section. Therefore, any change in permeability found in the present investigation might presumably be due to the flow-induced change in cytoskeletal structure, even though the permeability measurements and the flow exposure were not performed simultaneously.

Figure 3 summarizes values of the transendothelial permeability ($P_{\text{endothelial cell}}$) to each tracer under the static cultured control, shear-exposed PAECs and those treated with PMA. Figure 3-a represent data for permeability to BSA. Compared with the data for the static cultured control PAECs ($5.68 \pm 1.19 \times 10^{-6}$ cm s^{-1}, n=28), permeability after exposure to shear stress showed an increase to $1.70 \pm 0.70 \times 10^{-5}$ cm s^{-1} (n=8; $p < 0.05$ vs. static cultured control). Preincubation of the PAEC monolayer with PMA (0.5 μM, 30 min) resulted in an increase of the value to $1.67 \pm 0.13 \times 10^{-5}$ cm s^{-1} (n=6; $p < 0.05$ vs. static cultured control).

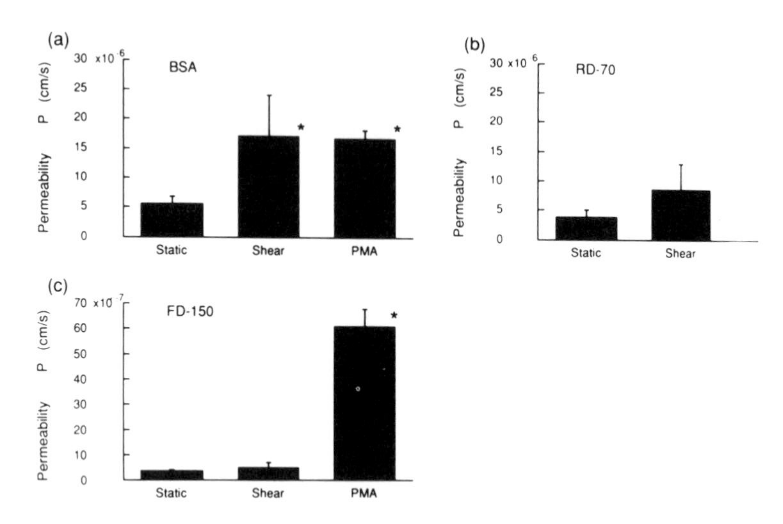

Figure 3. Transendothelial permeability characteristics to; (a) bovine serum albumin (BSA), (b) rhodamine-labeled dextran (MW = 70000, RD-70), and (c) FITC-labeled dextran (MW = 150000, FD-150). Mean ± SE. * $p < 0.05$ vs. static cultured control.

As shown in Fig. 3-b, permeability to RD-70 ($3.91 \pm 1.10 \times 10^{-6}$ cm/s, n=28) for the static cultured control PAECs was smaller than that of BSA. Although the value after exposure to shear stress showed 2-fold increase ($8.64 \pm 4.32 \times 10^{-6}$ cm s^{-1}, n=8), none of the treatments had statistically significant effects on transendothelial permeability to RD-70. Permeability to FD-150 ($3.77 \pm 0.44 \times 10^{-7}$ cm s^{-1}, n=21) for the static cultured control PAEC monolayers (Fig. 3-c) was 10 times smaller than that of RD-70 (Fig. 3-b). All these results seemed due to the direct effect of the bulkiness of the tracers, or the radius (Stokes-Einstein radius) of the molecule. The exposure to shear stress induced no significant permeability changes in the case of this tracer molecule ($5.20 \pm 1.75 \times 10^{-7}$ cm s^{-1}, n=3), while the treatment with PMA induced a marked increase in permeability as much as $6.09 \pm 0.72 \times 10^{-6}$ cm s^{-1} (n=6; $p < 0.05$ vs. static cultured control).

To maintain the integrity of the cell monolayers to be functioned as a barrier between vessels and tissues, cytoskeletal structure is considered to play a dominant role. A well-known permeability agonist, thrombin, known to affect transendothelial permeability (Killackey et al 1986), has been reported to induce a drastic changes in the distribution of cytoskeletal F-actin filaments in the cultured endothelial cells (Wong and Gotlieb 1990). Conversely, phalloidin that selectively binds to F-actin and inhibits the depolymerization of the filaments has been reported to reduce the inflammatory permeability increase by reinforcing the structures of the cultured endothelial cells (Alexander et al 1988). In view of these literature data and the fact that F-actin filament redistributed after exposure to shear stress, as demonstrated in Fig. 2 in the previous section, it was strongly suggested that shear-induced alterations in cytoskeletal structures was one of the factors that modulated the transendothelial permeability to macromolecules under the conditions studied.

The relationships between shear-induced changes of the endothelial cells and transendothelial permeability have recently attracted much attention in connection with the pathogenesis of such diseases as atherosclerosis or inflammatory diseases. Jo et al have recently reported that transendothelial permeability to BSA sensitively increased within 30 min after the onset of the medium flow and it showed 10-fold increase compared with the pre-shear value (Jo et al 1991). Since no apparent redistribution of cytoskeletal structure was reported before 2-3 h after the onset of shear stress application (Ookawa et al 1992), some fast-reacting messengers other than cytoskeletons might influence the permeation of solutes across the endothelial cells. Toward the thorough understanding of the permeability changes after shear exposure, further studies are still needed concerning the effects of other very early messengers responsible for the cellular signal transduction system.

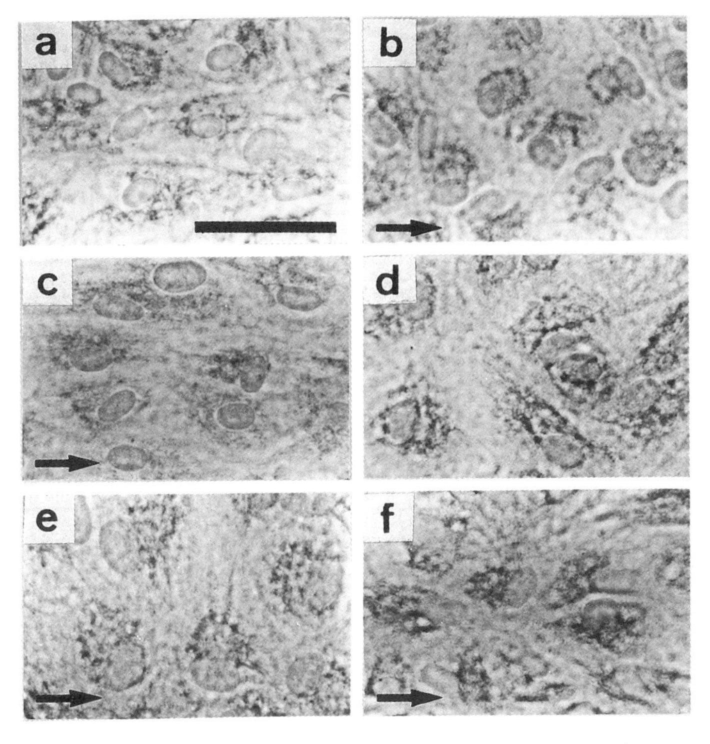

Figure 4. Typical photomicrographs of porcine aortic endothelial cells immunostained with antibodies against ELAM-1 (a-c) and VCAM-1 (d-f). Bar = 50 μm. Arrows = direction of the flow applied.

Adhesion molecules on endothelial cell surface

Under static culture conditions, the expression of ELAM-1 on PAEC was scarcely found (Fig. 4-a). After exposure to low-shear stress (ca. 2 dyn cm^{-2}), an enhanced and highly localized expression of ELAM-1 was observed around the nucleus of each cell (Fig. 4-b). In contrast, on the elongated PAECs under high-shear stress conditions (ca. 15 dyn cm^{-2}), a marked expression of ELAM-1 was observed (Fig. 4-c), while no apparent concentration or localization of the molecules were observed. As depicted by Fig. 4-d, VCAM-1 diffusely expressed on PAEC surface even under the static cultured conditions. The expression of VCAM-1 was highly localized after shear stress exposure (Figs. 4-e and f), although no apparent enhancement of the expression were not detected. The localization of these adhesion molecules were equally observed in both the upstream and downstream ends of the endothelial cells.

From our preliminary studies with guinea pig mesenteric microvasculature, VCAM-1 was mainly observed in venules while only weak expression of ELAM-1 was observed both in arterioles and venules. From these results, it is suggested that shear stress can act as a controlling factor of leukocyte-endothelial cell interactions in a similar way that affect morphology and permeability of the endothelial cells. Therefore, extensive studies on shear stress-induced functional changes in the light of other adhesion molecules known to interconnect adjacent cells are needed toward more detailed understanding of the endothelial barrier functions after exposed to shear stress.

Conclusions

Morphological and functional changes of cultured porcine aortic endothelial cells due to shear stress exposure were studied extensively in three different series of *in vitro* experiments. All data demonstrated that physiological level of shear stress is one of the controlling factors of structure and physiological functions of vascular endothelial cells. Hence, it is strongly suggested that fluid imposed shear stress modulate the function of vascular endothelial cells *in vivo*.

Acknowledgments

One of the authors (K. O.) are grateful to Dr. Masaaki Sato, presently with Faculty of Engineering, Tohoku University, for his continued guidance. This work was supported by research grants from University of Tsukuba Project Research, Grants-in-Aid from the Japanese Ministry of Education, Science and Culture (Nos. 63570380, 01570464, 05221211, 05454691), and in part supported by the Basic Research Core System of the Science and Technology Agency.

References

Alexander, J. S.; Hechtman, H. B.; Shepro, D. Phalloidin enhances endothelial barrier function and reduces inflammatory permeability in vitro. Microvasc. Res. 35:308-315; 1988.

Casnocha, S. A.; Eskin, S. G.; Hall, E. R.; McIntire, L. V. Permeability of human endothelial monolayers: effect of vasoactive agonists and cAMP. J. Appl. Physiol. 67:1997-2005; 1989.

Cooper, J. A.; Del Vecchio, P. J.; Minnear, F. L.; Burhop, K. E.; Selig, W. M.; Garcia, J. G. N.; Malik, A. B. Measurement of albumin permeability across the endothelial monolayers in vitro. J. Appl. Physiol. 62:1076-1083; 1987.

Cornhill, J. F.; Levesque, M. J.; Herderick, E. E.; Nerem, R. M.; Kilman, J. W.; Vasco, J. S. Quantitative study of the rabbit aortic endothelium using vascular casts. Atheroscleosis 35:321-337; 1980.

Franke, R. P.; Grafe, M.; Schnittler, H. M.; Seiffge, D.; Mittermayer, C.; Drenckhahn, D. Induction of human vascular endothelial stress fibers by fluid shear stress. Nature 307:648-649; 1984.

Ives, C. L.; Eskin, S. G.; McIntire, L. V. Mechanical effects on endothelial cell morphology: in vitro assessment. In Vitro Cell. Dev. Biol. 22:500-507; 1986.

Jo, H.; Dull, R. O.; Hollis, T. M.; Tarbell, J. M. Endothelial albumin permeability is shear dependent, time dependent, and reversible. Am. J. Physiol. 260:H1992-H1996; 1991.

Killackey, J. J. F.; Johnston, M. G.; Movat, H. Z. Increased permeability of microcarrier-cultured endothelial monolayers in response to histamine and thrombin. Am. J. Pathol. 122:50-61; 1986.

Ookawa, K.; Sato, M.; Ohshima, N. Changes in microstructure of cultured porcine aortic endothelial cells in the early stage after applying fluid-imposed shear stress. J. Biomech. 25:1321-1328; 1992.

Ookawa, K.; Sato, M.; Ohshima, N. Morphological changes of endothelial cells after exposure to fluid-imposed shear stress: Differential responses induced by extracellular matrices. Biorheology 30:131-140; 1993.

Sato, M.; Levesque, M. J.; Nerem, R. M.; Ohshima, N. Mechanical properties of cultured endothelial cells exposed to shear stress. Frontiers Med. Biol. Engng. 2:171-175; 1990.

Voyta, J. C.; Via, D. P.; Butterfield, C. E.; Zetter, B. R. Identification and isolation of endothelial cells based on their increased uptake of acetylated-low density lipoprotein. J. Cell Biol. 99:2034-2040; 1984.

Wechezak, A. R.; Wight, T. N.; Viggers, R. F.; Sauvage, L. R. Endothelial adherence under shear stress is dependent upon microfilament reorganization. J. Cell. Physiol. 139:136-146; 1989.

Wong, M. K. K.; Gotlieb, A. I. Endothelial monolayer integrity: perturbation of F-actin filaments and the dense peripheral band-vinculin network. Arteriosclerosis 10:76-84; 1990.

Part III

Mechanics and Biology of Cell-Substrate Interactions

9
Kinetics and Mechanics of Cell Adhesion under Hydrodynamic Flow: Two Cell Systems

D. A. Hammer, L. A. Tempelman, D. J. Goetz

Introduction

Transient, heterotypic adhesion between leukocytes and the endothelium is an essential part of many immune and inflammatory reactions (Stoolman, 1989; Osborn 1990; Springer, 1990; McEver, 1991; Lasky, 1992; Bevilacqua, 1993). It is generally believed adhesive interactions between leukocytes and endothelium are mediated by specific complementary binding between couterreceptors on both leukocytes and endothelium (Springer, 1990). Of particular interest are the adhesive interactions between neutrophils and endothelium near tissue sites of inflammation, where neutrophils are recruited to the inflammatory site to eliminate foreign invaders (Atherton and Born, 1972, 1973). A great deal of research is focused on understanding the molecular and physical requirements of adhesion molecules in this system which provide for local sequestration of neutrophils.

Neutrophils display a slow rolling over stimulated endothelium which appears to be important for their role as surveillance cells that monitor the health of nearby tissue. Rolling is defined as translation which is only some fraction of the translational velocity of particles of a similar size in the same hydrodynamic flow field. The reduced translational velocity is due largely to receptor-mediated adhesive interactions, which slow, but do not stop the neutrophil during its transit over the endothelial surface. Several lines of research suggest molecular interaction between receptors of the selectin family of adhesion molecules is principally responsible for this type of adhesion (Lawrence and Springer, 1991; Jones et al., 1993; Ley et al., 1991; von Adrian et al., 1991).

Although rolling has been seen in other cellular systems, such as tumor cells interacting with endothelium, lymphocytes interacting with endothelium, and monocytes interacting with laminin, rolling is only one dynamic manifestation of cell substrate interactions in hydrodynamic flow. There are many reports of receptor-mediated cell adhesion in hydrodynamic flow in which rolling is not observed (Wattenbarger et al., 1991; Tempelman and Hammer, 1994). We refer to these systems as binary, because only two states are observed; cells either translate at the unencumbered free stream velocity, or are firmly adherent. This leads us to ask, what are the properties of cell surface receptors and their binding which control the dynamics of adhesion? Is there anything unique about cell adhesion molecules which give rise to rolling? Are there ways of modulating systems between rolling and non-rolling states without changing the composition or structure of its cell surface receptors?

Although firm conclusions regarding differences between the molecules await direct measurement of the physical and chemical properties of individual molecules, others have tried to address this question indirectly by formulating mathematical models which contain mechanicochemical parameters for adhesion molecules (Hammer and Apte, 1992; Tozeren and Ley, 1992). These models, when applied to neutrophil rolling data, can delineate regions of parameter space where the observed dynamic behavior can be recreated.

In this paper, we describe experimental measurements of adhesion under flow with two separate cellular systems to elucidate the factors which control the dynamics of cell adhesion. In one of these systems, we use neutrophils interacting with stimulated, cultured endothelium. Clearly, this system displays the above mentioned rolling behavior. A big disadvantage in the neutrophil system is that it is not completely clear which neutrophil cell surface ligands bind to which endothelial ligands; as a result, it is not possible to systematically manipulate molecular chemistry and study the resulting effect on adhesion.

As an alternative, we have developed a second model system to study adhesion, the rat basophilic leukemia (RBL) cell. We selected the RBL cell system both because of morphological similarities to leukocytes (Oliver et al., 1988) and because a battery of well-characterized reagents have been developed for this system in the course of its use as a model system for studying the relationship between receptor crosslinking and secretion (Metzger et al., 1986). By using antibody as an intermediary between the RBL cell's $Fc_{\varepsilon}R1$ receptor and an antigen-coated surface, we can control both the number of binding sites on the cell and the kinetics of binding. In addition, many characteristics of the cell such as the number of its Fc_{ε} receptors are known. The cell has 1-3 x 10^5 high affinity Fc_{ε} receptors (Metzger, et al., 1986) which bind the Fc portion of IgE. Many well-characterized IgE antibody/antigen pairs are available to study the role of kinetics and affinity on adhesion. In our experiments thus far, we have used a single antigen, dinitrophenol (DNP) and an anti-DNP IgE antibody. The number of anti-DNP IgE surface molecules, and hence the binding sites for DNP, is changed by adjusting the proportion of anti-DNP IgE and another IgE (directed against a neutral antigen) which is used to fill the remaining Fc_{ε} receptors. Measured in solution, the affinity between the antibody and DNP is $K_A = 10^8$ M^{-1} and the forward rate constant is $k_f = 10^7$ $M^{-1}sec^{-1}$ (Erickson, et al., 1987). Even with this single IgE-antigen system, an alternate binding avenue exists, in which we prebind the antibody bivalently to the DNP and use the recognition between the $Fc_{\varepsilon}R1$ receptor and the Fc portion of IgE to mediate adhesion. The affinity between the Fc region of IgE and the Fc_{ε} receptor is 100 fold higher (10^{10} M^{-1}) but the forward rate constant is only 10^5 $M^{-1}sec^{-1}$ (Metzger, et al., 1986). Using IgE in these two separate configurations (IgE on the cell vs. IgE on the substrate) allows us to test the role of receptor recognition kinetics and affinity in adhesion.

There are advantages in using this second cell system for making a link between the physical chemistry of cell surface receptors and the dynamics of adhesion. First, the wide variety of antigen-IgE (ligand-receptor) pairs allow us to test hypotheses regarding the link between the kinetics and thermodynamics of ligand -receptor recognition and adhesion; in contrast, we often do not know the kinetics and thermodynamics of kinetic recognition in the neutrophil system unambiguously. Accordingly, there is a much greater spectrum of receptor-ligand chemistries available, given the wide variety of recognition that can be achieved with antigen-antibody pairs. Second, we can alter the number of IgE molecules on the RBL cell surface to test the effect of "receptor number" on adhesion; in contrast, we do not know the complete set of neutrophil surface receptors, nor do we know how many there are. Furthermore, although changing this number is possible (such as in the case of L-selectin) by stimulating cells to shed the receptor, adhesion experiments have to be carefully coordinated with the time after stimulation, a rather difficult procedure. Third, RBL cells can be grown in culture and large numbers of relatively homogenous cells can be easily obtained.

Here we report results of adhesion experiments with these two cell systems. With the neutrophil, we have measured rolling velocities at shear stresses between 1 and 4 dynes/cm^2 in an *in vitro* flow chamber. We show how the rolling velocity changes with shear stress, but also illustrate that the dynamics of neutrophil rolling is not completely described by this quantity. In the RBL cell system, we measured the effects of reaction rate and affinity between receptor and ligand, the number of IgE molecules on the cell surface, and the shear rate on adhesion. We show RBL cells will bind at various locations on a gel coated homogeneously with ligand, suggesting adhesion is a probabilistic process. Therefore, statistical factors might play a role in both types of adhesion, yet due to the different dynamics, might manifest themselves in different ways. These findings imply that a mathematical model must contain stochastic/probabilistic components in order to fully capture the dynamics of cell adhesion under hydrodynamic flow.

Materials and Methods

Neutrophil Experiments

Preparation of Neutrophils. Neutrophils were isolated from citrate-anticoagulated whole bovine blood using a modified version of Carlson and Kaneko (1973) as described by Gilbert et al. (1993). Recovered neutrophils were held on ice for less than four hours until they were used in experiments. Immediately prior to use, the supernatant was removed and the neutrophils were resuspended in 20 ml of DMEM supplemented with penicillin, streptomycin, 1mM $MgCl_2$, 1.2 mM $CaCl_2$, and 2.5 mM HEPES. The neutrophils were then placed in an incubator at 37 C, 5.0% CO_2, and 90% humidity for 30 minutes. An aliquot of the neutrophils was resuspended in supplemented DMEM to a final concentration of 1.7×10^5 cells/ml. The suspension was poured into a glass syringe which was then placed on a syringe pump in preparation for the experiment.

Preparation of Endothelial Cells. Bovine aortic endothelial cells (BAEC) were isolated from thoracic aortas of 18 month old bovines, essentially as described by Booyse et al. (Booyse et al., 1975) . BAEC were grown in Dulbecco's Modified Eagle Medium (DMEM) (Gibco, Grand Island, NY) supplemented with 10% heat inactivated fetal bovine serum (FBS) (Gibco, Grand Island, NY) and frozen in liquid nitrogen at passage 2 or 3. BAEC used in these experiments were recovered from liquid nitrogen and then passaged twice; once on tissue culture plastic and once on a glass slide pre-coated with 100 μg/ml of collagen type I (Collagen Corporation, Palo Alto, CA). Cells were cultured for 48 hours at standard culture conditions (37 ^{o}C, 5% CO_2 in air, 100% humidity). Four hours prior to the experiments, lipopolysaccharide (LPS) (Sigma, St. Louis, MO) was added to the endothelial cell media at a final concentration of 1 μg/ml.

Flow Chamber Assay: Neutrophils. An adapted version of the radial flow chamber described by Cozens-Roberts et al. (1990a) was used to measure the rolling velocity of the neutrophils. Our chamber consists of a lower plate and an upper plate separated by a gasket (Silastic Sheeting, Dow Corning, Midland, MI). The top plate has an inlet for the cell suspension. The bottom plate has a trough where a coverslip can be deposited. For flow rates and chamber dimensions used in these experiments, an axisymmetric laminar flow field is created in the horizontal gap between the plates (Cozens-Roberts et al., 1990a; Fowler and McKay, 1980). The shear stress at a radial position r on the bottom surface is given by $\tau(r) = 3q\mu/\pi r h^2$ where q is the volumetric flow rate, h is the thickness of the horizontal gap between the plates, and μ is the viscosity of the media (Fowler and McKay, 1980). For this particular application the flow rates used were 2.5 ml/min and 0.7 ml/min. The thickness of the horizontal gap was 0.20 mm and the viscosity of the media was 0.01 p. The flow chamber was placed on top of an inverted phase-contrast microscope (Diaphot-TMD, Nikon Inc., Garden City, NJ) connected to a video camera (70S camera, Dage-MTI, Inc., Michigan City, IN). The events were recorded on a VCR (BR-S611U, JVC, Elmwood Park, NJ) connected to a monitor (PVM-91, Sony, Park Ridge, NJ) and in line with a frame code generator (SA-F911U, JVC, Elmwood Park, NJ). The resolution of the recording was 30 frames per second. In line with the camera was a reticle which superimposed a grid of dimensions 250 μm x 250 μm onto the recorded image. Prints of the videoframes were made from a thermal printer (VideoGraphic Printer, UP-870MD, Sony, Park Ridge, NJ).

The experiments were performed at room temperature as follows. The glass slide covered with the LPS treated BAEC monolayer was placed in the bottom piece of the flow chamber. The flow chamber was assembled and the endothelial cell layer was washed with LPS-free supplemented DMEM for 10 minutes. Following the wash, the flow of the neutrophil suspension was initiated. Neutrophil trajectories were videotaped at radial positions corresponding to shear stresses of 4,3,2 and 1 dynes/cm^2.

Data Analysis. The average velocity of the cells at a given shear stress was determined from the time taken for a cell to cross a 250 μm section of the field of view defined by the reticle. To determine the instantaneous velocities, the tapes were played back on a VCR connected to an image analysis workstation (TCL-IMAGE Software, Multihouse TSI, Amsterdam, Holland, running on a Mac IIx computer, data translation frame grabber). Every two seconds, determined by the time codes, an image was sent to the image analysis work station. We then utilized the image analysis software to determine the position of the neutrophil in the field of view to a resolution of 0.76 μm.

The theoretical variance of the instantaneous velocity measurement was determined using standard error analysis. The velocity is defined by $v = d/t$ where d is the distance the cell moved in time t (2 seconds). The theoretical measurement variance was then determined from the equation $<\sigma_v> = (\delta d/t)^2 + (v\delta t/t)^2$ where δd is the theoretical error in the distance measurement (one half the smallest resolution, 0.38 μm) and δt is the theoretical error in the time measurement (one half the smallest time resolution, 1/60 second). In calculating the theoretical measurement variance, the average velocity at the given shear stress was used.

The experimental variance of the instantaneous velocities of cell j were determined from the equation $<\sigma v_j> = \Sigma\, (V_i - V_j)^2/(n-1)$, where the sum is taken from 1 to n where n is the number of instantaneous velocity measurements for cell j, V_j is the average instantaneous velocity for cell j, and V_i is the instantaneous velocity measurement corresponding to time i.

RBL cell experiments

Preparation of RBL cells. The RBL subline 2H3 (Barsumian et al., 1981) was a gift from Dr. C. M. S. Fewtrell. These cells were cultured and harvested as described by Taurog and coworkers (Taurog et al., 1979) for use between days 4 and 7. RBL cells are spherical, and our subline measured 13 μm diameter (measured using Coulter Counter™ Hileah, FL, and electron microscopy) and high microvilliated with 3000 microvilli (based on transmission electron microscopy; Oliver, et al., 1988) that are 0.5 μm long (Oliver, et al. 1988) with a cross sectional area of 0.01 μm^2 (Oliver et al., 1988; Bongrand and Bell, 1984). RBL cells possess on average 2 x 10^5 Fcε receptors (Metzger, et al., 1986; Ryan, 1989). Purified mouse monoclonal anti-2,4 dinitrophenol IgE from hybridoma H1 26.82 (Lui, et al., 1980) was a generous gift from Drs. B. Baird and D. Holowka. Mouse monoclonal anti-dansyl IgE from clone 2774 [PharMingen, San Diego, CA] was used as a neutral antibody that does not bind DNP. Cells (2 x 10^6/ml) were incubated in a 24 nM antibody mixture for at least 30 minutes. The anti-DNP to anti-dansyl ratio was adjusted to fill between 20 and 100% of a cell's 2 x 10^5 $Fc_\varepsilon RI$ receptors (Metzger, et al., 1986) with anti-DNP IgE (Weetal, 1992). In this way, we change the number of binding sites for DNP from 0, 20, 60 or 100% of 4 x 10^5 (IgE is bivalent). The cells were centrifuged and resuspended in a modified Tyrodes buffer (125 mM NaCl, 5 mM KCl, 10 mM HEPES, 1 mM $MgCl_2$, 1.8 mM $CaCl_2$, 0.05 % gelatin and 5.6 mM glucose in deionized water) at 3-4 x 10^5 cells/ml and held at 37 ^oC under gentle agitation in a waterbath for use within several hours. Control experiments and one experiment (described later in the text) were done without adding antibody to the cell suspension.

Antigen-Coated Gels. Gel preparation. Acrylamide, N, N'-methylene-bis-acrylamide, N,N,N',N'-tetramethylethylene diamine (TMED), and ammonium persulfate were from Bio-Rad Laboratories [Hercules, CA]. High percent acrylamide (21 g monomer/100 ml) gels which were also highly crosslinked [0.083 g crosslinker/ (g crosslinker +monomer)] were cast using 0.25 mm teflon spacers in a vertical gel caster [SE215 or 275 Hoeffer Scientific, San Francisco, CA]. The solution was buffered with 54 mM HEPES to pH 6.0. 162μl of TMED and 429 μl of 7% w/v ammonium persulfate per 100 ml are used as initializers. The monomer solution contained 1.25-20 mM of the N-succinimidyl ester of acrylamidohexanoic acid, a bifunctional linker, which copolymerizes with acrylamide. The linker was synthesized in our laboratory using the method of Pless and coworkers (Pless, et al., 1983). Dr. B. Brandley [Glycomed, Alameda CA] provided us with advice on the chemistry of linker synthesis and a sample of linker to use as a crystallization seed.

Derivatization. After casting for one hour and rinsing gels in cold water for 15 minutes, half the gel is inactivated by reacting the linker with ethanolamine at 1 μl/ml in 50 mM HEPES in deionized water with 10% ethanol buffer (pH 8.0) for 5 minutes. This step is performed by dipping the gel vertically in this capping solution and rinsing thoroughly with water. The entire gel is then treated horizontally with a saturated solution (5.3 mM) of N-ε-2,4 DNP-lysine hydrochloride [Sigma, St. Louis MO] in 50 mM HEPES with 10% ethanol buffer (pH 8.0) for 60-150 minutes. The DNP attaches to the active half of the gel by an amide linkage, displacing N-hydroxysuccinimide. A final thirty minute treatment with ethanolamine solution caps any unreacted linker sites with a methyl group, preventing the formation of charged carboxyl groups that would result from further hydrolysis of the linker. Reaction parameters were chosen to maximize aminolysis (attachment of DNP) and minimize hydrolysis (Schnaar and Lee, 1975; Pless, et al., 1983). Normally, the amount of DNP in a gel is controlled by the amount of linker in the gel and varies linearly with linker concentration (Chu, et al., 1994).

DNP Surface Density. Circles of gel were cut (1.45 cm dia, 0.025 cm height), covered with 5N NaOH and incubated at 37° C for 16-17 hours. This liberated the DNP from the gel, and the solution was assayed for DNP spectrophotometrically at 365 nm [Milton Roy Spectronic 1001 plus]. The assay was calibrated with known concentrations of DNP-lysine HCl which had also been treated with 5N NaOH. The results gave the efficiency of the derivatization reaction to be approximately 0.5 μmol DNP/μmol linker. The volumetric density was converted to a surface density using the assumption that DNP molecules within the top 10 Å of the gel would be accessible to antibody bound to cells moving across the surface. The surface density of DNP for gels with 1.25 μmol linker/ ml of gel which were used in the majority of experiments is approximately 3.6×10^{10} molecules/cm^2. Several saturation binding experiments of ^{125}I-anti-DNP IgE to DNP gels were performed to assess levels of DNP or IgE bound to the gels. These experiments did not give an absolute DNP surface concentration because in the time necessary for equilibrium binding between IgE and gel-bound DNP, the IgE in solution was able to penetrate beyond the surface of the gel. The level of IgE association with the gel far exceeded the density of linker in the gel's upper surface, suggesting deep (>100 Å) penetration of the IgE into the gel. Thus, saturation binding study results over-estimate the density of DNP available to cell-bound antibody. However, a saturation binding study on a 1.25 μmol/ml gel gave the same level of binding as a subsaturation binding (23.9 nM) to a 20 μmol/ml gel. Also, the characteristic time for dissociation of half the IgE bound at subsaturation on the 20 μmol/ml gel was over three hours, indicating that antibodies were bivalently bound to the gel. Details can be found in Tempelman (1993). This information is used to compare ligand density in experiments presented later in the text.

Flow Chamber Assay: RBL cells. The flow chamber is a parallel-plate chamber with a separate inlet for cells and accommodations to hold a thin gel. The channel for fluid flow has dimensions of 72 x 15.2 x 0.53 mm (L x W x H). The chamber is made from two blocks of transparent Lexan which are held together with plastic screws. The bottom piece has a double width microscope slide for visualization which is glued into the chamber with silicon. The slide is recessed from the chamber bottom to accommodate the gel. The top piece has the buffer inlet and outlet ports as well as the separate cell inlet port. The height of the channel is determined by a Silastic™ Sheeting gasket [Dow Corning, Midland MI] which is 0.021" thick. Buffer flow to the chamber is provided by a syringe pump [Sage Model 355, Orion Research Incorp., Boston MA] using a 500 ml glass syringe [Scientific Glass Engineering, Austin TX] and volumetric flow rates were measured daily and compared to a pump calibration curve. Cells are observed under 200X magnification on an inverted microscope [Nikon Diaphot-TMD] using phase contrast optics. The experiments are videorecorded [Dage-MTI 67S Newvicon videocamera; JVC BR-S611 videoplayer] for analysis at a later time. A thermal printer [Sony VideoGraphic Printer UP-870MD] was used to make prints of some videoframes.

Adhesion Experiment. The chamber is filled with modified Tyrodes buffer. In the absence of buffer flow, 20-40 μl of cell suspension is introduced from the cell inlet using a 500 μl syringe [Hamilton Corp., Reno NV]. Cells are allowed to settle for approximately 4 minutes onto the non-DNP portion of the gel. The inlet region is scanned spatially and videotaped in a systematic manner. The 20X microscope objective is then positioned downstream along the centerline of the chamber just within the adhesive region of the gel. Buffer flow is directed into the chamber for a period of time more than sufficient for all cells to travel the length of the chamber. Then the DNP region of the gel is scanned systematically. Negligible numbers of cells remain adherent to the region of the gel near the cell inlet despite being on the surface under static conditions for a significant length of time. Between trials, the flow chamber is cleared of cells by the high shear force of purging the chamber with water and air. Then a duplicate trial is performed. Analysis of duplicate and triplicate trials shows gel adhesiveness is not affected by repeated use of the gel.

Analysis. Three dimensional plots of the cell inlet region and the DNP region are prepared using WingZ™ [Informax Software, Lenexa, KS] software. The fraction of adhesion is calculated as the total number of cells adhered in the DNP region divided by the number of cells initially observed in the inlet region. Only individual cells are counted in the inlet and in the DNP region;

doublets, triplets and clumps are ignored. Our method of cell delivery ensures that the vast majority of cells are at the bottom of the chamber. This makes a comparison of the number of cells that attach to the DNP portion of the gel with the number of cells counted in the inlet area a particularly meaningful measure. If cells were present at all separation distances in the inlet fluid (as cells have been in previous researchers' adhesion systems), visualization of cells interacting with the substrate and determination of the number of cells available for binding would be more complicated and ambiguous.

Error in Cell Counting. While a counting error in any individual field of view may be + 5%, the errors are random and the total error over about 300 fields was measured to be only 1%. Error in scanning the gel systematically is about 5%. This is because scanning is done rather quickly both to speed the experiment and also to minimize the time bound cells remain on the gel. This can lead to as much as a + 10% error (calculated) in percent adhesion which is only 1% error for 10% adhesion, but 10% error for 100% adhesion. This accounts for reported values of adhesion greater than 100% adhesion in some experiments.

Methods of determining shear rate. Wall shear rate is calculated based on the measured volumetric flow rate and channel dimensions. Wall shear stress is proportional to wall shear rate; the proportionality factor is the viscosity, which we have measured to be 0.01 g/cm-s for modified Tyrodes buffer. Clearly, flow rate is not a unique indicator of shear rate, since in day-to-day assembly of the chamber, the thickness d may vary. Therefore, for some trials, we further confirmed shear rate by measuring the population average velocity of cells traveling across a 650 μm field of view (at the beginning of the DNP region of the gel), and comparing it to the average velocity for 14.5 μm diameter polystyrene beads [Coulter Corp., Hialeah, FL] (Tempelman, 1993). Because the dependence of bead velocity on shear rate agreed fairly well with theory (Goldman, et al., 1967), calibration curves could be constructed which related bead velocity to shear rate, and hence cell velocity to shear rate. This technique was used for an extensive study of RBL cell motion in simple shear (Tempelman et al., 1994); interested readers are directed there for further information.

Results

Neutrophils

Average Velocity vs. Shear Stress. To determine the average rolling velocity of the neutrophils we measured the time required for several cells to transverse the 250 x 250 μm^2 grid superimposed on the videotaped image. Figure 1 shows how the average velocity varies with shear stress. The rolling velocity increases with increasing shear stress up to about 3 dynes/cm^2, and appears to plateau in the vicinity of 4 dynes/cm^2. The average velocities at the three lower shear stresses are significantly different from one another ($p < 0.05$) while the average velocity at 4 dynes/cm^2 is not significantly different from that seen at 3 dynes/cm^2 ($p = 0.4$). From the work of Goldman et al. (1967), it is possible to predict the velocity at which a solid sphere having the same radius as the neutrophil would be moving if it were 500Å from the wall and in low Reynolds number linear shear flow (500Å being the approximate maximal separation distance for two selectin adhesion receptors to interact (Springer, 1990)). Table 1 shows this prediction, as well as the average velocities reported in Figure 1 as a function of shear stress. As shown, the neutrophils roll between 0.9% and 0.5% of the velocity predicted by Goldman and coworkers (Goldman et al., 1967). The translational velocities we measured for bovine neutrophils over LPS stimulated BAEC are quantitatively similar to those reported by Lawrence and Springer (1991) for human neutrophils rolling over a substrate consisting of reconstituted P-selectin at a surface density of 400 sites/μm^2.

Instantaneous Velocities. To further characterize the motion of the neutrophils, we measured the instantaneous velocities for several neutrophils at the four different shear stresses. This was done by importing an image from the videotape to an image analysis workstation. The position of the neutrophils in the image was then determined up to a theoretical resolution of 0.78 μm. This process

Table 1. Rolling velocity as a function of shear rate.

Shear stress dynes/cm^2	Rolling velocity (μm/sec)	Theoretical velocity* (μm/sec)	% of theory	P-selectin 400 /μm^2 #
1	2.0	200	0.98	1.8
2	2.5	400	0.63	3.4
3	3.6	600	0.60	3.9
4	3.9	800	0.49	4.8

*Goldman et al., 1967.
#Lawrence and Springer, 1991.

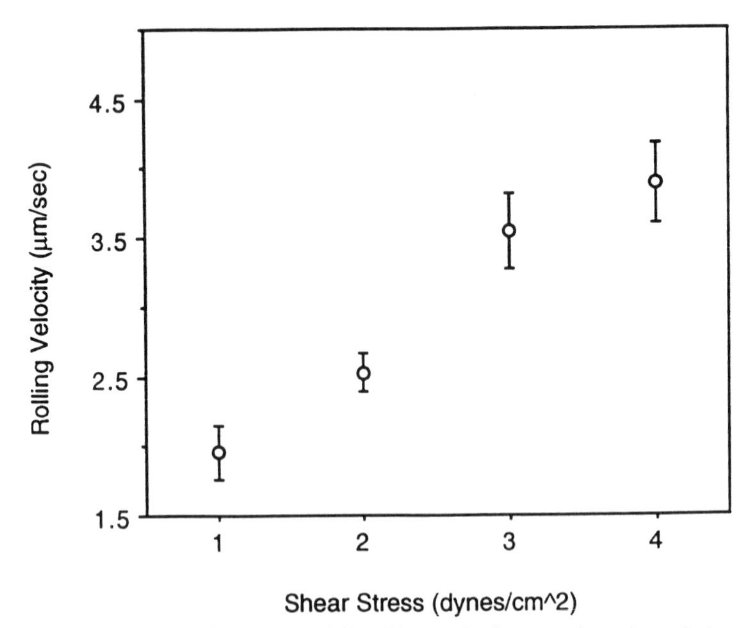

Figure 1. Average bovine neutrophil rolling velocity as a function of shear stress over LPS-stimulated bovine aortic endothelial cells.

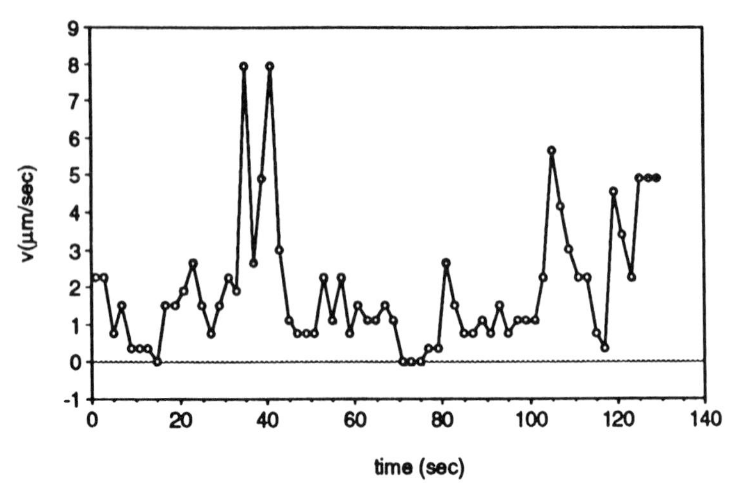

Figure 2. Example of neutrophil instantaneous velocity as a function of time at 2 dynes/cm^2.

was repeated many times for frames 2 seconds apart. Distance versus time plots were then converted to velocity versus time using finite difference. A typical result for a single neutrophil at 2 dynes/cm^2 is given in Figure 2, which shows the instantaneous velocity for a typical neutrophil varies considerably during its

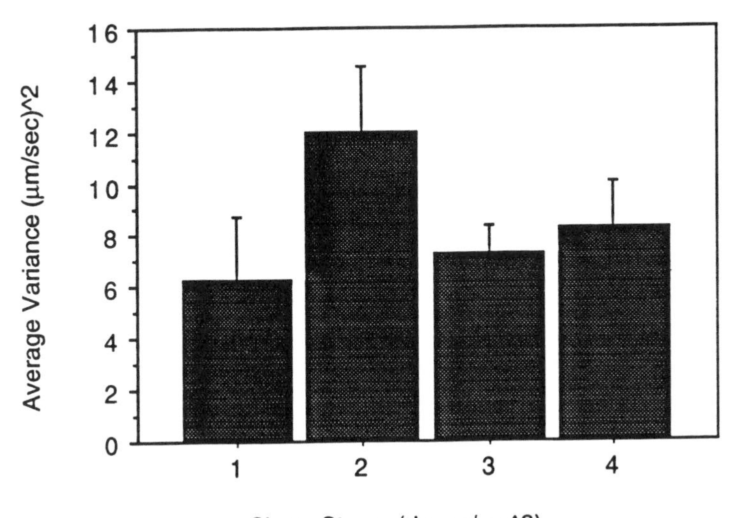

Figure 3. Variance in neutrophil rolling velocity as a function of shear stress.

transit over the field of view. This was true for all neutrophils observed at all shear stresses. To quantify the variability in the neutrophil motion, we calculated the variance in the instantaneous velocity for each individual neutrophil as described in Materials and Methods. We then averaged these variances and plotted the average variance as a function of shear stress (Figure 3). As expected the variance is non-zero at all shear stresses. However, even for perfectly smooth motion, the variance will be non-zero due to inherent measurement error. Hence, it is important to determine if the variance is significantly different from the theoretical measurement variance which arises from the measurement error. The theoretical measurement variance was calculated as shown in Material and Methods and compared with the observed variance at the four different shear stresses. The observed variance is over two orders of magnitude higher than the theoretical variance at all shear stresses, and this difference was found to be significant.

RBL cells

Control Experiments. To test for nonspecific adhesion under flow, we measured the adhesion of RBL cells without antibody to DNP-coated gels. Duplicate runs at 22 sec^{-1} showed less than 3% adhesion (21 bound out of 2014 and 124 out of 4373). A control at a higher shear rate of 99 sec^{-1} gave 0 out of 1804 cells bound. This control and others (cells with anti-dansyl IgE on DNP gels, cells with and without antibody on non-DNP gels) also showed low nonspecific adhesion in a static assay (Chu et al., 1994).

Location of Binding. At high DNP concentrations (>10 μmol linker/ml), it is possible to see the DNP portion of the gel because DNP absorbs in the yellow (as well as in the UV). We have observed that the start of the DNP region and the start of cell adhesion coincide and that cells adhere only on the DNP region of the gel. At lower DNP concentrations the yellow demarcation cannot be clearly seen, but the start of cell adhesion remains distinct and can be used to mark (infer) the border of the DNP region. Although there is an occasional cell adhering slightly before the interface or well inside the non-DNP portion, likely due to a slight jaggedness of the non-DNP/DNP interface or to nonspecific adhesion, there is well-defined separation between regions of the gel which support adhesion (DNP-derivatized) and regions which do not (DNP-free).

Bimodal Binding Behavior. A field of view was selected which spanned the boundary between a DNP and non-DNP region of the gel. The number of frames (1/30 sec) which it took for individual cells to travel the first 223 μm (non-DNP region) and last 223 μm (DNP region) of the 650 μm field were recorded for 57 cells. The average velocities in the non-DNP and DNP portions of the gel were 1.4 and 1.5 cm/min, respectively. The data demonstrate that cells coated with anti-DNP IgE do not travel at a reduced velocity over a DNP-coated surface before they bind. In fact, over the entire range of experimental conditions tested, there was no indication that RBL cells make transient bonds which reduce their velocity without leading to complete adhesion. We observed that cells move with a constant velocity right up until the last few frames (< 0.1 sec) before coming to a stop. Most cells stopped quite abruptly, but occasionally a cell would pause, turn over once and then stop.

Effect of Shear Rate. Figure 4 shows cell adhesion as a function of shear rate for two levels of anti-DNP IgE at a fixed ligand density of 3.6×10^{10} molecules/cm^2. For 20% anti-DNP IgE, adhesion is complete at about ~ 20 sec^{-1} and becomes quite sparse at ≥ 100 sec^{-1}. Two days' experiments with individual data points are shown. For 100% anti-DNP IgE, complete adhesion is supported at ≤ 70 sec^{-1}, but adhesion drops off rapidly to less than 10% adhesion at ~ 99 sec^{-1}. 60% anti-DNP IgE data is not shown, because it is similar to the 100% anti-DNP data. The fraction of

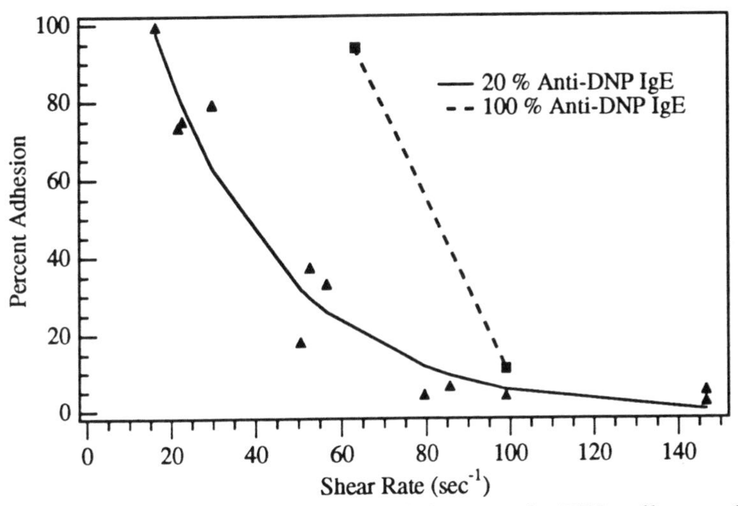

Figure 4. Percent adhesion as a function of shear rate for RBL cells coated with different densities of IgE.

cells adherent is linear with the inverse of shear rate (Tempelman and Hammer, 1994).

An example of a typical spatial pattern of adhesion is shown in Figure 5A. At this shear rate (64 sec^{-1}), adhesion is 22%, and the pattern of binding appears disperse, almost independent of position. Binding continues greater than 1500 cell diameters from the beginning of the DNP-gel and might continue further if the flow chamber were longer. This shows there is a small, nonzero probability of adhesion at all positions down the length of the chamber; this probabilistic nature of adhesion is considered in the Discussion.

Effect of Receptor Number. Total percent adhesion is also affected by the number of binding sites on the cell. Table 2 shows adhesion for different IgE coverages and three shear rates at a fixed ligand density of 3.6×10^{10} molecules/cm^2. For the lowest shear rate ($\gamma = 30$ sec^{-1}) adhesion is already very high when there is only 20% anti-DNP IgE, and we assume higher levels of anti-DNP IgE would also lead to nearly complete adhesion. For the highest shear rate ($\gamma = 99$ sec^{-1}), the maximum level of adhesion is 11%, even at the highest anti-DNP IgE surface density. At the middle shear rate, $\gamma = 64$ sec^{-1}, we see the most sensitivity to anti-DNP IgE surface density. When the number of binding sites is increased from 20 to 60 % anti-DNP IgE (a three fold increase) the amount of adhesion increases from an average of 23% to 101% (a 4.4 fold increase). A three dimensional plot of adhesion at $\gamma = 64$ sec^{-1} and 100% anti-

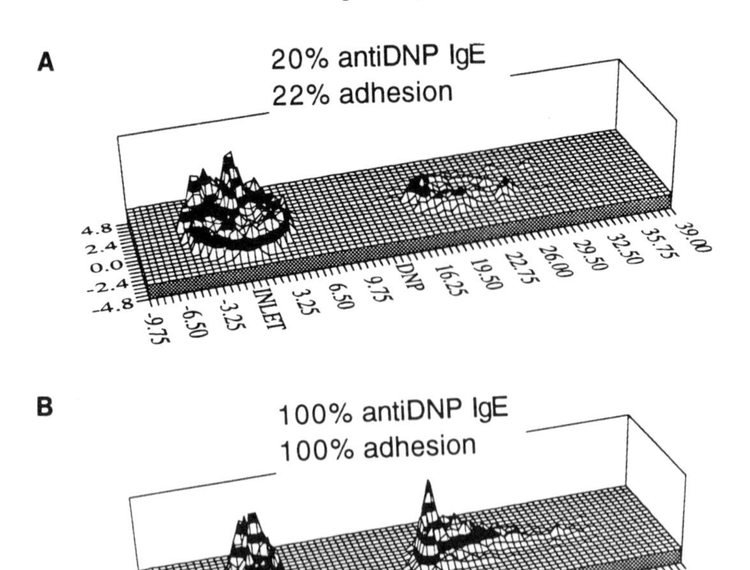

Figure 5. Spatial patterns of adhesion for RBL cells binding to DNP-coated gels. A. 20% anti-DNP IgE. B. 100% anti-DNP IgE. Shear rate is 64 sec^{-1}.

DNP IgE is shown in Figure 5B; comparison between Figures 5A and 5B illustrates the effect of increasing the number of IgE binding sites by a factor of 5. Figure 5B shows a distinct ridge of adhesion where the DNP portion of the gel begins, in contrast to the diffuse pattern of binding in Figure 5A.

Table 2. Fraction of adhesion as a function of shear rate and IgE density.

% anti-DNP	$\gamma = 30\ s^{-1}$	$\gamma = 64\ s^{-1}$	$\gamma = 99\ s^{-1}$
0	0	0	0
20	86 +/- 10	23 +/- 4	4 +/- 1
60	ND	101 +/- 2	11 +/- 4
100	ND	93 +/- 8	10 +/- 2

ND: not determined

Effect of Bond Kinetics. Table 3 shows the percent adhesion at various shear rates for two different surface chemistries which are known to exhibit different binding kinetics. The "high k_f" system is where a mixture of anti-DNP and anti-dansyl IgE is preincubated with the cells. The "high k_f" system displays a kinetic rate of binding of $k_f = 10^7\ M^{-1}sec^{-1}$ (Erickson et al., 1987). The "low k_f" system is where anti-DNP IgE was preincubated on high density DNP (20 μmol linker/ml) gels at 23.9 nM and the cells were free of antibody. The "low k_f" system displays a kinetic rate of recognition of $k_f = 10^5\ M^{-1}sec^{-1}$ (Metzger et al., 1986). Adhesion is very low for the "low k_f" system. For example, the "high k_f" system at 99 sec^{-1} can support more adhesion on average than the "low k_f" system at 30 sec^{-1}. Differences between "high k_f" and "low k_f" systems are not due to differences in ligand density; the density of surface binding sites is similar for both systems. That is, the number of DNP sites on a 1.25 μmol linker/ml gel (3.6×10^{10} molecules/cm^2) is similar to the number of Fc ends available on a 20 μmol linker/ml gel after a 2 hour incubation at 23.9 nM as measured by ^{125}I anti-DNP IgE binding studies (Tempelman, 1993).

A semi-static binding assay that further elucidates the differences in binding behavior between the "high k_f" and "low k_f" systems. In this assay, flow is turned on to deliver the cells to the DNP portion of the gel. Flow is turned off for the specified amount of time and then turned on to a low shear rate of 26 sec-1 to wash away unattached cells. Percent adhesion is based on about 10 cells in this assay and most trials were performed in quadruplet. After a 1 minute incubation the difference between the two systems is striking. In the "low k_f" system an average of only 12% of cells bind (range 0 to 30%), whereas in the "high k_f" system an average of 74% bind (range 50 to 89%). However, after 5 minutes the "low k_f" system exhibits near maximal binding similar to that

Table 3. Fraction adherent for different reaction and shear rates

Case	γ (sec^{-1})	Fraction adherent
high k_f	60	1.00
high k_f	99	0.12
low k_f	30	0.11
low k_f	60	0.01

seen in the "high k_f" system at 2 minutes. As a control, the adhesion of cells coated with 100% anti-dansyl IgE were used. Only one cell out of 68 cells in a series of time trials bound, again showing nonspecific adhesion is negligible. The semi-static binding assay demonstrates that there is sufficient IgE remaining on the 20 μmol/ml gels preincubated with IgE to support adhesion, provided there is sufficient time for reaction in the "low k_f" system.

Table 3 gives the extent of adhesion for combinations of k_f and shear rate. Clearly, attachment can be modulated by changes in either shear rate or intrinsic bond kinetics, and we consider whether there might be a unifying way of scaling these two parameters. Various models suggest a unifying parameter governing receptor-ligand recognition under flow is the ratio of the reaction rate to the shear rate (Hammer and Apte, 1992). One can compare reaction and shear rates with a parameter $\underline{\kappa}$ (units time) = $\nu\phi/\gamma$, where ν is the valence of the reaction (2 for "high k_f" and 1 for "low k_f"), and ϕ describes the extent of diffusion limitation for the reaction (1 for "high k_f" and 1/2 for "low k_f" systems). When one plots the extent of adhesion versus $\underline{\kappa}$ for the data shown in Table 3, one sees that all the combinations of extent and $\underline{\kappa}$ fall on a single, nonlinear curve, suggesting adhesion is uniquely related to this parameter (Figure 6).

Discussion

The two cell systems we have used in these experiments - neutrophils binding to stimulated endothelium and IgE sensitized RBL cells binding to antigen-coated substrates - display very different dynamic behavior. Under conditions of hydrodynamic flow, neutrophils "roll" - translate with a velocity well below the hydrodynamic velocity - over stimulated endothelium or substrates reconstituted with endothelial cell adhesion molecules (Lawrence et al., 1990, Lawrence and Springer, 1991). In contrast, RBL cells display binary adhesion behavior - they either move at the hydrodynamic free stream velocity, or are firmly adherent. In our experiments, we have explored a factor of five range in the number of cell surface IgE binding sites, a factor of five range in shear rates, a factor of 100 range in affinity, and a factor of five range in forward reaction rate. None of these changes alter the appearance of adhesion in the RBL cell system from "binary" to rolling.

The question remains, what inherent difference is there between these two systems that controls adhesive phenotype? Two possibilities are the mechanicochemical properties of the individual adhesion molecules, and the mechanics of cell surface microstructure. Using a microvilli hard sphere model of cell adhesive dynamics (Hammer and Apte, 1992), we have explored how two parameters which govern the mechanicochemical response of individual cell adhesion receptors to stress affect cell adhesion dynamics. The first is α, a dimensionless parameter which compares the stiffness of the spring-like adhesion molecule to the hydrodynamic force acting on the cell. It can be thought of as the bond's mechanical stiffness. The second is F_σ, a dimensionless parameter which describes the influence of stress on the rate of breakage of adhesion molecules. It can be viewed as the

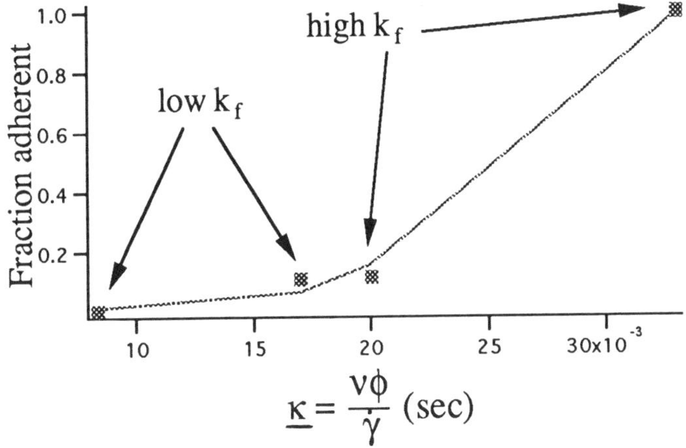

Figure 6. Adhesion for both low and high k_f systems can be plotted against a single scaled reaction rate, $\underline{\kappa}$. Adhesion monotonically increases with increasing $\underline{\kappa}$, suggesting reaction and flow rates are equally important determinant of adhesion.

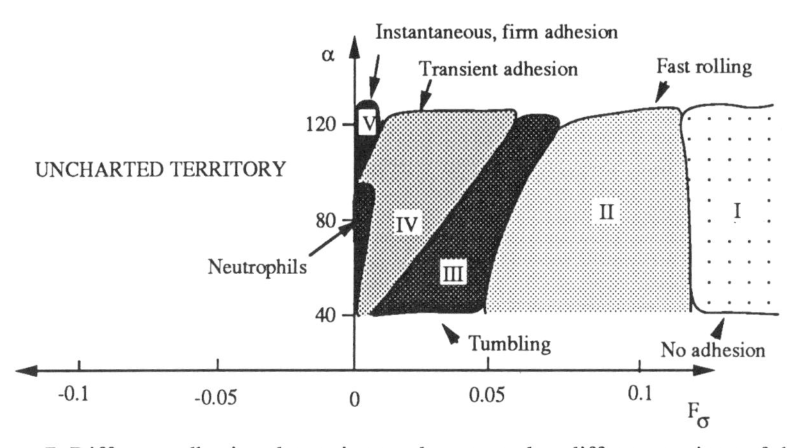

Figure 7. Different adhesion dynamics can be mapped to different regions of the α - F_σ plane, where α is bond stiffness and F_σ is bond reactive compliance. RBL cell behavior would correspond to region V.

reactive compliance; higher values of this compliance lead to faster rates of bond breakage under strain. One can concisely describe how adhesion varies with these two bond mechanical properties using a phase diagram in the α - F_σ plane (Figure 7), which illustrates how combinations of α and F_σ give rise to different forms of adhesion. Neutrophil and RBL cell adhesive dynamics map to different regions of the phase plane. In general, smaller values of F_σ and larger values of α are required to regenerate RBL cell adhesion dynamics. Direct measurement of the mechanicochemical properties of individual adhesion molecule bonds involved in each type of adhesion would be required to confirm the link between the macroscopic manifestation of adhesion and the mechanicochemical properties of the molecules.

An alternative explanation for the difference in adhesive dynamics between the two cellular systems is that there are different micromechanical compliances to the cell microstructure, even if the molecules have similar mechanical properties. If the RBL cell microstructure were quite compliant, RBL cells might not roll, since the microstructure would adsorb stresses, thus relieving the stress from the molecules. In contrast, neutrophils roll because the cell surface microstructure is stiff, and all stresses imparted on the cell are transmitted directly to the adhesion molecules, causing them to break, and forcing the cell to continue to translate. Our experimental data does not provide a sufficient test of this hypothesis. While we can impart stresses on RBL cells which permit little adhesion (100 sec^{-1}), we are not sure if the failure to adhere is due to breaking existing bonds, failure to form additional bonds, or both. This continues to be a viable hypothesis provided neutrophil reaction kinetics are faster than RBL cell reaction kinetics, either because of faster inherent binding between receptor and ligand, or because of the cell's intelligent placement of adhesion molecules (such as on the tips of microvilli (Picker et al., 1991).

Our experiments with the RBL cell line show conclusively that reaction rate, not affinity, controls adhesion under flow. The "low k_f" system displays weaker adhesion than the "high k_f" system, despite a two order of magnitude greater affinity of recognition. In fact, Figure 6 illustrates that the critical parameter that governs adhesion is the ratio of reaction rate to shear rate. This is the first experimental demonstration of this principle.

An additional comparison we can make between these two systems is in how probabilistic processes of recognition manifest themselves in the dynamics of adhesion. Neutrophils do not roll with a constant velocity over stimulated endothelium and that the variance in the velocity is more than two orders of magnitude higher than the theoretical variance generated by measurement error. Other groups have described the irregular motion of neutrophils in contact with endothelium (Lawrence and Springer, 1991; House and Lipowsky, 1991; Lipowsky et al., 1991). While the existence of nonzero variances in rolling velocity is indisputable, the underlying fundamental cause of that variance in unknown. Three possible causes are the irregular topography of the endothelial substrate, the uneven distribution of ligands on the endothelial substrate, or inherent stochasticity or probabilistic binding between cell and substrate. In our experiments, we cannot rule out any of these possibilities; however, in measurements of rolling of human neutrophils over reconstituted P-selectin, Lawrence and Springer (1991) showed neutrophil rolling was irregular. Although there is some possibility of the clustering of molecules on the substrate, the results show irregular rolling does not require an endothelial substrate. Certainly, probabilistic or stochastic binding remains an underlying mechanism which could explain data in both our system, as well as theirs. Whatever the direct cause of the variability, mathematical formulations which endeavor to fully describe the dynamic range of neutrophil motion must generate trajectories which display a variable velocity. The microvilli hard sphere model (Hammer and Apte, 1991) is a stochastic simulation in which the neutrophil velocity at any instant is determined from a force and torque balance on the neutrophil; molecular bonding and breakage causes fluctuations in levels of binding, which leads to fluctuations in the velocity itself. Hence, this model has the inherent ability to predict and simulate a non-zero variance in velocity.

We have shown examples of the three dimensional patterns of adhesion seen in RBL cell binding over homogeneous antigen substrates. One common feature of the patterns is that cells can adhere 20 mm, or about 1500 cell diameters, from the start of the DNP region. Two possible underlying mechanisms would give rise to these spatial patterns is an inherent probabilistic recognition between cell and substrate, or heterogeneity in the cell population (where cells endowed with more $Fc_{\varepsilon}R1$ receptors, and hence more IgE) would bind first.

Therefore, a self-consistent underlying explanation for adhesive dynamics in the two systems is that cell-substrate recognition is a probabilistic process, which manifests itself as irregular rolling in neutrophils or distributed spatial patterns of binding in the RBL cell system (Hammer and Apte, 1991; Cozens-Roberts et al., 1990b). This probabilistic recognition may be a fundamental, biophysical mechanism by which cells may sample the substrate - without committing unless necessary - using transient adhesive interactions. We will continue to explore the implications of probabilistic recognition in adhesion, using the experimental and theoretical methods described in this paper.

Acknowledgments

This work was supported by Public Health Service Grant PO1-Hl18208 from the National Institutes of Health and National Science Foundation grant EET-8808867 (Daniel A. Hammer).

References

Atherton, A.; Born, G.V.R. Quantitative investigations of the adhesiveness of circulating polymorphonuclear leucocytes to blood vessel walls. J. Physiol. 222:447-474; 1972.

Atherton, A.; Born, G.V.R. Relationship between the velocity of rolling granulocytes and that of the blood flow in venules. J. Physiol. 233:157-165, 1973.

Barsumian, E. L.; Isersky, E.; Petrino, M. G.; Siraganian, R. P. IgE-induced histamine release from rat basophilic leukemia cell lines: Isolation of releasing and non-releasing clones. Eur. J. Immunol. 11:317-323; 1981.

Bevilacqua, M.P. Endothelial-leukocyte adhesion molecules. Annual Review of Immunology. 11:767-804, 1993.

Bongrand, P.; Bell, G. I. Cell-cell adhesion: Parameters and possible mechanisms. In: Perelson, A. S.; DeLisi, C.; Wiegel, F. W. eds. Cell Surface Dynamics: Concepts and Models. New York: Marcel Dekker, Inc.; 1984: p. 459-493.

Booyse, F.M.; Sedlak, B. J.; Rafelson Jr, M. E. Culture of arterial endothelial cells. Thrombosis et Diathesis Haemorrhagica. 34, 825-839, 1975.

Carlson, G.P.; Kaneko, J.J. Isolation of leucocytes from bovine peripheral blood. Proceedings of the Society for Experimental Biology and Medicine 142:853-856, 1973.

Chu, L.; Tempelman, L.A.; Miller, C.; Hammer, D.A. Centrifugation assay of IgE-mediated rat basophilic leukemia cell adhesion to antigen-coated polyacrylamide gels. A.I.Ch.E. Journal, in press, 1994.

Cozens-Roberts, C.; Quinn, J. A.; Lauffenburger, D. A. Receptor-mediated adhesion phenomena: Model studies with the radial-flow detachment assay. Biophys. J. 58:107-125, 1990a.

Cozens-Roberts, C.; Lauffenburger, D. A.; Quinn, J. A. Receptor-mediated cell attachment and detachment kinetics: Part I- Probabilistic model and analysis. Biophys. J. 58:841-856, 1990b.

Erickson, J.; Goldstein, B.; Holowka, D.; Baird, B. The effect of receptor density on the forward rate constant for binding of ligands to cell surface receptors. Biophys. J. 52:657-662, 1987.

Fowler, H. W.; McKay, A. J. The measurement of microbial adhesion. In: Berkeley, R. C. W.; Lynch, J. M.; Melling, J.; Rutter, P. R.; Vincent, B., eds. Microbial Adhesion to Surfaces. Chichester, England: Ellis Horwood Limited; 1980: p.143-161.

Gilbert, R.O.; Grohn, Y.T. ; Guard, C.L.; Surman, V. Impaired post partum neutrophil function in cows which retain fetal membranes. Research in Veterinary Science 55:15-19, 1993.

Goldman, A. J.; Cox, R. G.; Brenner, H. Slow viscous motion of a sphere parallel to a plane wall - II couette flow. Chem. Eng. Sci. 22:653-660, 1967.

Hammer, D. A.; Apte, S. M. Simulation of cell rolling and adhesion on surfaces in shear flow: General results and analysis of selectin-mediated neutrophil adhesion. Biophys. J. 62:35-57, 1992.

House, S. D.; Lipowsky, H. H. Dynamics of leukocyte-endothelium interactions in the splanchnic microcirculation. Mircovas. Res. 42:288-304, 1991.

Jones, D.A.; Abbassi, O.; McIntire, L.V.; McEver, R.P.; Smith, C.W. P-selectin mediated neutrophil rolling on histamine-stimulated endothelial cells. Biophys. J. 65:1560-1569, 1993.

Lasky, L.A. Selectins: interpreters of cell-specific carbohydrate information during inflammation. Science 258:964-969, 1992.

Lawrence, M. B.; Smith, C. W. ; Eskin, S. G.; McIntire, L. V. Effect of venous shear stress on CD18-mediated neutrophil adhesion to cultured endothelium. Blood. 75(1):227-237, 1990.

Lawrence, M. B.; Springer, T. A. Leukocytes roll on a selectin at physiological flow rates: Distinction from and prerequisite for adhesion through integrins. Cell. 65:859-873, 1991.

Ley, K.; Gaehtgens, P.; Fennie, C.; Singer, M.S.; Lasky, L.A.; and Rosen, S.D. Lectin-like adhesion molecule 1 mediates leukocyte rolling in mesenteric venules in vivo. Blood 77:2553-2555, 1991.

Lipowsky, H.H.; Riedel, D.; Shan Shi, G. In vivo mechanical properties of leukocytes during adhesion to venular endothelium. Biorheology 28:53-64, 1991.

Lui, F. T.; Bohn, J. W.; Ferry, E. L., Yamamoto, H. ; Molinaro, C. A.; Klinman, N. R.; Katz, D. H. Monoclonal dinitrophenyl-specific murine IgE antibody: Preparation, isolation, and characterization. J. Immunol. 124: 2728-2736, 1980.

McEver, R. P. Leukocyte interactions mediated by selectins. Thrombosis and Haemostasis. 66:80-87, 1991.

Metzger, H.; Alcaraz, G. ; Hohman, R.; Kinet, J.-P.; Pribluda, V.; Quarto, R. The receptor with high affinity for immunoglobulin E. Ann. Rev. Immunol. 4:419-470, 1986.

Oliver, J. M.; Seagrave, J.-C.; Stump, R. F.; Pfeiffer, J. R.; Deanin, G. G. Signal transduction and cellular response in RBL-2H3 mast cells. Prog. Allergy. 42:185-245, 1988.

Osborn, L. Leukocyte adhesion to endothelium in inflammation. Cell. 62:3-6, 1990.

Picker, L.J.; Warnock, R.A.; Burns, A.R.; Doerschuk, C.M.; Berg, E.L.; Butcher, E.C. The neutrophil selectin LECAM-1 presents carbohydrate ligands to the vascular selectins ELAM-1 and GMP-140. Cell 66:921-934, 1991.

Pless, D. D.; Lee, Y.-C.; Roseman, S.; Schnaar, R. L. Specific cell adhesion to immobilized glycoproteins demonstrated using new reagents for protein and glycoprotein immobilization. J . Biol. Chem. 258:2340-2349, 1983.

Ryan, T. A. Signal Transduction of Immunoglobulin E Receptor Crosslinking. Ph.D. Thesis. Cornell University; 1989.

Schnaar, R. L.; Lee, Y.-C. Polyacrylamide gels copolymerized with active esters. A new medium for affinity systems. Biochemistry. 14:1535-1541, 1975.

Springer, T. A.Adhesion receptors of the immune system. Nature. 346:425-434, 1990.

Stoolman, L.M. Adhesion molecules controlling lymphocyte migration. Cell. 56:907-910, 1989.

Taurog, J. D.; Fewtrell, C.; Becker, E. IgE mediated triggering of rat basophilic leukemia cells: Lack of evidence for serine esterase activation. J. Immun. 122: 2150-2153, 1979.

Tempelman, L. A. Quantifying receptor-mediated cell adhesion under hydrodynamic flow using a model cell line. Ph.D. Thesis.Cornell University; 1993.

Tempelman, L. A.; Hammer, D. A. Quantitating receptor-mediated cell adhesion under flow using a model cell line. J. Cell Biol. 111:404a, 1990.

Tempelman, L.A.; Park, S.; Hammer, D.A. Motion of model leukocytes near a wall in simple shear flow. Biotechnology Progress, in press, 1994.

Tempelman, L.A.; Hammer, D.A. Receptor-mediated binding of IgE-sensitized rat basophilic leukemia cells to antigen-coated substrates under hydrodynamic flow. Biophysical Journal, in press, 1994.

Tozeren, A.; Ley, K. How do selectins mediate leukocyte rolling in venules? Biophys. J. 63:700-709, 1992.

von Andrian, U. H.; Chambers, J. D.; McEvoy, L. M.; Bargatze, R. F.; Arfors, K. E.; Butcher, E. C. Two-step model of leukocyte-endothelial cell interaction in inflammation: distinct roles for LECAM-1 and the leukocyte β_2 integrins in vivo. Proc. Natl. Acad. Sci. USA 88:7538-7542, 1991.

Wattenbarger, M. R.; Graves, D. J. ; Lauffenburger, D. A. Specific adhesion of glycophorin liposomes to a lectin surface in shear flow. Biophys. J. 57:765-777, 1990.

Weetal, M. Studies on the high affinity receptor for IgE ($Fc_\varepsilon RI$): Binding of chimeric IgE/IgG and desensitization of cellular responses. Ph. D. Thesis. Cornell University; 1992.

10
Initial Steps of Cell-Substrate Adhesion

A. Pierres, A. M. Benoliel, P. Bongrand

Introduction

The decisive step of receptor-mediated cell adhesion may well be the formation of the first bond following contact between a cell and an adhesive surface. Indeed, after the first receptor-ligand association occurred, two limiting situations may be considered. If the rate of bond formation between surfaces maintained in contact by the first linkage is higher than the rate of bond dissociation, the number of cell-surface bonds will increase, making adhesion irreversible in absence of significant disruptive forces or active cell detachment behavior. Conversely, if the rate of bond formation is lower than the rate of bond dissociation, most bonds will not be conducive to durable attachment. Indeed, when rat thymocytes were agglutinated by varying minimal amounts of concanavalin A, the adhesion efficiency was found to be proportional to the first power of the lectin concentration (Capo et al., 1982), thus supporting the view that the rate limiting step was the formation of the first bond.

Therefore, as a first approximation, the outcome of an encounter between a receptor-bearing cell and a ligand-coated surface may be modeled with three basic parameters (all expressed in second^{-1}) :
- k_o is the "on rate" of bond formation between ligands and receptors borne by kinetically independent surfaces. This parameter is not only dependent on the molecular properties of these ligands and receptors, but

it is also modulated by possible forces between approaching surfaces (e.g. electrostatic or steric repulsion, gravitational or centrifugal forces as well as hydrodynamic interactions, see Bongrand et al., 1982 ; Bongrand, 1988 for reviews).

- k_- is the rate of bond dissociation. This is obviously dependent on possible mechanical forces tending to detach adherent cells as well as the number and spatial distribution of existing bonds.
- k_+ is the rate of bond formation between surfaces maintained in contact by at least one bond. This is expected to be substantially higher than k_o, however, when the number of bonds increases, k_+ may be expected to decrease due to steric hindrance or limited availability of binding molecules.

There is little direct information on the value of parameters k_o, k_+ and k_- in situations of biological interest. A general theoretical framework was provided by George Bell (1978) who emphasized two important concepts. Firstly, he described a method for relating two- and three-dimensional rates of bond formation by splitting molecular interactions into two sequential steps, namely diffusion-driven approach of reagents and actual association. He was thus able to estimate at about $1.3 \times 10^{-2} \times N_1N_2$ µm²/sec the maximal rate of bond formation between surfaces bearing ligands and receptors with respective densities N_1 and N_2 (per µm²) and diffusion constants of 10^{-10} cm²/sec. Secondly, he used an empirical law derived by Zurkhov (1965) from experimental data on the mechanical resistance of solids to suggest a quantitative formula for the effect of stress on the dissociation rate of molecular bonds, namely :

$$k_- = k^\circ_- \exp(\gamma f/kT) \quad (1)$$

where k_- and k°_- are the stressed and unstressed dissociation rates, k is Boltzmann's constant, T is the absolute temperature, f is the force and γ is an empirical coefficient. This formula was used by many later authors (Dembo et al., 1988 ; Evans et al., 1991 ; Hammer and Apte, 1992 ; Tözeren and Ley, 1992 ; Tees et al., 1993).

In view of the aforementioned considerations, it seems of obvious interest to look at direct experimental information on the numerical value of parameters k_o, k_+ and k_-. The most promising way of achieving this goal was to measure the adhesion efficiency of cells encountering surfaces under conditions of flow. Thus, colloidal scientists derived theoretical frameworks to relate the rate of orthokinetic (brownian motion driven) or perikinetic (shear driven) particle coagulation to surface properties (see Bongrand et al., 1982, for a brief review) and this approach was applied to cells by Curtis (1969). More recently, Hammer and Lauffenburger (1987) elaborated a theoretical model of receptor-mediated cell adhesion to surfaces under conditions of flow and this model was refined or used in numerous further studies (Dembo et al., 1988 ; Cozens-Roberts et al., 1990 ; Hammer and Apte, 1992 ; Tözeren and Ley, 1992). The present limitation of these models is that they cannot be used to analyse experimental data without introducing several assumptions (such as cell-substrate contact area, stress dependence of bonds, behavior of cell-substrate repulsion, molecular crowding in the contact area). Although usually quite reasonable, these assumptions are often difficult to validate in detail.

The experimental approach we suggest consists of analysing the initial attachment of cells moving along a ligand-coated surface under laminar shear flow low enough that a single bond might induce a detectable arrest. Thus, it may be expected that the determination of the time distribution of cell arrests might give direct information on constants k_+ and k_-, and the frequency of cell arrests might reflect the numerical value of parameter k_0.

In this respect, it is useful to recall that the *order of magnitude* of the strength of molecular attachments was estimated at a few tens of piconewtons (Bell et al., 1978 ; Tha et al., 1986 ; Evans et al., 1991 ; Tees et al., 1993).

In the next section, we shall discuss the theoretical potential and limitations of this approach. We shall next describe some experimental results obtained by monitoring the motion of antibody-coated spherical beads of 2.8 µm diameter driven along antigen-bearing surfaces by a low shear flow.

Theoretical Background.

Motion of a sphere near a surface under flow

The motion of a neutrally buoyant sphere driven along a plane surface in a laminar shear flow was studied by Goldman et al. (1967). Two basic results are of interest.

First, when the sphere is close to the substrate, its translational velocity is inversely proportional to the logarithm of the width of the cell-substrate gap. This result is very difficult to apply rigorously to biological surfaces, since these are not smooth at the submicrometer level. However, numerous experimental studies performed on various cell types (Tissot et al., 1991 ; Lawrence and Springer, 1991 ; Tissot et al., 1992 ; Tissot et al., 1993 ; Kaplanski et al., 1993 ; Tempelman et al., 1994) and latex beads (Tempelman et al., 1994) showed that a convenient *order of magnitude* for the velocity of particles in apparent contact with the surface is about 75% of the product of the particle radius a and the wall shear rate G. Interestingly, cells rolled faster than artificial beads of similar diameter, suggesting an influence of cell surface microstructure on the motion (Tempelman et al., 1994).

Second, the hydrodynamic force experienced by a sphere bound to the substrate is :

$$F = 32\ \mu\ a^2\ G \qquad (2)$$

Where μ is the fluid viscosity, a is the cell radius and G is the shear rate. Using a viscosity of 0.0007 Pascal.second (corresponding to water and dilute aqueous solutions at 37°C) and a radius of 1.4 μm (corresponding to the radius of the beads used in the present work) we find :

$$F_{piconewton} = 0.044\ G_{1/second} \qquad (3)$$

Theoretical analysis of arrest duration

The model we shall use was described in a previous report (Kaplanski et al., 1993). We followed an approach suggested by Cozens-Roberts et al. (1991) who emphasized the expected peculiarities of attachments mediated by a very low number of molecular bonds.

We defined as $P_i(t)$ the probability that a particle be bound by i bonds at time t after formation of the first linkage. We define as k_- and k_+ the rates of bond dissociation and formation for adherent particles. We assume that the use of these constants is valid as far as the number of bonds remains low and the hydrodynamic drag is markedly lower than 10 piconewton. Since a cell will depart when the number of bonds is zero, we may write :

$$dP_1/dt = 2k_-\ P_2 - (k_+ + k_-)\ P_1 \qquad (4)$$
$$dP_i/dt = k_+\ P_{i-1} + (i+1)\ k_-\ P_{i+1} - (k_+ + ik_-)P_i \quad (\text{if } i>1)$$

The probability that a cell will remain bound at time t is simply

$$P_b(t) = \sum_{i>1} P_i(t) \qquad (5)$$

Equations (4) were solved numerically using $1/k_-$ as unit of time and varying the dimensionless ratio k_+/k_-. For this purpose, $P_i(t)$ was set at zero for i>10, and time remained low enough that $P_{10}(t)$ remained negligible. Results are shown on figure 1.

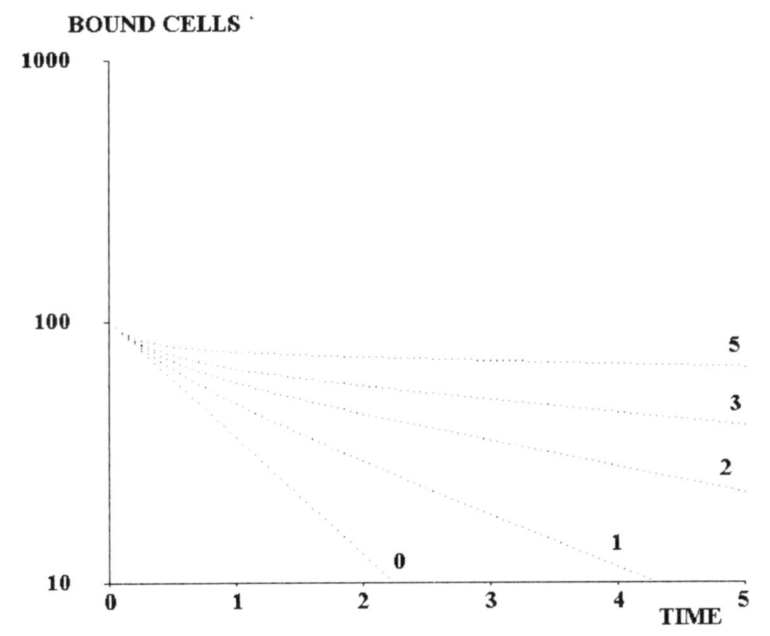

Figure 1. Theoretical distributions of arrest durations. Equations (4) were solved numerically with one bond at time zero et different values of the ratio k+/k- (0, 1, 2, 3 and 5 as shown). The number of bound cells was 100 at time zero. Time unit is 1/k-.

Two important points may be emphasized. First, if the rate of bond formation is low ($k_+/k_- \approx 0$), the logarithm of the number $N_b(t)$ of bound cells at time t must be linearly dependent on time. Second, all curves *become* fairly linear when time increases. This may be explained by noticing that the average number of bonds per adherent cell reaches a plateau (which must be equal to k_+/k_-) and the detachment probability per cell also approaches a limit depending on k_+/k_-. This point is illustrated on figure 2. It is concluded that the ratio k_+/k_- could in principle be derived from the distribution of arrest durations, the region of short times being expected to be the most informative. As described earlier (Kaplanski et al., 1993), a simple way of achieving this might be to determine the times $t_{3/4}$ and $t_{1/2}$ where respectively 3/4 and 1/2 of cells bound at time zero remain bound and use the theoretical curve shown on figure 3. This approach was used in the present work.

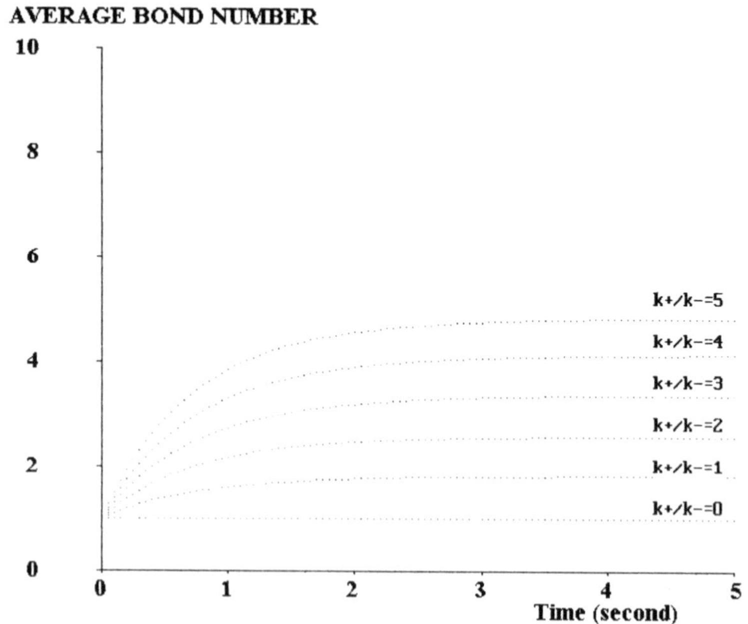

Figure 2. Evolution of the number of cell-substrate bonds after contact formation.

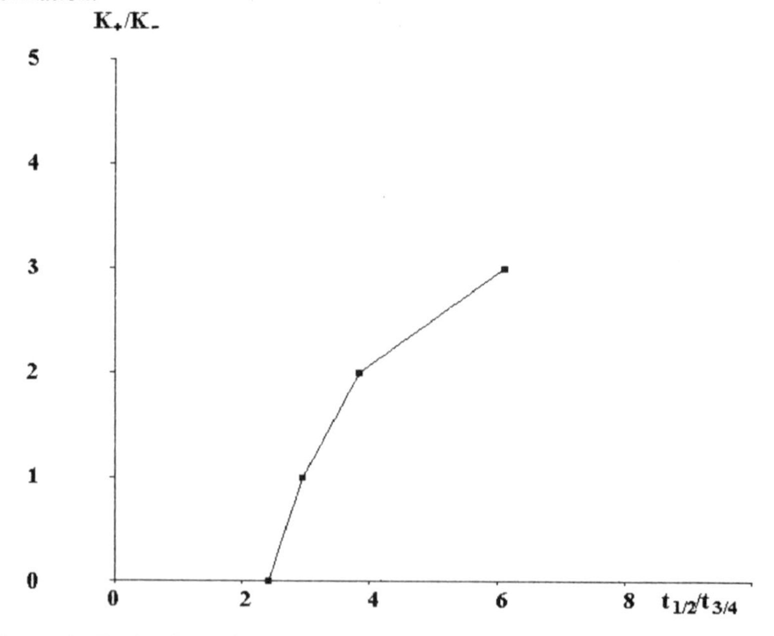

Figure 3. Derivation of the ratio k_+/k_- from the experimental distribution of arrest durations.

Materials and Methods

Particles and surface treatment.

Particles were 2.8 µm diameter spheres coated with streptavidin, a protein with high affinity for biotin (Dynabeads M280, Dynal, U.K.). They were supplied by Biosys (Compiègne, France). Before each experiment, 50 µl aliquots of bead suspensions (6-7 x 10^8 /ml) were incubated for 20 minutes at room temperature with 50 µl of a mixture of biotinylated murine monoclonal antibodies at a total concentration of 0.2 mg/ml. Antibodies were a monoclonal IgG2a specific for Fc fragment of rabbit IgG (clone IgL173, supplied by Immunotech, Marseille, France) and a monoclonal IgG2a specific for CD14 (clone UCHM1, supplied by Sigma, St Louis, MO). The latter antibody was used to dilute specific anti-rabbit immunoglobulin antibodies and will be refered to as "irrelevant" antibody. Beads were washed in phosphate buffered saline (pH 7.2) supplemented with 2mg/ml bovine albumine (in order to prevent nonspecific adhesion) and they were used at 10^7/ml in the same solution for adhesion assay. The surface density of immunoglobulins was performed as previously described (Pierres et al., 1994). Beads were incubated with an excess of fluorescein-labeled rabbit immunoglobulins (polyclonal rabbit anti-lactoferrin antibodies, Jackson Immunoresearch lab., Westgrove, PA). They were then examined with an confocal laser scanning microscope (CLSM, Leica, Heidelberg). Absolute calibration was achieved by measuring under similar conditions the fluorescence of small volumes (typically 5 µl) of known concentrations of the same immunoglobulin preparation. Each bead was found to bear an average of about 2.2 x 10^6 rabbit immunoglobulin molecules (i. e. about 90,000 molecules per squared micrometer).

Glass coverslips were coated with rabbit immunoglobulins with a modification of a method described by Michl et al. (1979). They were washed with sulphuric acid, then rinsed in distilled water and air dried. They were then incubated for 30 minutes at room temperature with 1 mg/ml polylysine (Molecular weight >300,000, Sigma) and washed in phosphate buffered saline. They were then incubated another thirty minutes in the dark with 16.8 mg/ml 2,4-dinitrobenzenesulfonic acid (Eastman Kodak, Rochester, NY) in pH 11.6 carbonate buffer. They were washed and incubated with 120 µg/ml rabbit anti-dinitrophenol antibodies (Sigma), then washed again in phosphate buffer containing 2 mg/ml bovine albumine before use.

Flow chamber.

We used a modification of a previously described apparatus (Tissot et al., 1991 & 1992). Briefly, a rectangular cavity of 17x6x0.2 mm^3 was cut into a plexiglas block. the bottom was a removable glass coverslip (about 22 x 10 mm^2) that was coated with rabbit immunoglobulins as previously described. The flow was generated with a plastic syringe (1 or 2 ml) mounted an an electric syringe holder (Razel scientific, Samford, CT, supplied by Bioblock, France, Ref. K88906) allowing digital setting of the flow. The wall shear rate was determined experimentally as previously described by monitoring the motion of 0.8 µm diameter latex beads (Tissot et al., 1991 & 1992).

Motion analysis.

The chamber was set on the stage of an inverted microscope (Olympus IM) bearing a 100X immersion lens. The microscope was equipped with a SIT videocamera (Model 4015, Lhesa, Cergy Pontoise, France) and all experiments were recorded with a Mitsubishi HS3398 tape recorder for delayed analysis.

In a typical experiment, a single microscope field (about 86 µm width) was monitored for a fixed period of 5 minutes. The experiment was replayed several times in order to allow individual monitoring of the motion of each visible particle. The duration of each arrest was determined with about 1/5 second accuracy.

In some cases, more refined trajectory analysis were achieved by subjecting the video output to real-time digitization with a PCVision + card (Imaging Technology, Bedford, MA) connected to a IBM-compatible desk computer. The motion of individual particles was followed with a cursor that was driven with the computer mouse and superimposed on live images. Digitized files representing small (32x32 pixel, 0.34 µm/pixel) fields were transferred to the host computer memory with a typical rate of 8/second for delayed determination of the accurate position of the examined particle.

Results

The duration of particle arrests is easily determined with 0.1-0.2 second accuracy.

When the shear rate was low (11 s^{-1}), it was fairly easy to discriminate between particles that were in contact with the chamber floor and faster particles that were flowing at distance (figure 4).

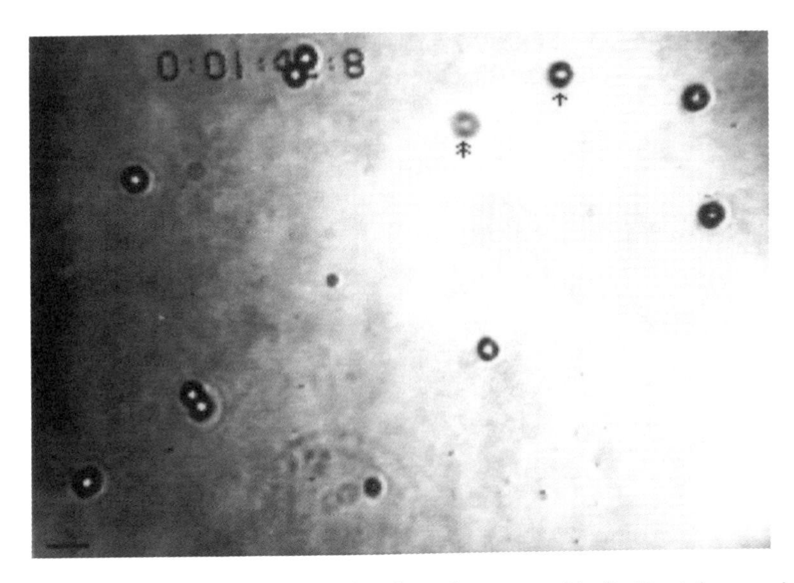

Figure 4. Motion of flowing particles. low shear rate ($11 s^{-1}$). Particles moving at distance from the chamber floor (double arrow) were easily discriminated from particles moving in contact with the coverlip (arrow) which were quite similar to immobile adherent particles. Bar is 5 µm.

When higher shear rates were used (up to 72 s^{-1}), moving particles were almost invisible and they appeared only when they stopped (figure 5).

In contrast with earlier study where irregularities of cell motion made very difficult a rigorous definition of cell arrests (Kaplanski et al., 1993 ; Pierres et al., 1994), spherical particles exhibited very smooth motion (probably due to the absence of surface protrusions) which allowed unambiguous detection of arrest and departure.

Particles exhibit numerous stops of widely varying duration

In a systematic study, the duration of cell arrests was determined for different values of the shear rate (11, 22, 44 and 72 s^{-1}) and specific antibody dilution (1, 1/10, 1/100, 1/1000, as defined in materials and methods section). A total number of 839 arrests were found and the distribution of arrest durations is shown on figure 6. The range of variation of these values was very wide, since 23 % of these arrest were shorter than 1 second and 46 % of them were longer than 60 seconds.

It was interesting to assess the importance of antibody dilution : when beads were coated with pure anti-rabbit immunoglobulin antibodies, 85

% of arrests (89/105) lasted more than 60 seconds, and a similar proportion was obtained whatever the shear rate. Conversely, with the highest

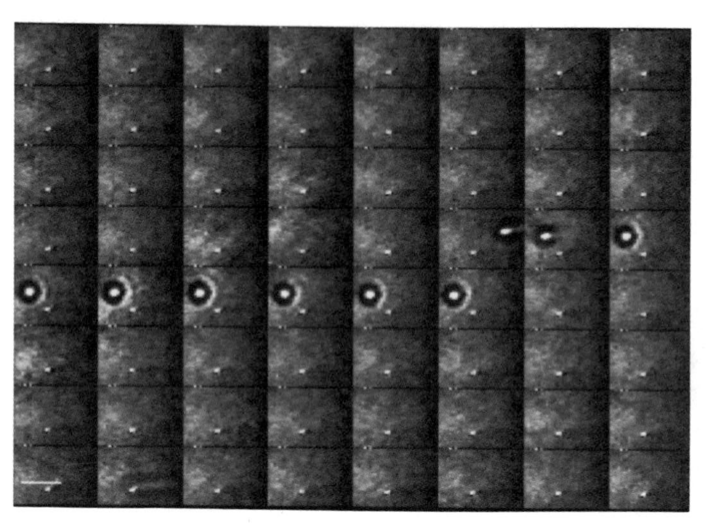

Figure 5. Detection of cell arrests. High shear (72 s^{-1}). The position of a particle arrest was determined in a first study of the record of an experiment. The field surrounding this position was then continously digitized and 64 sequential images are shown. A particle is clearly visible on frames 38-46. The interval between two sequential images is about 0.12 second. Bar is 5 µm.

dilution of specific antibodies (1/1000), only 36 % (44/122) lasted more than 60 seconds. This was similar to the proportion of longer arrests measured on particles coated with irrelevant antibodies (i.e. 37%, 16/43). However, arrests were much less frequent with the latter particles (not shown).

Tentative derivation of kinetic parameters of bond formation and dissociation.

We made use of the theoretical data described above as guidelines to analyse experimental data. The basic procedure is depicted on figure 7. The durations of arrests measured in a given experimental situation were ordered and a plot of the number of arrests longer than time t versus t was built on a logarithmic scale. Since arrests shorter than about 0.2 seconds were not detected with our method, curves were extrapolated to time zero in order to obtaine a "true" number of arrests. The initial rate of cell detachment was then calculated as the slope of the linear regression line calculated within the time interval of 0-1 second. Results are summarized on table 1.

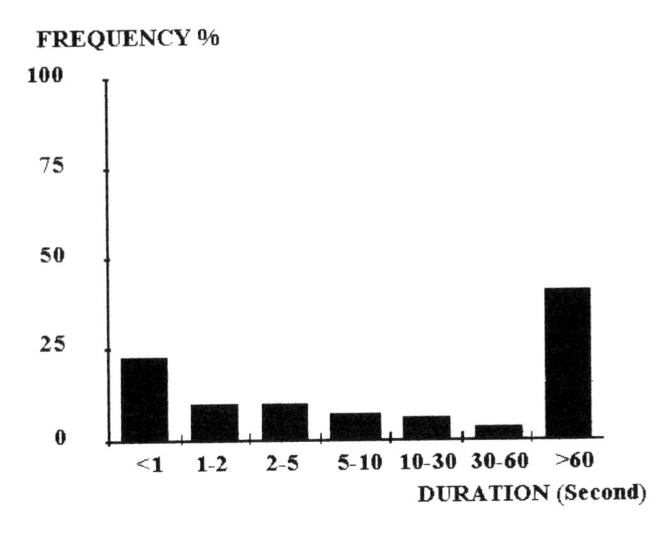

Figure 6. Distribution of arrest durations. Results obtained with different shear rates and antibody dilutions were pooled. A total number of 839 arrest durations were used to calculate the histogram shown above.

Table 1. Initial rate of cell detachment after an arrest (s^{-1})

Shear rate (s^{-1}) Antibody dilution	11	22	44	72
1/1	0.00	0.00	0.11	0.13
1/10	0.20	0.10	0.07	0.10
1/100	0.33	0.06	0.24	0.28
1/1000	0.39	0.15	N.D.	0.26

Each result on table 1 was obtained with a number of arrest durations ranging between 19 and 141. Despite fairly wide experimental variations, two trends were clearly observed. First, the rate of cell detachment increased when antibodies were diluted, but similar values were found with 1/100 and 1/1000 dilutions. Second, the rate of cell detachment was not substantially dependent on the shear rate in the studied domain.

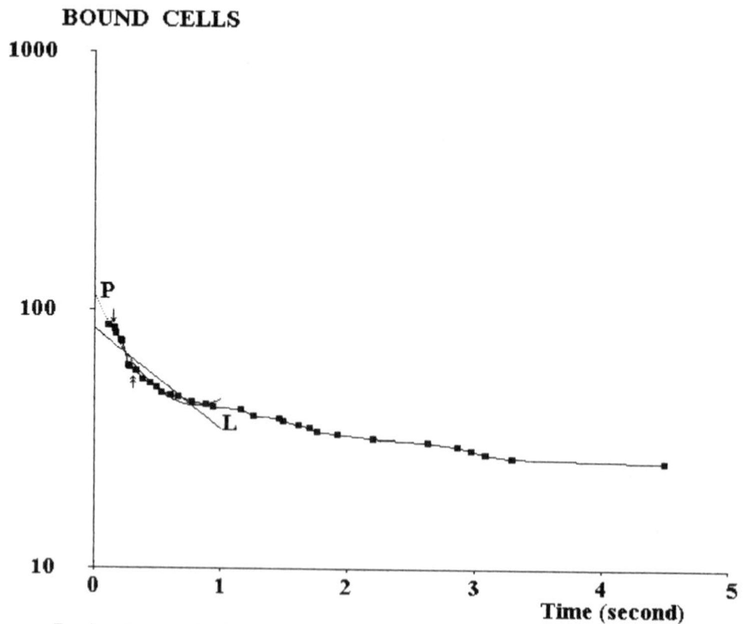

Figure 7. Analysis of distributions of arrest durations. Beads coated with a 1/1,000 dilution of specific anti-rabbit immunoglobulin antibodies were driven along a ligand-coated coverslip with a low shear rate (11 s^{-1}). A total number of 87 arrests were observed. The cumulative curve of arrest durations was subjected to linear (L) or parabolic (P) regression in order to obtain a "true" number of arrests by extrapolation to time zero. The slope of the linear regression line determined on the time interval (0,1) was used to calculate the initial rate of cell detachment (table 1).

The ratio between the periods of time required to achieve 50% and 25% cell detachment was used to estimate the ration k_+/k_- between the on- and off- rates of bond formation for adherent cells (Table 2).

Table 2. Estimated values of k_+/k_-.

Shear rate (s^{-1}) Antibody dilution	11	22	44	72
1/10	0.7	2.4	ND	>3
1/100	>3	ND	2.9	1.6
1/1000	0	>3	ND	3

A major problem in the interpretation of the data shown on Table 2 is that many more experimental values than available would be required to compare experimental and theoretical detachment curves over a sufficiently large period of time (e. g. 60 seconds). Indeed, it was felt that the crude model we used might be valid only over a limited period of time following cell arrest.

Discussion

The aim of this work was to determine some kinetic parameters of bond formation and dissociation between ligand and receptor molecules borne by interacting surfaces. The basic idea was to study the duration of arrests of particles driven along adhesive surfaces by a very low shear flow, weaker than the expected strength of a single molecular bond. Particles were spherical beads instead of actual cells for two reasons. First, it was thus possible to achieve a better control of the nature and density of adhesion molecules. Second, since these particles were very regular shaped, the motion irregularities observed during their displacement along adherent surfaces might be ascribed to molecular interactions rather than an influence of surface protrusions, thus alleviating some difficulties in the definition of adhesion-mediated arrests (Kaplanski et al., 1993 ; Pierres et al., 1994). Indeed, spherical particles were often used in recent quantitative studies of adhesion (Cozens-Roberts et al., 1990 ; Kuo et al., 1993 ; Tempelman et al., 1994).

The main conclusions of our study were that i) specific antigen-antibody bonds generated numerous arrests with a wide range of durations, i. e. from several tenths of a seconds to more than one minute. ii) While most arrests were durable (i.e. more than one minute) when beads were coated with a high antibody concentration, this duration was not markedly dependent on this concentration when antibody concentration was varied between 1/100 and 1/1000. This is precisely the expected behavior of single molecular attachments. iii) The duration of these arrests was essentially independent of the hydrodynamic shear force when this was lower than about 4 piconewtons. Thus, if initial arrests were induced by single molecular bonds, the lifetime of these bonds would not be affected by disrupting forces lower than 4 piconewtons. This is in line with previous estimates of individual bond strengths (Tha et al., 1986 ; Evans et al.. 1991 ; Tees et al., 1993). However, it is not possible at the present time to obtain more precise information on the force-dependence of the rate of bond dissociation, in order to perform an experimental check of Zurkhov's law.

Further, the validity of the above conclusions is hampered by a major problem that must be solved before more experiments are done in order to refine experimental results. Indeed, even the order of magnitude of the lifetime of unstressed molecular bonds is not known with certainty due to our lack of knowledge of the quantitative influence of particle structure and adhesive molecule behavior. Indeed, the natural dissociation rate of free and particle-bound molecules of similar structure may differ by a factor of ten thousand or more (Bell, 1978 ; Kaplanski et al., 1993). This point is important since our analysis of arrest durations is valid only if it is possible to exclude the occurrence of numerous undetectable arrests with a duration substantially smaller than a tenth of a second. Therefore, it seems important to use an improved experimental apparatus allowing better temporal resolution of the cell movement et check the validity of the extrapolation of experimental curves, as exemplified on figure 7. If this is achieved, the minimal duration Δt of detectable arrest will be given by :

$$\Delta t = \Delta x/v = 2\Delta x/(Ga) \qquad (6)$$

where Δx is the spatial resolution of our analysis system (i.e. at best about 0.33 µm), v is the rolling velocity, G is the wall shear rate and a is the particle radius. Thus, is conceivable to achieve a temporal resolution of about 20 ms with the highest flow used. This would require an improvement of the performance of our imaging system. This approach is presently considered in our laboratory.

References

Bell, G.I. Models for the specific adhesion of cells to cells. Science 200:618-627. 1978.

Bongrand, P. (editor). Physical Basis of Cell-Cell Adhesion. CRC press, Boca Raton, 1988.

Bongrand, P.; Capo, C.; Depieds, R. Physics of Cell Adhesion. Progress Surface Sci. 12:217-286. 1982.

Capo, C.; Garrouste, F.; Benoliel, A.M.; Bongrand, P.; Ryter, A.; Bell, G. I. Concanavalin A-mediated thymocyte agglutination : a model for a quantitative study of cell adhesion. J. Cell Sci. 56:21-48. 1982

Cozens-Roberts, C.; Lauffenburger, D.A.; Quinn, J.A. Receptor-mediated cell attachment and detachment kinetics. I. Probabilistic model and analysis. Biophys. J. 58:841-856. 1990a.

Cozens-Roberts, C.; Quinn, J.A; Lauffenburger, D.A. Receptor-mediated adhesion phenomena : model studies with the radial-flow detachment assay. Biophys. J. 58:107-125. 1990b.

Curtis, A.S.G. The measurement of cell adhesiveness by an absolute method. J. Embryol. Exp. Morphol. 23:305-325. 1969.

Dembo, M.; Torney, D.C.; Saxman, K.; Hammer, D. The reaction-limited kinetics of membrane-to-surface adhesion and detachment. Proc. R. Soc. Lond. B 234:55-83. 1988.

Evans, E.; Berk, D.; Leung, A. Detachment of agglutinin-bonded red blood cells. I. Forces to rupture molecular-point attachments. Biophys. J. 59:838-848. 1991.

Goldman, A.J.; Cox, R.G.; Brennner, H. Slow viscous motion of a sphere parallel to a plane wall. II. Couette flow. Chem. Engn. Sci. 22:653-660. 1967.

Hammer, D.; Apte, S.M. Simulation of cell rolling and adhesion on surfaces in shear flow : general results and analysis of selectin-mediated neutrophil adhesion. Biophys. J. 63:35-57. 1992.

Hammer, D.A.; Lauffenburger, D.A. A dynamical model for receptor-mediated cell adhesion to surfaces. Biophys. J. 52:475-487. 1987.

Kaplanski, G.; Farnarier, C.; Tissot, O.; Pierres, A.; Benoliel, A.M.; Alessi, M.C.; Kaplanski, S.; Bongrand, P. Granulocyte endothelium initial adhesion. Analysis of transient binding events mediated by E-selectin in a laminar shear flow. Biophys. J. 64:1922-1933. 1993.

Kuo, S.C.; Lauffenburger, D.A. Relationship between receptor/ligand binding affinity and adhesion. Biophys. J. 65:2191-2200. 1993.

Lawrence, M.B.; Springer, T.A.. Leukocytes roll an a selectin at physiological flow rates : distinction from and prerequisite for adhesion through integrins. Cell 65:859-873. 1991.

Michl, J.; Pieczonka, M. M.; Unkeless, J. C.; Silverstein, S.C. Effects of immobilized immune complexes on Fc- and complement-receptor function in resident and thioglycollate-elicited mouse peritoneal macrophages. J. Exp. Med. 150:607-621. 1979.

Pierres, A.; Tissot, O.; Malissen, B.; Bongrand, P. Dynamic adhesion of CD8-positive cells to antibody-coated surfaces. J. Cell Biol. in press.

Tees, D.F.J.; Coenen, O.; Goldsmith, H.L. Interaction forces between red cells agglutinated by antibody. IV. Time and force dependence of break-up. Biophys. J. 65:1318-1334. 1993.

Tempelman, L.A.; Park, S.; Hammer, D.A. Motion of model leukocytes near a wall in simple shear flow : deviation from hard sphere theory. Biotechnology Prog. in press. 1994.

Tha, S.P.; Shuster, J.; Goldsmith, H.L. Interaction forces between red cells agglutinated by antibody. II. Measurement of hydrodynamic force of breakup. Biophys. J. 50:1117-1126. 1986.

Tissot, O.; Foa, C.; Capo, C.; Brailly, H.; Delaage, M.; Bongrand, P. Influence of adhesive bonds and surface rugosity on the interaction between rat thymocytes and flat surfaces under laminar shear flow. J. Dispersion Sci. & Technol. (Biocolloids & Biosurface issue) 12:145-160. 1991.

Tissot, O.; Pierres, A.; Bongrand, P. Monitoring the formation of individual cell-surface bonds under low shear hydrodynamic flow. Life Sci. Adv. 12:71-82. 1993.

Tissot, O.; Pierres, A.; Foa, C.; Delaage, M.; Bongrand, P. Motion of cells sedimenting on a solid surface in a laminar shear flow. Biophys. J. 61:204-215. 1992.

Tözeren, A.; Ley,K. How do selectins mediate leukocyte rolling in venules ? Biophys. J. 63:700-709. 1992.

Zurkhov, S.N. Kinetic concept of the strength of solids. Int. J. Fract. Mech. 1:311-323. 1965.

11

A Cell-Cell Adhesion Model for the Analysis of Micropipette Experiments

C. Zhu, T. E. Williams, J. Delobel, D. Xia and M. K. Offermann

Introduction

Cell-cell adhesion is a fundamental biological phenomenon, and it plays an important role in the function of cells in many physiologic and pathologic processes (Alberts et al 1989). For about a decade, the research interest on cell adhesion has been rapidly increasing. More and more cell adhesion molecules (CAMs) (surface receptors by which specific cell-cell interactions are mediated) have been identified and characterized. The CAMs are expressed on the surface of specific cell types at variable densities, recognize and bind to their ligands on other cells with certain binding affinities, may move laterally on the cell membrane by diffusion, and are regulated by various biological response modifiers (Springer 1990a; Springer 1990b).

The mechanical strength of adhesiveness is an important parameter quantifying the adhesive interactions. A number of experimental techniques have been developed to quantify the adhesive strength between cells and/or to use adhesive strength as a measure to compare the effects of various factors influencing the adhesion. One of these is the micropipette manipulation technique. In this technique, a pair of cells are brought into contact by two micropipettes to allow adhesion to occur. The cell pair is then pulled apart by applying a sequence of incrementally increasing pipette suction pressures. The minimum aspiration pressure required to completely separate the cell pair is designated as the critical pressure. The maximum force the adhesion can sustain is computed as the critical pressure times the cross-sectional area of the pipette. This technique was initially used to measure the adhesive strength between membrane vesicles and between erythrocytes (Evans and Buxbaum 1981; Evans and Leung 1984; Evans 1988; Evans et al 1991a; Evans et al 1991b). Recently, it has been applied to the study of other cell types, including the conjugation of a T-lymphocyte and a target cell (Sung et al 1986), the adhesion of a helper T-lymphocyte to an antigen presenting cell mediated by ICAM-1/LFA-1 receptor-ligand pair (Sung et al 1992), the binding of a T-lymphocyte to a glass-supported lipid bilayer mediated respectively by CD2/LFA-3 (Tözeren et al 1992b) and by ICAM-1/LFA-1 molecules (Tözeren et al 1992a), the adhesion of two CHO cells mediated by fibrinogen and gpIIb-IIIa molecules (Sung et al 1993), and the binding of leukocyte cell lines to Kaposi's sarcoma cells (Delobel et al 1992; Delobel 1992; Zhu et al 1993).

The analysis of the micromanipulation data is greatly enhanced when coupled with mathematical modeling. Not only can a single end-point value of the critical force of detachment be found, but also the histories and distributions of a variety of physical quantities become determinable. More importantly, these quantities are interrelated quantitatively by fundamental principles. This helps to elucidate the mechanism of cell-cell adhesion and enables us to understand this biological phenomenon in physical, chemical, and molecular terms.

The physics of cell adhesion can be separately analyzed at two different length scales. At the scale of the cell body (microns), the analysis is based on continuum mechanics of the bulk cellular deformation; while at the scale of the intercellular spacing between the two cell surfaces at the contact edge (tens of nanometers), the analysis is based on molecular dynamics of the adhesive receptors. Such a separate treatment is possible because spatial variations of key physical quantities occur very rapidly within a thin zone at the edge of the adhesion area but relatively slowly outside (Zhu 1991; Dembo 1994). Therefore, the structure of the governing equations resembles that of the boundary layer problem, which greatly simplifies the mathematical analysis.

Previous work focusing on the "inner problem," i.e. on the peeling of the cell membrane under molecular bonding forces within the boundary layer, required the solution in the outer region as boundary conditions and known *a priori* (Evans 1985a; Evans 1985b; Dembo et al 1988; Dembo 1994). Among the published work on the "outer problem," the deformed shapes of erythrocyte rouleau in suspension has been computed using a finite element method (Skalak et al 1981). The detachment of adherent cell pairs by micropipette manipulation has been analyzed based on an approximate global force balance (Evans and Leung 1984; Tözeren et al 1989). Solutions to the micromanipulation problem were mentioned and drawn graphically by Tözeren et al (1992a), however, no detailed equations or solution procedures were presented in their paper.

In this paper, we present a mechanical analysis of cellular deformations due to adhesion and due to separation by externally applied forces. The objective was to evaluate the histories of adhesion force and energy during the entire processes of conjugation and detachment. Based on the idea that the cell itself could be used as a mechanical transducer, we sought a scheme to simultaneously calculate the adhesive properties, such as binding force and energy, and the mechanical properties, such as the viscoelastic constants, from the observed cellular deformation under controlled applied loading. The cortical shell-liquid core model was adapted for the cell (Dong et al 1988; Yeung and Evans 1989). A sequence of constitutive equations with increasing order of approximation and level of complexity were employed in the analysis. For the cortical shell, these included a uniform prestress tension, the shear and area elasticity, and the bending elasticity. For the liquid core, these included incompressible inviscid fluid and incompressible Newtonian fluid. After setting up the governing equations for the general problem, solutions were obtained for some particular cases. Depending on the complexity of the system, the boundary value problems were solved using an exact analytical method, an approximate analytical method based on small strain linearization, and a regular perturbation method based on the difference between the shear and area elastic moduli. These results provide the "outer solution" to be matched by the "inner solution" from the boundary layer analysis of the molecular dynamics of the cell adhesion molecules†.

Theory

Governing Equations

Cortical Shell. Fig. 1 shows the geometries and coordinate systems of the conjugation of a cell to a flat plane and its detachment from that plane by micropipette aspiration. Since the thickness of the cell cortex is much smaller than the two radii of principal curvature of the cell surface (except possibly at the corners), the mechanics is that of a thin shell problem. The internal stress resultants at every point on the plane of the cell cortex include the membrane tension T and bending moment M, which satisfy the following equations of equilibrium (Evans and Skalak 1980):

† The following abbreviations were used in the text:

AIDS	= acquired human immunodeficiency syndromes
CAM	= cell adhesion molecule
CD2	= cluster of differentiation 2
CHO	= Chinese hamster ovary
EDTA	= ethylenediamine tetraacetic acid
gpIIb-IIIa	= platelet membrane glycoprotein IIb-IIIa
HIV	= human immunodeficiency virus
ICAM-1	= intracellular adhesion molecule 1
KS	= Kaposi's sarcoma
LFA-1 (3)	= lymphocyte function-associated antigen 1 (3)

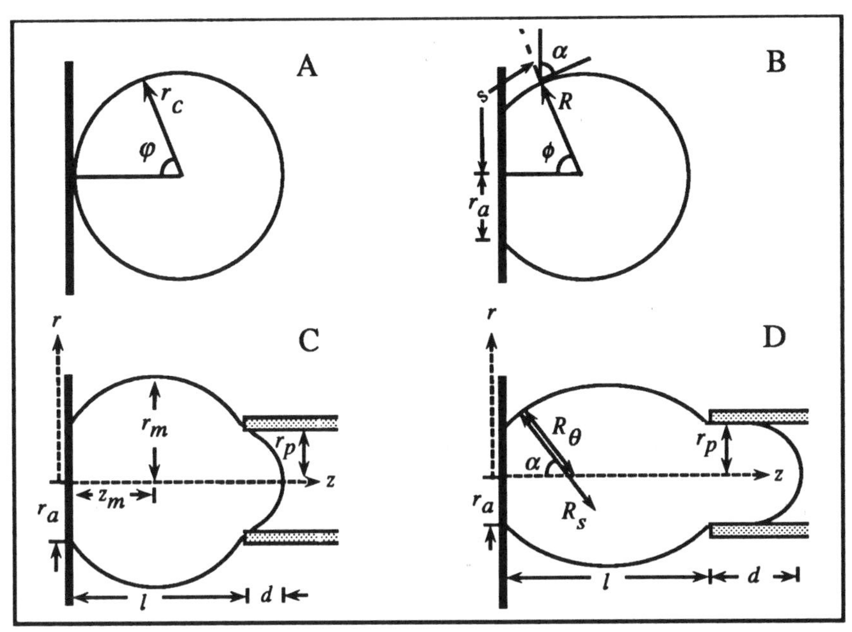

Fig. 1. (A) At the beginning of the contact, the originally undeformed shape of the cell is a sphere of radius r_c. In the meridional plane, the position of a point on the cell surface can be described by the angle φ. (B) After conjugation, the axisymmetrically deformed shape of the cell can be described either by $R(\phi)$ in the spherical coordinate system or by $\alpha(s)$ in the natural coordinate system. (C) and (D) The cell shape can also be described by $r(z)$ in the cylindrical coordinate system. Geometric parameters include the conjugation radius, r_a, the maximum latitude radius, r_m, and its position, z_m, the pipette radius, r_p, and its tip position, l, and the length of the cell tongue, d, being aspirated into the pipette. Also shown are the two principal radii of curvature, $R_s = |ds/d\alpha|$ and $R_\theta = |r/\sin\alpha|$, in meridional and circumferential directions, respectively. (C) represents phase I, or light aspiration, i.e. the case of $d < r_p$; and (D) represents phase II, or heavy aspiration, i.e. the case of $d > r_p$.

$$\frac{1}{rR_s}\left[\frac{d(rM_s)}{ds} - \frac{dr}{ds}M_\theta\right] + \frac{1}{r}\left[\frac{d(rT_s)}{ds} - \frac{dr}{ds}T_\theta\right] + p_s = 0 \tag{1a}$$

$$\frac{1}{r}\frac{d}{ds}\left[\frac{d(rM_s)}{ds} - \frac{dr}{ds}M_\theta\right] - \frac{T_s}{R_s} - \frac{T_\theta}{R_\theta} + p_n = 0 \tag{1b}$$

where the subscripts s and θ denote, respectively, the meridional arc length and circumferential directions. The externally applied loads have two components: p_n is normal to the membrane and is equal to the transcortical pressure difference, while p_s is tangential to the meridional contour and is to be matched by the wall shear stress of the cytoplasmic flow.

The cortical shell is assumed to be elastic, obeying the following tension-strain and moment-curvature relationships (Skalak et al 1973; Dong and Skalak 1992):

$$T_s = T_0 - E_s\left(\frac{1}{\lambda_s^2} - \frac{1}{\lambda_\theta^2}\right) + E_a\left(\lambda_s\lambda_\theta - 1\right) \tag{2a}$$

$$T_\theta = T_0 + E_s\left(\frac{1}{\lambda_s^2} - \frac{1}{\lambda_\theta^2}\right) + E_a\left(\lambda_s\lambda_\theta - 1\right) \tag{2b}$$

$$M_s = E_b\left(\frac{1}{R_s} - \frac{1}{r_c}\right) \tag{2c}$$

$$M_\theta = E_b\left(\frac{1}{R_\theta} - \frac{1}{r_c}\right) \tag{2d}$$

where λ_s and λ_θ are, respectively, the principal stretch ratios in the meridional and circumferential directions. T_0 is the prestress tension. E_s, E_a and E_b are, respectively, the shear, area, and bending elastic moduli. Since the stretch ratios and radii of curvature can be expressed in terms of the cell shape function $R(\phi)$ or $r(z)$, Eqs. 1 and 2 allow one to calculate the deformed shape of the cell provided that the external loads p_s and p_n are given.

Liquid Core. To determine the shear stress and pressure applied to the cortical shell one must solve for the cytoplasmic flow of the liquid core. In the present axisymmetric, incompressible case, a Stokes stream function ψ exists (Happel and Brenner 1986). For low Reynolds number flow of a Newtonian fluid, the equation of motion can be converted into a bi-Stokes equation on ψ. In spherical coordinates, the equation reads:

$$E^4\psi \equiv \left[\frac{\partial^2}{\partial R^2} + \frac{\sin\phi}{R^2}\frac{\partial}{\partial\phi}\left(\frac{1}{\sin\phi}\frac{\partial}{\partial\phi}\right)\right]^2 \psi = 0 \tag{3}$$

the solution of which can readily be obtained as an infinite series of products of the powers of R and associated Legendre functions of $\cos\phi$. The coefficients in the series can be determined using the no slip boundary conditions at the interface of the cortical shell (c) and the liquid core (ℓ), i.e. at the boundary:

$$\boldsymbol{v}|_\ell \equiv \frac{1}{R\sin\phi}\nabla\psi \times \boldsymbol{e}_\theta\Big|_{R=R(\phi)} = \boldsymbol{v}|_c \tag{4}$$

where ∇ is the gradient operator and $\boldsymbol{e}_\theta$ is the unit vector in the circumferential direction.

The normal and shear stresses, τ_{nn} and τ_{ns}, of the cytoplasm at the interface provides the normal and tangential loads for the cortex. Thus, at the boundary,

$$
\begin{aligned}
p_n = -\tau_{nn}\big|_\ell - p_0 = \mu \int & \left(\frac{dR}{R^2 \sin\phi} \frac{\partial}{\partial \phi} E^2 \psi - \frac{d\phi}{\sin\phi} \frac{\partial}{\partial R} E^2 \psi \right) \Bigg|_{R=R(\phi)} \\
& + \frac{2\mu}{\left[1 + (R'/R)^2\right] R \sin\phi} \left\{ \left(\frac{R'}{R}\right)^2 \left[\frac{\sin\phi}{R} \frac{\partial^2}{\partial R \partial \phi} \left(\frac{\psi}{\sin\phi} \right) - \frac{1}{R^2} \frac{\partial \psi}{\partial \phi} \right] \right. \\
& \left. + \left(\frac{R'}{R}\right) \left[E^2 \psi - 2R \frac{\partial}{\partial R} \left(\frac{1}{R} \frac{\partial \psi}{\partial R} \right) \right] - R \frac{\partial^2}{\partial R \partial \phi} \left(\frac{\psi}{R^2} \right) \right\} \Bigg|_{R=R(\phi)} - p_0
\end{aligned}
\tag{5a}
$$

$$
\begin{aligned}
p_s = \tau_{ns}\big|_\ell = & \frac{\mu}{R \sin\phi} \left[E^2 \psi - 2R \frac{\partial}{\partial R} \left(\frac{1}{R} \frac{\partial \psi}{\partial R} \right) \right] \Bigg|_{R=R(\phi)} \\
& - \frac{2\mu (R'/R)}{\left[1 + (R'/R)^2\right]} \left[\frac{1}{R^2} \frac{\partial}{\partial \phi} \left(\frac{1}{\sin\phi} \frac{\partial \psi}{\partial R} \right) + \frac{R}{\sin\phi} \frac{\partial}{\partial R} \left(\frac{1}{R^3} \frac{\partial \psi}{\partial \phi} \right) \right] \Bigg|_{R=R(\phi)}
\end{aligned}
\tag{5b}
$$

where μ is the cytoplasmic viscosity. $E^2(\bullet)$ is the Stokes operator (cf. Eq. 3) (Happel and Brenner 1986). R' stands for the derivative of the cell shape function $R(\phi)$ with respect to ϕ. The pressure p_0 outside the cortical shell may either be the pressure of the suspending medium (assumed to be zero), the pressure inside the aspirating micropipette, or the contact pressure inside the conjugation region. The tangential traction applied from outside the cortical shell is negligible except possibly at the pipette tip and inside the contact area.

The cell volume remains constant during the adhesion process if one assumes that the cytoplasm is incompressible and that the mass exchange across the cell membrane is negligible. The equation for the volume conservation is:

$$
\pi \int_0^{l+d} r^2 dz = \frac{4\pi}{3} r_c^3 \tag{6}
$$

where the geometric parameters r_c, l, and d are defined in Fig. 1. Eq. 6 allows for determination of the integration constant resulting from Eq. 5a.

Solutions

Uniform Tension Cortex - Uniform Pressure Core Model. The mechanics of the cortical shell can be decoupled from that of the liquid core if the cytoplasm is in static equilibrium. For slow deformations, such a quasi-static assumption could be a realistic approximation because the pressure gradient and shear stress of the liquid core may be negligible. The contribution of the liquid core can then be represented by a uniform intracellular pressure, p_c, so that p_s becomes zero and p_n becomes a piecewise constant function.

If the bending elasticity E_b is neglected, the tensions of the cortical shell under a piecewise constant normal loading p_n become statically determined. Direct integration of Eq. 1 (after setting $M_s = M_\theta = 0$) in the cylindrical coordinate system yields:

$$T_s = \frac{1}{2}\sqrt{1+r'^2}\left(rp_n + \frac{f}{\pi r}\right) \tag{7a}$$

$$T_\theta = \frac{1}{2}\frac{rr''}{\sqrt{1+r'^2}}\left(rp_n + \frac{f}{\pi r}\right) + rp_n\sqrt{1+r'^2} \tag{7b}$$

where the primes indicate differentiation with respect to z. The integration constant f has the physical meaning of the axial force resultant at any cross section normal to the z axis. This can easily be shown by considering a force balance at the cross section through $z = z_m$, at which r reaches its maximum, r_m, and $r'(z_m) = 0$ (Fig. 2).

The cortical tensions must be specified in order to further integrate Eq. 7. A simplest model is isotropic or uniform tensions, as suggested by Tözeren et al (1992a). However, the assumption of a uniform nonzero meridional tension and a vanishing circumferential tension (Tözeren et al 1992a) is incompatible with the equilibrium equation. It can be shown from Eq. 1 or 7 that isotropic tension implies uniform tensions, that uniform meridional tension implies isotropic tension, but that uniform circumferential tension is not sufficient either for the tensions to be isotropic or for the meridional tension to be uniform. In essence, the uniform or isotropic tension model assumes that the tensions are independent of the spatial variation in local strains, which can be obtained from, but not limited to, the assumption of negligible shear and area elastic moduli, i.e. $T_s = T_\theta = T_0$ (after setting $E_s = E_a = 0$ in Eqs. 2a and 2b). Using this model, the deformed shape of the cell can be expressed in terms of an elliptic integral,

$$\left|z - z_m\right| = \int_r^{r_m}\left[\left(\frac{rp_n}{2T_0} + \frac{f}{2\pi rT_0}\right)^{-2} - 1\right]^{-1/2} dr \tag{8a}$$

where

$$z_m = \int_{r_a}^{r_m}\left[\left(\frac{rp_n}{2T_0} + \frac{f}{2\pi rT_0}\right)^{-2} - 1\right]^{-1/2} dr \tag{8b}$$

and

$$r_m = \frac{T_0}{p_n} + \left[\left(\frac{T_0}{p_n}\right)^2 - \frac{f}{\pi p_n}\right]^{1/2} \tag{8c}$$

For the case of spontaneous adhesion without externally applied loads (Fig. 1B), Eq. 8 reduces to an indented sphere of radius $2T_0/p_c$

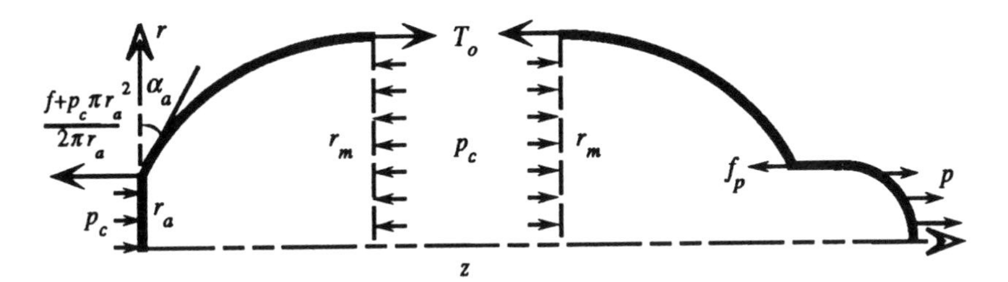

Fig. 2. Free body diagram of parts of an adherent cell subject to micropipette aspiration. Only the balance of axial forces is shown because the radial forces are self-balanced for the present axisymmetric problem. Also due to axisymmetry, only the upper half of the diagram is shown. The normal stress distribution on the adhesion area is drawn according to our model prediction, i.e. $\sigma_z = -p_c$ for $r < r_a$ and $\sigma_z \rightarrow [(f + \pi r_a^2 p_c)/(2\pi r_a)]\,\delta(r - r_a)$ as $r \rightarrow r_a$. What prevents one from being able to directly measure the adhesion force f is that the indentation force f_p by the pipette tip is usually not known. Only when the cell is about to slip out from the pipette tip will f_p vanish and hence can f be calculated from the aspiration pressure through simple force balance.

the value of which can be determined from the volume conservation condition of Eq. 6. The result is:

$$\frac{2T_0}{p_c} = r_c\left[2 + \frac{1}{8}\left(\frac{r_a}{r_c}\right)^6\right]^{\frac{1}{3}}\left[\left(1 + \sqrt{1+\beta}\right)^{\frac{1}{3}} + \left(1 - \sqrt{1+\beta}\right)^{\frac{1}{3}}\right]^{-1} \tag{9a}$$

where

$$\beta = \left(\frac{r_a}{r_c}\right)^{12}\left[2^{10} + 2^6\left(\frac{r_a}{r_c}\right)^6\right]^{-1} \tag{9b}$$

For the case of detachment of an adherent cell from a flat plane by micropipette (Figs. 1C and 1D), Eq. 8 describes the deformed shape of the cell between the adhesion plane and the pipette tip as well as the shape of the "cell tongue" inside the pipette. The latter is either a partial sphere (Fig. 1C) or a circular cylinder plus a hemisphere (Fig. 1D), so that the radius of the sphere, the cortical tension, and the transcortical pressure difference are related by the law of LaPlace,

$$\frac{2T_0}{p_c - p} = \begin{cases} \left(r_p^2 + d^2\right)/2d & d \leq r_p \\ r_p & d > r_p \end{cases} \tag{10}$$

where p is the aspiration pressure inside the pipette.

The surface adhesion energy density, γ, defined as the mechanical work done to detach a unit area of adhesive contact, can be calculated using the Young equation when bending is neglected (Adamson 1976; Evans 1994), as follows:

$$\gamma = T_s\left(1 - \cos\alpha_a\right) = T_0\left(1 - \frac{r'}{\sqrt{1 + r'^2}}\right)\Bigg|_{z=0}$$
$$= T_0\left[1 - \text{sign}\left(r'|_{z=0}\right)\sqrt{1 - \left(\frac{f + \pi r_a^2 p_c}{2\pi r_a T_0}\right)^2}\right] \tag{11}$$

where α_a is the angle between the r axis and the tangent of the meridional contour at the edge of the adhesion area ($z = 0$, $r = r_a$), as shown in Fig. 2, and sign(•) denotes the sign function. Note that Eq. 11 applies to both cases of spontaneous adhesion and detachment by micropipette aspiration. For the former case, the net axial force $f = 0$, while for the latter case, $f > 0$.

The distribution of stress normal to the contact disk (Fig. 2) is compressive (due to indentation from the original spherical shape) except within a thin zone at the edge where it is tensile (to provide adhesive forces) (Zhu 1991). The more flexible the cell cortex, the thinner the tensile zone (boundary layer). When the bending rigidity of the cortical shell is neglected, sudden jumps in the surface curvatures are allowed, the boundary layer becomes infinitesimally thin, and the tensile stress distribution approaches a Dirac delta function. The strength of the stress concentration can be derived by considering the force balance at the conjugation area (cf. Fig. 2). It has the physical meaning of adhesive force per unit length of boundary. Thus,

$$\sigma_z = -p_c + \frac{f + \pi r_a^2 p_c}{2\pi r_a}\delta(r - r_a) \tag{12}$$

where δ(•) denotes the Dirac delta function.

Elastic Cortex - Uniform Pressure Core Model. To examine the effect of elasticity, the problem of spontaneous formation of adhesion (Figs. 1A and 1B) was solved using the model of elastic cortical membrane and liquid core with uniform intracellular pressure. In contrast to the model of uniform prestress cortical tension in which the solution outside the adhesion region is independent of that inside, for the model of elastic cortex, the cell shape outside the conjugation area is coupled to the membrane deformation inside the adhesion region. Since the governing equations become highly nonlinear once elasticity is included, approximate analytical methods were used. These included small strain linearization to solve for the cell shape outside the conjugation area and a regular perturbation method to solve for the deformation inside the adhesion region.

For the outer region, two functions were introduced: $R(\phi) = r_c[1 + \varepsilon(\phi)]$ to describe the deformed shape of the cell in spherical coordinates and $\varphi(\phi) = \phi[1 + e(\phi)]$ to specify the mapping between the deformed and undeformed coordinates. Substituting Eqs. 2a and 2b into Eqs. 7a and 7b respectively and expressing λ_s, λ_θ, r, and z in terms of ε, e and their derivatives with respect to ϕ, two coupled ordinary differential equations can be obtained. Assuming that ε and e are small and hence nonlinear terms on ε and e can be neglected, solutions to the linearized equations can be found. The result is:

$$\varepsilon = \left[\left(1 - \frac{2E_a}{T_0}\right) c\cos\phi - \left(1 - \frac{p_c r_c}{2T_0}\right)\right] \Big/ \left(\frac{2E_a}{T_0} - \frac{p_c r_c}{2T_0}\right) \quad (13a)$$

$$e = c\frac{\sin\phi}{\phi}\left(1 - \frac{2E_a}{T_0}\right) \Big/ \left(\frac{2E_a}{T_0} - \frac{p_c r_c}{2T_0}\right) \quad (13b)$$

Upon substitution of Eq. 13a into $R(\phi) = r_c(1 + \varepsilon)$, it can be shown that this is a linearized elliptic equation in the polar coordinate system. Thus, the deformed cell shape is approximately an indented ellipsoid rather than an indented sphere when the membrane elasticity is included in the model. The integration constant c has the geometric meaning of the eccentricity. The semilatus rectum of the ellipse, $r_c[(2E_a/T_0 - 1)/(2E_a/T_0 - p_c r_c/2T_0)]$, depends on the ratio of the area elastic modulus, E_a, to the prestress tension, T_0. The cortical tension is still uniform and isotropic, but has a value different from T_0 and depends also on the ratio E_a/T_0, as follows:

$$\frac{T_s}{T_0} = \frac{T_\theta}{T_0} = \left(\frac{2E_a}{T_0} - 1\right) \Big/ \left(\frac{4E_a}{p_c r_c} - 1\right) \quad (14)$$

To determine the integration constant c from Eqs. 13a and 13b, one must solve the deformation within the contact area. Only one equation of equilibrium, Eq. 1a, suffices for determination of the axisymmetric deformations of a flat membrane, because Eq. 1b is automatically satisfied. Eqs. 1a (after setting $M_s = M_\theta = 0$ and $s = r$), 2a and 2b were solved by a perturbation method using the ratio of shear to area elastic moduli or its reciprocal as the perturbation parameter. For the case of $E_s/E_a \ll 1$, the zeroth order solution for the mapping between the deformed coordinate r and undeformed coordinate φ is:

$$\varphi = \cos^{-1}\left[1 + \frac{1}{2}\left(\frac{r}{r_c}\right)^2 \left(\frac{p_c r_c}{2T_0} - \frac{2E_a}{T_0}\right) \Big/ \left(\frac{p_c r_c}{2T_0} + \frac{2E_a}{T_0} - 2\right)\right] \quad (15a)$$

For the case of $E_s/E_a \gg 1$, it is:

$$\varphi = 2\tan^{-1}\left\{\frac{r}{r_a}\tan\left[\frac{1}{2}\sin^{-1}\left(\frac{r_a}{r_c}\sqrt{\frac{2E_a/T_0 - p_c r_c/2T_0}{2E_a/T_0 + p_c r_c/2T_0 - 2}}\right)\right]\right\} \quad (15b)$$

The boundary condition that the meridional tension be continuous at the edge of the contact ($r = r_a$ or $\phi = \phi_a$) has already been imposed to obtain the solutions given by Eqs. 15a and 15b. The geometric boundary conditions at the contact edge, i.e., $\varphi(r_a) = \phi_a[1 + e(\phi_a)]$ and $r_a = r_c[1 + \varepsilon(\phi_a)]\sin\phi_a$, plus the volume conservation condition of Eq. 6, allow for determination of the eccentricity c and the intracellular pressure p_c for any given configuration of adhesion (as specified by the radius of the adhesion area r_a). The surface adhesion energy density, γ, can be calculated using Eq. 11 (after setting $f = 0$ and replacing T_0 by T_s given by Eq. 14).

Experiment

Cells

Kaposi's sarcoma (KS) cells were obtained from punch biopsies of skin lesions of Kaposi's sarcoma in HIV-infected AIDS patients (whose informed consents were obtained) and cultured using conditions that favored endothelial cell growth (Yang et al 1994). KS cells (passage 4 - 10) were detached from tissue culture flasks by 0.5 m*M* EDTA (Gibco, Grand Island, NY) and washed twice before being used in the adhesion experiment. The promyelocytic leukemia cell line HL60 was obtained from American Type Culture Collection (Rockville, MD). HL60 cells were used as a model because many of their properties resemble those of granulocytes and because they adhered avidly to KS cells (Zhu et al 1993). As far as the bulk cellular deformation is concerned, the KS cell - HL60 cell system appears to behave similarly to many other cell systems whose adhesion has been studied using the micropipette method (e.g., Sung et al 1986; Sung et al 1992; Sung et al 1993; Tözeren et al 1992b; Tözeren et al 1992a), although the cell adhesion molecules underlying the interactions are different in different cell systems. The molecular mechanisms mediating the strong KS cell - HL60 cell adhesion are currently being explored.

Experimental Protocol

Our micropipette apparatus has been described (Delobel 1992; Delobel et al 1992; Zhu et al 1993). The methods used to manipulate adhesion of individual cell pairs are illustrated by the sequential photomicrographs in Fig. 3. Briefly, two micropipettes with inner diameters of 4 - 6 μm at the tip were prepared using a pipette puller (Kopf, Tujunga, CA) and a microforge (built in-house). The wide end of each pipette was connected to a hydraulic pressure regulation system (built in-house). The movement of each pipette was controlled by a hydraulic micromanipulator (Narishige, Tokyo). Cells were suspended in a small chamber mounted on the stage of an inverted

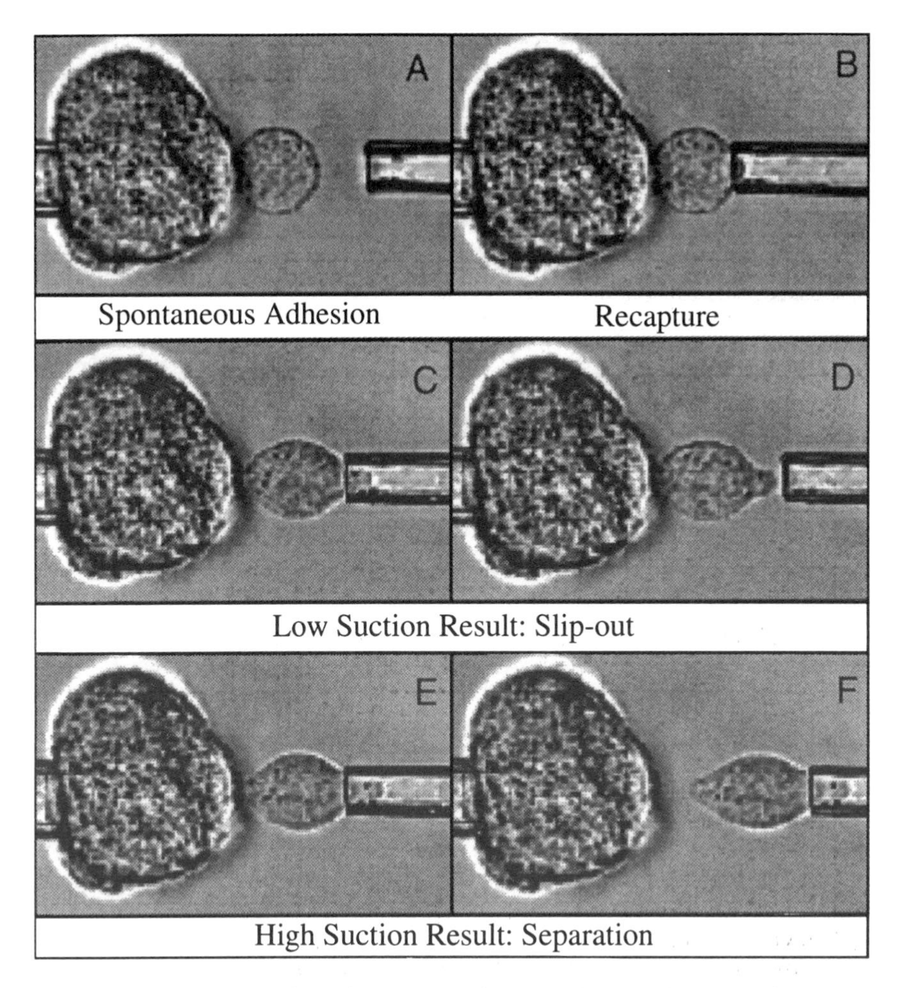

Fig. 3. Six sequential photomicrographs of a typical micropipette manipulation experiment. The tape-recorded images were digitized and processed using the Macintosh Quadra 950 to allow for direct printing from a laserwriter. The larger cell on the left was a KS cell held by the holding pipette. The smaller cell on the right was a HL60 cell held by the manipulating pipette. The two cells were aligned and brought to contact. The HL60 cell first was released (A) to allow for adhesion to the KS cell free of applied forces and then recaptured (B) at a later time. The two cells then were gradually pulled away from each other (C). The HL60 cell slipped out from the manipulating pipette when the suction pressure was small (D). At larger aspiration pressure, the HL60 cell was stretched to elongate (E). As the manipulating pipette was slowly maneuvered to the right, the two cell bodies were separated (F).

microscope (Olympus, Tokyo) which was placed on an anti-vibration table (Kinetic Systems, Boston, MA). A KS cell and a HL60 cell were respectively captured by two pipettes, aligned, and brought to close proximity. The HL60 cell was released by removal of the aspiration pressure from its pipette (manipulating pipette) and allowed to interact freely with the KS cell which was held in a fixed position for observation by its pipette (holding pipette) (Fig. 3A). After a predetermined period (10 - 30 min) of incubation, the HL60 cell was recaptured with a small suction pressure (~ 2 mm H_2O) (Fig. 3B) which was increased stepwise in increments of 0.25 - 1 mm H_2O. At each level of aspiration pressure, the HL60 cell was pulled away gradually by maneuvering the micropipette (Figs. 3C and 3E). To ensure the detachment process to be quasi-static as required by the model, the pipette was moved at an extremely low velocity of about 0.5 μm/sec. At low levels of aspiration pressure, the HL60 cell slipped out of the pipette tip when the manipulating pipette was moved too far right (Fig. 3D). Above a certain level of suction pressure, the body of the HL60 cell was detached from the KS cell body (Fig. 3F). As can be seen in Fig. 3, the HL60 cell deformed into a series of elongated shapes in response to aspiration and micromanipulation. By contrast, the KS cell showed no detectable deformation suggesting much more rigid mechanical properties. The manipulation process was viewed through the bottom of the cell chamber with an 100× objective (numerical aperture of 1.25, oil immersion) and a 20× eyepiece. The moving images with digital time were continuously recorded through a charge coupled device camera (Dage MTI, Michigan City, IN), a video timer (For.A, Boston, MA), and a Super-VHS video cassette recorder (Mitsubishi, Cypress, CA) and displayed on a video monitor (Panasonic, Secaucus, NJ) at a final magnification of approximately 2,500× as calibrated by a stage micrometer.

Image Processing and Data Acquisition

At a later time, the recorded images were played back in a computer-controlled video cassette recorder (NEC Technologies, Wood Dale, IL) and processed using a Macintosh Quadra 950 equipped with an image grabber (Neotech, United Kingdom). At various levels of aspiration pressure p, and at time intervals of 1 - 2 sec, the following geometric measurements were made from the processed images: the contact radius, r_a, the inner radius, r_p, and the location, l, of the pipette, the cell tongue length, d, the radius of the maximum latitude circle, r_m, and its location, z_m, and a set of coordinates on the cell meridional contour, (z_i, r_i).

Results and Discussion

We shall illustrate the results for the problem of detachment of adhesion by micropipette aspiration (Figs. 1C and 1D) using the simplest mechanical property model: a thin shell with uniform cortical tension and a liquid core with uniform intracellular pressure (i.e. quasi-static process). The effect of cortical elasticity will be examined using the results for the problem of spontaneous formation of adhesion (Figs. 1A and 1B).

Model Validation

For the problem of detachment of adhesion by micropipette aspiration, two approaches were utilized to test the validity of our model by comparing the observed with the predicted shapes of the deformed cells. In the first approach (Fig. 4), the adhesion radius, r_a,

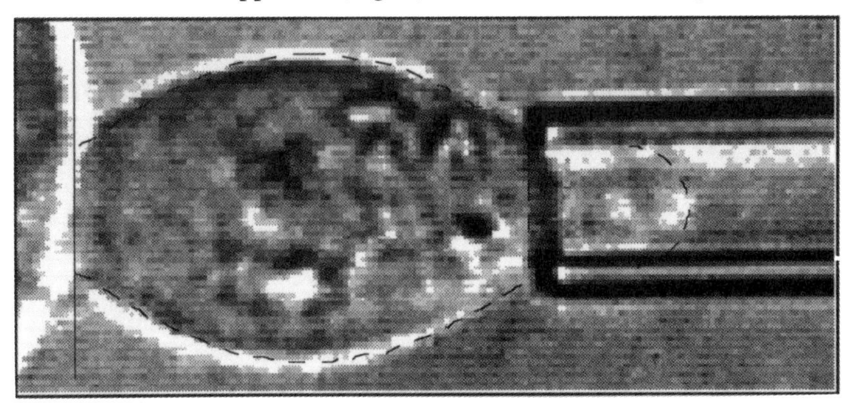

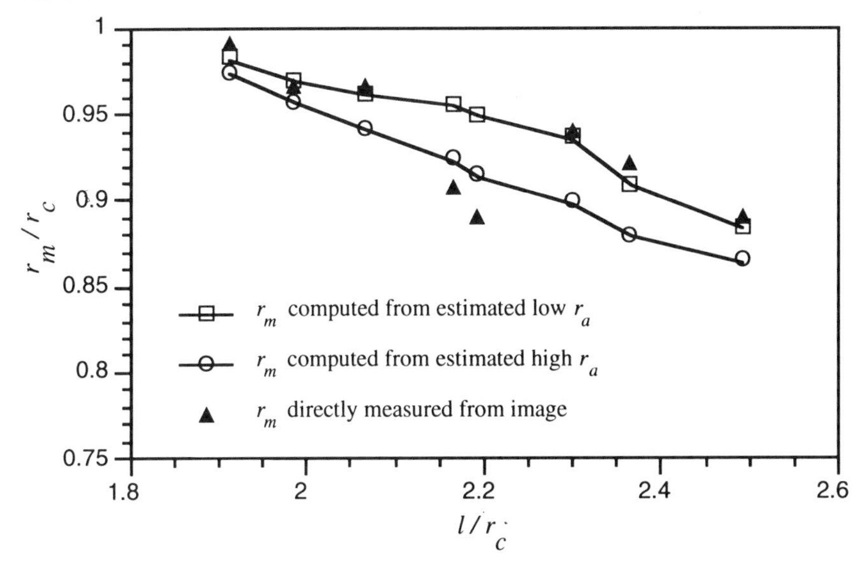

Fig. 4. (A) Computer-processed image of an adherent HL60 cell recorded from a typical experiment overlain by the outline of the cell shape predicted from our model (dashed curve). Measured parameters used as input for the calculation of the deformed shape of the cell were the initial radius of the undeformed spherical cell, r_c, the radius of the circular adhesion area, r_a, the inner radius, r_p, the location, l, and the aspiration pressure, p, of the pipette, and the cell tongue length, d, being aspirated into the pipette. (B) Comparison of the theoretically predicted and experimentally measured radius of the maximum latitude circle, r_m, as a function of the pipette location, l, at aspiration pressure $p = -7$ mm H_2O for the HL60 cell shown in (A). The curves connecting the open symbols are predicted r_m calculated from either the upper or lower bounds of the r_a values estimated to the best of our ability from the diffraction band of the image, whereas the closed symbols are the directly measured r_m.

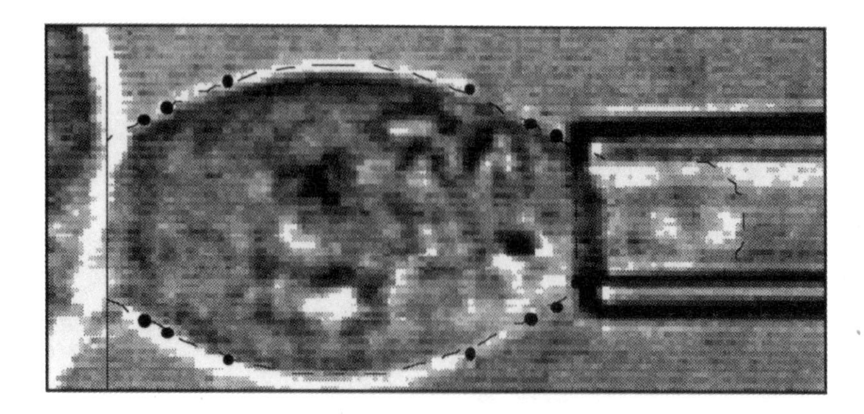

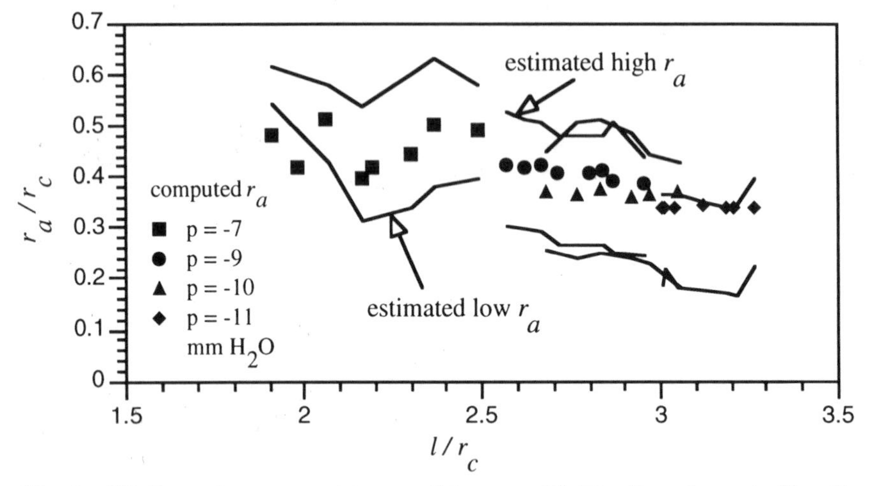

Fig. 5. (A) Computer-processed image of the same HL60 cell as shown in Fig. 4A overlain by the computed cell shape (dashed curve) to best fit a set of measured cylindrical coordinates, (z_i, r_i) $(i = 1 - 6)$, of the cell outline (dots). Additional measured parameters used as model input included the initial radius of the undeformed spherical cell, r_c, the maximum latitude radius, r_m, the inner radius, r_p, the location, l, and the aspiration pressure, p, of the pipette, and the cell tongue length, d, being aspirated into the pipette. (B) Comparison of the theoretically predicted and experimentally measured radius of the adhesion area, r_a, for four levels of aspiration pressure, p, and various values of the pipette location, l, for the HL60 cell shown in (A). The solid curves represent upper and lower bounds measured to the best of our ability from the diffraction band of the image.

and the cell tongue length, d, were measured for each controlled pipette pressure, p, and location, l. These were used as model inputs to calculate the adhesive properties (f and γ) and mechanical properties (T_0 and p_c). The predicted cell shape was then compared with the experimental observation.

In the second approach (Fig. 5), the deformed shape of the cell portion between the adhesion plane and the pipette tip (i.e. $0 < z < l$) was measured by a set of cylindrical coordinates, (z_i, r_i). T_0, p_c, f, and γ were computed for various values of p and l. The computation was based on a nonlinear regression method such that a best fit was obtained between the theoretically predicted and the experimentally observed cell shape. It should be noted that, in contrast to the first approach, the contact radius, r_a, was not required as a model input, but rather, was a computed result from the model.

It can be seen from Figs. 4 and 5 that the model predictions compare fairly well with experimental data, both visually (Figs. 4A and 5A) and numerically (Figs. 4B and 5B). The second approach is deemed to be more reliable because the best fit method utilizes more data points to minimize the errors between the predicted and measured cell shape, which helps to smooth out some of the observational noise inherent in a single measurement. The first approach requires less data points, but one of the required data points, the adhesion radius r_a, is usually difficult to measure and the measurements are of larger uncertainty (see Figs. 4 and 5). Also, the exact fit method used by the first approach is often more susceptible to the accuracy (or the lack thereof) of the input data. In addition, for higher aspiration pressures and larger deformations of the cell, the numerical scheme used by the exact fit method becomes ill-conditioned resulting in a very slow rate of convergence.

Model Predictions

The theoretically predicted axial force resultant, f (scaled by $2\pi r_c T_0$), and the surface adhesion energy density, γ (scaled by T_0) are respectively shown in Figs. 6A and 6B as functions of the pipette location, l (scaled by r_c), for several levels of aspiration pressure, p. Note that the present model works for both cases of pulling ($f > 0$) the cell away and pushing ($f < 0$) the cell toward the adhesion plane (Fig. 6A). Combining Figs. 4B and 6B, it can be seen that, during the process of detachment by micropipette aspiration, the surface adhesion energy density γ increases while the conjugation radius r_a decreases. By contrast, γ increases with the increasing r_a during the process of spontaneous adhesion (see Fig. 7 below). Moreover, the adhesion energy is much larger during the detachment process than the conjugation process. These results support the notion that cell-cell adhesion is a thermodynamically irreversible process.

To predict f and γ for the entire processes of adhesion and detachment requires the use of the cell itself as a force and energy transducer. The validity of such an application, and the accuracy of the mechanical transducer, depends on an appropriate mechanical model for the cell and reliable values of the mechanical property constants. In the present study, the mechanical property constants of each individual cell were solved simultaneously with the adhesive properties of that cell in order to minimize the influence of variations in mechanical properties among cells on the evaluation of the adhesion force and energy. Fig. 7 shows our model predictions of the cortical tension T_0 and the intracellular pressure p_c. The predicted

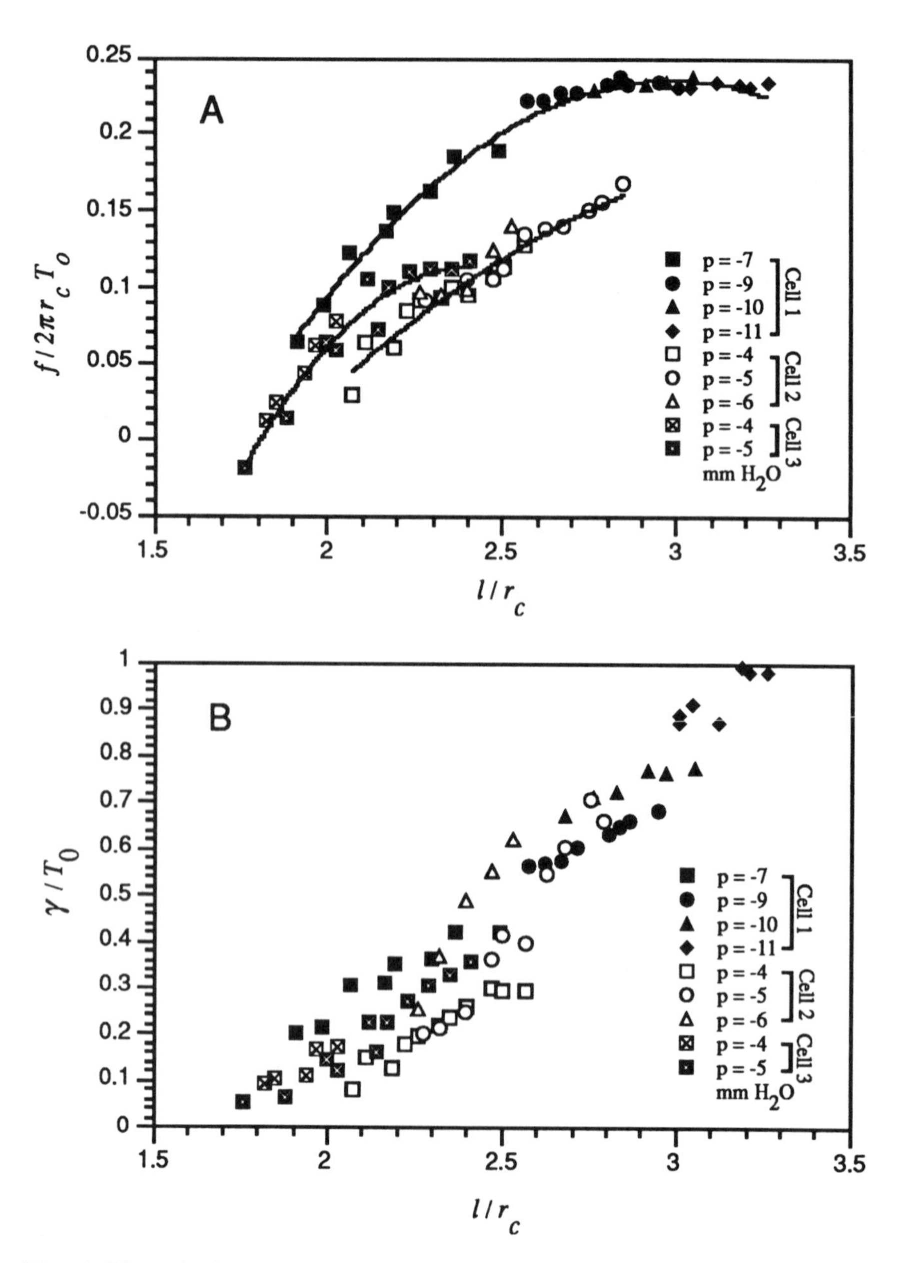

Fig. 6. Dimensionless axial force resultant, $f/(2\pi r_c T_0)$ (A), and surface adhesion energy density, γ/T_0 (B), as functions of the dimensionless pipette location, l/r_c, at various levels of aspiration pressure, p (in mm H_2O). Each datum point in Fig. 6A or 6B was computed from measurements from one image of a deformed cell, such as those shown in Fig. 4A or 5A. Data for three cells were shown, and tens of deformation states were analyzed for each cell. The continuous curves in (A) represent quadratic curves to best fit the data for each cell.

cortical tension agrees well with those reported by others for leukocytes (Dong et al 1988; Yeung and Evans 1989; Needham and Hochmuth 1992; Dong and Skalak 1992).

In the studies cited above, a single micropipette was utilized either to aspirate the cell into a cylindrical pipette (Dong et al 1988; Yeung and Evans 1989; Dong and Skalak 1992) or to push the cell along a tapered pipette (Needham and Hochmuth 1992). Only could a single variable, the pipette pressure, be controlled by the experimentalist. In the present work, the HL60 cell was allowed to adhere to a KS cell on one side and to be aspirated by a micropipette on the other side. Such an experimental design enabled us to deform the cell into various configurations by independently controlling two variables: the pipette position and the aspiration pressure. The cortical tension, T_0, can be calculated from each equilibrium configuration as a function of the pipette location, l, and the aspiration pressure, p, as shown in Fig. 7.

According to Eq. 8 and the preceding discussion, T_0 in Eq. 8 must be the uniform and isotropic cortical tension independent of the spatial strain variations. However, since Eq. 8 is derived for static equilibrium, it allows for different T_0 values for different equilibrium configurations. In such a case, T_0 may not be a true constant but may depend on the bulk deformation of the cell, as has been suggested by Dong and Skalak (1992) and Needham and Hochmuth (1992). The increase of the computed T_0 values with increasing p (Fig. 7) might reflect the limitation of the present model rather than the real change

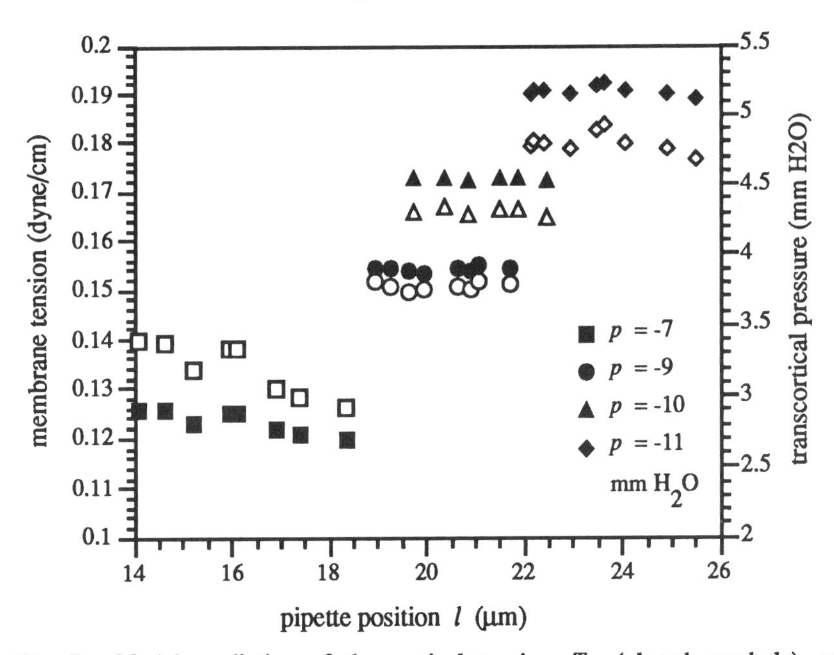

Fig. 7. Model prediction of the cortical tension, T_0 (closed symbols), and intracellular pressure, p_c (open symbols), as functions of the pipette position, l, at several levels of aspiration pressure, p.

of the mechanical properties. The reason is that, from direct observation, it was seen that changing l caused much more bulk deformation of the cell than changing p. One of the possible model limitations may result from the assumption of negligible tangential load, p_s, acting on the cell surface. This assumption is unlikely to be valid at the ring where the tip of the pipette indents onto the cell surface. If there exists a concentrated tangential load there, then there will be a discontinuity in the cortical tension. An increase in the aspiration pressure would induce an increase in the tension of the cell tongue inside the pipette but probably has little effect on the cortical tension of the cell portion outside. Another possible model limitation may result from the quasi-static assumption. Although the time rate of change in l was controlled in our experiment to be sufficiently small, after a sudden change in the aspiration pressure, there might not be enough time to allow the cytoplasmic flow into the pipette to reach the equilibrium state. We are currently examining these issues.

Effects of Elasticity

The effects of shear and area elasticity of the cortical shell on adhesion energy were examined using the problem of spontaneous adhesion (Figs. 1A and 1B). Compared in Fig. 8 are γ/T_0 vs. r_a/r_c curves computed based on the models without (prestress tension alone) and with cortical elasticity. For the latter model, two extreme cases were solved: the case in which the shear elastic modulus was much smaller than the area modulus and the case in which the shear modulus was much larger than the area modulus.

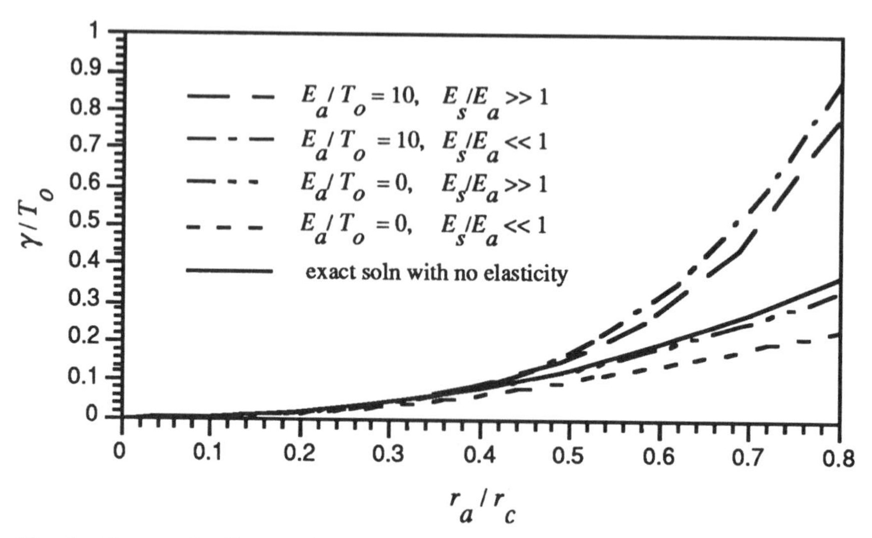

Fig. 8. Computed γ/T_0 vs. r_a/r_c curves for the problem of spontaneous formation of adhesion. The predictions of two models are compared: one with and the other without cortical elasticity. For the elastic model, both cases of $E_s/E_a << 1$ and $E_s/E_a >> 1$ are shown for two values of E_a/T_0: 0 and 10.

In order to ensure the reliability of the approximate solutions, their accuracy must first be examined, since only linear terms were kept in the governing equation for the region outside the conjugation area and since only the leading term of the perturbation series was used to approximate the solutions inside the adhesion region. This was done by setting the elastic moduli to zero in the approximate solutions (obtained from the equations with elasticity) and comparing them with the exact solution (obtained from the equation without elasticity). As can be seen in Fig. 8 (curves 3, 4 and 5), the two approximate solutions coincide very closely with the exact solution.

As can easily be understood intuitively, the effects of cortical elasticity increase with the degree of deformation and with the ratio of the elastic modulus E_a to the prestress tension T_0. The only reported values of cortical elastic moduli for leukocytes that we know of are $E_a \approx 0.01$ dyne/cm, $E_s \approx 0.14$ dyne/cm, and $T_0 \approx 0.12$ dyne/cm (Dong and Skalak 1992). We have examined the elasticity effects for a much wider variety of E_a/T_0 values. The case of $E_a/T_0 = 10$ is shown in Fig. 8 (curves 1 and 2). For smaller E_a/T_0 values, the effects of elasticity on the adhesion energy are minimal.

Conclusion

Of all existing methods for the measurement of the physical strength of cell-cell adhesion, the micropipette manipulation technique is probably the most quantitative one, because it allows for determination of the absolute value of the force required to separate a single pair of adherent cells. However, because of the unknown indentation force exerted to the cell by the pipette tip (cf. Fig. 2), only can the end-point value of the force required to separate the cell pair be directly measured using a sequence of pulls. In the present work, a model has been developed to analyze the deformation of an adherent cell in the micropipette experiment. The model allows us to obtain detailed information on the adhesion force and energy, namely, their histories during the entire processes of adhesion and detachment. Moreover, the model allows us to obtain such information using less experimental manipulations, say, a single pull of the adherent cell at a moderately-high aspiration pressure. This will help to minimize the undesirable mechanical disturbances to the cell during repeated pulling at multiple levels of aspiration pressure. The information obtained from the present model can be used as input for the modeling of the dynamics of the cell adhesion molecules.

Acknowledgment

This work was supported by NSF Grants Nos. 9210684 and 9350370, and Whitaker Foundation Biomedical Engineering Research Grant to CZ, and Emory/Georgia Tech Biomedical Technology Research Grant to CZ and MKO.

References

Adamson, A. W. Physical chemistry of surfaces. New York: John Wiley & Sons, Inc.; 1976.

Alberts, B.; Bray, D.; Lewis, J.; Raff, M.; Roberts, K.; Watson, J. D. Molecular biology of the cell. New York: Garland Publishing, Inc.; 1989.

Delobel, J. Quantification of the adhesion force between individual promyelocytic cells and Kaposi's sarcoma cells using a micropipette technique. Gerogia Institute of Technology. 1992.

Delobel, J.; Yang, J.; Offermann, M. K.; Zhu, C. Mechanical properties of membrane tethers mediating the cell adhesion of Kaposi's sarcoma. Bidez, M. W., ed. 1992 Advances in Bioengineering: proceedings of ASME Winter Annual Meeting; BED-Vol. 22; November 8 - 13, 1992; Anaheim, CA. American Society of Mechanical Engineers; 1992: 391-394.

Dembo, M. On peeling an adherent cell from a surface. In: Goldstein, D.; Wossy, C., eds. Some mathematical problems in biology. Providence, RI: American Mathematical Society; 1994, (in press).

Dembo, M.; Torney, D. C.; Saxman, K.; Hammer, D. A. The reaction-limited kinetics of membrane-to-surface adhesion and detachment. Proc. R. Soc. Lond. B. 234:55-83; 1988.

Dong, C.; Skalak, R. Leukocyte deformability: Finite element modeling of large viscoelastic deformation. J. Theor. Biol. 158:173-193; 1992.

Dong, C.; Skalak, R.; Sung, K.-L. P.; Schmid-Schönbein, G. W.; Chien, S. Passive deformation analysis of human leukocytes. J. Biomech. Eng. 110:27-36; 1988.

Evans, E. A. Detailed mechanics of membrane-membrane adhesion and separation. I. Continuum of molecular cross-bridges. Biophys. J. 48:175-183; 1985a.

Evans, E. A. Detailed mechanics of membrane-membrane adhesion and separation. II. Discrete kinetically trapped molecular cross-bridges. Biophys. J. 48:185-192; 1985b.

Evans, E. A. Micropipette studies of cell and vesicle adhesion. In: Bongrand, P., ed. Physical basis of cell-cell adhesion. Boca Raton: CRC Press; 1988:p. 173-189.

Evans, E. A. Physical actions in biological adhesion. In: Lipowski, R.; Sackmann, E., eds. Handbook of biophysics. Amsterdam: Elsevier; 1994, (in press).

Evans, E. A.; Berk, D.; Leung, A. Detachment of agglutinin-bonded red blood cells: I. Forces to rupture molecular-point attachments. Biophys. J. 59:838-848; 1991a.

Evans, E. A.; Berk, D.; Leung, A.; Mohandas, N. Detachment of agglutinin-bonded red blood cells: II. Mechanical Energies to separate large contact areas. Biophys. J. 59:849-860; 1991b.

Evans, E. A.; Buxbaum, K. Affinity of red blood cell membrane for particle surfaces measured by the extent of particle encapsulation. Biophys. J. 34:1-12; 1981.

Evans, E. A.; Leung, A. Adhesivity and rigidity of erythrocyte membrane in relation to wheat germ agglutinin binding. J. Cell Biol. 98:1201-1208; 1984.

Evans, E. A.; Skalak, R. Mechanics and thermodynamics of biomembranes. Boca Raton, FL: CRC Press; 1980.

Happel, J.; Brenner, H. Low Reynolds number hydrodynamics. Dordrecht: Martinus Nijhoff; 1986.

Needham, D.; Hochmuth, R. M. A sensitive measure of surface stress in the resting neutrophil. Biophys. J. 1992.

Skalak, R.; Tözeren, A.; Zarda, P. R.; Chien, S. Strain energy function of red blood cell membranes. Biophys. J. 13:245-264; 1973.

Skalak, R.; Zarda, P. R.; Jan, K.-M.; Chien, S. Mechanics of rouleau formation. Biophys. J. 35:771-781; 1981.

Springer, T. A. Adhesion receptors of the immune system. Nature. 346:425-434; 1990a.
Springer, T. A. The sensation and regulation of interactions with the extracellular environment: The cell biology of lymphocyte adhesion receptors. Annu. Rev. Cell Biol. 6:359-402; 1990b.
Sung, K.-L. P.; Frojmovic, M. M.; O'Toole, T. E.; Zhu, C.; Ginsberg, M. H.; Chien, S. Determination of adhesion force between single cell pairs generated by activated GpIIb-IIIa receptors. Blood. 81:419-423; 1993.
Sung, K.-L. P.; Kuhlman, P.; Maldonado, F.; Lollo, B. A.; Chien, S.; Brian, A. A. Force contribution of the LFA-1/ICAM-1 complex to T cell adhesion. J. Cell Sci. 103:259-266; 1992.
Sung, K.-L. P.; Sung, L. A.; Crimmins, M.; Burakoff, S. J.; Chien, S. Determination of junction avidity of cytolytic T cell and target cell. Science. 234:1405-1408; 1986.
Tözeren, A.; Mackie, L. H.; Lawrence, M. B.; Chan, P.-Y.; Dustin, M. L.; Springer, T. A. Micromanipulation of adhesion of phorbol 12-myristate-13-acetate-stimulated T lymphocytes to planar membranes containing intracellular adhesion molecule-1. Biophys. J. 63:247-258; 1992a.
Tözeren, A.; Sung, K.-L. P.; Chien, S. Theoretical and experimental studies on cross-bridge migration during cell disaggregation. Biophys. J. 55:479-487; 1989.
Tözeren, A.; Sung, K.-L. P.; Sung, L. A.; Dustin, M. L.; Chan, P.-Y.; Springer, T. A.; Chien, S. Micromanipulation of adhesion of a Jurkat cell to a planar bilayer membrane containing lymphocyte function-associated antigen 3 molecules. J. Cell Biol. 116:997-1006; 1992b.
Yang, J.; Xu, Y.; Zhu, C.; Hagan, M. K.; Lawley, T.; Offermann, M. K. Regulation of adhesion molecules expressed in Kaposi's sarcoma cells. J. Immunol. 152:361-373; 1994.
Yeung, A.; Evans, E. A. Cortical shell-liquid core model for passive flow of liquid-like spherical cells into micropipette. Biophys. J. 56:139-149; 1989.
Zhu, C. A thermodynamic and biomechanical theory of cell adhesion. Part I: General formulism. J. Theor. Biol. 150:27-50; 1991.
Zhu, C.; Delobel, J.; Ferguson, L. A.; Offermann, M. K. Quantitation of adhesion forces of cultured Kaposi's sarcoma cells for leukocyte cell lines. Tarbell, J. M., ed. 1993 Advances in Bioengineering: proceedings of ASME Winter Annual Meeting; BED-Vol. 26; November 28 - December 3, 1993; New Orleans, LA. American Society of Mechanical Engineers; 1993: 351-354.

Part IV

Cell-Matrix Interactions and Adhesion Molecules

12

Biphasic Theory and *In Vitro* Assays of Cell-Fibril Mechanical Interactions in Tissue-Equivalent Gels

V. H. Barocas and R. T. Tranquillo

Introduction

The mechanical interaction of motile cells with fibers in the surrounding extracellular matrix (ECM) is fundamental to cell behavior in soft tissues and tissue-equivalent reconstituted gels, and thus to many biomedical problems and tissue engineering applications. Examples are wound healing (Ehrlich 1988), tissue regeneration (Madri and Pratt 1986), and bioartificial skin and organs (Weinberg and Bell 1986; Yannas et al. 1982). A key phenomenon in these interactions is the traction exerted by cells on local collagen fibers that typically constitute the solid network of these tissues and gels and impart gross mechanical integrity. Two important consequences of cells exerting traction on such collagen networks are first, when the cells coordinate their traction, resulting in cell migration, and second, when their traction is sufficient to deform the network. Such cell-collagen network interactions are coupled in a number of ways. Network deformation, for example, can result in orientation of collagen fibers, inducing cell contact guidance, wherein cells orient and migrate with bi-directional bias along the direction of fibril orientation. This may govern cell accumulation in a wound and

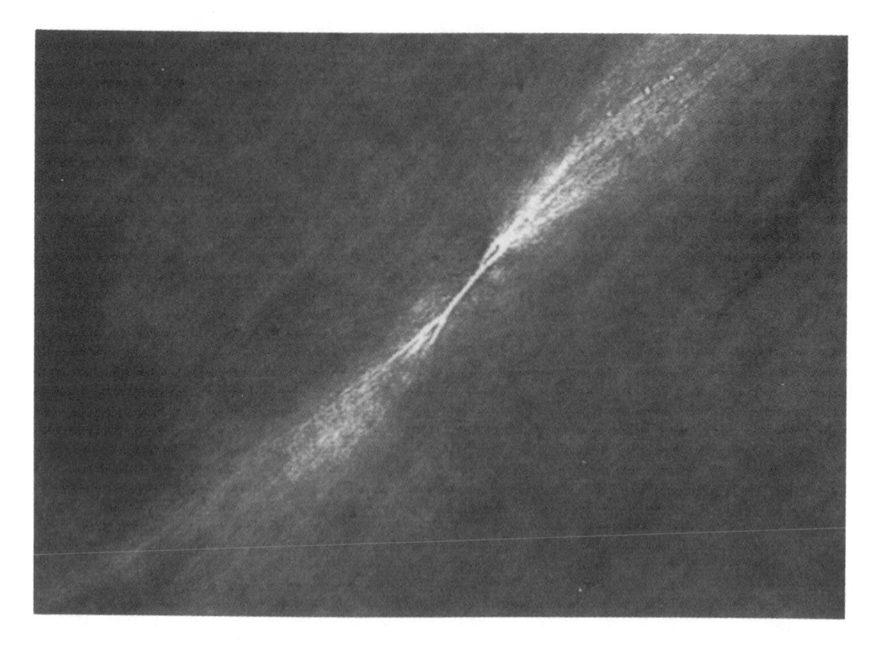

Figure 1. Polarized light micrograph of a fibroblast in collagen gel. Bright regions extending outward from both ends of the cell indicate fibril orientation due to "tractional structuring." Reproduced from Moon and Tranquillo (1993) by permission of the American Institute of Chemical Engineers. © 1993 AIChE - all rights reserved.

thereby its rate and extent of contraction, and be exploited to control the organization of bioartificial tissues and organs and thereby their functionality. We summarize here a biphasic isotropic theory which we have used to model traction-driven compaction of cell-populated collagen gel microspheres as *in vitro* assays of fibroblast traction and dermal wound contraction. An anisotropic extension of the theory that accounts for cell contact guidance is subsequently presented. Further, we make comparisons between fibroblast-populated collagen gel and cartilage, and between our biphasic theory for the former with that developed by Mow, Lai, and coworkers for the latter. First, we introduce the nature and consequences of cell traction through a description of *in vitro* assays used to measure it (see Tranquillo et al. 1992 for a review of the role of cell traction in dermal wound contraction), and present properties of collagen gel relevant for our theory.

Cell-Populated Collagen Gel Assays of Cell Traction

Harris and co-workers first documented the extensive reorganization of collagen fibrils associated with the extension and retraction of pseudopods by cells dispersed in a type I collagen gel (e.g. Stopak and Harris 1982), clearly evident in the polarized light micrograph of Figure 1. The macroscopic manifestation of this "tractional structuring" was documented earlier by Bell et al. (1979), who conducted a study based on their seminal fibroblast-populated collagen lattice (FPCL) assay. They exploited the ability to assess cell behavior in the physiologically-relevant environment of a type I collagen gel, made by restoring a dilute aqueous solution of collagen to physiological conditions in order to initiate collagen fibrillogenesis. They observed that fibroblasts cultured in a small floating disk of collagen gel (made by pouring cold neutralized cell-containing collagen solution into a well of a tissue culture dish and then warming to 37 C) can dramatically compact the gel, displacing tissue culture medium from the gel while organizing and concentrating the collagen fibrils - a cell-induced syneresis of the gel.

The FPCL assay and its variants quickly became established as the primary means of investigating cell traction (reviewed in Tranquillo et al. 1992). Grinnell and co-workers conducted a series of investigations to elucidate the mechanism of fibril reorganization and gel compaction (Grinnell and Lamke 1984; Guidry and Grinnell 1986). Several important observations and conclusions were made: fibrils in the gel interior are rearranged even when the cells reside only at the surface, and disrupting the network connectivity inhibits compaction, both implying the transmission of traction force through a connected fibrillar network of collagen; only 5% of the collagen is degraded even though the gel volume may decrease by 85% or more, implying compaction involves primarily a rearrangement of existing collagen fibrils rather than their degradation and replacement; few covalent modifications of the collagen occur; a partial reexpansion of compacted gels occurs after treatment of the cells with cytochalasin D or removal of the cells with detergent; and, cell-free gel compacted under centrifugal force exhibits a partial reexpansion similar to fibroblast-compacted gel. Based on these observations, a two-step mechanism was hypothesized for the mechanical stabilization of collagen fibrils during gel compaction: cells pull collagen fibrils into proximity via traction-exerting pseudopods, and over a longer time scale, the fibrils become noncovalently crosslinked, independent of cell-secreted factors (Guidry and Grinnell 1986).

We have devised a protocol for a spherical version of the FPCL assay based on pipetting cold cell-containing collagen solution into warm (immiscible) silicone fluid (Moon and Tranquillo 1993). The spherical geometry has several key advantages over the traditional disk geometry (e.g. affording the assumption of spherical symmetry in modeling equations) for only a modest increase in experimental effort. We have measured the change in diameter of fibroblast-populated collagen microspheres (FPCM) with time (Figure 2), confirming the increase in the rate of compaction with increasing cell concentration reported by Bell et al. (1979) and finding no dependence on the initial FPCM diameter (see later).

Despite the significant understanding which has been obtained

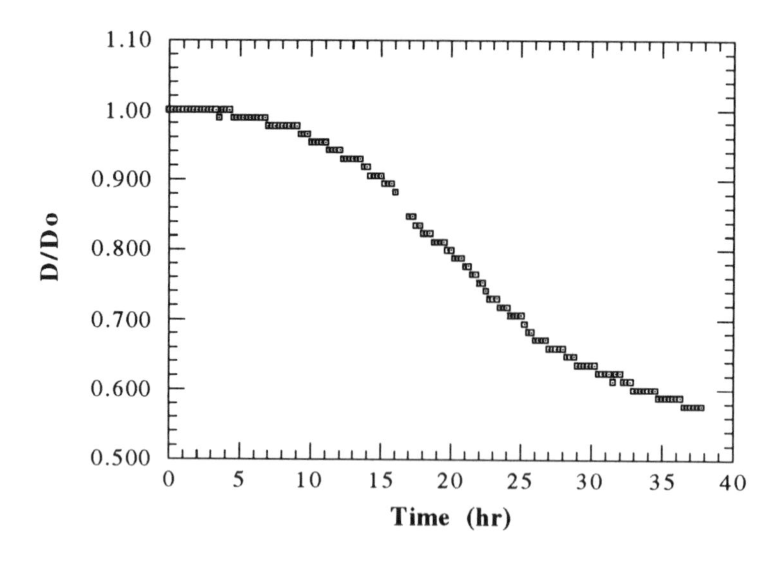

Figure 2. Compaction in the FPCM Traction Assay. Reproduced from Moon and Tranquillo (1993) by permission of the American Institute of Chemical Engineers. © 1993 AIChE - all rights reserved.

from the FPCL assay about the mechanical interactions between cells and the ECM, there is a fundamental limitation in characterizing compaction simply in terms of rate or extent of diameter reduction. Both of these measures of compaction are dependent on assay properties which complicate the interpretation of the traction being exerted by the cells, such as the initial cell and collagen concentrations (the viscoelastic properties of the gel, for example, are highly dependent on the collagen concentration, so the measured traction-driven compaction is also directly dependent on the collagen concentration). Nor do they provide a basis for predicting the compaction that will occur in other geometries relevant for tissue engineering applications, such as smooth muscle cell-populated tubes that serve as the basis for bioartificial arteries (Weinberg and Bell 1986). In order to improve this unsatisfactory situation, we applied the Oster-Murray monophasic theory of cell-ECM mechanical interactions (Oster et al. 1983) to model our FPCM assay and determine a value for the cell traction parameter that appears in the mechanical force balance for the cell/gel composite (Barocas et al. 1993). Monophasic refers to the fact that collagen gel, which is intrinsically a biphasic material composed of a collagen network and interstitial solution, is treated as a single homogeneous phase. As will be seen, this monophasic theory is related to a limiting case of our biphasic theory. First, relevant information about type I collagen gel is presented.

Structure and Properties of Type I Collagen Gel

Type I collagen gel consists of a hydrated network of native collagen fibrils. Self-assembly of collagen monomers into native fibrils is induced by raising the ionic strength, pH, and temperature of cold acid-stabilized collagen solution to physiological levels. Gel prepared from commercially available sources (e.g. Vitrogen 100, Celtrix Labs) usually contains 0.1-0.2 wt % of type I collagen initially present in a random network of long entangled fibrils ranging from 90 to 400 nm in diameter and having no apparent free ends, with segments of fibrils appearing quite linear between points of apparent entanglement (Allen et al. 1984). Fibril-fibril associations are mainly non-covalent in nature (Guidry & Grinnell 1987).

The mechanical properties of collagen gel are a consequence of the organization of the collagen network and the interaction of the fibrils with the surrounding liquid. The collagen network entraps the liquid and keeps it from flowing away, while the liquid prevents the network from collapsing. The mechanical role of the network is to withstand tensile stresses and to recover from finite deformations of shape when applied stresses are removed. The contribution of the fibrils is strength and rigidity, the fibrils being much weaker in torsion and flexure than in tension. The combination of the biphasic nature of the gel (collagen network with entrapped liquid) with the complex microstructure of the collagen network determine the effective rheological and mechanical properties of the gel, which are very dependent on the time scale of the applied load (Moon 1992).

We have measured the small shear strain viscoelastic properties (dynamic and creep) of type I collagen gel using the Rheometrics, Inc. Fluids Spectrometer (RFSII) and Dynamic Stress Rheometer at a single collagen concentration. At long times, the gel exhibits sustained creep characteristic of a linear viscoelastic liquid (for sufficiently small shear strain) with a single relaxation time of approximately 13 hr. The shear behavior can thus be satisfactorily described by the Maxwell fluid constitutive equation (Barocas et al. 1993):

$$\frac{1}{2G}\frac{D\boldsymbol{\sigma}_n}{Dt} + \frac{1}{2\mu}\boldsymbol{\sigma}_n = \frac{1}{2}\left[\nabla\mathbf{v}_n + (\nabla\mathbf{v}_n)^T\right] - \frac{1}{3}(\nabla\cdot\mathbf{v}_n)\mathbf{I} \qquad (1)$$

where $\boldsymbol{\sigma}_n$ is the Cauchy stress tensor, $\mathbf{v}_n$ is the velocity, G is the shear modulus, and μ the shear viscosity (all with reference to the collagen network), and $\mathbf{I}$ is the unit tensor. $D(\cdot)/Dt$ denotes the substantial time derivative with respect to the network velocity, $\mathbf{v}_n$, throughout this paper. An improved fit of the short time response can be made with an additional viscous term; however, the short time response (10 min) is negligible on the long time course of compaction in the FPCM assay (10 hr) and we thus forego it for the purpose of modeling the FPCM assay. We have not yet performed compression studies, so we assume that the relaxation times in compression and shear are equal.

Therefore, we use the Maxwell constitutive equation for both shear and bulk deformation with a prescribed value for the network Poisson's ratio υ (with $\upsilon << 0.5$, reflecting the large compressibility of a highly porous network of entangled fibrils):

$$\frac{1}{2G}\frac{D\boldsymbol{\sigma}_n}{Dt} + \frac{1}{2\mu}\boldsymbol{\sigma}_n = \frac{1}{2}\left[\nabla \mathbf{v}_n + (\nabla \mathbf{v}_n)^T\right] + \frac{\upsilon}{1-2\upsilon}(\nabla \cdot \mathbf{v}_n)\mathbf{I} \quad (2)$$

We also have not yet performed permeability studies on collagen gel, but given the great similarity of rheological properties with those reported for fibrin gel (Ferry 1988), we assume that the collagen gel permeability will be comparable to that of fibrin gel with the same network concentration (van Gelder 1993).

Biphasic Theory of Cell-Populated Biological Gel

There is considerable literature concerning the flow of fluid through connective tissue as it is deformed, largely due to investigations of articular cartilage mechanics. Mow, Lai, and co-workers (Mow et al. 1980 1986; Kwan et al. 1990) and subsequently others (Farquhar et al. 1990; Mak 1986) have developed a class of theories for describing cartilage mechanics based on the theory of mixtures (Craine et al. 1970), the seminal theory being known as the KLM theory. These theories account (at a minumum) for conservation of mass and momentum for both collagen network and interstitial fluid phases, including momentum transfer between the two phases that occurs due to relative motion. A central interest is to predict the stress field in cartilage due to applied loads. There is no account for cell traction due to chondrocytes that reside in cartilage. Thus, these theories cannot be used directly to model the FPCM assay.

One approach would be to add a cell traction stress in the mechanical force balance for the network in the same way a swelling stress was added in the triphasic mechano-electrochemical theory proposed by Mow, Lai, and co-workers (Lai et al. 1991). Rather, we adopt the biphasic theory developed by Dembo and Harlow (1986) for cytomechanics based on the multi-phase averaging theory of Drew and Segel (1971), since a contractive stress is inherent in that theory. Averaging theory involves averaging variable values by two spatial and two time integrations, converting the locally discrete phase variables into continuous average variables; thus, every point in a multi-phase mixture must contain only one phase, but all phases coexist continuously on average. The model of Dembo and Harlow accounts for contractive forces due to actomyosin complexes distributed within an actin filament network and their activation of the sliding filament mechanism. This is analogous to the contractive forces generated by cells distributed within a collagen fibril network and their tractional structuring activity. The effect of the finite size of the cells on the mechanics of the collagen gel has been shown to be negligible at FPCM cell concentrations (Jain et al. 1988), so cells are treated as a component of the network phase rather than as a separate phase. A summary of the biphasic theory developed by Dembo and Harlow (1986) follows.

Defining the average fractional volumes of network and solution as θ_n and θ_n, the excluded volume relation requires:

$$\theta_n + \theta_s = 1 \tag{3}$$

Defining the average velocities of network and solution as $\mathbf{v}_n$ and $\mathbf{v}_s$, mass conservation can be expressed in terms of volume fractions when the intrinsic phase densities are constant and nearly equal (a good approximation for protein networks and biological solutions, as Dembo and Harlow note):

$$\frac{D\theta_n}{Dt} = -\theta_n \nabla \cdot \mathbf{v}_n + R_n \tag{4}$$

$$\frac{D\theta_s}{Dt} = -\theta_s \nabla \cdot \mathbf{v}_s - R_n \tag{5}$$

where R_n is the net rate at which volume is transferred from solution to network phase due to all reactions; $R_n = 0$ is appropriate for the FPCM assay since fibrillogenesis is complete before the cells begin to exert traction and there is negligible synthesis or degradation of collagen due to the cells during the assay (Nusgens et al. 1984). The assumption of equal intrinsic phase densities is unnecessary for writing (4) & (5) in terms of volume fractions when $R_n = 0$. Their sum yields the overall incompressibility relation:

$$\nabla \cdot (\theta_n \mathbf{v}_n + \theta_s \mathbf{v}_s) = 0 \tag{6}$$

Of course, only two of the three forms of mass conservation equations are independent.

The key result of the averaging theory of Drew and Segel as adapted by Dembo and Harlow, after omitting inertial terms which are negligible for cell/ECM mechanics (Odell et al. 1981), is the following set of coupled mechanical force balances:

$$\nabla \cdot (\theta_n \boldsymbol{\sigma}_n{}^*) + P\nabla\theta_n + \varphi(\mathbf{v}_s - \mathbf{v}_n) + \rho_n\theta_n\mathbf{f} = \mathbf{0} \tag{7}$$

$$\nabla \cdot (\theta_s \boldsymbol{\sigma}_s{}^*) + P\nabla\theta_s + \varphi(\mathbf{v}_n - \mathbf{v}_s) + \rho_s\theta_s\mathbf{f} = \mathbf{0} \tag{8}$$

where

$$\boldsymbol{\sigma}_n{}^* = \boldsymbol{\sigma}_n - P_n\mathbf{I}, \quad \boldsymbol{\sigma}_s{}^* = \boldsymbol{\sigma}_s - P_s\mathbf{I}$$

and ρ_n and ρ_s are the intrinsic phase densities, and $\boldsymbol{\sigma}_s$ is the stress tensor for the solution phase. The terms involving P and φ represent interphase forces. P is the "interphase pressure" existing at the hypothetical interface between the interspersed phases, and is generally not equal to the intraphase pressures, P_n and P_s. P_n is associated with the traction-induced tension in the network, and P_s is associated with hydrostatic pressure of the solution (a component of P_n associated with a reaction force due to dilation of the network phase is retained in the form of $\boldsymbol{\sigma}_n{}^*$). $\varphi(\bullet)$ is an interphase friction coefficient, which in general depends on some local measure of porosity of the network. Using (3), note that the total interphase forces defined by (7) & (8) are equal in magnitude and of opposite sign, as required. Defining

$$\Psi = P_s - P_n \equiv \text{traction stress}$$
$$\Gamma = P_s - P \equiv \text{solvation stress}$$
$$P_f = P_s + \Gamma \ln \theta_s \equiv \text{effective pressure}$$

then (7) & (8) can be written

$$\nabla \cdot (\theta_n \boldsymbol{\sigma}_n) - \theta_n \nabla P_f + \Gamma \nabla \ln \theta_s + \nabla(\theta_n \Psi) + \varphi(\mathbf{v}_s - \mathbf{v}_n) + \rho_n \theta_n \mathbf{f} = \mathbf{0} \tag{9}$$

$$\nabla \cdot (\theta_s \boldsymbol{\sigma}_s) + \theta_s \nabla P_f + \varphi(\mathbf{v}_n - \mathbf{v}_s) + \rho_s \theta_s \mathbf{f} = \mathbf{0} \tag{10}$$

by assuming that Γ is a constant independent of θ_n (justified by Dembo and Harlow on the basis that solvation forces are of nonspecific nature) but that $\Psi(\bullet)$ and $\varphi(\bullet)$ may depend on other variables.

We now make assumptions to simplify (9) & (10) in applying the biphasic theory to the FPCM assay. Since collagen gel microspheres (without cells) are not observed to spontaneously swell or contract, we take $\Gamma = 0$, so that $P_f = P_s$ (= P). Since the FPCM is almost neutrally buoyant, we take $\mathbf{f} = \mathbf{0}$. Given the relatively small viscosity of the solution phase (tissue culture medium), we take $\boldsymbol{\sigma}_s = \mathbf{0}$. We take $\boldsymbol{\sigma}_n$ to be given by (2). In the absence of guiding data, we take $\varphi(\bullet) = \varphi\theta_n\theta_s$ as proposed by Dembo and Harlow, which makes the friction coefficient take on a maximum value when both phases are present in equal amounts and a value of zero when either phase is absent. (9) & (10) thus become:

$$\nabla \cdot (\theta_n \boldsymbol{\sigma}_n) - \theta_n \nabla P + \nabla(\theta_n \Psi) + \varphi\theta_n\theta_s(\mathbf{v}_s - \mathbf{v}_n) = \mathbf{0} \tag{11}$$

$$\theta_s \nabla P + \varphi\theta_n\theta_s(\mathbf{v}_n - \mathbf{v}_s) = \mathbf{0} \tag{12}$$

It remains to define a form for the traction stress, $\Psi(\bullet)$, which is a contractive stress assumed to act isotropically and is thus equivalent to a negative pressure. The form proposed by Oster, Murray, and co-workers for the traction stress in their monophasic theory is bilinear in cell and gel concentrations. Since our biphasic formulation inherently accounts for the network concentration (note $\Psi(\bullet)$ is multiplied by θ_n in (11)), we will use the equivalent form:

$$\Psi = \tau_0 n \tag{13}$$

where n is the cell concentration and τ_0 is the cell traction parameter, a measure of the intrinsic traction capacity of the cells. It may be necessary to account for inhibition of traction at high cell and collagen concentrations by modifying (13), for example (Oster et al. 1983):

$$\Psi = \frac{\tau_0 n}{(1 + \lambda n^{\lambda'})(1 + \beta \theta_n^{\beta'})} \tag{14}$$

although without guiding data, we take $\lambda = \beta = 0$ (see later). In the FPCM (and FPCL) assay, the cells are initially spherical and nonmotile. They develop a polar morphology and begin to exhibit motility indicative of traction during a "lag period" that precedes and overlaps initial FPCM compaction. We have measured the longest cell dimension, D(t), relative to the initial cell diameter, D_0, to develop a cell spreading curve (Moon and Tranquillo 1993), which we fit with the following function

$$\frac{D - D_0}{D_0} = \frac{t^\gamma}{t^\gamma + t_{1/2}^\gamma} \tag{15}$$

where $t_{1/2}$ is the half-time for spreading and γ is a constant that depends on n_0 (possibly by determining the accumulation rate of autocrine growth factors that stimulate cell spreading). We assume that τ_0 is time-dependent being proportional to the right-hand side of (15) (Barocas et al. 1993).

Equation (13) requires that a cell conservation equation defining n be included in the theory. Fibroblasts exhibit limited migration through the collagen network and limited cell division during the FPCM assay under typical conditions. Consistent with the isotropic assumption implicit in the biphasic theory presented thus far, we assume the cells migrate randomly, which on a continuum scale can be modeled as Fickian diffusion (Gail and Boone 1970). We have found that exponential growth is sufficient to model cell division (Barocas et al. 1993). The cell conservation equation is thus

$$\frac{Dn}{Dt} = \mathcal{D}_0 \nabla^2 n + kn \tag{16}$$

where $\mathcal{D}_0$ is the random migration coefficient (analogous to a molecular diffusivity), and k is the exponential growth constant. Since the cells crawl through the network, the network velocity implicit in the substantial time derivative is appropriate for their convective flux.

Comparison to the Oster-Murray Theory and KLM Theory

Equations (11) & (12) are functionally very similar to the equations of the KLM theory (disregarding the traction stress), with the exception that there is an explicit weighting of the intrinsic phase stress tensors by the local volume fractions of the phases, that is, a phase carries a fraction of the total load equal to its volume fraction; less precisely, the moduli have a linear dependence on volume fraction. (Note that other applications of mixture theory (e.g. Johnson et al. 1991) have retained such explicit weightings.) The traction stress appearing in our network force balance is a contractive stress on the network and is functionally analogous to the swelling stress added to the KLM network force balance in the triphasic theory (Lai et al. 1991). The analogy further extends by the need for a cell conservation equation to define the traction stress and the need for ion conservation equations to define the swelling stress.

We can reduce our biphasic theory to a (pseudo-)monophasic theory by setting $\varphi = 0$ in (11) & (12), making (12) trivial (there is still solution flow in general, but since there is no frictional loss there is no pressure gradient); (11) so modified is similar to the Oster-Murray theory, with the exceptions that the stress tensor is well-defined, being that of the network phase and not the ill-defined stress tensor of a gel phase, as well as that noted above in comparison to the KLM theory.

Biphasic Theory Predictions for the FPCM Traction Assay

The collagen fibril orientation appears quite isotropic initially in the FPCM. Thus, we assume that (2), (4), (6), (11), (12), (13), & (16) apply initially. Because of the approximating spherical symmetry of the assay, only the radial components of $\mathbf{v}_n$ and $\mathbf{v}_s$, v_n and v_s, need to be considered (omitting the subscript "r" for notational simplicity). The equation set is reduced by substituting $\theta_s = 1 - \theta_n$ from (3) (the subscript n for θ_n is omitted hereafter), and (11) & (12) are combined by eliminating P. Finally, we make the observation that (6) is equivalent to

$$r^2\left[\theta v_n + (1-\theta) v_s\right] = f(t) \tag{17}$$

Since the value of the left hand side of the equation is zero at r = 0 for all t, we can set the quantity inside the brackets equal to zero and solve for v_s and then substitute for v_s in the combined mechanical force balance in terms of v_n (the subscript n for v_n is omitted hereafter). We are then left with three balance equations, namely combined mechanical force, network, and cell, and two auxiliary constitutive equations for the radial and transverse network stresses. Using the following scalings:

$$\tilde{r} = r / R_0 \qquad \tilde{n} = n / n_0$$
$$\tilde{t} = t / (\mu / G) \qquad \tilde{\theta} = \theta / \theta_0$$
$$\tilde{v} = v / (R_0 G / \mu) \qquad \tilde{\sigma}_{ii} = \sigma_{ii} / 2G$$

where R_0 is the initial FPCM radius, n_0 and θ_0 are the initial (uniform) cell concentration and network volume fraction, and dimensionless parameters

$$\tilde{\tau}_0 = \tau_0 n_0 / 2G \qquad \tilde{\varphi} = \varphi R_0^2 / 2\mu$$
$$\tilde{\mathcal{D}}_0 = \mathcal{D}_0 \mu_0 / R_0^2 G \qquad \tilde{k} = k\mu / G$$

these equations in dimensionless form are (tildes omitted):

$$\frac{\partial}{\partial r}(\theta \sigma_{rr}) + \frac{2\theta}{r}(\sigma_{rr} - \sigma_{\theta\theta}) + \tau_0 \frac{\partial}{\partial r}(n\theta) - \frac{\varphi \theta v}{1 - \theta_0 \theta} = 0 \qquad (18)$$

$$\frac{D\theta}{Dt} + \theta\left(\frac{\partial v}{\partial r} + \frac{2v}{r}\right) = 0 \qquad (19)$$

$$\frac{Dn}{Dt} + n\left(\frac{\partial v}{\partial r} + \frac{2v}{r}\right) - \mathcal{D}_0\left(\frac{\partial^2 n}{\partial r^2} + \frac{2}{r}\frac{\partial n}{\partial r}\right) - kn = 0 \qquad (20)$$

$$\frac{D\sigma_{rr}}{Dt} + \sigma_{rr} - \left(\frac{1-\upsilon}{1-2\upsilon}\right)\frac{\partial v}{\partial r} - \left(\frac{2\upsilon}{1-2\upsilon}\right)\frac{v}{r} = 0 \qquad (21)$$

$$\frac{D\sigma_{\theta\theta}}{Dt} + \sigma_{\theta\theta} - \left(\frac{\upsilon}{1-2\upsilon}\right)\frac{\partial v}{\partial r} - \left(\frac{1}{1-2\upsilon}\right)\frac{v}{r} = 0 \qquad (22)$$

It is necessary to prescribe initial and boundary conditions on the variables θ, v, and n. Symmetry at the origin requires that there be zero velocity of network (and solution), the stresses be isotropic, and there be zero cell flux. At the outer (moving) boundary, we prescribe zero net radial stress at the boundary since there is negligible interfacial tension between the network and the surrounding medium, and zero cell flux since cells are not observed to leave the FPCM. We assume that the pressure, P, at the FPCM surface equals the hydrostatic pressure surrounding the FPCM, the same assumption typically applied in modeling unconfined compression of articular cartilage (e.g., Kenyon 1979). As indicated above the initial conditions for n and θ are (spatially uniform) constants. The initial condition for v is generally obtained by evaluating the combined mechanical force balance (18) at $t = 0$. However, our chosen form for the time-dependence of τ_0 (15) imposes $v = 0$ initially.

The values of the dimensionless parameters used in the computed model predictions are given in Table 1. The values are typical for the assay (Barocas et al. 1993). From examination of the forms of the parameters, we can distinguish between changes in the conditions of the assay (e.g. making a larger FPCM, R_0, or using a higher collagen concentration, θ_0) and changes in the intrinsic cell properties (traction, τ_0, migration, $\mathcal{D}_0$, or division, k) that might be modulated

Table 1. Dimensionless Parameter Values

Parameter	Value
τ_0	0.100
φ	0.002
$\mathcal{D}_0$	0.005
k	0.380

with exogenous chemical factors. Noting the small value of the dimensionless interphase friction, $\tilde{\varphi}$, for values of R_0 we have used to date, we could use a (pseudo-)monophasic approximation (i.e. $\tilde{\varphi} = 0$), in which case the system of equations and boundary conditions admit a "homogeneous solution" which is of the form

$$\begin{array}{ll} n(r,t) = n(t) & u(r,t) = h(t) \cdot r \\ \theta(r,t) = \theta(t) & \sigma(r,t) = \sigma(t)\mathbf{I} \end{array} \tag{23}$$

where u is the radial displacement of the network. This solution is obtained by numerical integration of a system of ordinary differential equations, similar to the model based on the monophasic theory of Oster and Murray (Moon and Tranquillo 1993; Barocas et al. 1993). However, we present computed results using the standard Galerkin finite element method with piecewise quadratic basis functions (100 elements), solving the resulting differential-algebraic system with the subroutine SDASRT (Brenan 1989).

In Figure 3 we show the predicted profiles of u, θ, n, σ_{rr}, and $\sigma_{\theta\theta}$ for the typical FPCM Traction Assay parameter values based on an initial FPCM radius, $R_0 = 0.05$ cm, and for $R_0 = 1$ for comparison. The expected "homogeneous solution" is obtained for $R_0 = 0.05$ cm, for which $\tilde{\varphi} = 0.002$ and interphase drag is negligible, but a spatially inhomogeneous solution is obtained for $R_0 = 1$ cm, for which $\tilde{\varphi} = 0.8$ and interphase drag is significant. In the latter case, cells and network are predicted to accumulate near the inward moving FPCM surface where displacement is enhanced. Also, the stress is dissipated inward from the FPCM surface by interphase drag. In Figure 4 we show both the biphasic and (Oster-Murray) monophasic model predictions of the time-course of compaction along with representative data from an FPCM Traction Assay. In comparing the biphasic and monophasic predictions, the biphasic prediction yields an acceptable fit for a longer time, because the θ_n weighting factor effectively makes the network stiffer and more viscous as it compacts and thereby able to balance the increasing traction stress. In comparing the predictions and the data, both predictions fit the data well over an initial period, but then predict an increasing rate of compaction in contradiction to the data, because no stress term increases with compaction as much as the traction stress owing to its bilinear dependence (effectively) on cell and network concentrations.

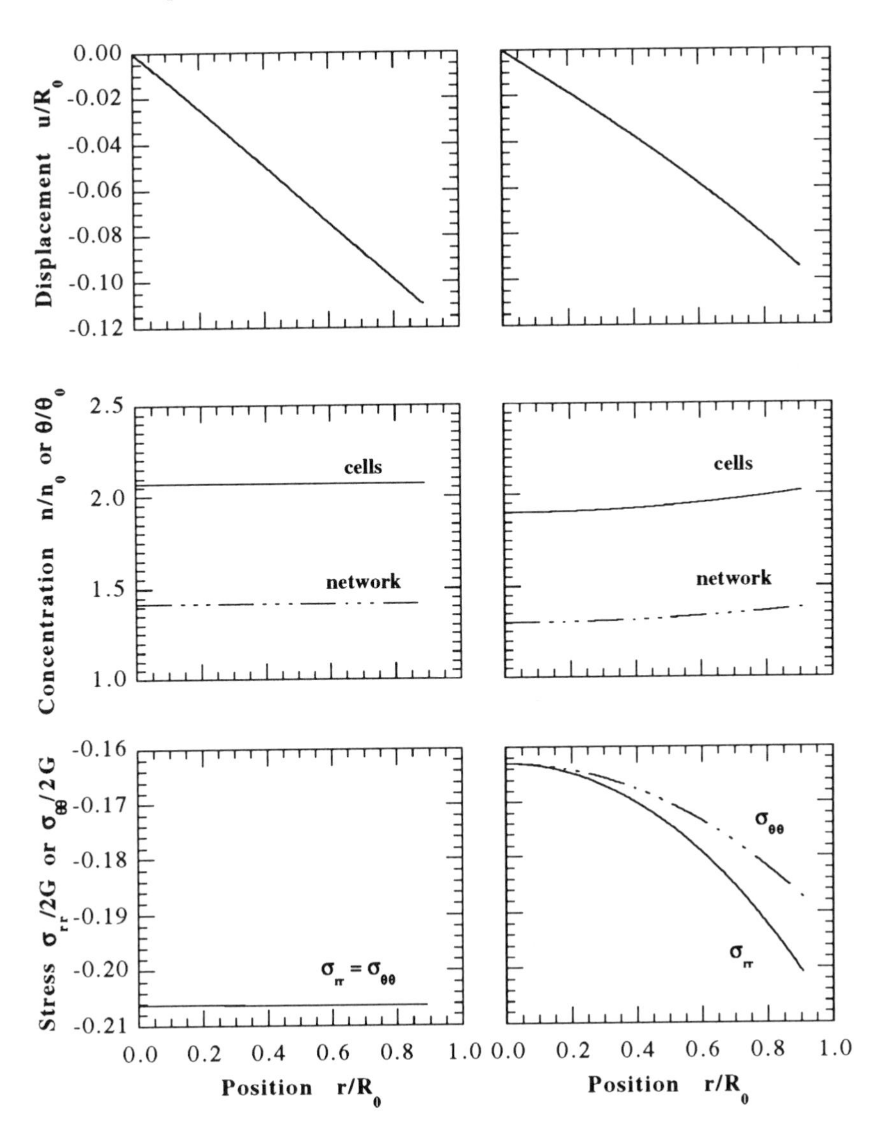

Figure 3. Predicted displacement, cell concentration, network concentration, and radial and transverse viscoelastic stress profiles for the FPCM Traction Assay at dimensionless time 1.0 (~13 hr). The graphs on the left are for our typical assay with $R_0 = 0.05$ cm; those on the left are for $R_0 = 1.0$ cm.

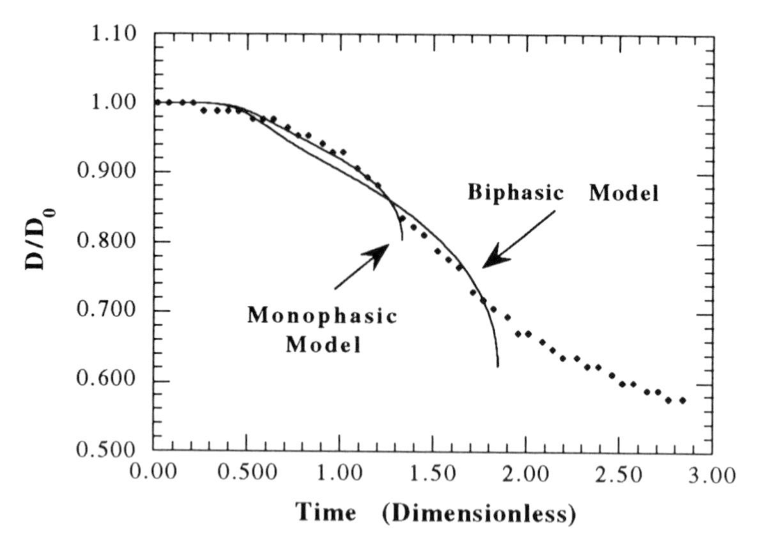

Figure 4. Predicted and experimental compaction for the FPCM Traction Assay

If a form for Ψ like (14) were used that includes inhibition of traction at high cell and/or network concentrations, then the qualitative discrepancy apparent in Figure 4 at longer times could be resolved. Before considering an anisotropic extension of our biphasic theory, we make a comparison of collagen gel and cartilage mechanics based on the isotropic biphasic theory.

The comparison is based on calculating values of the dimensionless interphase friction, $\tilde{\varphi}$, for both materials. As stated above, the value is 0.002 for our typical initial FPCM radius, R_0 = 0.05 cm and, therefore, the consequences of interstitial flow are negligible (our finding that the compaction profiles are independent of R_0 in a range 0.045 - 0.055 is thus consistent with this). The form of $\tilde{\varphi}$ appropriate for typical rheological tests performed on cartilage is given by

$$\tilde{\varphi} = \frac{\varphi LV}{M} \tag{24}$$

where L and V are a characteristic length and velocity, and M is a characteristic modulus. If we use reported data for an oscillatory compression test on cartilage (Mow et al. 1986), we calculate $\tilde{\varphi}$ = 0.2, so we are not surprised that interstitial flow plays a role in such a test.

Anisotropic Biphasic Theory: Contact Guidance

Contact guidance refers to the biased orientation (and possibly biased migration) of cells in or on a substratum with structural or textural anisotropy. The relevant example is the bi-directional orientation (Figure 5) and migration (Figure 6) of fibroblasts along the axis of fibril orientation in oriented collagen gel (Guido and Tranquillo 1993 and Dickinson et al. 1994 - the fibril orientation was induced by a magnetic field during fibrillogenesis in these studies). Although we have not observed any contact guidance in our typical FPCM Traction Assay (in fact, Figure 3 predicts that it should not be observed except for much larger values of R_0), we have observed it in our FPCM Wound Assay. In this assay, the core of the FPCM is initially devoid of cells, representing the absence of cells in the fibrin clot that initially fills the wound space (Tranquillo et al. 1992; Bromberek and Tranquillo 1994). An important consequence of this initial nonuniform cell distribution is that compaction is inhomogeneous for all values of R_0 (Figure 7). We infer from the videomicrograph of the FPCM Wound Assay in Figure 8 that the FPCM compaction is inducing circumferential orientation of fibrils that is manifested by cell contact guidance. Assuming that a polarized cell exerts traction anisotropically, as supported by Figure 1, the consequences of anisotropic fibril orientation are anisotropic traction stress as well as anisotropic cell migration. There is an inherent mechanical/ biological feedback, then, in the FPCM Wound Assay (which we believe is ubiquitous in cell-tissue systems): cell traction drives compaction inducing anisotropic fibril orientation, which induces contact guidance, which induces anisotropic cell migration and orientation (thereby an inhomogeneous cell distribution and anisotropic traction stress), which modulates subsequent compaction.

In modeling this anisotropic composite of cells and fibrillar network, we follow three steps. First, we define a variable to characterize the orientation of the network. Second, we describe how the network orientation evolves with time (i.e. we write an equation of change for the orientation variable). Finally, we incorporate the orientation variable into the equations we developed for the isotropic biphasic theory, in particular, generalizing the random migration coefficient, $\mathcal{D}_0$, and the traction stress, Ψ, to variable tensors rather than scalar constants.

The development of orientation of collagen fibrils under stress has been observed and characterized qualitatively by several investigators (e.g., Grinnell and Lemke 1984, Klebe et al. 1989). Stress can develop from two sources: internally, from cell traction, and externally, from an applied load. Although the two sources of stress are fundamentally different, our initial treatment of traction-induced fibril orientation assumes that anisotropy developing from cell traction can be modeled as if it developed from macroscopic deformation of the network.

Before proceeding, we note that orientation of fibrils affects the mechanics of the collagen gel itself as well as the cell-gel composite via contact guidance. However, the effect on the network rheology will be small as long as the network strains are within the LVE limit (in contrast to articular cartilage, *cf.* Farquhar et al. 1990), since any effect of orientation would result in a deviation from linearity. Further, the effect on the gel permeability will be small for typical FPCM assay values of $\tilde{\varphi}$: Sangani and Yao (1988) used numerical

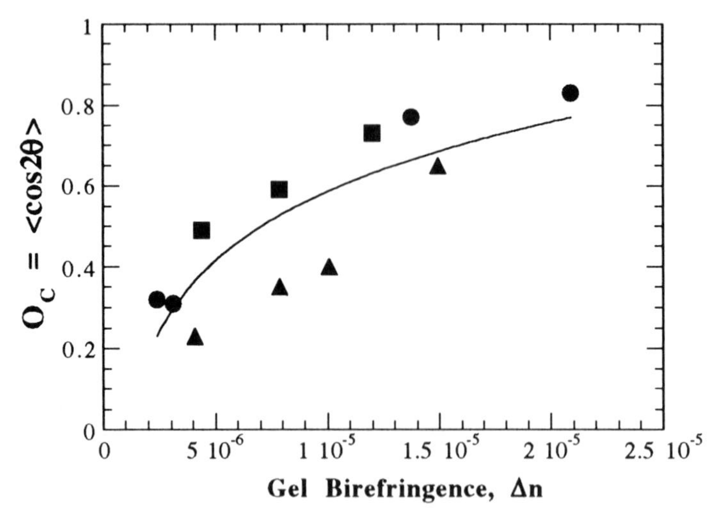

Figure 5. Cell orientation vs. birefringence for cells in magnetically oriented collagen gel. Reproduced from Guido and Tranquillo (1993) by permission of The Company of Biologists Ltd. © 1993 The Company of Biologists Ltd. - all rights reserved.

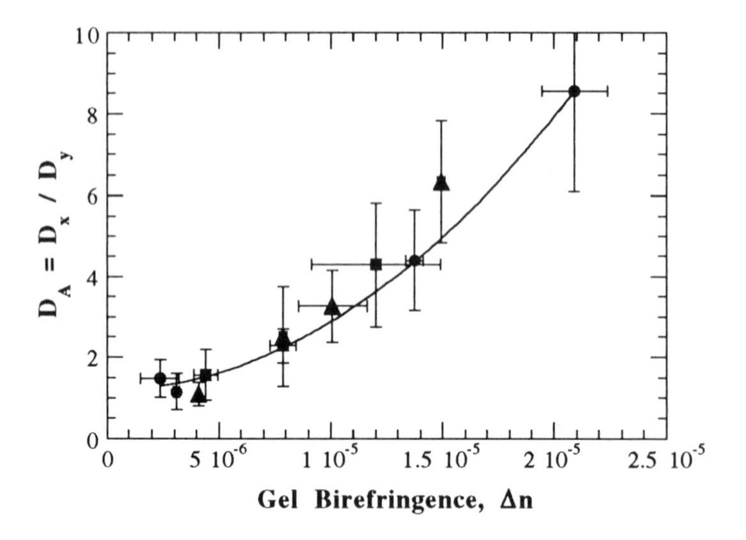

Figure 6. Anisotropic migration vs. birefringence for cells in magnetically oriented collagen gel. Reproduced from Dickinson et al. (1994) by permission of the Biomedical Engineering Society. © 1993 BMES - all rights reserved.

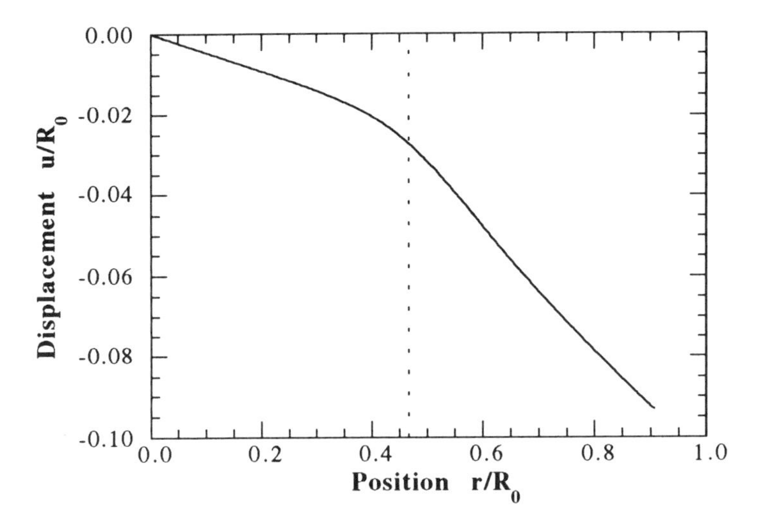

Figure 7. Predicted displacement profile in the FPCM Wound Assay at dimensionless time 1.0 (~13 hr). The dotted line marks the location of the boundary between the initially cell-free inner region, which undergoes relatively little compaction, and cell-rich outer region, which shows more compaction.

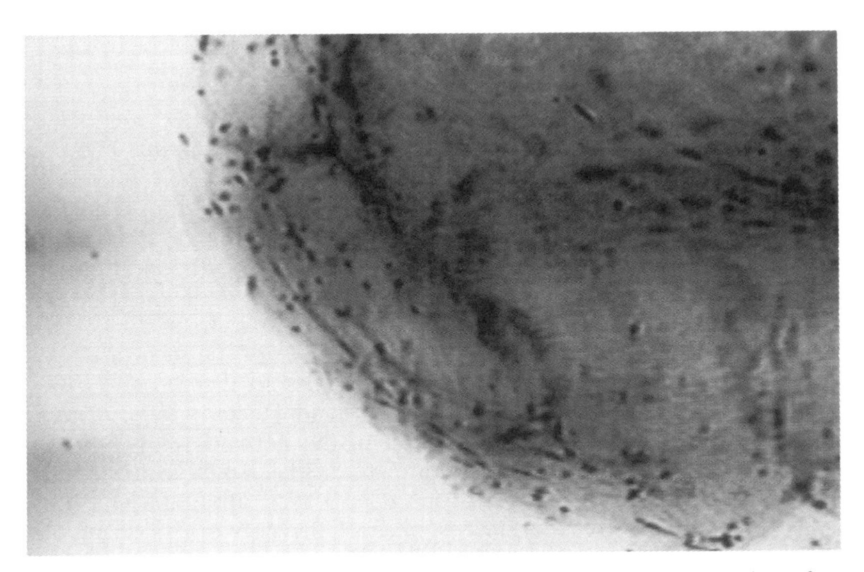

Figure 8. Videomicrograph of the FPCM Wound Assay. A cross-section of one quadrant of the FPCM is presented. Circumferential orientation of cells is observed in the outer region of the FPCM.

simulations to predict a four-fold change in permeability between transverse and longitudinal flow in a randomly arranged uniaxial array of cylinders; since our analysis has shown that $\tilde{\varphi}$ would have to increase by two orders of magnitude to become significant, anisotropic permeability effects are negligible for our typical assays.

Rather than attempt to model the detailed network microstructure, we introduce the concept of a model fiber as a representation of the actual fibrillar network. Mathematically, the fiber is a unit vector that captures the fibril orientation state at a point in the gel (the fiber exists at a point and occupies no volume). Since we are concerned only with fiber direction, fiber stretching is not considered. Further, the reorientation of a fiber is a function only of the local macroscopic deformation and does not depend on other fibers. The response of the cells to fibril orientation is thus modeled as a function of fiber orientation.

The orientation of a fiber with respect to a characteristic direction can be described completely by an angle α away from the characteristic direction and an angle ϑ around the characteristic direction. For the spherically symmetric FPCM assays, the orientation of a fiber at a point in the microsphere can be completely characterized by α, the angle between the fiber and the radius of the sphere. The continuum assumption made in our isotropic biphasic theory implies that there is a continuous distribution of fibers at every point in space, so we consider the distribution $P(\alpha,\vartheta)$ such that the probability of finding a fiber with orientation between (α,ϑ) and $(\alpha+d\alpha,\vartheta+d\vartheta)$ is given by

$$\frac{P(\alpha,\vartheta)\sin(\alpha)}{2\pi}d\alpha d\vartheta, \alpha \in \left[0,\frac{\pi}{2}\right], \vartheta \in [0,2\pi) \qquad (25)$$

Of course, for the spherically symmetric case, $P(\alpha,\vartheta)$ is just $P(\alpha)$ and the integral over ϑ removes the factor of 2π.

There are three classes of methods to define $P(\alpha,\vartheta)$. The most complex of these attempt to characterize P using balance laws for the orientation of individual fibers. The intermediate methods attempt to capture the basic nature of P in a representative tensor (typically second-order). The simplest methods postulate a form for P with parameters that depend on the macroscopic state of the network.

The most direct of the complex methods is to discretize P into a finite number of values, $P_{ij}(\alpha_i,\vartheta_j)$ (Advani and Tucker 1987). Since P must be defined at every spatial point in the network, however, this approach becomes unwieldy for more than just a few values of P_{ij}. Another method consists of a Monte-Carlo type approach (Farquhar et al. 1990). The stress-strain response of the network is defined by selecting random fiber orientations and modeling the response of fibers with those orientations. The sum of the calculated fiber responses, weighted by P, gives the macroscopic network response as well as the updated P. Although computationally-intensive, this method has the advantage that no assumptions about the form of P need be made *a priori*.

The class of methods of intermediate complexity consists of characterizing P with a tensor, then manipulating this tensor. The fabric tensor of Cowin (1985) is any tensor, which captures the anisotropy of the material, used to determine the necessary properties. P can also be approximated by a series of even-order orientation tensors (Advani and Tucker 1987). The higher-order tensors can be approximated in terms of a symmetric second-order tensor, allowing determination of material properties based on only five independent terms. Further, the changes in P are defined by those for an individual fiber. This method, like that of Farquhar et al. (1990), rests heavily on the assumption that the fibers reorient independently.

The final class of methods, postulating a form for P, has the advantage of simplicity but the disadvantage of being unable to incorporate single-fiber mechanics. One possibility for transversely isotropic problems is the bimodal form of the Von Mises distribution (Tranquillo et al. 1992). Postulating that the Von Mises concentration parameter is proportional to the magnitude of the strain then gives a straightforward way to account for the orientation of the fibers.

A more general method, which we use here for an anisotropic extension of our biphasic theory, is to define P based on the diagonalized Finger deformation tensor, $\mathbf{B} = (\nabla \mathbf{u})(\nabla \mathbf{u}^T)$ and its associated ellipsoid (i.e. the ellipsoid whose axes have lengths equal to the eigenvalues of $\mathbf{B}$ and directions equal to the eigenvectors of $\mathbf{B}$). We define the probability of finding a fiber at an orientation (α,ϑ) to be proportional to the differential surface area of the ellipsoid at that position. Mathematically, this is expressed by

$$P(\alpha, \vartheta) = \frac{4\pi r^2(\alpha, \vartheta)}{A} \tag{26}$$

where $r(\alpha,\vartheta)$ is the radius of the ellipsoid at angles α and ϑ from the axes of the ellipsoid, and A is the total surface area of the ellipsoid. For isotropic strain, r is a constant, and $P(\alpha,\vartheta)$ is identically 1. The use of a probability distribution based on the deformation ellipsoid has two important advantages. First, there is no introduction of new parameters, since the orientation is determined by the macroscopic deformation. Second, unlike the Von Mises distribution approach, the method readily extends to problems without transverse isotropy. We therefore choose to use the deformation-based distribution as a preliminary model of fiber orientation.

Rather than use P directly in the equations, it is convenient to introduce a fiber orientation tensor, $\mathbf{\Omega}_f$, defined by

$$\Omega_f = 3\int_0^{\frac{\pi}{2}}\int_0^{2\pi} \mathbf{n}(\alpha, \vartheta) \otimes \mathbf{n}(\alpha, \vartheta) \frac{P(\alpha, \vartheta)\sin(\alpha)}{2\pi} d\vartheta d\alpha \tag{27}$$

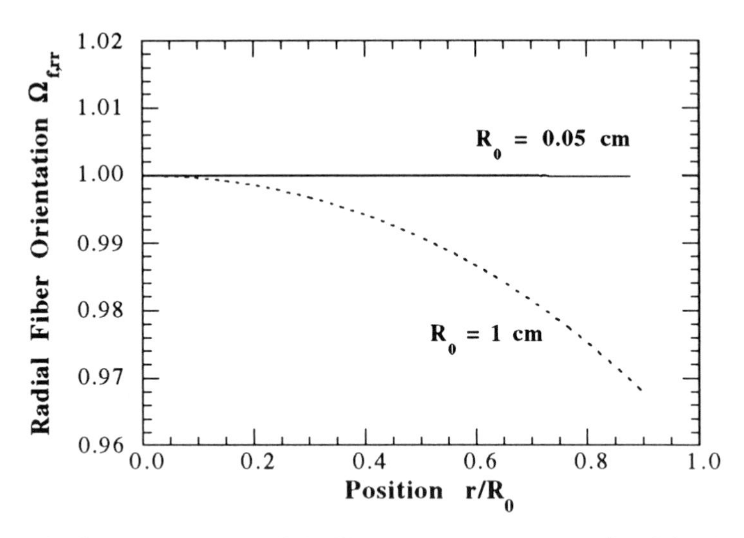

Figure 9. Radial component of the fiber orientation tensor predicted for the FPCM Traction Assay at dimensionless time 1.0 (~13 hr). A value of $\Omega_{f,rr}$ less than one indicates circumferential alignment. There is no orientation ($\mathbf{\Omega}_f = \mathbf{I}$) for R_0 = 0.05 cm, but circumferential orientation is predicted for R_0 = 1.0 cm.

where **n** is the unit vector in the direction of a fiber, and the symbol $\otimes$ denotes dyad product. The relative orientation of the fibers in the direction of an arbitrary vector **v** is defined by ($\mathbf{v} \cdot \mathbf{\Omega}_f \cdot \mathbf{v}$), with a value greater than one indicating that the direction of **v** is generally aligned with the direction of oriented fibers. The factor 3 ensures that $\mathbf{\Omega}_f = \mathbf{I}$ when the network is isotropic (rather than having tr $\mathbf{\Omega}_f = 1$). This is a continuous analog of the discrete definition of the orientation tensor by Farquhar et al. (1990).

Although we have not yet included the feedback effects of fiber orientation via contact guidance into the model equations (see below), we have predicted the evolution of fiber orientation due to traction-driven compaction in the FPCM Traction Assay. Figure 9 shows the profiles of $\Omega_{f,rr}$ corresponding to the cases of Figure 3 ($\Omega_{f,\theta\theta} = (3-\Omega_{f,rr})/2$ is not shown). For the case of R_0 = 0.05 cm where compaction is homogeneous, no fiber orientation is observed, consistent with our observation that cell contact guidance does not occur. For the case of R_0 = 1.0 cm where compaction is inhomogeneous, circumferential orientation of fibers is predicted. Although we have not performed the FPCM Traction Assay to verify this prediction of orientation, we have performed our FPCM Wound Assay where cell contact guidance is observed (Figure 8). The model based on the anisotropic theory predicts significant circumferential orientation of fibers (Figure 10), consistent with the observation.

In order to account for the contact guidance that is induced by such fiber orientation, the functional forms for the cell migration and

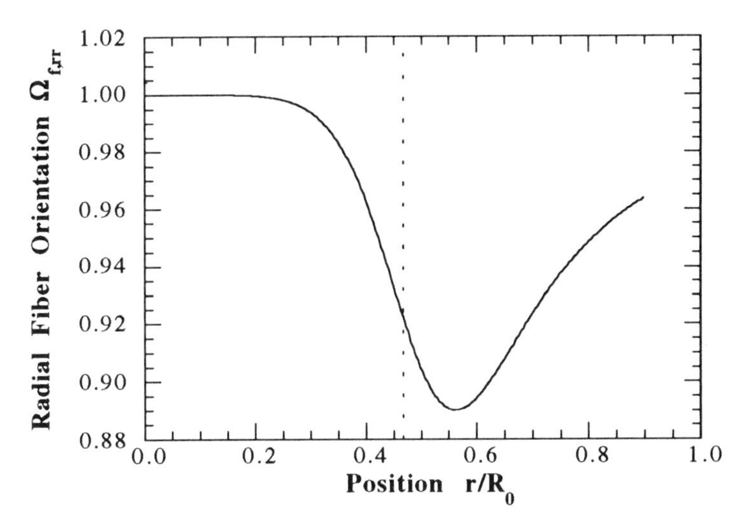

Figure 10. Radial component of the fiber orientation tensor predicted for the FPCM Wound Assay at dimensionless time 1.0 (~13 hr). The dotted line marks the boundary between the initially cell-free inner region and cell-rich outer region. A value of $\Omega_{f,rr}$ less than one indicates circumferential alignment, so the orientation is most pronounced in the region just outside the boundary.

cell traction terms must be appropriately modified. Two separate but related questions must be answered to make this modification: how does the cell orientation depend on the fiber orientation (we make the initial assumption that the fiber and fibril orientation are equivalent), and how do the anisotropic cell migration and traction stress depend on cell orientation?

Although we do not yet know the explicit relation between cell orientation and fibril orientation, we can comment on the expected nature of the relation. Observations suggest that cells align with respect to the same principal axes as the fibrils, and cell orientation is generally more pronounced than fibril orientation (Guido and Tranquillo 1993). These observations, along with the constraint that the cells are isotropically oriented when the fibrils are, suggest that the cell orientation tensor, $\mathbf{\Omega}_c$, can be written as a monotonically increasing function of the fiber orientation tensor, $\mathbf{\Omega}_f$, for example

$$\mathbf{\Omega}_c = (\mathbf{\Omega}_f)^{\kappa} \tag{28}$$

where $\kappa > 1$ is a fitting parameter. The cell orientation vs. birefringence data in Figure 5 (note that the cell orientation parameter $O_c = 2\Omega_{c,xx}-1$) will yield the necessary relation between cell and fiber orientation once fibril orientation and birefringence have been correlated. This correlation is currently under investigation.

Our answer to the second question at present is to modify (13) and (16) by generalizing the traction stress coefficient, τ_0, and the random motility coefficient, $\mathcal{D}_0$, to tensors $\boldsymbol{\tau}$ and $\mathbf{D}$, in particular

$$\begin{aligned} \boldsymbol{\tau} &= \tau_0 \boldsymbol{\Omega}_c \\ \mathbf{D} &= \mathcal{D}_0 \boldsymbol{\Omega}_c \end{aligned} \tag{29}$$

The assertion that $\mathbf{D}$ is proportional to $\boldsymbol{\Omega}_c$ is supported by the following result for a transversely isotropic gel (Dickinson et al. 1994):

$$\frac{D_{xx}}{D_{yy}} \approx \frac{\Omega_{c,xx}}{\Omega_{c,yy}} \left(\text{or } D_A \approx O_A \text{ in Figure 14 of Dickinson et al.}\right) \tag{30}$$

The assertion that $\boldsymbol{\tau}$ is proportional to $\boldsymbol{\Omega}_c$ is supported by the typical observation, portrayed in Figure 1, that the direction of oriented fibrils associated with tractional structuring is coincident with the cell orientation.

Thus, although we are still in the process of validating the specific functionalities relating fiber orientation and contact guidance, those proposed above comprise a complete anisotropic biphasic theory that accounts for this biomechanical feedback phenomenon. Model predictions for the FPCM assays based on this theory are presented elsewhere (Barocas and Tranquillo 1994).

Discussion

We have examined the mechanical interactions of cells with tissue-equivalent fibrillar gels using a biphasic theory to describe the mechanics of the gel. An isotropic theory has been used to model the FPCM Traction Assay. Modifications have been proposed to account for fibril orientation associated with deformation of the gel due to cell traction and the resulting anisotropic behavior of the cells, allowing us to model contact guidance observed in the FPCM Wound Assay. There remain, however, some important issues to be addressed.

Syneresis processes are often modeled by treating the process as a change in the stress-free state of the system leading to an apparent strain rate (Scherer 1989). This approach can be incorporated into our biphasic theory by replacing the traction stress term with an analogous strain rate term. For the simple geometry of the FPCM assays, the two approaches yield similar results (Barocas et al. 1993), but there could be significant differences for FPCL assays that have a constrained boundary (e.g. Grinnell 1991).

Just as the compaction can be treated as a stress-based or a strain-based process, the reorientation of the fibrils can be based on stress or on strain. In externally loaded systems, such as cartilage or fiber-reinforced composites, it is common to predict fiber reorientation based on the strain field (Schwartz et al. 1991), and we have presented here a simple method for predicting fiber reorientation based on the strain in the collagen network. Sherratt and Lewis (1993), in the modeling of actin cytogel, have suggested that reorientation can be predicted based on stress rather than strain (their treatment is also restricted to a planar medium in contrast to ours). Although this

Table 2. Contrasts Between Fibroblast-Populated Collagen Gel and Articular Cartilage

	Collagen Gel	Cartilage
collagen concn	low	high
proteoglycan concn	zero	high
cell concn	high	low
network rheology	VE liquid	elastic solid
network permeability	high	low
network swelling	zero	high
cell traction	significant	insignificant
fiber orientation	cell traction-induced	inherent
cell contact guidance	can be significant	insignificant
anisotropic network mechanical properties	insignificant	significant

approach would lead to similar results as the strain-based theory for the FPCM assays, there are significant differences in other cases (notably, the stress-based theory predicts spontaneous alignment of fibers in a constrained cell-populated gel, whereas the strain-based theory does not). We are currently investigating these alternative theories.

There are obvious similarities between fibroblast-populated collagen gel and articular cartilage, since both are composed primarily of a hydrated collagen network, but there are three major differences that lead to very different mechanical responses (Table 2). First, the presence of proteoglycans and associated fixed charges in cartilage leads to an ionic swelling pressure (Maroudas and Bannon 1981) that is not observed in collagen gel. Second, the tenfold lower collagen content in the gel produces a network that exhibits the characteristics of a viscoelastic liquid rather than an elastic or viscoelastic solid (Mak 1986); the lower collagen concentration also greatly increases the permeability of the network to interstitial fluid. The third difference, and the most important one for us, is that cartilage is essentially acellular, but collagen gel is prepared with cells specifically to study cell traction forces. The manner in which cells restructure the surrounding collagen network is central to the modeling of such tissue-equivalent systems.

Acknowledgments

This work has been supported by a National Institutes of Health FIRST Award to RTT (1R29-GM46052-01). A grant from the Minnesota Supercomputer Institute is acknowledged.

References

Advani, S. G.; Tucker C. L. III. The Use of Tensors to Describe and Predict Fiber Orientation in Short Fiber Composites. J. Rheol. 31:751-784; 1987.

Allen, T. D.; Schor S. L.; Schor A. M. An ultrastructural review of collagen gels, a model system for cell-matrix, cell-basement membrane and cell-cell interactions. Scan. Electron Microsc. 375-90; 1984.
Barocas, V. H.; Moon A. G.; Tranquillo R. T. The fibroblast-populated collagen microsphere assay of cell traction force - Part 2. Measurement of the cell traction parameter. J. Biomech. E. (submitted):1993.
Barocas, V. H.; Tranquillo R. T. Anisotropic biphasic theory of cell-fibril mechanical interactions in tissue-equivalent gels (in preparation) 1994.
Bell, E.; Ivarsson B.; Merrill C. Production of a tissue-like structure by contraction of collagen lattices by human fibroblasts of different proliferative potential *in vitro*. Proc. Natl. Acad. Sci. USA 76:1274-1278; 1979.
Brenan, K. E.; Campbell S. L.; Petzold L. R. Numerical solution of initial-value problems in differential-algebraic equations. New York: Elsevier; 1989.
Bromberek, B. A.; Tranquillo R. T. A novel in vitro wound healing and contraction assay (in preparation) 1994.
Cowin, S. C. The relationship between the elasticity tensor and the fabric tensor. Mechanics of Materials 4:137-147; 1985.
Craine, R. E.; Green A. E.; Naghdi P. M. A mixture of viscous elastic materials with different constituent temperatures. Quart. Journ. Mech. and Applied Math 23:171-184; 1970.
Dembo, M.; Harlow F. Cell motion, contractile networks, and the physics of interpenetrating reactive flow. Biophys. J. 50:109-21; 1986.
Dickinson, R. B.; Guido, S.; Tranquillo, R. T. Biased cell migration of fibroblasts exhibiting contact guidance in oriented collagen gels. Ann. Biomed. Eng. (in press).
Drew, D. A.; Segel L. A. Averaged equations for two-phase flows. Studies in Applied Mathematics 1:205-231; 1971.
Ehrlich, H. P. Wound closure: evidence of cooperation between fibroblasts and collagen matrix. Eye 2:149-157; 1988.
Farquhar, T.; Dawson P. R.; Torzilli P. A. A microstructural model for the anisotropic drained stiffness of articular cartilage. J. Biomech. E. 112:414-25; 1990.
Ferry, J. D. Biological and synthetic polymer networks. New York: Elsevier Applied Science; 1988.
Gail, M. H.; Boone, C. W. The locomotion of mouse fibroblasts in tissue culture. Biophys J. 10:980-993; 1970.
Grinnell, F. Fibroblast reorganization of three-dimensional collagen gels and regulation of cell biosynthetic function. In: Okamura, S.;S. Tsuruta;Y. Imanishi;J. Sunamoto, eds. Fundamental investigations on the creation of biofunctional materials. Kyoto: Kagaku-Dojin; 1991:p. 33-43.
Grinnell, F.; Lamke C. R. Reorganization of hydrated collagen lattices by human skin fibroblasts. J. Cell Sci. 66:51-63; 1984.
Guido, S.; Tranquillo R. T. A methodology for the systematic and quantitative study of cell contact guidance in oriented collagen gels: correlation of fibroblast orientation and gel birefringence. J. Cell Sci. 105:317-331; 1993.
Guidry, C.; Grinnell F. Contraction of hydrated collagen gels by fibroblasts: evidence for two mechanisms by which collagen fibrils are stabilized. Collagen Rel. Res. 6:515-529; 1986.
Jain, M. K.; Chernomorsky A.; Silver F. H.; Berg R. A. Material properties of living soft tissue composites. J. Biomed. Mater. Res. 22:311-326; 1988.
Johnson, G.; Massoudi M.; Rajagopal R. J. Flow of a fluid-solid mixture between flat plates. Chem. Eng. Sci. 46:1713-1723; 1991.
Kenyon, D. E. Consolidation of transversely isotropic solids. J. Appl. Mech. 46:65-70; 1979.

Klebe, R. J., Caldwell, H. & Milam, S. Cells transmit spatial information by orienting collagen fibers. Matrix 9:451-458; 1989.

Kwan, M. K.; Lai W. M.; Mow V. C. A finite deformation theory for cartilage and other soft hydrated connective tissues--I. Equilibrium results. J Biomech 23:145-55; 1990.

Lai, W. M.; Hou J. S.; Mow V. C. A triphasic theory for the swelling and deformation behaviors of articular cartilage. J. Biomech. E. 113:245-258; 1991.

Madri, J. A.; Pratt B. M. Endothelial cell-matrix interactions: *in vitro* models of angiogenesis. J. Histochem. Cytochem. 34:85-91; 1986.

Mak, A. F. The apparent viscoelastic behavior of articular cartilage - the contributions from the intrinsic matrix viscoelasticity and interstitial fluid flows. J Biomech Eng 108:123-130; 1986.

Maroudas, A.; Bannon C. Measurement of swelling pressure in cartilage and comparison with the osmotic pressure of constituent proteoglycans. Biorheology 18:619-632; 1981.

Moon, A. G. Cell traction forces exerted on the extracellular matrix: Modeling and Measurement. Ph.D. Thesis: University of Minnesota; 1992.

Moon, A. G.; Tranquillo R. T. The fibroblast-populated collagen microsphere assay of cell traction force - Part 1. Continuum model. AIChE J. 39:163-177; 1993.

Mow, V. C.; Kuei S. C.; Lai W. M.; Armstrong C. G. Biphasic creep and stress relaxation of articular cartilage in compression: theory and experiments. J. Biomech. E. 102:73-84; 1980.

Mow, V. C.; Kwan M. K.; Lai W. M.; Holmes M. H. A finite deformation theory for nonlinearly permeable soft hydrated biological tissues. In: Schmid-Schonbein, G. W.;S. L.-Y. Woo;B. W. Zweifach, eds. Frontiers in biomechanics. New York: Springer-Verlag; 1986:p. 153-179.

Nusgens, B.; Merrill C.; Lapiere C.; Bell E. Collagen biosynthesis by cells in a tissue equivalent matrix in vitro. Collagen Relat. Res. 4:351-63; 1984.

Odell, G. M.; Oster G.; Alberch P.; Burnside B. The mechanical basis of morphogenesis. I. Epithelial folding and invagination. Dev. Biol. 85:446-62; 1981.

Oster, G. F.; Murray J. D.; Harris A. K. Mechanical Aspects of Mesenchymal Morphogenesis. J. of Embryol. Exp. Res. 78:83-125; 1983.

Sangani, A. S.; Yao C. Transport processes in random arrays of cylinders. II. Viscous flow. Physics of Fluids 31:2435-2444; 1988.

Scherer, G. W. Mechanics of syneresis, I. Theory. J. of Non-Crystalline Solids 108:18-27; 1989.

Schwartz, M.; Leo P. H.; Lewis J. L. A Microstuctural model of articular cartilage. J. Biomech. 25 (in press); 1994.

Sherratt, J. A.; Lewis J. Stress-induced alignment of actin filaments and the mechanics of cytogel. Bull. Math. Biol. 55:637-654; 1993.

Stopak, D.; Harris A. K. Connective tissue morphogenesis by fibroblast traction. I. Tissue culture observations. Dev. Biol. 90:383-398; 1982.

Tranquillo, R. T.; Durrani M. A.; Moon A. G. Tissue engineering science: consequences of cell traction force. Cytotechnology 10:225-250; 1992.

van Gelder, J.M.; Nair, C.H.; Dhall, D.P. Colloid determination of fibrin network permeability. Biorheology (in press); 1994.

Weinberg, C. B.; Bell E. A blood vessel model constructed from collagen and cultured vascular cells. Science 231:397-400; 1986.

Yannas, I. V.; Burke J. F.; Orgill D. P.; Skrabut E. M. Wound tissue can utilize a polymeric template to synthesize a functional extension of skin. Science 215:174-6; 1982.

13

Mechanical Load ± Growth Factors Induce $[Ca^{2+}]_i$ Release, Cyclin D_1 Expression and DNA Synthesis in Avian Tendon Cells

A.J. Banes, M. Sanderson, S. Boitano, P. Hu,
B. Brigman, M. Tsuzaki, T. Fischer, W.T. Lawrence

Introduction

The Tendon Cell Model

Tendons are fibrous connective tissues designed to transmit the force of muscle contraction to bone to effect limb movement. To accomplish the latter task, tendons mandate a more complex architecture than is generally appreciated: the origin is spread over muscle in a trellis-like epimycium to permit maximum surface area for contractile input. The principle bulk of tendon is comprised of highly aligned matrix containing 70-80% type I collagen to provide tensile strength, 10-40% elastin, yielding compliance and elasticity, proteoglycans as pulse dampeners, as well as lipids, whose presence in the tendon epitenon may reduce shear stress-induced friction (Oakes and Bialkower 1977; Vogel and Evanko 1988; Banes et al. 1988; Tsuzaki et al. 1993; Brigman et al. In press). There are at least two cell populations represented in the major anatomical compartments of tendon (Riederer-Henderson et al. 1983; Banes et al. 1988; Tsuzaki et al. 1993). The epitenon contains a large, polygonal to round cell (tendon epitenon synovial cell, TSC) embedded in a lipid and proteoglycan-rich matrix containing only 25% collagen, while the internal portion of tendon contains fibroblasts (tendon internal fibroblasts, TIF) in tightly packed rows amidst linear and branching collagen fascicles and bundles (Riederer-Henderson et al. 1983; Banes et al. 1988). TSC occupy the surface of tendon in a 2-8 cell-thick border contiguous with cells in the endotenon that partition collagen fascicles from one another (Greenlee and Ross 1967; Rowe 1985a, b). TSC produce and store in the epitenon, types I and III collagens, fibronectin, TGF-β, as well as positive (IGF-I) and negative modulators (unknown etiology) of cell division (Banes et al. 1988a, b; Tsuzaki et al. 1993; Brigman et al. In press). It is possible that some of the negative modulators are IGF binding proteins that reduce bioavailability of IGF (Clemmons 1991; Tsuzaki personal communication). TIF in mature tendon are positioned in arrays, intimately connected to each other, but less in communication with other arrays. This arrangement appears optimal to promote communication through connexin-43 gap junctions to propagate a calcium ion signal via IP_3 for intercellular signaling of mechanical load conditions (Sanderson et al. 1988, 1990; Boitano et al. 1992; Beyer et al. 1992).

The Single Cell Mechanical Challenge Model

Sanderson and coworkers have pioneered a model system to explore how cells communicate a mechanical signal from a target cell to neighboring cells (Sanderson et al. 1986). The cartoon in Fig. 1 depicts a longitudinal cross-section of a tendon that has received a notch wound. The epitenon cells are shown on the tendon upper and lower aspects as a continuous array that presents as a circumferential sheet that covers the tendon surface. Tendon internal fibroblasts within the structural matrix compartment of tendon appear in discrete, linear arrays of 10-20 cells in a row with less obvious connectivity to other rows of TIF (circled row). Cells are loaded with fura-2AM, the acetoxymethylester of the calcium ion binding fura-2 compound, either in tissue or in cultured cells, then subjected to fluorescence imaging and mechanical challenge. A glass micropipet tip is brought to bear on the surface of the plasma membrane of a target cell then rapidly advanced two microns to indent the cell. The target cell responds with an instantaneous release of intracellular calcium stores, $[Ca^{2+}]_i$. This effect occurs in the presence or absence of calcium in the medium, but in some cells can be prevented if the poke is performed in

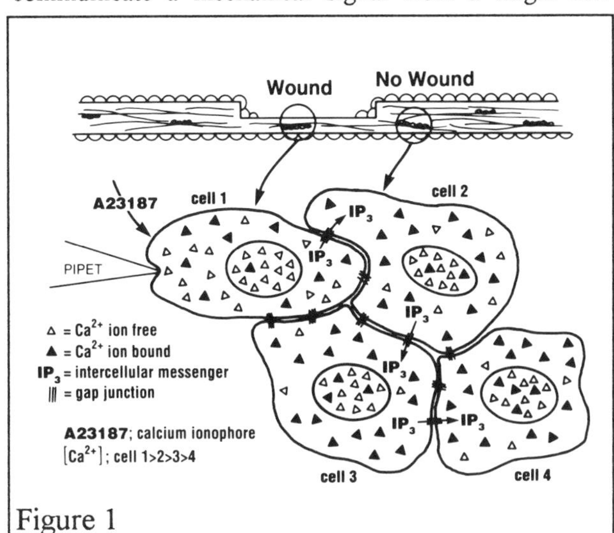

Figure 1

calcium-free medium. A calcium-induced calcium release mechanism may be operative but Dantrolene, a CICR inhibitor, abolishes calcium oscillations but not the rate of calcium wave propagation (Charles et al. 1992). The calcium wave is propagated to neighboring cells in a time-dependent fashion. Calcium wave propagation can require seconds to reach cells furthest from the target before the signal stops. The calcium release phenomenon usually reaches cells in a 7-10 cell radius, but the rate of transmission may vary with time in culture, degree of confluency and intrinsic ability of the cells to form gap junctions (Charles et al. 1992). If cells are treated with halothane, heptanol or other gap junction blockers and poked, only the target cell released Ca^{2+} (Boitano et al. 1992). These results indicate that a chemical mediator induced by mechanical challenge transits the cell from the point of pipet contact to gap junction conduits with adjoining cells. Results of experiments with cells that have had heparin (an IP_3 receptor blocker) forced into the cell interior by electroporation, then are challenged with a poke, reveal that only the target cell releases $[Ca^{2+}]_i$. These results indicate that inositol phosphate 3 is required for calcium wave propagation, but not initial release in the target cell, and that the signal is propagated through gap junctions. Proteins that comprise gap junctions are members of the *connexin family* (Beyer et al. 1990). Gap junctions are assembled from 6 identical, connexin subunits to form a hexamer channel termed a *connexon* (Robertson 1963). Connexons from adjacent cells contact at a 2 nm spacing to form a *functional gap junction channel* (Saez et al. 1989). It is through these apposed channels that IP_3 is postulated to move to propagate the mechanical load signal as a calcium wave.

Model for Strain Application to Cultured Cells

Cyclic strain is applied to cultured tendon cells by growing cells on a flexible, rubber surface bonded with collagen and deforming the membrane by stretching (Banes et al. 1985, 1990). The mechanical load regime involves a half sinusoidal waveform with 0.2 maximal strain and 1 Hz frequency as one type of deformation that is a gradient of strain, maximal at the periphery and minimal at the culture well center (Flex I culture plate, 2.3 mm thick membrane) (Banes et al. 1991; Pederson et al. 1992; Gilbert et al. In press). A second, flexible membrane substrate is a thin (0.020 inch thick) membrane that has 85% homogeneous, biaxial strain over the culture surface. The microprocessor-driven controller (Flexercell Strain Unit, FSU) regulates the deformation regime to the culture plates from 1 to 200% elongation and a frequency up to 5 Hz (Banes et al. 1985, 1991; Pederson et al., 1993; Gilbert et al., in press). Deformation of fura-2-loaded cells can be performed directly on the microscope stage with a jig that allows dynamic flexing of the substratum controlled by an FSU. Cells can also be flexed, then mechanically challenged with the micropipet to test for changes in rate of intercellular transmission of the calcium wave.

Cyclins Regulate Cell Cycle Phase Entry and Division

A family of 34 kD proteins termed cyclins regulate S phase entry as well as exit from G2 phase (Xiong et al. 1992). In $_{G1}$ phase, D cyclins and a cyclin-dependent kinase (CDK2) are expressed when quiescent cells are stimulated with serum, PDGF-BB, IGF-I the growth factor combination, or cyclic load and growth factors (Xiong et al. 1992; Xiao et al. 1993; Winston et al. 1993; Hu et al. 1993, 1994). Tendon cells express novel D1 cyclins that are induced by mitogens, including serum, PDGF-BB, IGF-I and load (Hu et al. 1993, 1994). However, TIF express more D1 cyclin clone 203a mRNA upon serum stimulation than do TSC. Since both cell types synthesize DNA from quiescence in response to serum, TSC probably express another cyclin during $_{G1}$ phase.

Methods

Cell Culture

Tendon epitenon synovial cells (TSC) and internal fibroblasts (TIF) were isolated from avian flexor tendons by the method of Banes and coworkers (Banes et al. 1988). Briefly, the method involved sequential trypsin and collagenase treatment of the dissected flexor tendons of the legs to first isolate TSC, followed by mechanical scraping with a rubber policeman to remove residual epitenon. The remaining tendon samples were minced and treated with 1% collagenase to free TIF. Cells were washed in complete medium (10% fetal calf serum in DMEM-H with 0.5 mM ascorbate and antibiotics then plated at 25 k cells/cm^2 in 83 mm diameter collagen bonded culture plates for cell expansion. Cells were passaged in 0.01% trypsin in PBS, pH 7.2, after washing in 0.1 M phosphate buffered saline (PBS) pH 7.2. Cells were used from passes two to three.

For experiments involving cyclic mechanical loading, cells were plated in complete medium at 25 k cells/cm^2 onto type I collagen bonded Flex I™, flexible bottomed culture plates (Flexcell Intl. Corp. McKeesport, PA).

For experiments involving mechanical challenge with a glass micropipet, cells were plated either on collagen-coated glass cover slips or BiFlex™ collagen bonded rubber membranes (0.020 inch thick, with 85% homogeneous, biaxial load when subjected to cyclic strain) at 25 k cells/cm^2 in complete medium (Pederson et al. 1992, 1993; Gilbert et al. In press). Cells grown on the BiFlex membranes were used for the pipet poke, mechanical challenge experiments.

Micropipet Cell Membrane Indentation Model

The membrane indentation model of Sanderson and coworkers was utilized to challenge single cells in an array with a mechanical stimulus (Sanderson et al. 1986, 1988, 1989, 1990; Boitano et al. 1992). Briefly, TSC and TIF were plated on collagen-coated glass coverslips or BiFlex, 0.020 inch thick, collagen bonded rubber membranes in complete medium at 25 k cells/cm^2 and grown to confluence and quiescence for 7 days without a medium change (Pledger et al. 1978). Medium was aspirated, the cells were washed free of medium in DMEM with HEPES pH 7.2, then loaded with 5 μM fura-2AM in DMEM for 45 min at RT. Cells were washed in DMEM without fura-2AM for 5-10 min, then placed in a holding jig on the stage of an Olympus OM-2 upright, fluorescence microscope equipped with a long working distance, water immersion, 40X, high transmission 340 nm UV objective (NA 0.7, WD 3.1 mm, W Plan, cat. # 1-LB450, Olympus Corp. Tokyo, Japan), filter wheel system with 340 and 380 nm filters, 510 nm and above cutoff filter, CCD camera, intensifier screen, and Image-1 image analysis hardware with fura-2 signal collection software (Universal Imaging West Chester, PA). A target cell in a monolayer was selected at random for mechanical challenge with a 0.5-1 μm diameter tip glass micropipet controlled by a Narishige micromanipulator. The pipet was positioned to a point just touching the plasma membrane then the pipet was advanced rapidly, 2 μm, with the micromanipulator (Sanderson et al. 1988). Intracellular calcium ion release, $[Ca^{2+}]_i$, was quantitated by ratio imaging the 340 and 380 nm excitation fields with 510 nm emission. Absolute calcium quantitation was performed by comparison to intensities of standard calcium solutions. Electroporation of heparin (an IP_3 receptor blocker) and Texas Red dextran (a positive control label to assure entry of heparin into the cell) was accomplished by pulsed field electroporation (Boitano et al. 1992). Halothane at 4.3 mM in DMEM was incubated with cultures for 15 minutes prior to testing response to a mechanical challenge. Data were expressed as the rise in $[Ca^{2+}]_i$ from a basal level of 35-50 nM to over 1000 nM in 5 different cell arrays per culture. Cultures from three different isolations were used in the experiments.

Cyclic Mechanical Loading with a Flexercell Strain Unit

Cells were plated on Flex I, collagen bonded, flexible bottomed culture plates as described above, grown to confluence and quiescence for 7 days, then washed with DMEM and the following groups tested ± 1 Hz , 0.2 strain (20% elongation) for 24 h: 1. No medium change, 2. DMEM only, 3. PDGF-BB 3.33 ng/ml, 4. IGF-I 6.67 ng/ml, 5. PDGF-BB 3.33 ng/ml + IGF-I 6.67 ng/ml.

Optimal growth factor doses were determined previously (Banes et al. Submitted). Each 25 mm diameter well of the Flex plates contained 1.5 ml DMEM. Serum-containing medium (10% FCS) was used as a positive control, DMEM alone served as a negative control. Culture medium included 0.5 μCi ^{3}H-thymidine/well during the 24 h incubation period. After 24 h, culture plates were placed on ice, medium was aspirated, and cells were treated with 5% trichloroacetic acid (TCA), scraped from the wells, and washed exhaustively in 5% TCA and radioactivity determined in a scintillation counter. Replicate wells were used for cell counts for each group. Replicate determinations were performed for each group (n=5/group). Data were expressed as DPM ^{3}H-thymidine/cell.

Detection of Cyclin D1 mRNA by Reverse-Transcription Polymerase Chain Reaction and Northern Analysis

Total RNA was collected from quiescent TSC and TIF at 8 h post-serum treatment to induce maximal cyclin D1 messenger RNA (Berk and Sharp 1977; Favoloro et al. 1980; Gilman 1989). Sufficient total RNA was collected from two 83 mm diameter culture plates or two 6 well Flex I plates for a single determination by Northern analysis (Lermach et al. 1977; Thomas 1980; Hassouna et al. 1984; Raynal et al. 1984). The 1.3 kb cyclin D1 203a mRNA was detected using a tail labeled, 203 base pair cDNA probe (Hu et al. 1994). We have sequenced the PCR cDNA product and verified that the DNA sequence is homologous to that of the published sequence for human cyclin D1 message (Xiong et al. 1992; Hu et al. 1993, 1994).

The Cetus-Perkin Elmer GeneAmp RNA PCR kit was used to perform RT-PCR of TSC and TIF mRNAs (Mullis and Faloona 1987). Some substitutions in reagents were made as kit components were utilized: Promega taq DNA polymerase and AMV reverse transcriptase were used to supplement other reagents in the kit. Equal amounts of sample total RNA were used for the reverse transcription reaction. Replicate samples were separated on an agarose gel, stained with ethidium bromide and the band densities of the 28 and 18s ribosomal RNAs quantitated to assure that equal portions of RNA samples were assayed. TSC and TIF cyclin D1 mRNAs were reverse transcribed, then the cDNAs were amplified for 25 cycles, an amplification regime we have documented as within a linear, quantitative region of the curve (Brigman et al. In press). cDNA band densities were determined by image analysis of the ethidium bromide-stained products. Quantitation of the amount of amplified product from the TSC and TIF samples was performed by hybridizing a ^{32}P-labeled 203 base pair probe (cyc D1 203b) to a blot of the separated cDNAs using stringent conditions (Hu et al. 1994).

Results

Data in Figures 2 and 3 depict intracellular calcium ion, $[Ca^{2+}]_i$, basal states as well as calcium ion release in TSC and TIF stimulated with a micropipet. The figures are gray tone conversions of pseudocolor picture montages. A dark background indicates resting levels of calcium (35 to 50 nM Ca^{2+}). A light background indicates calcium levels up to 1000 nM. For the purposes of describing results in these figures, expression of quantitation will be limited to demonstrating intracellular release rather than quantitation. Figure 2A shows a brightfield frame with a 0.5-1.0 μm diameter tip of a glass micropipet touching the plasma membrane of a target TSC prior to a membrane "poke". The resting calcium ion level is approximately 35-50 nM. At 1 sec after membrane indentation, a target TSC has released intracellular calcium stores with maximal $[Ca^{2+}]_i$ values over the nucleus of greater than 1000 nM. At 4 sec, several adjacent cells have released intracellular calcium. Calcium wave propagation continues in the next frames at 7 and 10 seconds with more cells involved in intracellular calcium release, but with the calcium concentration decreasing with distance from the target cell and time. Data in the Figure 2B series shows the reaction of a TSC target cell and adjacent cells that have been infused with heparin, an IP_3 receptor blocker (Boitano et al. 1992). The cells infused with heparin are gray (using a filter to visualize Texas Red dextran) in the first column of 2B. The target cell released intracellular calcium stores beginning at 1 sec, whose intensity increased from 4 to 7 sec then diminished by 10 sec. No $[Ca^{2+}]_i$ release was detected in adjacent cells, indicating that the signal was not propagated from the target cell to nearest neighbors. Blockade of calcium wave propagation by heparin indicates that intercellular signaling occurs via an IP_3-dependent system, since heparin inhibits activity of the IP_3 receptor (Boitano et al. 1992). The series in Figure 2C shows a border of the Texas Red dextran and heparin to the left of the white line. A target cell on the right was indented with a micropipet, intracellular calcium was released and a wave propagated to adjacent cells in the area distant from the heparin. The signal was not propagated to heparin-loaded cells to the left of the line. These results demonstrate that heparin specifically blocked signal transduction from nontreated to heparin-treated cells.

Results in Figure 3 pertain to TIF as in TSC for Figure 2. TIF responded to the mechanical challenge (series A) as well as heparin (series B) in the same way as did TSC. In series C, cells to the left of the white line were electroporated with heparin and Texas Red dextran. A target cell to the left of the line, within the heparin zone was poked and released intracellular calcium stores. The signal did not propagate to neighboring cells due to inhibition by heparin of the IP_3 receptor. Both TSC and TIF treated with 4.3 mM halothane, a gap junction blocker, released intracellular calcium stores in response to a poke, but failed to propagate a wave to adjacent cells. These results indicate that signal propagation elicited by a mechanical stimulus is transmitted through gap junctions. Results in Figure 4 show the plot of elapsed time to intracellular calcium release in neighboring cells vs. cell diameters from the target cell.

The signal heralding intracellular calcium release was transmitted from cell

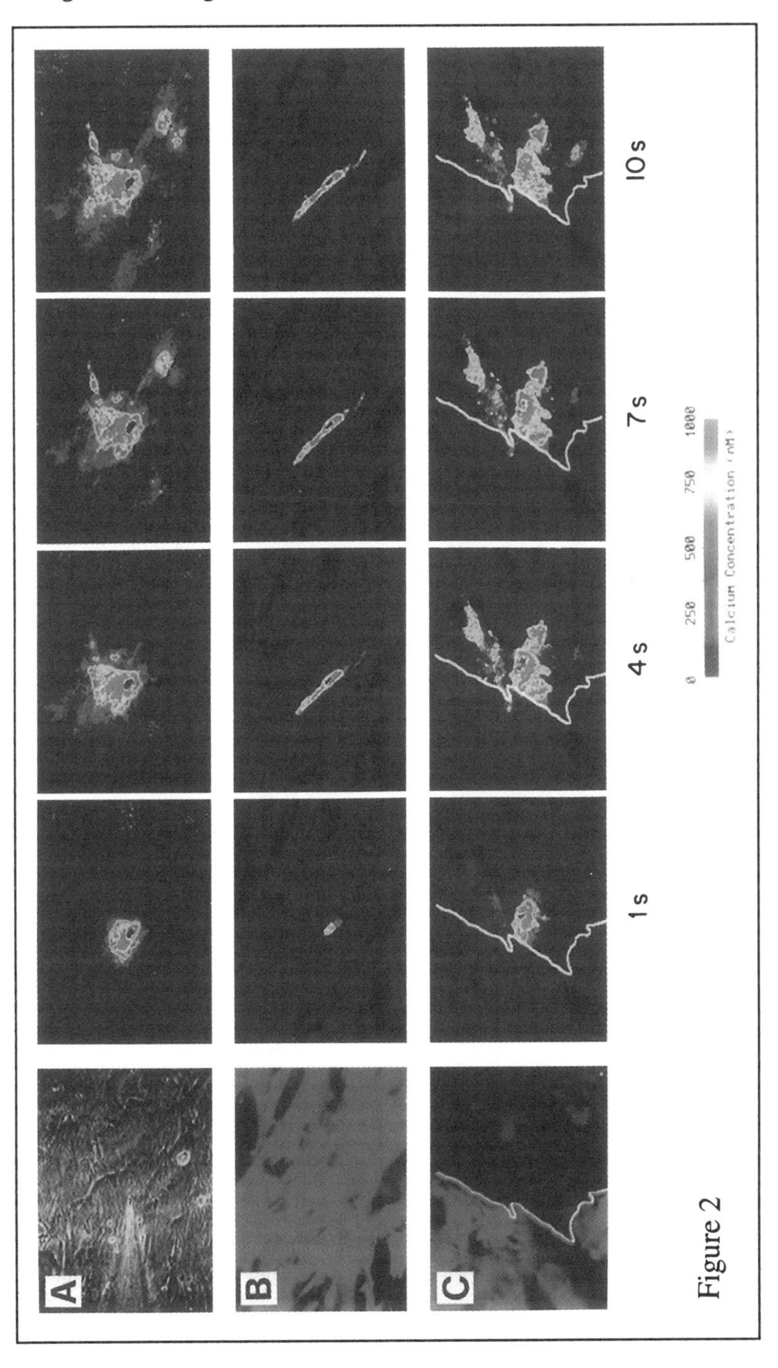

Figure 2

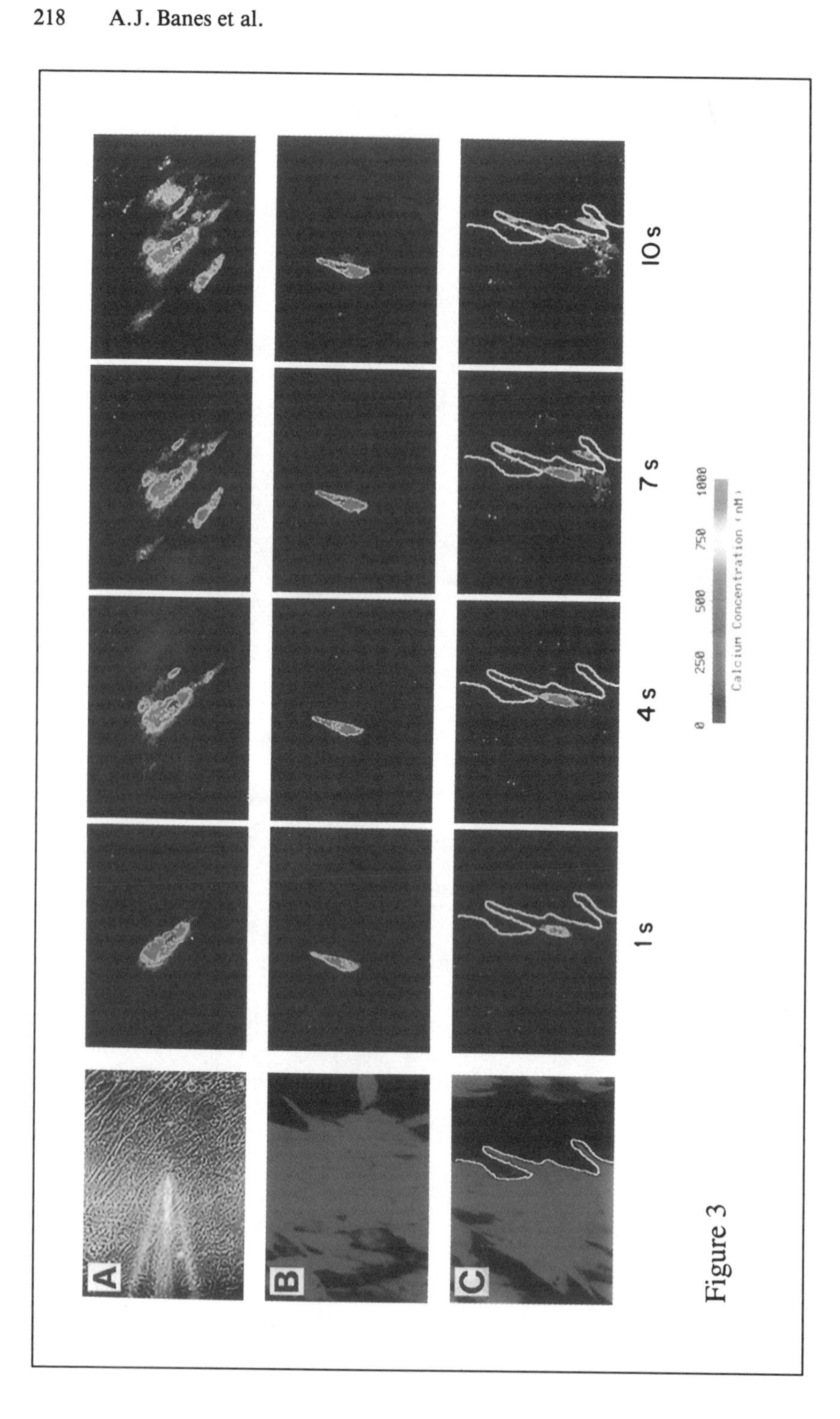

Figure 3

to cell in linear fashion at approximately two sec/cell. This number is only approximate since the actual rate of $[Ca^{2+}]_i$ increase is much higher within the cell than between cells. These results indicate that tendon cells from either the surface or interior of tendon that reside in different mechanical environments *in vivo*, respond to single cell mechanical challenge *in vitro* in like manner.

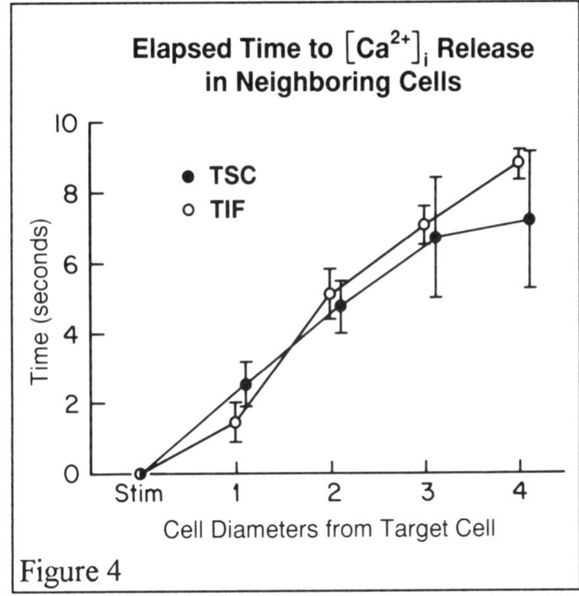

Figure 4

Results in Figure 5a, b show the effects on DNA synthesis in quiescent TSC and TIF treated with growth factors and/or 1 Hz, 0.2 strain, cyclic mechanical load for 24 h. For both TSC and TIF, IGF-I was able to stimulate cells grown on collagen coated rubber to synthesize DNA with or without applied cyclic load. In TSC without load, DNA synthesis was stimulated 10.3 fold ($p<0.0001$) in the IGF-I and 8.3 fold in the combination group, but was only 23% of control in the PDGF-BB alone group, compared to the DMEM control value. Treatment of non loaded TSC with 3.33 ng/ml PDGF-BB and 6.67 ng/ml IGF-I stimulated DNA synthesis 8.3 fold ($p<0.0001$) compared to the DMEM control value. In cyclically loaded cells, DNA synthesis in TSC was increased 4 ($p<0.025$),

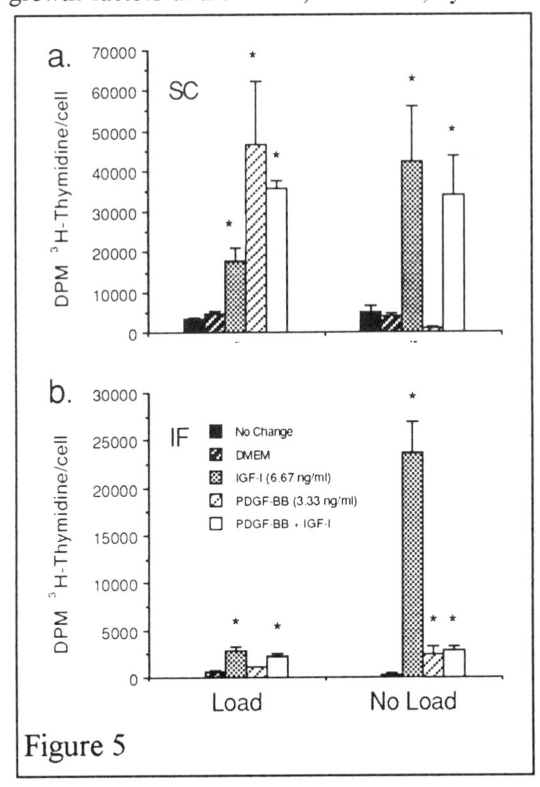

Figure 5

7.9 ($p<0.0001$) and 10.4 fold ($p<0.0001$), in IGF-I, PDGF-BB and the combination group respectively. DMEM and no medium change values were similar for TSC in both the load and no load groups. Load and IGF-I were half as stimulatory as were IGF-I and no load, but treatment of TSC with PDGF-BB alone increased DNA synthesis almost 49 fold ($p<0.0001$). In a dose response experiment where [IGF-I] was constant at 6.74 ng/ml and PDGF-BB was given at 0.0067, 0.067, 0.67, 3.37 or 6.74 ng/ml, nonloaded cells reached an equivalent maximum of DNA synthesis at 0.67 but loaded cells were maximal at 6.74 ng/ml. However, at the lowest dose, loaded TSC had a 2.3 fold greater DNA synthetic response than the control ($p<0.001$).

In TIF without load, IGF-I treatment was the major mitogen, stimulating DNA synthesis 76 fold ($p<0.0001$) compared to the DMEM value (DMEM alone value was extremely low due to the presence of a proposed inhibitor of DNA synthesis). DNA synthesis increased in TIF treated with PDGF-BB or the IGF-I + PDGF-BB combination were 7.9 ($p<0.04$) and 9.2 fold ($p<0.019$). By comparison, growth factor stimulation in loaded cells was less dramatic. DNA synthesis in IGF-I-treated TIF subjected to load increased only 5.4 fold ($p<0.049$). PDGF-BB had no significant effect on DNA synthesis in loaded TIF, whereas treatment with PDGF-BB and IGF-I increased DNA synthesis 4.3 fold ($p<0.016$). Therefore, the increase in DNA synthesis in loaded TIF was attributable to IGF-I, the principle mitogen for TIF that were not loaded. Moreover, DNA synthesis was increased in TIF incubated in DMEM vs. the group that received no medium change 13.6 fold in loaded TIF and 8.3 fold in nonloaded cells. The latter result indicated that TIF secreted a powerful substance in their conditioned medium that was inhibitory for DNA synthesis.

Data in Figure 6 show a Northern blot for a 1.3 kb messenger RNA for cyclin D1 in quiescent avian TSC and TIF stimulated with serum-containing medium and harvested at 8 h post-stimulation. The messenger size is identical to that published for the human analog isolated from WI38 human diploid fibroblasts (Xiong et al. 1992). Xiong and coworkers also identified a higher molecular weight messenger species that was cyclin-associated (4.5 kb). Both the 1.3 and 4.5 kb species were identified in the avian samples.

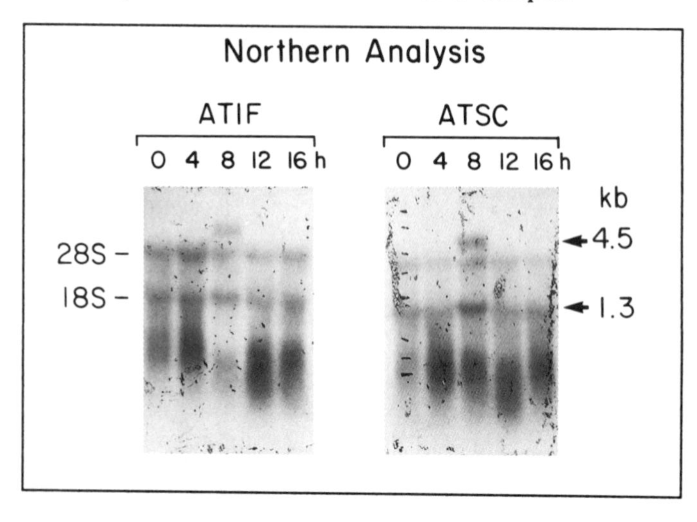

Data in Figure 7 show the results of 25 cycles of amplification of cyclin D1 clone 203a cDNA from equal amounts of total RNA from quiescent avian TSC and TIF after 8 h of treatment with serum (positive controls, lanes 1, 5) or ± PDGF-BB, IGF-I or the combination and ± load. The most obvious effect on detection of cyclin D1 mRNA was that of inhibition by PDGF-BB in TSC but not TIF (lanes 2, 6 PDGF-BB; 3, 7 IGF-I; 4, 8 PDGF-BB + IGF-I). Load + IGF-I may reduce the inhibitory effect of PDGF-BB on cyclin D1 mRNA expression in TSC (lane 4), but this conclusion is only speculative. Cyclin D1 mRNA expression was not inhibited by PDGF-BB in TIF. These results indicate that entry into S phase is regulated by different cyclins that may be inducible by different growth factors in TSC and TIF. Cyclin D1 clone 203a mRNA is under less stringent control in TIF than in TSC, since its expression in TIF is equivalent with the various growth conditions tested.

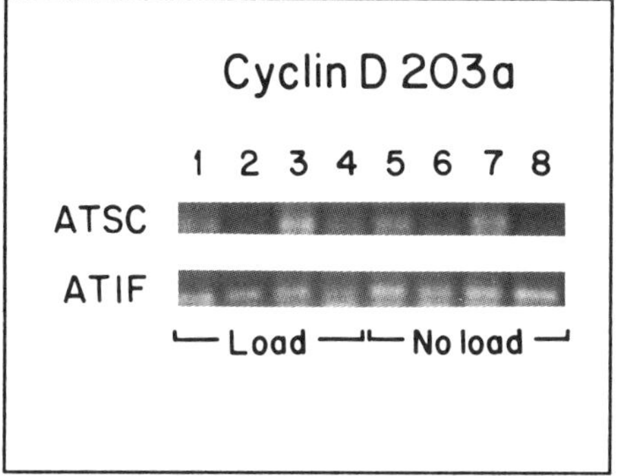

Discussion

Tendon functions to transmit the contractile force of muscle to bone to facilitate limb movement. The tendon matrix bears the structural load but the resident cells are subjected to strains as the collagen fibrils are placed in tension and shear stress is created by tendon gliding (Walker 1964; Abrahams 1967; Herrick et al. 1978; Walker 1978). Evolutionary pressures likely selected for cells that could withstand forces that became more focused and thus increased as limb development progressed (Rodbard 1970). It has been suggested that tendons develop by tractional structuring at a load intensive locale (Stopak and Harris 1982). Although flexor tendons in the horse can withstand strains of up to 18%, cyclic elongation in the area of a few percent is more common. Muscle clearly responds to increased loading by rapid hypertrophy (Banes 1993). Tendons also hypertrophy in response to load. A mechanism to stimulate division in tendon cells may be the production, release and activation of IGF-I, particularly in conjunction with mechanical loading. It has been reported that tendons hypertrophy in laborers who use vibration-intensive tools (Hasson et al. 1988). How are the mechanical signals transduced and interpreted by cells that are routinely subjected to complex, cyclic loads from muscle contraction?

Signal Transduction in Cultured Tendon Cells in Response to a Mechanical Challenge: Connectivity of cell arrays

We have shown that cell types isolated from the two principle compartments of tendon, the surface, epitenon synovial cells (TSC) and the internal fibroblasts (TIF), responded to a single mechanical signal by releasing intracellular calcium stores, $[Ca^{2+}]_i$. Activation of the target cell resulted in immediate $[Ca^{2+}]_i$ release whose signal was transmitted to neighboring cells up to 7 diameters in radius, through gap junctions by an IP_3-mediated mechanism. This mechanism is active in many diverse cell types such as airway epithelial cells, endothelial cells, smooth muscle cells, osteoblasts and glioma cells (Sanderson et al. 1986, 1988, 1989, 1990; Boitano et al. 1992; Charles et al. 1992; Demer et al. 1993; Baird et al. Submitted). In each cell type studied to date, the signaling response was limited to an area in a monolayer that was 14-20 cells in diameter. The intensity of the intracellular calcium wave diminished as distance from the target cell increased. This action has been explained for airway epithelial cells in terms of a local increase in ciliary beat frequency in the trachea to clear mucus that has captured particles or bacteria that have activated the response system (Sanderson et al. 1988). This local intercellular signaling system differs from the more global nerve-response driven autonomic system responsible for gross signaling in connective tissues. In tendon, cells in the epitenon reside in a gelatinous, lipid matrix designed for ease of cell migration and mechanical pulse dampening (Tsuzaki et al. 1993). *In vivo*, epitenon cells are plump and rounded without extensive interconnecting pseudopods, but juxtaposed to each other in a circumferential sheet between layers of matrix above and below (Greenlee and Ross 1967, Chaplin and Greenlee 1975). Conversely, internal fibroblasts are aligned in well organized, linear arrays of 10 or more cells with intervening collagen bundles segregating arrays by approximately 10 cell diameters. The TIF are arranged in succession, as boxcars in a train, and would appear to be ideal for electrical and chemical coupling. Results of unpublished experiments indicate that both epitenon and internal cells of whole, living *ex vivo* tendon, loaded with fura-2, release intracellular calcium stores when treated with a calcium ionophore such as 4-BrA23187 (Banes, unpublished results). However, preliminary experiments with cells, *in situ*, in whole tendon, have not yet shown that resident cells release $[Ca^{2+}]_i$ when challenged with a mechanical poke. The ability of tendon cells, *in vitro*, to respond to a single cell mechanical challenge indicates that the local signaling system is active in these cells. Lack of a similar, detectable response in cells in *ex vivo* tissue may indicate that, 1. the response is active but below the 35-50 nM $[Ca^{2+}]_i$ detectable range, 2. that the cells require some additional, priming factor before signaling can commence, 3. the system is inoperative *in vivo*.

Cells from diverse mechanically active tissues have responded in similar fashion to the single cell mechanical poke challenge. Target cells uniformly release intracellular calcium and propagate a calcium wave to neighboring cells (Sanderson et al. 1988; Boitano et al. 1992; Demer et al. 1993). Interestingly, osteoblasts (primary avian calvarial cells), which are purported to respond to as little as 100 μstrain in a stretch paradigm *in vitro* (Brighton et al. 1991), respond to a cell poke with great rapidity (Banes, personal observation). However, the transmission distance of the signal to adjacent cells is in the same order of magnitude as in all other cells tested - 5 to 10 cell diameters from the target cell. Nevertheless, the single cell mechanical testing regime may prove invaluable in assessing the limits of individual cells to receive and process a mechanical signal and the mechanism(s) with which the response is achieved.

The results of experiments with the inositol phosphate 3 receptor blocker, heparin, clearly demonstrate that the target cell can mount an intracellular calcium release, but that the propagation reaction is blocked. If the poke test is performed in calcium-free medium, often, but not always, the target cell fails to release calcium as well. The latter finding implicates calcium influx from the medium as an initiator of the calcium ion release reaction (CICR, calcium-induced calcium release) however, Dantrolene, an inhibitor of CICR, diminishes the $[Ca^{2+}]_i$ but not the rate of wave propagation in poked cells (Charles et al. 1992). What is the evidence that IP_3, and not calcium ion, is the mediator of calcium wave propagation between cells? Data from Allbritton and coworkers, with diffusion rates of IP_3 and Ca^{2+} in cytosolic extracts, indicate that calcium ion is a short range (0.2 μm), but that IP_3 is a long range signal molecule (up to 24 μm) (Albritton et al. 1992). The biological half-life of IP_3 is brief (seconds) but its diffusion rate is high, therefore it is capable of spanning the average cell, and of passage through a gap junction. Ca^{2+}, on the other hand, has a short diffusion distance and is bound by many molecules. It does not have the characteristics of a signal messenger vital to span a cell in the proper time frame to explain the spread of a calcium ion wave across a cell, nor the communication between cells. Therefore, one must assume that after a cell receives a mechanical signal, an IP_3 wave traverses the cell and arrives at a gap junction prior to that of the calcium ion wave. If so, then IP_3 is most likely the signal molecule that crosses through connexon hemichannels that form the gap junction, recognizable in transmission electron microscopy photos of cell junctions. If the limit of diffusion distance of the average IP_3 molecule is about 24 μm, then inositol, yielding IP_3, that originates near the plasma membrane of the first point of contact of a micropipet, should be hydrolyzed by the time it reaches a gap junction or shortly after it crosses over into the adjacent cell. What then is the mechanism by which IP_3 is regenerated to cause propagation of a calcium ion wave from cell to cell? Phospholipase C action , activated by Ca^{2+}, may provide more IP_3 in the second cell to regenerate the signal, but the precise mechanism by which this may occur is speculative (Eberhard and Holz 1988; Whitaker 1989). However, it is clear from an assessment of diverse cell types that respond to a poke by calcium release, that the response is self-limiting with a range of 5-10 cell diameters from the target cell in airway epithelial cells (Sanderson et al. 1988), glial cells (Charles et al. 1991), endothelial cells (Demer et al. 1993), osteoblasts (Geist et al. 1993), tendon cells (Banes et al. 1993), and smooth muscle cells (Christ et al. 1993; Baird et al. Submitted). The self-limiting calcium release phenomenon in cells that act as an effective syncytium defined a local reaction to a mechanical stimulus. Although more than 10 cells can be shown to be connected by gap junctions and are electrically connected, the propagation of the initial mechanical stimulus stops! Why? If the tensegrity theory of cell signaling is taken to an upper limit, then a mechanical signal sufficient to invoke a signal response, such as intracellular calcium release, in a target cell (cell 1) perturbs the matrix around the cell, as well as the cytoskeleton and nucleus (Ingber 1991, 1993; Wang et al. 1993). Cell 1 is mechanically linked to cell 2, therefore, mechanically, every cell that is physically connected could, theoretically, receive a remnant of the original mechanical perturbation. Wang and coworkers performed an interesting experiment with ligand-coated, ferrous beads that could be magnetized in a uniform direction on the cell surface prior to ligand-receptor binding (Wang et al. 1993). Beads that were coated with ligands that had fixed receptors tethered to the cytoskeleton, such as collagen or fibronectin, offered a resistance to movement in a direction opposed to the original direction of polarization. This resistance to movement from the preferred direction could be measured and quantitated. Beads which had been coated with albumin or lipoprotein which did not have receptors fixed to the cytoskeleton rotated freely to a direction 90 degrees to the direction of initial polarization. These results confirm that integrins are bound to the cytoskeleton, but not that mechanical signals are transmitted through the cytoskeleton. Other indirect evidence that integrins respond to mechanical load is the work of Davies and coworkers who showed that integrins cluster at the leading edge of endothelial cells subjected to shear stress (Robotewskyj et al. 1993). Cells are largely membrane bound aquifers with buoyant and fixed organelles that not only support cell structure, but simultaneously, are capable of transmitting force, from the point of reception at attachment points (integrins, cadherins, gap junctions, others), through talin, paxillin, α-actinin, vinculin, vimentin and actin directly to the cell nucleus and beyond. The metaphor that the cytoskeletal system is akin to a finely tuned string instrument and as such, can convey information by altering tension on the strings is an interesting concept (Ingber 1993, personal communication). There is evidence that shear stress or strain can upregulate integrin expression, but definitive evidence that mechanical signals are transmitted along cytoskeletal routes and result in action at a distance remains forthcoming (Ingber 1991, 1993; Mitsumata et al. 1993; Wang et al. 1993; Robotewskyj et al. 1993; Okamoto et al. 1993). The cells themselves and the transmission lines may also dampen the signal and limit its transmission. Clearly the signal has a defined range with respect to eliciting $[Ca^{2+}]_i$ release. It is also clear that gap junctions act as conduits for cell-cell signaling because gap junction blockers such as the anesthetics, heptanol, octanol and halothane block mechanically induced calcium wave propagation (Peracchia 1991; Boitano et al. 1992). The electron microscopy evidence is that gap junctions do not have demonstrable cytoskeletal connections (Robertson 1963; Revel and Karnofsky 1967; Saez 1989; Bennett 1991). On the other hand, one report indicates that antibody to the gap junction protein, connexin 43, coprecipitates an actin, that can be removed from the menstrum with anti-actin antibody (Laird et al. 1991). If gap junctions are not connected to cytoskeletal proteins, and gap junctions are a bona fide conduit for cell-cell signaling, then the theory concerning transmission of mechanical signals by direct mechanical linkage (tensegrity model) represents only one signal transmission mechanism cells may use to communicate between and among themselves. That gap junctions are vital for intercellular signaling has been shown by a multitude of experiments involving dye and electrical coupling of cells (Crow 1990). A precedent report however, is that of Charles and coworkers, who showed that C6 glioma cells, which are not strongly electrically or dye coupled, express only low levels of connexin-43, a principal gap junction protein (Charles et al. 1992). Ability to propagate a calcium wave in response to a poke was restored after C6 cells were transfected with a cDNA for CXN-43, expressed both the mRNA and the protein and regained the ability to pass current among themselves and respond to a poke with a propagated calcium wave. In addition, the uncontrolled rate of division slowed to a more normal rate, lending credence to the thought that intercellular signaling via $[Ca^{2+}]_i$ release and IP_3 diffusion through gap junctions is a key part of the cell division regulatory system. These results add evidence to the theory that cells use gap junctions as key conduits through which to rapidly signal each other upon receipt of a mechanical signal. If the tensegrity mechanism were the sole player in signaling through intracellular calcium release, then gap junction inhibitors would have no impact on wave propagation-but they block it!

Quiescent TSC and TIF Respond to Cyclic, Mechanical Load Differently When PDGF-BB or IGF-I are Present

TSC grown on collagen-bonded, flexible bottomed plates and subjected to 1 Hz 0.2 maximum, cyclic strain in the presence of PDGF-BB were stimulated to synthesize DNA whereas PDGF-BB without load was not stimulatory for TSC. TSC were responsive to IGF-I alone without load, but only half as well with load. We know that both TSC and TIF produce IGF-I binding proteins (primarily BP-3, Tsuzaki and Banes, unpublished data). Therefore, we hypothesize that load may induce production and secretion of IGF binding proteins that reduce IGF-I action, since nonloaded cells respond to IGF-I and incorporate thymidine. Alternatively, PDGF-BB may have a suppressive effect on IGF binding protein production and secretion. These hypotheses remain to be tested. TIF are much more *refractory* to load than are TSC. IGF-I is the principle mitogen for TIF (Banes et al. Submitted; Tsuzaki et al. Submitted).

However, as with TSC, when load was applied to TIF, even in the presence of IGF-I, only a slight mitogenic response was mounted. Results of previous studies in tendon cells have shown that TSC grown on a positively charged polystyrene responded mitogenically to both PDGF-BB or IGF-I or the combination, whereas TIF responded maximally only to IGF-I (Banes et al. Submitted). It is now clear that TSC and TIF respond differently to growth factors depending on whether they are cultured on collagen or plastic, a finding supported in the literature for other cell types. Moreover, growth on a malleable surface, such as rubber, is more akin to a native matrix *in vivo* rather than an extremely stiff material such as plastic. An analogy is the response of cells grown on or in a collagen gel whose collagen fibers they can reorganize and from which they can exclude water (Bell et al. 1979; Ehrlich et al. 1988; Greco and Ehrlich 1992). Normal human dermal fibroblasts divide readily on a plastic or collagen coated, plastic surface, but not in a collagen gel (Greco and Ehrlich 1992).

D1 Cyclin is Expressed by TSC and TIF Treated with Serum or IGF-I But is Inhibited by PDGF-BB in TSC (But Not TIF)

In serum-treated, quiescent TSC or TIF, S phase begins at 12 h and continues through 18 h (Banes et al. Submitted). Serum treatment of quiescent TSC or TIF induced cyclin D1 mRNA which was maximal at 8 h post-treatment. However cyclin D1 expression was greater in TIF. In TSC, IGF-I also induced cyclin D1 message with the same kinetics as serum; however, PDGF-BB ablated its expression. Preliminary data from PCR experiments indicates that TIF with or without load express D1 cyclin message. Load and IGF-I in TSC treated with PDGF may reduce the inhibition of cyclin D expression, but these results must be confirmed. The results suggest that TSC may express another cyclin between G_0 and onset of S phase. Therefore, TSC and TIF may regulate entry into S phase using different D1 cyclins.

A Model for Growth Regulation in Tendon

We hypothesize that TSC and TIF respond to mechanical signals but may regulate DNA synthesis and cell division in different ways. *In vivo*, the majority of both tendon cell types is in G_0 (Banes et al. Submitted). A model for growth regulation in tendon is shown in Fig. 8. During trauma to tendon, bleeding occurs from local capillaries or extratendinous sources, providing PDGF and TGF-β from platelets, IGF-I from plasma, as well as proteases that may activate local growth factor stores. TSC migration to the site is stimulated first, followed by activation and utilization of endogenous IGF-I and TGF-β, then TSC cell division. TIF appear to be less involved in an initial hyperplastic reaction but may activate growth factor stores present in the internal tendon compartment. Cell division is stimulated for two to three rounds of replication by external and local stores of growth factors, then division slows as matrix synthesis increases. Production of IGF binding proteins by both TSC and TIF may be one mechanism by which mitogenesis is regulated and reduced. Differential expression of D cyclins and inhibition by PDGF-BB in TSC may also provide additional regulation of cell division.

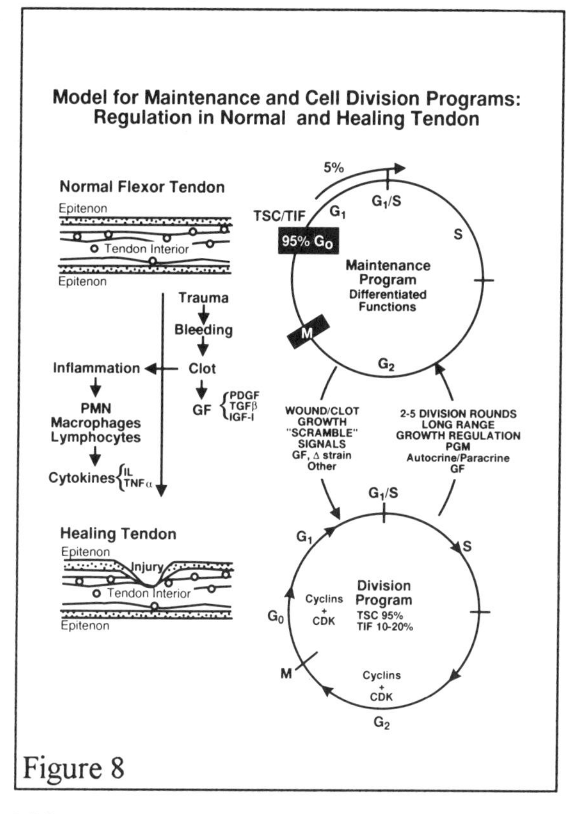

Figure 8

Summary

Isolated tendon cells respond to mechanical load signals by releasing intracellular calcium stores and propagating a calcium wave by IP_3 transmission to cells connected by gap junctions. The response is confined to a local region, 5-10 cells in diameter. The surface cells of tendon, TSC, are more mechanosensitive with respect to stimulation of cell division, than are the TIF. TSC treated with PDGF-BB and cyclic load resulted in mitogenesis. TIF were refractory to load and growth factors. Likewise, TSC and TIF expression of D cyclins is regulated differently by PDGF-BB: it is unaffected in TIF but suppressed in TSC. We conclude that in tendon cells, mechanical load signals are received routinely, are processed, but do not lead necessarily to global cell division. Load in conjunction with growth factors such as PDGF-BB and/or IGF-I can stimulate cell division, principally in the pivotal cell known to populate healing wounds *in vivo*-the epitenon cell. Differential responses in $_{G1}$ regulatory molecules such as cyclins and cyclin-dependent kinases may provide some insight into cell division control in tendon, but expression of growth factor receptors and growth factor binding proteins may also play a regulatory role. Supported by NIH-AR38121.

References

Abrahams, M. Mechanical behavior of tendon in vitro. Med. Biol. Engin. 5:433; 1967.

Allbritton, N. L.; Meyer, T.; Stryer, L. Range of messenger action of calcium ion and inositol 1,4,5-trisphosphate. Science 258:1812-1815; 1992.

Baird, C. W.; Boitano, S.; Brigman, B.; Sanderson, M.; Keagy, B. A.; Banes, A. J. Heparin inhibits mechanically induced calcium wave propagation in smooth muscle cells. Submitted.

Banes, A. J.; et al. Cell cycle kinetics and stimulation of DNA synthesis in tendon synovial and internal fibroblasts released from quiescence by serum or growth factors. J. Orthop. Res. Submitted.

Banes, A. J.; Brigman, B. E.; Baird, C.; Yin, H.; Tsuzaki, M.; Almekinders, L.; Lawrence, W. T. Stimulation of DNA synthesis in tendon synovial and internal fibroblasts released from quiescence by serum, growth factors or mechanical loading. Orthop. Trans. In Press.

Banes, A. J. Mechanical strain and the mammalian cell. In: Frangos, J., ed. Physical forces and the mammalian cell. Orlando: Academic Press; 1993.

Banes, A. J.; Baird, C.; Tsuzaki, M.; Brigman, B.; Lawrence, W. T.; Keagy, B.; Sanderson, M.; Boitano, S. Tendon synovial and internal fibroblasts respond rapidly to mechanical load with IP and Ca^{2+} bursts communicated to adjacent cells, followed by DNA synthesis. Abstracts of the 38th Plastic Surgery Research Council Meeting: 25-29; 1993.

Banes, A. J.; Link, G. W.; Gilbert, J. W.; Monbureau, O. Culturing cells in a mechanically active environment: I. The Flexercell Strain Unit can apply cyclic or static tension or compression to cells in culture. Am. Biotech. Lab. 8:12-22; 1990.

Banes, A. J.; Donlon, K.; Link, G. W.; Gillespie, G. Y.; Bevin, A. G.; Peterson, H. D.; Bynum, D.; Watts, S.; Dahners, L. A simplified method for isolation of tendon synovial cells and internal fibroblasts: Conformation of origin and biological properties. J. Orthop. Res. 6:83-94; 1988a.

Banes, A. J.; Link, G. W.; Bevin, A. G.; Peterson, H. D.; Gillespie, G. Y.; Bynum, D.; Watts, S.; Dahners, L. Tendon synovial cells secrete fibronectin in vivo and in vitro. J. Orthop. Res. 6:73-82; 1988b.

Banes, A. J.; Gilbert, J.; Taylor, D.; Monbureau, O. A new vacuum-operated stress-providing instrument that applies static or variable duration cyclic tension or compression to cells in vitro. J. Cell Sci. 75:35-42; 1985.

Bennett, M. V. L.; Barrio, L. C.; Bargiello, T. A.; Spray, D. C.; Hertzberg, E.; Saez, J. C. Gap junctions: New tools, new answers, new questions. Neuron 6:305-320; 1991.

Berk, A. J.; Sharp, P. A. Sizing and mapping of early adenovirus mRNAs by gel electrophoresis of S1 endonuclease-digested hybrids. Cell 12:721-732; 1977.

Beyer, E. C.; Paul, D. L.; Goodenough, D. A. Connexin family of gap junction proteins. J. Mem. Biol. 116:187-194; 1990.

Boitano, S.; Dirksen, E. R.; Sanderson, M. Intercellular propagation of calcium waves mediated by inositol trisphosphate. Science 258:292-295; 1992.

Brighton, C. T.; Strafford, B.; Gross, S. B.; Leatherwood, D. F.; Williams, U. L.; Pollack, S. R. The proliferative and synthetic response of isolated calvarial bone cells of rats to cyclic biaxial mechanical strain. J. Bone Joint Surg. Am 73:320-331; 1991.

Brigman, B. E.; Yin, H.; Tsuzaki, M.; Lawrence, W. T.; Banes, A. J. Fibronectin in the tendon-synovial complex: Quantitation in vivo and in vitro by ELISA and relative mRNA levels by quantitative PCR and Northern analysis. J. Orthop. Res. In Press.

Chaplin, D. M.; Greenlee, T. K. The development of human digital tendons. J. Anat. 120:253-274; 1975.

Charles, A. C.; Merrill, J. E.; Dirksen, E. R.; Sanderson, M. J. Intercellular calcium signaling via gap junctions in glioma cells. J. Cell. Bio. 118:195-201; 1992.

Christ, G. J.; Moreno, A. P.; Melman, A.; Spray, D. C. Gap junction-mediated intercellular diffusion of Ca^{2+} in cultured human corporal smooth muscle cells. Am. J. Phys. C373-C383; 1992.

Clemmons, D. R. Insulin-like growth factor binding protein control secretion and mechanisms of action. Adv Exp Med Biol 293:113-123; 1991.

Crow, D. S.; Beyer, E. C.; Paul, D. L.; Kobe, S. S.; Lau, A. F. Phosphorylation of connexin43 gap junction protein in uninfected and Rous sarcoma virus-transformed mammalian fibroblasts. Mol. Cell Biol. 10:1754-1763; 1990.

Demer, L. L.; Wortham, C. M.; Dirksen, E. R.; Sanderson, M. J. Mechanical stimulation induces intercellular calcium signaling in bovine aortic endothelial cells. Am. J. Physiol. 264:H2094-H2102; 1993.

Eberhard, D.A. and Holz, R.W. Intracellular Ca2+ activates phospholipase C. Trends Neurosci. 11: 517-520.

Ehrlich, H. P.; Buttle, D. J.; Bernanke, D. H. Physiological variables affecting collagen lattice contraction by human dermal fibroblasts. Exp. Mol. Pathol. 50:220-229; 1989.

Ehrlich, H.P. The modulation of contraction of fibroblast populated collagen lattices by types I, II, and III collagen. Tissue-Cell. 20: 47-50; 1988.

Greco, R.M., Ehrlich, H.P. Differences in cell division and thymidine incorporation with rat and primate fibroblasts in collagen lattices. Tissue-Cell. 24: 843-851; 1992.

Favoloro, J.; Treisman, R.; Kannen, R. Transcription maps of polyoma virus-specific RNA: Analysis by two-dimensional nuclease S1 gel mapping. Meth. Enzymol. 65:718-749; 1980.

Geist, S. T.; Civitelli, R.; Beyer, E. C.; Steinberg. Calcium waves in osteoblastic cells. Mol. Biol. Cell 4:218a; 1993.

Gilbert, J. A.; Weinhold, P. S.; Banes, A. J.; Link, G. W.; Jones, G. L. Strain profiles for circular cell culture plates containing flexible surfaces employed to mechanically deform cells in vitro. J. Biomech. In Press.

Gilman, M. Preparation of cytoplasmic RNA from tissue culture cells. In: Ausubel, F. M.; et al. eds. Current protocols in molecular biology. New York: Greene.; 1989: pp 4.1.2-4.1.6.

Greco, R. M.; Ehrlich, H. P. Differences in cell division and thymidine incorporation with rat and primate fibroblasts in collagen lattices. Tiss. Cell. 24:843-851; 1992.

Greenlee, T. K.; Ross, R. The development of rat flexor digital tendon, a fine structure study. J. Ultrastr. Res. 18:354-376; 1967.

Hasson, H. A.; Engstrom, A. M. C.; Holm, S.; Rosenqvist, A. L. Somatomedin C immunoreactivity in the Achilles tendon varies in a dynamic manner with the mechanical load. Acta Physiol. Scand. 134: 199-208; 1988.

Hassouna, N.; Michot, B.; Bachellerie, J.-P. The complete nucleotide sequence of mouse 28 S rRNA gene. Implications for the process of size increase of the large subunit rRNA in higher eukaryotes. Nucl. Acids Res. 12:3563; 1984.

Herrick, W. C.; Kingsbury, H. B.; Lou, D. Y. S. A study of the normal range of strain, strain rate and stiffness of tendon. J. Biomed. Matls. Res. 12:877-894; 1978.

Hu, P.; Xiao, H.; Brigman, B.; Lawrence, W. T.; Banes, A. J. $_{G1}$ D cyclins are differentially regulated in tendon epitenon and internal fibroblasts. Trans. Orthop Res. Soc. In Press.

Hu, P.; Xiao, H.; Brigman, B.; Lawrence, W. T.; Banes, A. J. Serum, growth factors, and mechanical load stimulate cyclin D1 mRNA and protein in quiescent tendon cells. Molec. Bio. Cell 4:239a; 1993.

Ingber, D. E. Cellular tensegrity: defining new rules of biological design that govern the cytoskeleton. J. Cell Sci. 104:613-627; 1993.

Ingber, D. E. Control of capillary growth and differentiation by extracellular matrix. Use of a tensegrity (tensional integrity) mechanism for signal processing. Chest 99:34s-40s; 1991.

Laird, D. W.; Puranam, K. L.; Revel, J. P. Turnover and phosphorylation dynamics of connexin 43 gap junction protein in cultured cardiac myocytes. Biochem. J. 273:67-72; 1991.

Lehrach, H.; Diamond, D.; Wozney, J. M.; Boedtker, H. RNA molecular weight determinations by gel electrophoresis under denaturing conditions: a critical reexamination. Biochem. 16:4743; 1977.

Mitsumata, M.; Fishel, R. S.; Nerem, R. M.; Alexander, R. W.; Berk, B. C. Fluid shear stress stimulates platelet-derived growth factor expression in endothelial cells. Am. J. Physiol. 265:H3-H8; 1993.

Mullis, K. B.; Faloona, F. A. Specific synthesis of DNA in vitro via a polymerase-catalyzed chain reaction. In Wu, R., ed. Methods in Enzymology 155:335-350; 1987.

Oakes, B. W.; Bialkower, B. Biomechanical and ultrastructural studies on the elastic wing tendon from the domestic fowl. J. Anat. 123:369-387; 1977.

Okamoto, T.; Dorofi, D.; Hu, P.; Tsuzaki, M.; Keagy, B. A.; Banes, A. J. Mechanical load stimulates $\alpha v\beta 3$ mRNA and protein in human umbilical endothelial cells. Mol. Bio. Cell 4:406a; 1993.

Pedersen, D. R.; Bottlang, M.; Brown, T. D.; Banes, A. J. Hyperelastic constitutive properties of polydimethyl siloxane cell cuture membranes. Meeting of the American Society of Mechanical Engineering (ASME); 1993.

Pedersen, D. R.; Brown, T. D.; Banes, A. J. Mechanical behavior of a new substratum for strain-mediated cell culture experiments. NACOB II:The Second North American Congress on Biomechanics. Chicago, IL, Aug 24-28, 1992.

Peracchia, C. Effects of the anesthetics heptanol, halothane and isoflurane on gap junction conductance in crayfish septate axons: a calcium- and hydrogen-independent phenomenon potentiated by caffeine and theophylline, and inhibited by 4-aminopyridine. J. Mem. Bio. 121:67-78; 1991.

Pledger, W. J.; Stiles, C. D.; Antoniades, H. N.; Scher, C. D. An ordered sequence of events is required before BALB/c-3T3 cells become committed to DNA synthesis. Proc. Natl. Acad. Sci. U. S. A. 75:2839-2843; 1978.

Raynal, F.; Michot, B.; Bachellerie, J. P. Complete nucleotide sequence of mouse 18S rRNA gene: Comparisons with other available homologs. FEBS Lett. 167:263; 1984.

Revel, J. P.; Karnovsky, J. J. Hexagonal array of subunits in intercellular junctions of the mouse heart and liver. J. Cell Biol. 33:C7-C12; 1967.

Riederer-Henderson, M. A.; Gauger, A.; Olson, L.; Robertson, C.; Greenlee, T.K. Jr. Attachment and extracellular matrix differences between tendon and synovial fibroblastic cells. In Vitro 19:127-133; 1983.

Robertson, J. D. The occurrence of a subunit pattern in the unit membrane of club ending in Mauthner cell synapses in goldfish brains. J. Cell Biol. 19:201-221; 1963.

Robotewskyj, A.; Griem, M. L.; Davies, P. F. Remodeling of endothelial focal adhesion sites in response to flow: quantitative studies by confocal image analysis. Faseb J. 7(3):A54; 1993.

Rodbard, S. Negative feedback mechanisms in the architecture and function of the connective and cardiovascular tissues. Perspect. on Med. Biol. 13: 507-727; 1970.

Rowe, R. W. D. The structure of the rat tail tendon. Conn. Tiss. Res. 14:9-20; 1985a.

Rowe, R. W. D. The structure of rat tail tendon fasicles. Conn. Tiss. Res. 14:21-30; 1985b.

Saez, J. C.; Connor, J. A.; Spray, D. C.; Bennett, M. V. Hepatocyte gap junctions are permeable to the second messenger, inositol 1,4,5-trisphosphate, and to calcium ions. Proc. Natl. Acad. Sci. U. S. A. 86:2708-2712; 1989.

Sanderson, M. J.; Charles, A. C.; Dirksen, E. R. Mechanical stimulation and intracellular communication increases intracellular Ca^{2+} in epithelial cells. Cell Regulation 1:585-596; 1990.

Sanderson, M. J.; Dirksen, E. R. Inositol trisphosphate mediates intercellular communication between ciliated epithelial cells. J. Cell Biol. 109:304a; 1989.

Sanderson, M. J.; Chow, I.; Dirksen, E. R. Intercellular communictaion between ciliated cells in culture. Am. J. Physiol. 254:C63-C74; 1988.

Sanderson, M. J.; Dirksen, E. R. Mechanosensitivity of cultured ciliated cells from the mammalian respiratory tract: implications for the regulationof mucociliary transport. Proc. Natl. Acad. Sci. U. S. A. 83:7302-7306; 1986.

Stopak, D.; Harris, A. Connective tissue morphogenesis by fibroblast traction. Devel. Biol. 90: 383-398; 1982.

Thomas, P. S. Hybridization of denatured RNA and small DNA fragments transferred to nitrocellulose. Proc. Natl. Acad. Sci. U. S. A 77:5201; 1980.

Tsuzaki, M.; Xiao, H.; Brigman, B.; Lawrence, W. T.; Van Wyk, J.; Banes, A. J. Growth stimulatory and inhibitory factors in tendon: autocrine/paracrine functions of IGF-1 and TGF-b in maintenance and repair. J. Orthop. Res. Submitted.

Tsuzaki, M.; Yamauchi, M.; Banes, A. J. Tendon Collagens: Extracellular matrix composition in shear stress and tensile components of flexor tendon. Conn. Tiss. Res. 29:141-152; 1993.

Vogel, K.; Evanko, S. P. Proteoglycans of fetal bovine tendon. Trans. Orthop. Res. Soc. 13:182; 1988.

Walker, L. B.; Harris, E. H.; Benedict, J. V. Stress-strain relationaship in human cadaveric plantaris tendon: A preliminary study. Med. Elect. Bio. Engin. 2:31; 1964.

Walker, P.; Amstutz, H. C.; Rubinfeld, M. J. Canine tendon studies. II. Biomechanical evaluation of normal and regrown canine tendons. Biomed. Matls. Res. 10:61; 1976.

Wang, N.; Butler, J. P.; Ingber, D. E. Mechanotransduction across the cell surface and through the cytoskeleton. Science 260:1124-1127; 1993.

Whitaker, M. Phosphoinositide second mesengers in eggs and oocytes. In: Inositol lipids in cell signalling, ed. R.H. Michell, A.H. Drummond and C.P. Downes. London: Academic Press, 459-483.

Winston, J. T.; Pledger, W. J. Growth factor regulation of cyclin D1 mRNA expression through protein synthesis dependent and independent mechanisms. Mol. Bio. Cell 4:1133-1144; 1993.

Xiao, H.; Hu, P.; Banes, A. J.; Lawrence, W. T. Cyclin D1 mRNA is serum inducible in both normal human dermal and keloid fibroblasts but temporally later in keloid cells. Mol. Cell. Bio. 4:240a; 1993.

Xiong, Y.; Menninger, J.; Beach, D.; Ward, D. C. Molecular cloning and chromosomal mapping of CCND genes encoding human D-type cyclins. Genomics 13:575-584; 1992.

14
Cytomechanics of Transdifferentiation

M. Opas

Introduction

To differentiate, epithelial cells withdraw from cell cycle and reorganize their cytoarchitecture from a "spread" to a "round" phenotype. These structural changes activate the expression of tissue-specific genes by a process in which the cell structure is both the signal and its medium. The transmembrane linkage complexes integrating extra- and intracellular environments as well as soluble (growth) factor receptors cooperate in transducing signals *via* activation of protein kinases (Ginsberg et al 1992; Hynes, 1992; Humphries et al 1993). Phosphorylation often starts a chain of events leading to DNA synthesis (Zelenka, 1990; Birchmeier et al 1993). In turn, protein kinases exert extensive control over the cell shape, cell contacts and the cytoskeleton. The involvement of a similar circle of inductive steps in transdifferentiation seems very likely as I regard transdifferentiation as a cell's "second chance" for differentiation (Chandebois, 1981; Nathanson, 1986). Both soluble factors and attachment factors affect the ability of cells to adhere, and consequently, to control their shape and proliferate or differentiate (Ingber and Folkman, 1989; Nakagawa et al 1989; Ingber, 1991a; Schwartz et al 1991; Schubert and Kimura, 1991; Sutton et al 1991). Hence it is possible to envisage a scenario in which, in a transdifferentiating cell system, after the decision to switch the genomic programme has been induced by a soluble factor, attachment factors, such as extracellular matrix (ECM) proteins, may further control its realization. Conversely, it is equally possible that action of an attachment factor would reside at the beginning of a chain of events that is further regulated by soluble factors.

Cell Adhesions

Strong cell-substratum adhesions of spread cells *in vitro* are known as focal contacts (Izzard and Lochner, 1976; Opas, 1985; Verschueren, 1985; Burridge et al 1988). Focal contacts are specialized transmembrane cytoskeleton-ECM linkage complexes typically associated with the ends of microfilament bundles (Singer, 1979; Burridge et al 1987; Opas, 1987). The prominent microfilament bundles (stress fibres) are contractile (Kreis and Birchmeier, 1980; Burridge, 1981). Thus, focal contacts are the structures which transmit the tractional forces generated in the cytoplasm to the substratum (Chen et al 1985; Opas, 1987; Aubin and Opas, 1988; Ingber et al 1993) and it has actually been proposed that the arrangement of stress fibres in a spread cell reflects the lines of a tension field generated by cellular contractile activity and spatially restricted by adhesions to the substratum (Greenspan and Folkman, 1977). Focal contacts also comprise a set of specific proteins (Burridge et al 1988), some of which, e.g., vinculin and paxillin, are targets of tyrosine (Tyr) kinases (Sefton and Hunter, 1981; Turner et al 1990; Burridge et al 1992). Focal contacts link the cytoskeleton to proteins of the ECM (Singer, 1979; Singer et al 1984; Hynes et al 1982; Burridge and Fath, 1989) *via* the integrins (Hynes, 1992; Reichardt and Tomaselli, 1991; Ruoslahti, 1991). The focal contact-mediated adhesion is itself regulated by the activation of protein kinases (Hunter, 1989; Chen, 1990; Kellie, 1988; Kellie et al 1991; Kornberg and Juliano, 1992). For example, activation of protein kinase C by growth factors and tumour promoters is accompanied by changes in cell adhesion and affects focal contacts (Woods and Couchman, 1992; Herman et al 1987; Martini and Schachner, 1986; Meigs and Wang, 1986; Turner et al 1989; Zhou et al 1993); oncogenic viruses which encode Tyr kinases (e.g., *src*) have dramatic effects on cell shape and cell adhesion (Rohrschneider et al 1982; Burridge et al 1988; Kellie, 1988; Chen, 1990; Kellie et al 1991). Furthermore, several kinases such as an isoform of protein kinase C (Hyatt et al 1990; Jaken et al 1989) and, most notably, a focal contact-specific Tyr kinase, $pp125^{FAK}$ (Hanks et al 1992; Schaller et al 1992; Schaller and Parsons, 1993), localize to focal contacts. The clustering and immobilization of integrins in focal contacts appears to participate in cell signalling (Guan et al 1991; Kornberg et al 1991; Schwartz et al 1991; Lipp and Niggli, 1993; McNamee et al 1993) *via* Tyr kinase activation (Kornberg and Juliano, 1992; Juliano and Haskill, 1993).

Strong intercellular adhesions of cells packed into sheets are mediated by zonulae adherens. Zonulae adherens associate with circumferential rings of microfilaments which actively contract and maintain epithelial cell sheet under tension (Owaribe et al 1981; Owaribe and Masuda, 1982). It appears that in nonmotile cells both the stress fibres and the circumferential rings of microfilaments are contracting isometrically (Kreis and Birchmeier, 1980; Owaribe et al 1981). Hence, zonulae adherens have been postulated to act as mechanical equivalents of focal contacts (Opas, 1987). The difference between the linear arrangement of stress fibres in spread cells and the circumferential arrangement of microfilament rings in packed cells is likely to be caused by different spatial distribution of adhesions in these cells. While in the spread cells the strongest adhesion is realized *via* a planar arrangement of focal contacts, in the packed cells the strongest adhesion is realized *via* the belt of zonulae adherens circumscribing the apices of the cells. Zonulae adherens and focal contacts share some components (e.g., vinculin), but also have components unique for each adhesion type (Geiger et al 1985; Geiger et al 1987). The cytoplasmic protein talin (Burridge and Feramisco, 1980) and transmembrane integrins (Damsky et al 1985; Horwitz et al 1985; Tamkun et al 1986; Buck and Horwitz, 1987) are specific for focal contacts, while the cytoplasmic proteins, plakoglobin (Franke et al 1987) and catenins (Kemler, 1992; Piepenhagen and Nelson, 1993; Stappert and Kemler, 1993), and transmembrane cadherins (Hirano et al 1987; Takeichi, 1990; Tsukita et al 1992) are specific for zonulae adherens. In epithelial cells the proteins of zonulae adherens are major targets of protein Tyr kinases (Maher et al 1985; Volberg et al 1992; Volberg et al 1991; Matsuyoshi et al 1992; Tsukita et al 1991). The products of c-*src* and c-*yes* are concentrated in the adherens junctions (Tsukita et al 1991) and the junctions have elevated levels of Tyr-phosphoproteins compared to non-junctional areas (Shriver and Rohrschneider, 1981; Maher et al 1985). The level of Tyr phosphorylation in adherens junctions increases after inhibition of protein-Tyr phosphatases with sodium orthovanadate (Hecht and Zick, 1992), suggesting a role of Tyr phosphatases in their function (Tsukita et al 1991; Volberg et al 1991; 1992).

Because many of the regulatory proteins localize to "adhesive" areas of a cell, such as focal contacts and zonulae adherens, these structures have been postulated to be involved not only in adhesion but also in signal transduction (Ben-Ze'ev, 1991).

Adhesion-Related Non-Receptor Tyrosine Kinases

Proto-oncogenes that encode non-receptor Tyr kinases have dramatic effects on cell shape and cell adhesion (Rohrschneider et al 1982; Burridge et al 1988; Kellie, 1988; Chen, 1990; Kellie et al 1991), and indeed the association of these kinases with the cell surface allows them to interact with transmembrane receptors, such as integrins (Ingber, 1991b; Kornberg and Juliano, 1992; Juliano and Haskill, 1993). The product of c-*src* gene, $pp60^{c\text{-}src}$, is widespread (Kellie et al 1991; Brickell, 1992) and is particularly abundant in nervous tissues (Sobue, 1990; Maness and Cox, 1992). $pp60^{c\text{-}src}$ has been localized in the retina throughout development (Sorge et al 1984; Biscardi et al 1991) and is also present in chick RPE both *in vitro* and *in vivo* (Koh, 1992; Moszczynska and Opas, 1994b). At the cellular level, the viral and cellular *src* proteins are associated with plasma membranes, endocytotic vesicles, secretory granules in variety of cell types (David-Pfeuty and Nouvian-Dooghe, 1990; Kaplan et al 1990; 1992). In developing neurons *src* proteins are found in association with motile structures (e.g. filopodia) of neural growth cones (Maness et al 1988;

Sobue, 1990). $pp60^{v\text{-}src}$ associates particularly abundantly with the detergent-insoluble cytoskeletal matrix of adhesion plaques, in "rosettes" (Shriver and Rohrschneider, 1981; Tarone et al 1985; Marchisio et al 1987; Gavazzi et al 1989) and in point cell-substratum contacts (Nermut et al 1991) in transformed cells. In osteoclasts, specialized, "rosette"-like structures, known as podosomes (Marchisio et al 1984; Marchisio et al 1987; Turksen et al 1988) are in part responsible for the formation of the "sealing zone" separating extracellular space into that directly underneath the cell and the one besides it (Aubin, 1992). Interestingly, although it has been assumed that podosomes in osteoclasts and in RSV-transformed cells are similar structures, $pp60^{c\text{-}src}$ in osteoclasts, unlike $pp60^{v\text{-}src}$ in RSV-transformed cells, does not appear in podosomes, but instead it associates with intracellular membranes (Horne et al 1992) and is particularly abundant in the motile (ruffled) cell periphery (Tanaka et al 1992). Similarly to the viral protein, $pp60^{c\text{-}src}$ localizes to cell-cell contacts of several cell types in culture (Tsukita et al 1991) and to intracellular vesicles in c-*src* overexpresser cells (David-Pfeuty and Nouvian-Dooghe, 1990; Kaplan et al 1992). In keratinocytes and retinal cells in culture the c-*src* product is also present intranuclearly (Zhao et al 1992; Moszczynska and Opas, 1994b).

Focal contacts besides focal contact-specific structural proteins such as actin, α-actinin, talin, and vinculin (which itself is a major target of Tyr kinases (Sefton and Hunter, 1981)), contain proteins which may have regulatory functions such as paxillin, which is a major substrate for Tyr kinases and interacts with vinculin *in vitro* (Turner et al 1990; Burridge et al 1992), tensin, which has the SH_2 domain (Davis et al 1991) and is Tyr-phosphorylated in adhesion-dependent manner, zyxin, an α-actinin-binding protein which has proline-rich motifs that are also present in $pp125^{FAK}$ (Crawford and Beckerle, 1991; Crawford et al 1992; Sadler et al 1992), regulatory proteins such as protein kinase C (Jaken et al 1989; Hyatt et al 1990), and most likely others (Maher et al 1985; Beckerle et al 1987). Integrin clustering by a ligand could physically bring together proteins present in focal contact and allow them to interact. The mechanisms of signalling *via* integrins is likely to be unique because short cytoplasmic domains of integrins have no enzyme activity. Attachment of cells to ECM-coated substrata and clustering of integrins causes the enhanced Tyr phosphorylation of a 115-130 kDa complex of proteins (Guan et al 1991; Kornberg et al 1991; Kornberg et al 1992; Burridge et al 1992; Hanks et al 1992) of which a protein with MW of ~ 120 kDa (Guan and Shalloway, 1992; Hanks et al 1992), has been designated as focal adhesion-associated protein Tyr kinase, $pp125^{FAK}$. This adhesion-related Tyr kinase is neither a transmembrane nor a membrane-associated protein (Schaller et al 1992). Because $pp125^{FAK}$ is a substrate for $pp60^{v\text{-}src}$ (Kanner et al 1990; Guan and Shalloway, 1992; Schaller et al 1992) it is plausible that in normal cells $pp60^{c\text{-}src}$ might phosphorylate the $pp125^{FAK}$ rendering it sensitive to cell adhesion (*via* integrins) and to factors that activate Tyr kinases (Zachary and Rozengurt, 1992; Schaller and Parsons, 1993). $pp125^{FAK}$ has been found in every tissue examined so far. During chick development the kinase and its phosphorylation level are regulated. Both the protein expression and its phosphorylation level are upregulated in the first half of embryogenesis and then decline (Turner et al 1993).

Transdifferentiation

The retinal pigment epithelium (RPE) *in vivo* is a single cell-thick sheet of tightly adherent cells that rests on its basement membrane (BM). Embryonic RPE cells differentiate *in vitro*, i.e., pack into an epithelial sheet that actively contracts (Crawford, 1979; Owaribe et al 1981), acquire specific RPE markers (Chu and Grunwald, 1990) and heavy pigmentation (Crawford, 1979; Opas et al 1985; Opas and Dziak, 1988). During their differentiation *in vitro* the RPE cells display a differentiation-dependent organization of the cytoskeleton, adhesiveness and ECM production (Crawford, 1979; 1980; Turksen et al 1983; 1984; 1987; Crawford and Vielkind, 1985; Opas and Kalnins, 1985; Opas et al 1985; Opas, 1989; Owaribe, 1990; Rizzolo, 1991). Adhesion of RPE cells to fibronectin is β_1 integrin-mediated and involves the RGD signal sequence (Philp and Nachmias, 1987; Anderson et al 1990; Chu and Grunwald, 1991a; 1991b), and while RPE cells also recognize the YIGSR signal sequence in laminin, their attachment to laminin and to BMs is a multi-receptor process (Opas and Dziak, 1991; Zhou et al 1993). The differentiation-promoting effects of BM proteins are greatly enhanced if RPE cell spreading is prevented by low rigidity of the substratum (Opas, 1989).

The neural retina (NR), when grown *in vitro* forms epithelial sheets comprising flat cells and neuroblastic cells growing on top of the flat cells (Combes et al 1977; Li and Sheffield, 1986a; 1986b). Under standard culture conditions the neuroblastic cells are lost within a few weeks *in vitro*. The NR cell population which flattens out, proliferates and gives rise to the epithelial sheets is glial and derived from the same progenitor cells as the Müller cells (Li and Sheffield, 1984; Moyer et al 1990).

Transdifferentiation is the process by which the differentiated cells alter their identity from one to another distinct cell type (Okada, 1986; Eguchi and Kodama, 1993). Transdifferentiation has been described in several systems (Parker et al 1980; Moscona and Linser, 1983; Lopashov and Zviadadze, 1984; Eguchi, 1986; 1988; Okada, 1986; Schmid and Alder, 1986; McDevitt, 1989; Beresford, 1990; Watt, 1991; Schmid, 1992). The embryonic retina has been a good model system for transdifferentiation studies because at early stages of development of the retina any of the retinal cell subpopulations may transdifferentiate (Okada et al 1979; Okada, 1980; Moscona, 1986; Eguchi and Kodama, 1993; Okada and Yasuda, 1993). Barr-Nea and Barishak in their early interesting report (1972) described formation of a stratified squamous epithelium from RPE. Eguchi and Okada (1973) provided first unequivocal evidence that a clonal population of RPE cells can transdifferentiate into lens. Subsequently, it has been shown that the potential to transdifferentiate into lens resides also in iris epithelium (Eguchi et al 1974; Yasuda et al 1978; Eguchi, 1979). The RPE is at least bipotential as it can transdifferentiate into either lens epithelium (Eguchi, 1986; Itoh and Eguchi, 1986a; Kosaka et al 1992; Agata et al 1993; Okada and Yasuda, 1993) or NR (Coulombre and Coulombre, 1965; Okada, 1980; Tsunematsu and Coulombre, 1981). Transdifferentiation of the RPE into the NR has been induced by basic fibroblast growth factor (bFGF) after retinectomy *in situ* (Park and Hollenberg, 1989; 1991) as well as *in vitro* (Pittack et al 1991; Guillemot and Cepko, 1992; Opas and Dziak, 1994) and it was shown recently that bFGF also enhances the transdifferentiation of RPE into lens (Hyuga et al 1993). The NR usually transdifferentiates directly into lens epithelium (Okada et al 1979; de Pomerai and Clayton, 1980; Okada, 1980; Moscona and Degenstein, 1981; de Pomerai et al 1982; Moscona, 1986; Okada and Yasuda, 1993). The flat glial cells of NR may, however, convert into RPE cells (Okada et al 1979; Okada, 1980; Pritchard, 1981). Interestingly, this represents complete phenotypic conversion as the RPE cells derived by transdifferentiation from the NR can undergo a secondary transdifferentiation into the lens epithelium (Okada, 1980). The transdifferentiation pathways in the retina might, in a simplistic manner, be depicted as:

RPE ⇄ NR
↘ ↙
Lens

However, it has been shown that transdifferentiation of RPE into lens proceeds *via* an intermediate, dedifferentiated cell "type" in which neither the RPE- nor lens-specific genes are expressed (Agata et al 1993). It is more than likely that intermediate cell "types" for each of the transdifferentiation pathways will be unique and "pathway-specific", that is:

✣ → Lens ← ✻ ← **RPE** → ★ → *NR* → ✤ → *Lens* (?)
↑ ↑
NR → ✪ → *RPE*

where the stars depict different intermediate stages and *italics* denote the cell types obtained by secondary transdifferentiation.

Epithelial phenotype *in vitro* is affected by cell adhesiveness which, in turn, is regulated by the surface properties of its neighbours and by the nature of the ECM (Adams and Watt, 1993; Hay, 1993). Hence, in the RPE culture systems it is possible, by manipulating environmental conditions, to control the proliferative behaviour and differentiation of RPE (Yasuda, 1979; Itoh and Eguchi, 1986a; 1986b; Reh et al 1987; Kosaka et al 1992; Hyuga et al 1993; Opas and Dziak, 1994). These regulatory effects of adhesiveness have been also demonstrated during transdifferentiation of a variety of cell types (Pritchard et al 1978; Moscona et al 1983; Ophir et al 1985; Moscona, 1986; Boukamp and Fusenig, 1993; Schmid et al 1993), including RPE (Yasuda, 1979; Eguchi et al 1982; Reh et al 1987; Opas and Dziak, 1994). As I indicated above, the soluble factor, FGF, is instrumental in transdifferentiation of chick RPE into the NR both *in ovo* and *in vitro*. The abundance of FGFs, their mRNAs and their receptors in retina and RPE (Jeanny et al 1987; Baudouin et al 1990; Cirillo et al 1990; Fayein et al 1990; Heuer et al 1990; Jacquemin et al 1990; Mascarelli et al 1991; Wanaka et al 1991; Gao and Hollyfield, 1992; Ishigooka et al 1992) argues for FGF importance in the retinal cell biology although the precise role of FGFs there is not clear (Campochiaro, 1993). FGFs are bound to heparan sulfate proteoglycans in a variety of BMs (Moscatelli et al 1991), including the BM of the RPE (Jeanny et al 1987). Taken together, these data suggest that both soluble factors (e.g., FGFs) and adhesion *via* either attachment factors (e.g., ECM) or cell-cell interactions, play a pivotal role in determining cell fate choice during retinal differentiation.

The phenotypic expression of epithelial cells *in vitro* is adhesion-dependent in that development of strong cell-substratum adhesion is usually associated with the loss of differentiated traits while strong cell-cell adhesion accompanies expression of the differentiated phenotype. In both RPE and NR, strong cell-substratum and cell-cell adhesions are realized by two subclasses of adherens junctions: focal contacts and zonulae adherens, respectively. Protein phosphorylation on Tyr appears to regulate the function of focal contacts (Rohrschneider et al 1982; Kellie, 1988; Kornberg and Juliano, 1992). In epithelial cells, phosphorylation reciprocally modulates the stability of the two types of adherens junctions: increased phosphorylation degrades the cell-cell junctions, while dephosphorylation stabilizes them (or restores previously degraded junctions). The opposite seems to be true for focal contacts (Volberg et al 1991; 1992). Consequently, because of the importance of adhesion in epithelial cell function, and of the role of Tyr phosphorylation in junctional stability, non-receptor Tyr kinases have been implicated in the regulation of expression of the differentiated epithelial phenotype (Ellis et al 1987; Vardimon et al 1991; Schmidt et al 1992; Zhao et al 1992).

The role of protein phosphorylation on Tyr in regulation of focal contacts (Burridge et al 1992) and the existence of the focal contact specific kinase (Schaller and Parsons, 1993) allow some inferences as to their role in cell phenotypic expression. Dedifferentiated flat RPE cells in culture attach to the substratum with prominent focal contacts (Opas, 1985). Our immunolocalization studies (Moszczynska and Opas, 1994b) show that these focal contacts are highly enriched in $pp125^{FAK}$. When, during their differentiation *in vitro*, RPE cells reach an intermediate stage they start to express weaker cell-substratum adhesiveness. This has been demonstrated directly with interference reflection microscopy as a decrease in number and size of focal contacts (Turksen et al 1983; Opas et al 1985). Fully *in vitro* differentiated RPE cells form sheets of tightly packed cuboidal pigmented cells which express strong cell-cell adhesiveness and are only loosely attached to the underlying pad of ECM. The focal contacts are absent in the fully differentiated RPE cells (Turksen et al 1983; Opas et al 1985), hence $pp125^{FAK}$ cannot be detected there either. Thus, it appears that the expression of $pp125^{FAK}$ is regulated in a coordinated manner with a state of cell-substratum adhesiveness in that the $pp125^{FAK}$ protein is abundant in the cells strongly adhering to the substratum, is limited in distribution to a few sparse focal contacts in RPE cells at intermediate stages of differentiation, and is not detectable in the fully differentiated cells. pp60$^{c\text{-}src}$ localizes to submembranous cortex and vesicles in the flat undifferentiated cells, then it progressively becomes more intranuclear as the cells start to differentiate and pack more closely together. In fully packed differentiated cells the pp60$^{c\text{-}src}$ is predominately intranuclear. Interestingly, the same pattern of distribution and translocation of the kinase is observed the NR and the RPE cells during their differentiation *in vitro*, and translocation of pp60$^{c\text{-}src}$ to the nucleus also occurs during transdifferentiation of NR into the RPE *in vitro* (Moszczynska and Opas, 1994a). Although both pp60$^{v\text{-}src}$ and pp60$^{c\text{-}src}$ associate with the nuclear membranes (David-Pfeuty and Nouvian-Dooghe, 1990; Kaplan et al 1990), the intranuclear localization of pp60$^{c\text{-}src}$ is rather unusual. The redistribution of pp60$^{c\text{-}src}$ in our retinal cell cultures resembles, however, the redistribution of pp60$^{c\text{-}src}$ from the cell surface to the nucleus during differentiation of keratinocytes (Zhao et al 1992).

In summary, in the retinal cultures, the differentiation-associated redistribution of pp60$^{c\text{-}src}$ is accompanied by downregulation of pp125^{FAK} and a switch from predominantly cell-substratum adhesion associated with focal contacts to cell-cell adhesion mediated by zonulae adherens.

As far as zonulae adherens are concerned, the role of Tyr phosphorylation levels in their differentiation-related functions is not clear. An increase in Tyr-phosphorylation of several major proteins during differentiation has been detected in junction-rich areas of the retina during development (Shores and Maness, 1989). The junctional complexes are enriched in pp60$^{c\text{-}src}$ in the RPE cells at the intermediate-to-late stage of differentiation *in vitro* (Moszczynska and Opas, 1994b). This enrichment occurs irrespective of whether the RPE cells are derived by redifferentiation of dedifferentiated RPE or by transdifferentiation of NR. pp60$^{c\text{-}src}$ protein is also elevated in transdifferentiating chick NR where it was found in the cells committed to lens fate (Ellis et al 1987). The data of Ellis *et al.* (1987) parallel our findings in that the highest levels of pp60$^{c\text{-}src}$ are found in cells which are at intermediate-to-late stages of transdifferentiation. The pp60$^{v\text{-}src}$ kinase phosphorylates the catenin-cadherin protein complex thus causing degradation of adherens junctions (Matsuyoshi et al 1992; Behrens et al 1993; Hamaguchi et al 1993). However, while the expression of pp60$^{v\text{-}src}$ in epithelial cells degrades the adherens junctions (Volberg et al 1991; Volberg et al 1992; Behrens et al 1993), the overexpression of pp60$^{c\text{-}src}$ does not (Warren et al 1988; Behrens et al 1993). Circumstantial evidence indicates, however, that epithelial cells expressing elevated levels of pp60$^{c\text{-}src}$ are less rigid than their normal counterparts (Warren et al 1988). The interesting conjecture can therefore be made that cell rigidity is affected by the level of phosphorylation of the cytoskeletal protein, cortactin, a major target of *src* genes (Wu et al 1991). Cortactin is effectively phosphorylated by the activated *src* and is an actin-binding protein localized in the cell cortex of a variety of cell types (Wu and Parsons, 1993). The discovery of cortactin together with the observation of Warren *et al.* (1988) that pp60$^{c\text{-}src}$ "softens" cells poses an interesting question: Does the transient increase of pp60$^{c\text{-}src}$ abundance in cell-cell junctions of differentiating and/or transdifferentiating cells play a mechanical role?

Interplay between mechanical properties of cells and of their extracellular environment (i.e., cytomechanics (Opas, 1987) together with the biochemistry of the environment play important role in regulating the phenotypic expression in transdifferentiating cell systems (Opas, 1989; 1994; Schmid, 1992; Schmid et al 1993; Opas and Dziak, 1994). We have shown (Opas and Dziak, 1994) that while the soluble factor, bFGF, provides the stimulus necessary to redirect the choice of fate of the presumptive RPE from the RPE pathway into the neural pathway, the mechanical properties of the substratum determine the extent to which a neural phenotype is expressed by the transdifferentiating cells. And thus, the RPE transdifferentiates into a pleomorphic neuroepithelium on rigid, two-dimensional BM carpets, into a pseudostratified neuroepithelium on highly malleable BM gels, and into a stratified, NR-like neuroepithelium on its native basement membrane which is of intermediate rigidity. We have noticed recently that, during spontaneous transdifferentiation of RPE into NR that occurs during extended culture (Zhou and Opas, 1994) the transdifferentiating cells upregulate N-CAM expression and always protrude above the cell sheet as if squeezed up by the mechanical constriction exerted on them by the neighbouring RPE cells. The transdifferentiating cells have lower resting intracellular pH than cells that have maintained the RPE phenotype. In addition to the difference in the cell shape and resting pH, the transdifferentiating cells respond to bFGF with profound and sustained alkalinization. In contrast, packed differentiated RPE cells do not respond to bFGF with pH changes. This is consistent with the finding that RPE cells do not proliferate in response to bFGF (Guillemot and Cepko, 1992; Opas and Dziak, 1994) while NR cells respond to bFGF with a stimulation of growth. Our results regarding the bFGF-dependent pH responsiveness show that bFGF acts not on the fully differentiated RPE but only on those cells which have changed shape and already entered some, albeit very initial, stages of transdifferentiation. bFGF provides transdifferentiating N-CAM-positive cells which transdifferentiated from the N-CAM-negative RPE with a signal necessary to transdifferentiate and histodifferentiate further. As our data show that neither that the cuboidal RPE cells tightly packed in a cell sheet nor single round RPE cells respond to bFGF with pH changes, it is obvious that it is the cell shape, which determines RPE cell responsiveness to bFGF. What induces early phenotypic changes of RPE cells, such as the commencement of N-CAM expression, is not clear. However, the transient increase in junctional pp60$^{c\text{-}src}$ abundance leading to alterations in mechanical properties of the cell cortex ("softening") might be a step just preceding commencement of N-CAM expression and acquisition of the FGF responsiveness by round cells.

Acknowledgements

Support by a grant from RP Eye Research Foundation and by a grant MA-9713 from the Medical Research Council of Canada is gratefully acknowledged.

References

Adams, J. C.; Watt, F. M. Regulation of development and differentiation by the extracellular matrix. Development 117:1183-1198; 1993.

Agata, K.; Kobayashi, H.; Itoh, Y.; Mochii, M.; Sawada, K.; Eguchi, G. Genetic characterization of the multipotent dedifferentiated state of pigmented epithelial cells in vitro. Development 118:1025-1030; 1993.

Anderson, D. H.; Guérin, C. J.; Matsumoto, B.; Pfeffer, B. A. Identification and localization of a beta-1 receptor from the integrin family in mammalian retinal pigment epithelial cells. Invest. Ophthalmol. Vis. Sci. 31:81-93; 1990.

Aubin, J. E. Osteoclast adhesion and resorption: The role of podosomes. J. Bone Miner. Res. 7:365-368; 1992.

Aubin, J. E.; Opas, M. Cell adhesion and contractility. In: Davidovitch, Z., ed. The biological mechanisms of tooth eruption and root resorption. Birmingham: EBSCO Media; 1988:43-51.

Barr-Nea, L.; Barishak, Y. R. Behavior of retinal pigment epithelium in organ culture conditions. Ophthalmic Res. 4:321-327; 1972.

Baudouin, C.; Fredj-Reygrobellet, D.; Caruelle, J. -P.; Barritault, D.; Gastaud, P.; Lapalus, P. Acidic fibroblast growth factor distribution in normal human eye and possible implications in ocular pathogenesis. Ophthalmic Res. 22:73-81; 1990.

Beckerle, M. C.; Burridge, K.; DeMartino, G. N.; Croall, D. E. Colocalization of calcium-dependent protease II and one of its substrates at sites of cell adhesion. Cell 51:569-577; 1987.

Behrens, J.; Vakaet, L.; Friis, R.; Winterhager, E.; Van Roy, F.; Mareel, M. M.; Birchmeier, W. Loss of epithelial differentiation and gain of invasiveness correlates with tyrosine phosphorylation of the E-cadherin/β-catenin complex in cells transformed with a temperature-sensitive v-SRC gene. J. Cell Biol. 120:757-766; 1993.

Ben-Ze'ev, A. Animal cell shape changes and gene expression. BioEssays 13:207-212; 1991.

Beresford, W. A. Direct transdifferentiation: can cells change their phenotype without dividing? Cell Differ. Dev. 29:81-93; 1990.

Birchmeier, C.; Sonnenberg, E.; Weidner, K. M.; Walter, B. Tyrosine kinase receptors in the control of epithelial growth and morphogenesis during development. BioEssays 15:185-190; 1993.

Biscardi, J. S.; Shores, C. G.; Maness, P. F. Elevated protein tyrosine phosphorylation in the optic tract of the chick embryo. Curr. Eye Res. 10:1121-1128; 1991.

Boukamp, P.; Fusenig, N. E. "Trans-differentiation" from epidermal to mesenchymal/myogenic phenotype is associated with a drastic change in cell-cell and cell-matrix adhesion molecules. J. Cell Biol. 120:981-993; 1993.

Brickell, P. M. The p60c-src family of protein-tyrosine kinases: structure, regulation, and function. Crit. Rev. Oncog. 3:401-446; 1992.

Buck, C. A.; Horwitz, A. F. Cell surface receptors for extracellular matrix molecules. Annu. Rev. Cell Biol. 3:179-205; 1987.

Burridge, K. Are stress fibers contractile. Nature 294:691-692; 1981.
Burridge, K.; Molony, L.; Kelly, T. Adhesion plaques: sites of transmembrane interaction between the extracellular matrix and the actin cytoskeleton. J. Cell Sci. Suppl. 8:211-229; 1987.
Burridge, K.; Fath, K.; Kelly, T.; Nuckolls, G.; Turner, C. Focal adhesions: transmembrane junctions between the extracellular matrix and the cytoskeleton. Annu. Rev. Cell Biol. 4:487-525; 1988.
Burridge, K.; Turner, C. E.; Romer, L. H. Tyrosine phosphorylation of paxillin and $pp125^{FAK}$ accompanies cell adhesion to extracellular matrix: A role in cytoskeletal assembly. J. Cell Biol. 119:893-903; 1992.
Burridge, K.; Fath, K. Focal contacts: Transmembrane links between the extracellular matrix and the cytoskeleton. BioEssays 10:104-108; 1989.
Burridge, K.; Feramisco, J. R. Microinjection and localization of a 130K protein in living fibroblasts: A relationship to actin and fibronectin. Cell 19:587-595; 1980.
Campochiaro, P. A. Cytokine production by retinal pigmented epithelial cells. Int. Rev. Cytol. 146:75-82; 1993.
Chandebois, R. The problem of automation in animal development: confrontation of the concept of cell sociology with biochemical data. Acta Biotheor. 30:143-169; 1981.
Chen, W. -T.; Hasegawa, E.; Hasegawa, T.; Weinstock, C.; Yamada, K. M. Development of cell surface linkage complexes in cultured fibroblasts. J. Cell Biol. 100:1103-1114; 1985.
Chen, W. -T. Transmembrane interactions at cell adhesion and invasion sites. Cell Differ. Dev. 32:329-336; 1990.
Chu, P.; Grunwald, G. B. Generation and characterization of monoclonal antibodies specific for the retinal pigment epithelium. Invest. Ophthalmol. Vis. Sci. 31:856-862; 1990.
Chu, P.; Grunwald, G. B. Identification of the 2A10 antigen of retinal pigment epithelium as a β1 subunit of integrin. Invest. Ophthalmol. Vis. Sci. 32:1757-1762; 1991a.
Chu, P.; Grunwald, G. B. Functional inhibition of retinal pigment epithelial cell-substrate adhesion with a monoclonal antibody against the β1 subunit of integrin. Invest. Ophthalmol. Vis. Sci. 32:1763-1769; 1991b.
Cirillo, A.; Arruti, C.; Courtois, Y.; Jeanny, J. -C. Localization of basic fibroblast growth factor binding sites in the chick embryonic neural retina. Differentiation 45:161-167; 1990.
Combes, P. C.; Privat, A.; Pessac, B.; Calothy, G. Differentiation of chick embryo neuroretina cells in monolayer cultures. An ultrastructural study. I. Seven-day retina. Cell Tissue Res. 185:159-173; 1977.
Coulombre, J. L.; Coulombre, A. J. Regeneration of neural retina from the pigmented epithelium in the chick embryo. Dev. Biol. 12:79-92; 1965.
Crawford, A. W.; Michelsen, J. W.; Beckerle, M. C. An interaction between zyxin and α-actinin. J. Cell Biol. 116:1381-1393; 1992.
Crawford, A. W.; Beckerle, M. C. Purification and characterization of zyxin, an 82,000-dalton component of adherens junctions. J. Biol. Chem. 266:5847-5853; 1991.
Crawford, B. Cloned pigmented retinal epithelium: The role of microfilaments in the differentiation of cell shape. J. Cell Biol. 81:301-315; 1979.
Crawford, B. J. Development of the junctional complex during differentiation of chick pigmented epithelial cells in clonal culture. Invest. Ophthalmol. Vis. Sci. 19:223-237; 1980.

Crawford, B. J.; Vielkind, U. Location and possible function of fibronectin and laminin in clones of chick retinal pigmented epithelial cells. In Vitro Cell. Dev. Biol. 21:79-87; 1985.

Damsky, C. H.; Knudsen, K. A.; Bradley, D.; Buck, C. A.; Horwitz, A. Distribution of the cell substratum attachment (CSAT) antigen on myogenic and fibroblastic cells in culture. J. Cell Biol. 100:1528-1539; 1985.

David-Pfeuty, T.; Nouvian-Dooghe, Y. Immunolocalization of the cellular *src* protein in interphase and mitotic NIH c-*src* overexpresser cells. J. Cell Biol. 111:3097-3116; 1990.

Davis, S.; Lu, M. L.; Lo, S. H.; Lin, S.; Butler, J. A.; Druker, B. J.; Roberts, T. M.; An, Q.; Chen, L. B. Presence of an SH2 domain in the actin-binding protein tensin. Science 252:712-715; 1991.

de Pomerai, D. I.; Carr, A.; Soranson, J. A.; Gali, M. A. Pathways of differentiation in chick embryo neuroretinal cultures. Differentiation 22:6-11; 1982.

de Pomerai, D. I.; Clayton, R. M. The influence of growth-inhibiting and growth-promoting medium conditions on crystallin accumulation in transdifferentiating cultures of embryonic chick neural retina. Dev. Growth Diff. 22:49-60; 1980.

Eguchi, G.; Abe, S. -I.; Watanabe, K. Differentiation of lens-like structures from newt iris epithelial cells in vitro. Proc. Natl. Acad. Sci. USA 71:5052-5056; 1974.

Eguchi, G. Transdifferentiation in pigmented epithelial cells of vertebrate eyes in vitro. In: Ebert, J.D.; Okada, T.S., eds. Mechanisms of Cell Change. New York: John Wiley and Sons; 1979:273-291.

Eguchi, G.; Masuda, A.; Karasawa, Y.; Kodama, R.; Itoh, Y. Microenvironments controlling the transdifferentiation of vertebrate pigmented epithelial cells in in vitro culture. Adv. Exp. Med. Biol. 158:209-221; 1982.

Eguchi, G. Instability in cell commitment of vertebrate pigmented epithelial cells and their transdifferentiation into lens cells. Curr. Top. Dev. Biol. 20:21-37; 1986.

Eguchi, G. Cellular and molecular background of wolffian lens regeneration. Cell Differ. Dev. 25 Suppl:147-158; 1988.

Eguchi, G.; Kodama, R. Transdifferentiation. Curr. Opin. Cell Biol. 5:1023-1028; 1993.

Eguchi, G.; Okada, T. S. Differentiation of lens tissue from the progeny of chick retinal pigment cells cultured in vitro: A demonstration of a switch of cell types in clonal culture. Proc. Natl. Acad. Sci. USA 70:1495-1499; 1973.

Ellis, D. K.; Carr, A.; de Pomerai, D. I. pp60c-src expression in transdifferentiating cultures of embryonic chick neural retina cells. Development 101:847-856; 1987.

Fayein, N. A.; Courtois, Y.; Jeanny, J. C. Ontogeny of basic fibroblast growth factor binding sites in mouse ocular tissues. Exp. Cell Res. 188:75-88; 1990.

Franke, W. W.; Kapprell, H.; Cowin, P. Immunolocalization of plakoglobin in endothelial junctions: identification as a special type of zonulae adhaerentes. Biol. Cell 59:205-218; 1987.

Gao, H.; Hollyfield, J. G. Basic fibroblast growth factor (bFGF) immunolocalization in the rodent outer retina demonstrated with an anti-rodent bFGF antibody. Brain Res. 585:355-360; 1992.

Gavazzi, I.; Nermut, M. V.; Marchisio, P. C. Ultrastructure and gold-immunolabelling of cell-substratum adhesions (podosomes) in RSV-transformed BHK cells. J. Cell Sci. 94:85-99; 1989.

Geiger, B.; Volk, T.; Volberg, T. Molecular heterogeneity of adherens junctions. J. Cell Biol. 101:1523-1531; 1985.

Geiger, B.; Volk, T.; Volberg, T.; Bendori, R. Molecular interactions in adherens-type contacts. J. Cell Sci. Suppl. 8:251-272; 1987.

Ginsberg, M. H.; Du, X.; Plow, E. F. Inside-out integrin signalling. Curr. Opin. Cell Biol. 4:766-771; 1992.

Greenspan, H. P.; Folkman, J. Hypotheses on cell adhesion and actin cables. J. Theor. Biol. 65:397-398; 1977.

Guan, J. -L.; Trevithick, J. E.; Hynes, R. O. Fibronectin/integrin interaction induces tyrosine phosphorylation of a 120-kDa protein. Cell Regul. 2:951-964; 1991.

Guan, J. -L.; Shalloway, D. Regulation of focal adhesion-associated protein tyrosine kinase by both cellular adhesion and oncogenic transformation. Nature 358:690-692; 1992.

Guillemot, F.; Cepko, C. L. Retinal fate and ganglion cell differentiation are potentiated by acidic FGF in an *in vitro* assay of early retinal development. Development 114:743-754; 1992.

Hamaguchi, M.; Matsuyoshi, N.; Ohnishi, Y.; Gotoh, B.; Takeichi, M.; Nagai, Y. $p60^{v\text{-}src}$ causes tyrosine phosphorylation and inactivation of the N-cadherin-catenin cell adhesion system. EMBO J. 12:307-314; 1993.

Hanks, S. K.; Calalb, M. B.; Harper, M. C.; Patel, S. K. Focal adhesion protein-tyrosine kinase phosphorylated in response to cell attachment to fibronectin. Proc. Natl. Acad. Sci. USA 89:8487-8491; 1992.

Hay, E. D. Extracellular matrix alters epithelial differentiation. Curr. Opin. Cell Biol. 5:1029-1035; 1993.

Hecht, D.; Zick, Y. Selective inhibition of protein tyrosine phosphatase activities by H_2O_2 and vanadate *in vitro*. Biochem. Biophys. Res. Commun. 188:773-779; 1992.

Herman, B.; Roe, M. W.; Harris, C.; Wray, B.; Clemmons, D. Platelet-derived growth factor-induced alterations in vinculin distribution in porcine vascular smooth muscle cells. Cell Motil. Cytoskeleton 8:91-105; 1987.

Heuer, J. G.; von Bartheld, C. S.; Kinoshita, Y.; Evers, P. C.; Bothwell, M. Alternating phases of FGF receptor and NGF receptor expression in the developing chicken nervous system. Neuron 5:283-296; 1990.

Hirano, S.; Nose, A.; Hatta, K.; Kawakami, A.; Takeichi, M. Calcium-dependent cell-cell adhesion molecules (cadherins): Subclass specificities and possible involvement of actin bundles. J. Cell Biol. 105:2501-2510; 1987.

Horne, W. C.; Neff, L.; Chatterjee, D.; Lomri, A.; Levy, J. B.; Baron, R. Osteoclasts express high levels of $pp60^{c\text{-}src}$ in association with intracellular membranes. J. Cell Biol. 119:1003-1013; 1992.

Horwitz, A.; Duggan, K.; Greggs, R.; Decker, C.; Buck, C. The cell substrate attachment (CSAT) antigen has properties of a receptor for laminin and fibronectin. J. Cell Biol. 101:2134-2144; 1985.

Humphries, M. J.; Mould, A. P.; Tuckwell, D. S. Dynamic aspects of adhesion receptor function - Integrins both twist and shout. BioEssays 15:391-397; 1993.

Hunter, T. Protein modification: phosphorylation on tyrosine residues. Curr. Opin. Cell Biol. 1:1168-1181; 1989.

Hyatt, S. L.; Klauck, T.; Jaken, S. Protein kinase C is localized in focal contacts of normal but not transformed fibroblasts. Mol. Carcinog. 3:45-53; 1990.

Hynes, R. O.; Destree, A. T.; Wagner, D. D. Relationships between microfilaments, cell-substratum adhesion, and fibronectin. Cold Spring Harbor Symp. Quant. Biol. 46:659-670; 1982.

Hynes, R. O. Integrins: Versatility, modulation, and signaling in cell adhesion. Cell 69:11-25; 1992.

Hyuga, M.; Kodama, R.; Eguchi, G. Basic fibroblast growth factor as one of the essential factors regulating lens transdifferentiation of pigmented epithelial cells. Int. J. Dev. Biol. 37:319-326; 1993.

Ingber, D. E. Extracellular matrix and cell shape: Potential control points for inhibition of angiogenesis. J. Cell. Biochem. 47:236-241; 1991a.

Ingber, D. E. Integrins as mechanochemical transducers. Curr. Opin. Cell Biol. 3:841-848; 1991b.

Ingber, D. E.; Karp, S.; Plopper, G.; Hansen, L.; Mooney, D. Mechanochemical transduction across extracellular matrix and through the cytoskeleton. In: Frangos, J.A.; Ives, C.L., eds. Physical forces and the mammalian cell. San Diego: Academic Press; 1993:61-79.

Ingber, D. E.; Folkman, J. Mechanochemical switching between growth and differentiation during fibroblast growth factor-stimulated angiogenesis in vitro: role of extracellular matrix. J. Cell Biol. 109:317-330; 1989.

Ishigooka, H.; Aotaki-Keen, A. E.; Hjelmeland, L. M. Subcellular localization of bFGF in human retinal pigment epithelium in vitro. Exp. Eye Res. 55:203-214; 1992.

Itoh, Y.; Eguchi, G. In vitro analysis of cellular metaplasia from pigmented epithelial cells to lens phenotypes: a unique model system for studying cellular and molecular mechanisms of "transdifferentiation". Dev. Biol. 115:353-362; 1986a.

Itoh, Y.; Eguchi, G. Enhancement of expression of lens phenotype in cultures of pigmented epithelial cells by hyaluronidase in the presence of phenylthiourea. Cell Differ. 18:173-182; 1986b.

Izzard, C. S.; Lochner, L. R. Cell-to-substrate contacts in living fibroblasts: an interference reflexion study with an evaluation of the technique. J. Cell Sci. 21:129-159; 1976.

Jacquemin, E.; Halley, C.; Alterio, J.; Laurent, M.; Courtois, Y.; Jeanny, J. C. Localization of acidic fibroblast growth factor (aFGF) mRNA in mouse and bovine retina by in situ hybridization. Neurosci. Lett. 116:23-28; 1990.

Jaken, S.; Leach, K.; Klauck, T. Association of type 3 protein kinase C with focal contacts in rat embryo fibroblasts. J. Cell Biol. 109:697-704; 1989.

Jeanny, J. C.; Fayein, N.; Moenner, M.; Chevallier, B.; Barritault, D.; Courtois, Y. Specific fixation of bovine brain and retinal acidic and basic fibroblast growth factors to mouse embryonic eye basement membranes. Exp. Cell Res. 171:63-75; 1987.

Juliano, R. L.; Haskill, S. Signal transduction from the extracellular matrix. J. Cell Biol. 120:577-585; 1993.

Kanner, S. B.; Reynolds, A. B.; Vines, R. R.; Parsons, J. T. Monoclonal antibodies to individual tyrosine-phosphorylated protein substrates of oncogene-encoded tyrosine kinases. Proc. Natl. Acad. Sci. USA 87:3328-3332; 1990.

Kaplan, J. M.; Varmus, H. E.; Bishop, J. M. The *src* protein contains multiple domains for specific attachment to membranes. Mol. Cell. Biol. 10:1000-1009; 1990.

Kaplan, K. B.; Swedlow, J. R.; Varmus, H. E.; Morgan, D. O. Association of $p60^{c-src}$ with endosomal membranes in mammalian fibroblasts. J. Cell Biol. 118:321-333; 1992.

Kellie, S. Cellular transformation, tyrosine kinase oncogenes, and the cellular adhesion plaque. BioEssays 8:25-30; 1988.

Kellie, S.; Horvath, A. R.; Elmore, M. A. Cytoskeletal targets for oncogenic tyrosine kinases. J. Cell Sci. 99:207-211; 1991.

Kemler, R. Classical cadherins. Semin. Cell Biol. 3:149-155; 1992.

Koh, S. -W. M. The pp60$^{c\text{-}src}$ in retinal pigment epithelium and its modulation by vasoactive intestinal peptide. Cell Biol. Int. Rep. 16:1003-1014; 1992.

Kornberg, L.; Earp, H. S.; Parsons, J. T.; Schaller, M.; Juliano, R. L. Cell adhesion or integrin clustering increases phosphorylation of a focal adhesion-associated tyrosine kinase. J. Biol. Chem. 267:23439-23442; 1992.

Kornberg, L.; Juliano, R. L. Signal transduction from the extracellular matrix: The integrin-tyrosine kinase connection. Trends Pharmacol. Sci. 13:93-95; 1992.

Kornberg, L. J.; Earp, H. S.; Turner, C. E.; Prockop, C.; Juliano, R. L. Signal transduction by integrins: Increased protein tyrosine phosphorylation caused by clustering of β_1 integrins. Proc. Natl. Acad. Sci. USA 88:8392-8396; 1991.

Kosaka, J.; Watanabe, K.; Eguchi, G. Transdifferentiation of chicken retinal pigmented epithelial cells in serum-free culture. Exp. Eye Res. 55:261-267; 1992.

Kreis, T. E.; Birchmeier, W. Stress fiber sarcomeres of fibroblasts are contractile. Cell 22:555-561; 1980.

Li, H. -P.; Sheffield, J. B. Isolation and characterization of flat cells, a subpopulation of the embryonic chick retina. Tissue Cell 16:843-857; 1984.

Li, H. -P.; Sheffield, J. B. Retinal flat cells participate in the formation of fibers by retinal neuroblasts in vitro. Time lapse video studies. Invest. Ophthalmol. Vis. Sci. 27:307-315; 1986a.

Li, H. -P.; Sheffield, J. B. Retinal flat cells are a substrate that facilitates retinal neuron growth and fiber formation. Invest. Ophthalmol. Vis. Sci. 27:296-306; 1986b.

Lipp, P.; Niggli, E. Ratiometric confocal Ca^{2+}-measurements with visible wavelength indicators in isolated cardiac myocytes. Cell Calcium 14:359-372; 1993.

Lopashov, G. V.; Zviadadze, K. G. Mechanisms of transdifferentiation and their relation to induction and competence [Rus]. Ontogenez 15:339-347; 1984.

Maher, P. A.; Pasquale, E. B.; Wang, J. Y.; Singer, S. J. Phosphotyrosine-containing proteins are concentrated in focal adhesions and intercellular junctions in normal cells. Proc. Natl. Acad. Sci. USA 82:6576-6580; 1985.

Maness, P. F.; Aubry, M.; Shores, C. G.; Frame, L.; Pfenninger, K. H. c-src gene product in developing rat brain is enriched in nerve growth cone membranes. Proc. Natl. Acad. Sci. USA 85:5001-5005; 1988.

Maness, P. F.; Cox, M. E. Protein tyrosine kinases in nervous system development. Semin. Cell Biol. 3:117-126; 1992.

Marchisio, P. C.; Cirillo, D.; Naldini, L.; Primavera, M. V.; Teti, A.; Zambonin-Zallone, A. Cell-substratum interaction of cultured avian osteoclasts is mediated by specific adhesion structures. J. Cell Biol. 99:1696-1705; 1984.

Marchisio, P. C.; Cirillo, D.; Teti, A.; Zambonin-Zallone, A.; Tarone, G. Rous sarcoma virus-transformed fibroblasts and cells of monocytic origin display a peculiar dot-like organization of cytoskeletal proteins involved in microfilament-membrane interactions. Exp. Cell Res. 169:202-214; 1987.

Martini, R.; Schachner, M. Immunoelectron microscopic localization of neural cell adhesion molecules (L1, N-CAM, and MAG) and their shared carbohydrate epitope and myelin basic protein in developing sciatic nerve. J. Cell Biol. 103:2439-2448; 1986.

Mascarelli, F.; Tassin, J.; Courtois, Y. Effect of FGFs on adult bovine Muller cells: proliferation, binding and internalization. Growth Factors 4:81-95; 1991.

Matsuyoshi, N.; Hamaguchi, M.; Taniguchi, S.; Nagafuchi, A.; Tsukita, S.; Takeichi, M. Cadherin-mediated cell-cell adhesion is perturbed by v-*src* tyrosine phosphorylation in metastatic fibroblasts. J. Cell Biol. 118:703-714; 1992.

McDevitt, D. S. Transdifferentiation in animals. A model for differentiation control. Dev. Biol. 6:149-173; 1989.

McNamee, H. P.; Ingber, D. E.; Schwartz, M. A. Adhesion to fibronectin stimulates inositol lipid synthesis and enhances PDGF-induced inositol lipid breakdown. J. Cell Biol. 121:673-678; 1993.

Meigs, J. B.; Wang, Y. -L. Reorganization of alpha-actinin and vinculin induced by a phorbol ester in living cells. J. Cell Biol. 102:1430-1438; 1986.

Moscatelli, D.; Flaumenhaft, R.; Saksela, O. Interaction of basic fibroblast growth factor with extracellular matrix and receptors. Ann. NY Acad. Sci. 638:177-181; 1991.

Moscona, A. A.; Brown, M.; Degenstein, L.; Fox, L.; Soh, B. M. Transformation of retinal glia cells into lens phenotype: expression of MP26, a lens plasma membrane antigen. Proc. Natl. Acad. Sci. USA 80:7239-7243; 1983.

Moscona, A. A. Conversion of retina glia cells into lenslike phenotype following disruption of normal cell contacts. Curr. Top. Dev. Biol. 20:1-19; 1986.

Moscona, A. A.; Degenstein, L. Lentoids in aggregates of embryonic neural retina cells. Cell Differ. 10:39-46; 1981.

Moscona, A. A.; Linser, P. Developmental and experimental changes in retinal glia cells: cell interactions and control of phenotype expression and stability. Curr. Top. Dev. Biol. 18:155-188; 1983.

Moszczynska, A.; Opas, M. Regulation of adhesion-related protein tyrosine kinases during in vitro differentiation of retinal pigment epithelial cells: Translocation of $pp60^{c\text{-}src}$ to the nucleus is accompanied by downregulation of $pp125^{FAK}$. Biochem. Cell Biol. 1994a. (in press).

Moszczynska, A.; Opas, M. Involvement of non-receptor protein tyrosine kinases in phenotypic expression by cells of retinal origin. 1994b. Int. J. Dev. Biol. (in press).

Moyer, M.; Bullrich, F.; Sheffield, J. B. Emergence of flat cells from glia in stationary cultures of embryonic chick neural retina. In Vitro Cell. Dev. Biol. 26:1073-1078; 1990.

Nakagawa, S.; Pawelek, P.; Grinnell, F. Extracellular matrix organization modulates fibroblast growth and growth factor responsiveness. Exp. Cell Res. 182:572-582; 1989.

Nathanson, M. A. Transdifferentiation of skeletal muscle into cartilage: transformation or differentiation? Curr. Top. Dev. Biol. 20:39-62; 1986.

Nermut, M. V.; Eason, P.; Hirst, E. M. A.; Kellie, S. Cell/substratum adhesions in RSV-transformed rat fibroblasts. Exp. Cell Res. 193:382-397; 1991.

Okada, T. S.; Yasuda, K.; Araki, M.; Eguchi, G. Possible demonstration of multipotential nature of embryonic neural retina by clonal cell culture. Dev. Biol. 68:600-617; 1979.

Okada, T. S. Cellular metaplasia or transdifferentiation as a model for retinal cell differentiation. Curr. Top. Dev. Biol. 16:349-380; 1980.

Okada, T. S. Transdifferentiation in animal cells: Fact or artifact. Dev. Growth Diff. 28:213-221; 1986.

Okada, T. S.; Yasuda, K. How are non-lenticular cells ready for transdifferentiation. Dev. Dyn. 196:273-275; 1993.

Opas, M. The focal adhesions of chick retinal pigmented epithelial cells. Can. J. Biochem. Cell Biol. 63:553-563; 1985.

Opas, M.; Turksen, K.; Kalnins, V. I. Adhesiveness and distribution of vinculin and spectrin in retinal pigmented epithelial cells during growth and differentiation in vitro. Dev. Biol. 107:269-280; 1985.

Opas, M. The transmission of forces between cells and their environment. In: Bereiter-Hahn, J.; Anderson, O.R.; Reif, W.-E., eds. Cytomechanics. Berlin: Springer-Verlag; 1987:273-285.

Opas, M. Expression of the differentiated phenotype by epithelial cells in vitro is regulated by both biochemistry and mechanics of the substratum. Dev. Biol. 131:281-293; 1989.

Opas, M. Substratum mechanics and cell differentiation. Int. Rev. Cytol. 150:119-137; 1994.

Opas, M.; Dziak, E. Effects of substrata and method of tissue dissociation on adhesion, cytoskeleton, and growth of chick retinal pigmented epithelium in vitro. In Vitro Cell. Dev. Biol. 24:885-892; 1988.

Opas, M.; Dziak, E. Adhesion, spreading, and proliferation of cells on protein carpets: effects of stability of a carpet. In Vitro Cell. Dev. Biol. 27A:878-885; 1991.

Opas, M.; Dziak, E. bFGF-induced transdifferentiation of RPE to neuronal progenitors is regulated by the mechanical properties of the substratum. Dev. Biol. 1994. (in press).

Opas, M.; Kalnins, V. I. Spatial distribution of cortical proteins in cells of epithelial sheets. Cell Tissue Res. 239:451-454; 1985.

Ophir, I.; Moscona, A. A.; Loya, N.; Ben-Shaul, Y. Formation of lentoids from retina gliocytes: ultrastructural study. Cell Differ. 17:149-157; 1985.

Owaribe, K.; Kodama, R.; Eguchi, G. Demonstration of contractility of circumferential actin bundles and its morphogenic significance in pigmented epithelium in vitro and in vivo. J. Cell Biol. 90:507-514; 1981.

Owaribe, K. The cytoskeleton of retinal pigment epithelial cells. In: Osborne, N.; Chader, J., eds. Progress in retinal research. Pergamon Press; 1990:23-49.

Owaribe, K.; Masuda, H. Isolation and characterization of circumferential microfilament bundles from retinal pigment epithelial cells. J. Cell Biol. 95:310-315; 1982.

Park, C. M.; Hollenberg, M. J. Basic fibroblast growth factor induces retinal regeneration in vivo. Dev. Biol. 134:201-205; 1989.

Park, C. M.; Hollenberg, M. J. Induction of retinal regeneration in vivo by growth factors. Dev. Biol. 148:322-333; 1991.

Parker, K. K.; Norenberg, M. D.; Vernadakis, A. "Transdifferentiation" of C6 glial cells in culture. Science 208:179-181; 1980.

Philp, N. J.; Nachmias, V. T. Polarized distribution of integrin and fibronectin in retinal pigment epithelium. Invest. Ophthalmol. Vis. Sci. 28:1275-1280; 1987.

Piepenhagen, P. A.; Nelson, W. J. Defining E-cadherin-associated protein complexes in epithelial cells: Plakoglobin, β- and gamma-catenin are distinct components. J. Cell Sci. 104:751-762; 1993.

Pittack, C.; Jones, M.; Reh, T. A. Basic fibroblast growth factor induces retinal pigment epithelium to generate neural retina in vitro. Development 113:577-588; 1991.

Pritchard, D. J.; Clayton, R. M.; de Pomerai, D. I. 'Transdifferentiation' of chicken neural retina into lens and pigment epithelium in culture: controlling influences. J. Embryol. Exp. Morphol. 48:1-21; 1978.

Pritchard, D. J. Transdifferentiation of chicken embryo neural retina into pigment epithelium: indications of its biochemical basis. J. Embryol. Exp. Morphol. 62:47-62; 1981.

Reh, T. A.; Nagy, T.; Gretton, H. Retinal pigmented epithelial cells induced to transdifferentiate to neurons by laminin. Nature 330:68-71; 1987.

Reichardt, L. F.; Tomaselli, K. J. Extracellular matrix molecules and their receptors: Functions in neural development. Annu. Rev. Neurosci. 14:531-570; 1991.

Rizzolo, L. J. Basement membrane stimulates the polarized distribution of integrins but not the Na,K-ATPase in the retinal pigment epithelium. Cell Regul. 2:939-949; 1991.

Rohrschneider, L.; Rosok, M.; Shriver, K. Mechanism of transformation by rous sarcoma virus: Events within adhesion plaques. Cold Spring Harbor Symp. Quant. Biol. 46:953-968; 1982.

Ruoslahti, E. Integrins. J. Clin. Invest. 87:1-5; 1991.

Sadler, I.; Crawford, A. W.; Michelsen, J. W.; Beckerle, M. C. Zyxin and cCRP: Two interactive LIM domain proteins associated with the cytoskeleton. J. Cell Biol. 119:1573-1587; 1992.

Schaller, M. D.; Borgman, C. A.; Cobb, B. S.; Vines, R. R.; Reynolds, A. B.; Parsons, J. T. $pp125^{FAK}$, A structurally distinctive protein-tyrosine kinase associated with focal adhesions. Proc. Natl. Acad. Sci. USA 89:5192-5196; 1992.

Schaller, M. D.; Parsons, J. T. Focal adhesion kinase: An integrin-linked protein tyrosine kinase. Trends Cell Biol. 3:258-262; 1993.

Schmid, V. Transdifferentiation in medusae. Int. Rev. Cytol. 142:213-261; 1992.

Schmid, V.; Baader, C.; Bucciarelli, A.; Reber-Müller, S. Mechanochemical interactions between striated muscle cells of jellyfish and grafted extracellular matrix can induce and inhibit DNA replication and transdifferentiation *in vitro*. Dev. Biol. 155:483-496; 1993.

Schmid, V.; Alder, H. The potential for transdifferentiation of differentiated medusa tissues in vitro. Curr. Top. Dev. Biol. 20:117-135; 1986.

Schmidt, J. W.; Brugge, J. S.; Nelson, W. J. pp60src tyrosine kinase modulates P19 embryonal carcinoma cell fate by inhibiting neuronal but not epithelial differentiation. J. Cell Biol. 116:1019-1033; 1992.

Schubert, D.; Kimura, H. Substratum-growth factor collaborations are required for the mitogenic activities of activin and FGF on embryonal carcinoma cells. J. Cell Biol. 114:841-846; 1991.

Schwartz, M. A.; Lechene, C.; Ingber, D. E. Insoluble fibronectin activates the Na/H antiporter by clustering and immobilizing integrin $\alpha_5\beta_1$, independent of cell shape. Proc. Natl. Acad. Sci. USA 88:7849-7853; 1991.

Sefton, B. M.; Hunter, T. Vinculin: A cytoskeletal target of the transforming protein of Rous Sarcoma Virus. Cell 24:165-174; 1981.

Shores, C. G.; Maness, P. F. Tyrosine phosphorylated proteins accumulate in junctional regions of the developing chick neural retina. J. Neurosci. Res. 24:59-66; 1989.

Shriver, K.; Rohrschneider, L. Organization of $pp60^{src}$ and selected cytoskeletal proteins within adhesion plaques and junctions of Rous Sarcoma Virus-transformed rat cells. J. Cell Biol. 89:525-535; 1981.

Singer, I. I. The fibronexus: A transmembrane association of fibronectin-containing fibers and bundles of 5nm microfilaments in hamster and human fibroblasts. Cell 16:675-685; 1979.

Singer, I. I.; Kawka, D. W.; Kazazis, M.; Clark, R. A. F. In vivo co-distribution of fibronectin and actin fibers in granulation tissue: Immunofluorescence and electron microscope studies of the fibronexus at the myofibroblast surface. J. Cell Biol. 98:2091-2106; 1984.

Sobue, K. Involvement of the membrane cytoskeletal proteins and the *src* gene product in growth cone adhesion and movement. Neurosci. Res. 8 Suppl. 13:S80-S91; 1990.

Sorge, L. K.; Levy, B. T.; Maness, P. F. pp60$^{c\text{-}src}$ is developmentally regulated in the neural retina. Cell 36:249-257; 1984.

Stappert, J.; Kemler, R. Intracellular associations of adhesion molecules. Curr. Opin. Neurobiol. 3:60-66; 1993.

Sutton, A. B.; Canfield, A. E.; Schor, S. L.; Grant, M. E.; Schor, A. M. The response of endothelial cells to TGFβ-1 is dependent upon cell shape, proliferative state and the nature of the substratum. J. Cell Sci. 99:777-787; 1991.

Takeichi, M. Cadherins: A molecular family important in selective cell- cell adhesion. Annu. Rev. Biochem. 59:237-252; 1990.

Tamkun, J. W.; DeSimone, D. W.; Fonda, D.; Patel, R. S.; Buck, C.; Horwitz, A. F.; Hynes, R. O. Structure of integrin, a glycoprotein involved in the transmembrane linkage between fibronectin and actin. Cell 46:271-282; 1986.

Tanaka, S.; Takahashi, N.; Udagawa, N.; Sasaki, T.; Fukui, Y.; Kurokawa, T.; Suda, T. Osteoclasts express high levels of p60$^{c\text{-}src}$, preferentially on ruffled border membranes. FEBS Lett. 313:85-89; 1992.

Tarone, G.; Cirillo, D.; Giancotti, F. G.; Comoglio, P. M.; Marchisio, P. C. Rous sarcoma virus-transformed fibroblasts adhere primarily at discrete protrusions of the ventral membrane called podosomes. Exp. Cell Res. 159:141-157; 1985.

Tsukita, S.; Oishi, K.; Akiyama, T.; Yamanashi, Y.; Yamamoto, T. Specific proto-oncogenic tyrosine kinases of src family are enriched in cell-to-cell adherens junctions where the level of tyrosine phosphorylation is elevated. J. Cell Biol. 113:867-879; 1991.

Tsukita, S.; Nagafuchi, A.; Yonemura, S. Molecular linkage between cadherins and actin filaments in cell-cell adherens junctions. Curr. Opin. Cell Biol. 4:834-839; 1992.

Tsunematsu, Y.; Coulombre, A. J. Demonstration of transdifferentiation of neural retina from pigmented retina in culture. Dev. Growth Diff. 23:297-311; 1981.

Turksen, K.; Opas, M.; Aubin, J. E.; Kalnins, V. I. Microtubules, microfilaments and adhesion patterns in differentiating chick retinal pigment epithelial (RPE) cells in vitro. Exp. Cell Res. 147:379-391; 1983.

Turksen, K.; Aubin, J. E.; Sodek, J.; Kalnins, V. I. Changes in the distribution of laminin, fibronectin type IV collagen and heparan sulphate proteoglycan during colony formation by chick retinal pigment epithelial cells in vitro. Coll. Relat. Res. 4:413-426; 1984.

Turksen, K.; Opas, M.; Kalnins, V. I. Preliminary characterization of cell surface-extracellular matrix linkage complexes in cultured retinal pigmented epithelial cells. Exp. Cell Res. 171:259-264; 1987.

Turksen, K.; Kanehisa, J.; Opas, M.; Heersche, J. N.; Aubin, J. E. Adhesion patterns and cytoskeleton of rabbit osteoclasts on bone slices and glass. J. Bone Miner. Res. 3:389-400; 1988.

Turner, C. E.; Pavalko, F. M.; Burridge, K. The role of phosphorylation and limited proteolytic cleavage of talin and vinculin in the disruption of focal adhesion integrity. J. Biol. Chem. 264:11938-11944; 1989.

Turner, C. E.; Glenney, J. R.,Jr.; Burridge, K. Paxillin: a new vinculin-binding protein present in focal adhesions. J. Cell Biol. 111:1059-1068; 1990.

Turner, C. E.; Schaller, M. D.; Parsons, J. T. Tyrosine phosphorylation of the focal adhesion kinase pp125^{FAK} during development: Relation to paxillin. J. Cell Sci. 105:637-645; 1993.

Vardimon, L.; Fox, L. E.; Cohen-Kupiec, R.; Degenstein, L.; Moscona, A. A. Expression of v-src in embryonic neural retina alters cell adhesion, inhibits histogenesis, and prevents induction of glutamine synthetase. Mol. Cell. Biol. 11:5275-5284; 1991.

Verschueren, H. Interference reflection microscopy in cell biology: methodology and applications. J. Cell Sci. 75:279-301; 1985.

Volberg, T.; Geiger, B.; Dror, R.; Zick, Y. Modulation of intercellular adherens-type junctions and tyrosine phosphorylation of their components in RSV-transformed cultured chick lens cells. Cell Regul. 2:105-120; 1991.

Volberg, T.; Zick, Y.; Dror, R.; Sabanay, I.; Gilon, C.; Levitzki, A.; Geiger, B. The effect of tyrosine-specific protein phosphorylation on the assembly of adherens-type junctions. EMBO J. 11:1733-1742; 1992.

Wanaka, A.; Milbrandt, J.; Johnson, E. M.,Jr. Expression of FGF receptor gene in rat development. Development 111:455-468; 1991.

Warren, S. L.; Handel, L. M.; Nelson, W. J. Elevated expression of $pp60^{c-src}$ alters a selective morphogenetic property of epithelial cells in vitro without a mitogenic effect. Mol. Cell. Biol. 8:632-646; 1988.

Watt, F. M. Cell culture models of differentiation. FASEB J. 5:287-294; 1991.

Woods, A.; Couchman, J. R. Protein kinase C involvement in focal adhesion formation. J. Cell Sci. 101:277-290; 1992.

Wu, H.; Reynolds, A. B.; Kanner, S. B.; Vines, R. R.; Parsons, J. T. Identification and characterization of a novel cytoskeleton-associated pp60src substrate. Mol. Cell. Biol. 11:5113-5124; 1991.

Wu, H.; Parsons, J. T. Cortactin, an 80/85-kilodalton $pp60^{src}$ substrate, is a filamentous actin-binding protein enriched in the cell cortex. J. Cell Biol. 120:1417-1426; 1993.

Yasuda, K.; Okada, T. S.; Eguchi, G.; Hayashi, M. A demonstration of a switch of cell type in human fetal eye tissues in vitro: Pigmented cells of the iris or the retina can transdifferentiate into lens. Exp. Eye Res. 26:591-595; 1978.

Yasuda, K. Transdifferentiation of "lentoid" structures in cultures derived from pigmented epithelium was inhibited by collagen. Dev. Biol. 68:618-623; 1979.

Zachary, I.; Rozengurt, E. Focal adhesion kinase ($p125^{FAK}$): A point of convergence in the action of neuropeptides, integrins, and oncogenes. Cell 71:891-894; 1992.

Zelenka, P. S. Proto-oncogenes in cell differentiation. BioEssays 12:22-26; 1990.

Zhao, Y.; Sudol, M.; Hanafusa, H.; Krueger, J. Increased tyrosine kinase activity of c-Src during calcium-induced keratinocyte differentiation. Proc. Natl. Acad. Sci. USA 89:8298-8302; 1992.

Zhou, Y.; Dziak, E.; Opas, M. Adhesiveness and proliferation of epithelial cells are differentially modulated by activation and inhibition of protein kinase C in substratum-dependent manner. J. Cell. Physiol. 155:14-26; 1993.

Zhou, Y.; Opas, M. The cell shape, intracellular pH and FGF responsiveness during transdifferentiation of RPE into neuroepithelium. 1994. (submitted).

Part V

Molecular and Biophysical Mechanisms of Mechanical Signal Transduction

15

Signal Transduction Cascades Involved in Mechanoresponsive Changes in Gene Expression

P. A. Watson, J. Krupinski

Introduction

Over the past few years it has been recognized that a broad spectrum of extracellular forces which are imposed upon cells and tissue of the body have the ability to influence their biology. This is true of forces as seemingly insignificant as those generated when integrin receptors in the membranes of cells bind to extracellular matrix proteins, to forces as seemingly overwhelming as those placed upon skeletal muscles, tendons, and bones during intensive resistance training. One constant in this scenario is that cell's must have in their structure systems through which they can monitor their size, shape, and the forces imposed upon them by the external environment. Additionally, systems must be in place which convert such information into regulatory signals allowing the cell to react and adapt to changes in these parameters. Such adaptations and reactions are essential to the survival and function of these cells in an altered mechanical environment. As many of these cells serve to regulate functions of other cells, tissues, and organ systems, such responses to changes in the mechanical environment are critically important to the survival of the organism. In addition, a number of pathological states which are manifest following long term exposure to altered mechanical environments, such as congestive failure of the decompensated hypertrophic heart in hypertensive patients, appear to stem from the attempts of cells to adapt to this altered environment. It would therefore seem to be of significant value to understand how cells perceive the status of their mechanical environment, convert such information into regulatory signals, and subsequently adapt such that function can be maintained in the face of new mechanical challenges.

In this chapter, I will attempt to give an overview of our knowledge regarding how cells respond to mechanical challenges, and how these responses are regulated with regard to signal generation. I will also attempt to synthesize from the available data a speculative model which may address issues and gaps in our current understanding of these processes. I will examine our knowledge regarding the types of genes which comprise the genetic programs activated in certain cells which constitute both initial signaling events occurring immediately following mechanical challenge (0-2 hours) as well as the genetic programs involved in adaptive responses following mechanical challenges. Additionally, I will attempt to characterize the types of signals which may be generated in response to mechanical challenges and identify which cellular structures and systems may be involved in coupling morphological information to signal generation. It should be noted that while much energy has recently been devoted to these topics, this field is still in its infancy and much of the information we seek has yet to be obtained.

Mechanoresponsive Gene Expression

The cell types and tissues whose function and structure are influenced in some manner by the dynamics their mechanical environment are as diverse as the nature of the mechanical forces which are imposed upon them. Structures involved in locomotion and stasis including bones, skeletal muscles, and tendons all respond to forces generated with use and due to gravity. The tissues of the cardiovascular system which pump and conduct blood are subjected to the systemic pressure of containing this fluid mass, the pulsatile pressures generated by the action of the heart, as well as pressure and flow resulting from blood circulation. Chronic alterations in the mechanical environment results in adaptive responses in tissues and cells in an attempt to resist these forces and maintain cellular homeostasis in terms of structure and function. This chain of response requires that cells "sense" alterations in their mechanical environment and convert this information into regulatory signals which stimulate changes in gene expression for both further signaling as well as for adaptive remodeling of the cell and tissue. Thus, 2 general genetic programs must be activated to complete the adaptive cycle. The first genetic program consists of genes activated very early following imposition of a mechanical challenge which are involved in conveying signaling information and subsequently coordinating the adaptive response. Such genes have classically been referred to as Immediate Early Genes (IEGs) and are characterized by a plethora of oncogenes such as *c-myc*, *c-fos*, and members of the *ras* family of small GTP-binding proteins. The second program consists of genes whose expression is required for maintaining the structural and functional integrity of the cells and tissue in the face of an altered mechanical environment. These include genes for cellular proteins in the cytoskeleton as well as proteins involved in the specific function of the tissue or cell. Studies have documented changes in the expression of genes in both the early signaling program and later adaptive program of mechanoresponsive gene expression in a variety of cell types and with various mechanical challenges. However, little is known regarding how components of the early gene program may regulate the expression of the later adaptive program. Additionally, despite significant information regarding the various signaling pathways which respond to mechanical challenge in cells, very little is known regarding the role of specific signals in regulating expression of genes in either of these programs.

Expression of Early Reactive Signaling Genes: A number of cellular oncogenes have been described which serve as early markers for anabolic responses in several tissues and cell types following mechanical challenge. The role the products of these genes play in the actual adaptive response has yet to be elucidated. However, studies into the regulation of the expression these genes may offer the best opportunity to delineate signaling pathways which are both activated by mechanical challenge *and* play a significant role in downstream reactive and adaptive changes in gene expression.

The heart has long been a model for adaptive responses by tissues to mechanical challenge (Morgan, et al., 1989). Chronic hypertension often results in decompensated hypertrophy of the heart with subsequent failure being a major cause of mortality in these patients. Studies in isolated perfused hearts and, more recently, isolated cardiac myocytes and nonmyocytes have addressed the signaling events and patterns of gene expression which characterize and define this mechanosensitive response. Rapid increases in the messenger RNA for the cellular oncogenes *c-fos*, *c-jun*, *c-myc*, JE, and Egr-1 have been observed in cultured neonatal cardiac myocytes following mechanical challenge in cultured cell models (Komura, et al., 1990, Komura, et al, 1991, Sadoshima, et al. 1992a). Induction of *c-fos* has been used as a marker to define initial events which are presumed to lead to anabolic/hypertrophic responses in cultured cardiac cell models (Komura, et al., 1991). Exposure of human umbilical vein endothelial cells to shear stress also has been shown to induce both expression and nuclear localization of the *c-fos* gene product in the first hour following mechanical challange (Ranjan and Diamond, 1993). Upregulation of *c-fos* in both cultured cardiac myocytes and aortic endothelial cells following mechanical challenge has been presumed to occur downstream of activation of protein kinase C (Yazaki, et al. 1993, Sadoshima and Izumo, 1993a).

In addition to the increased expression of IEGs, genes which are involved in early reactive changes in cell behavior are also induced immediately following mechanical challenge in a number of cell types. The products of these genes are, for the most part, involved in regulating the functions of other cells in response to mechanical challenges. These include release of autocrine/paracrine factors involved in anabolic responses in both cardiac myocytes and smooth muscle cells in the vasculature. Release of angiotensin II from stretched cardiac myocytes plays a role in stretched-induced anabolic responses in these cells (Sadoshima, et al., 1993, Sadoshima and Izumo, 1993b). Likewise, shear and stretch of bovine aortic endothelial cells in culture results in the release of endothelin-1 (Morita, et al., 1993, Malek, et al. 1993a) tissue plasminogen activator (Diamond, et al. 1990, Iba, et al., 1991), basic fibroblast growth factor (Malek, et al., 1993b) and platelet-derived growth factor (PDGF; Hsieh, et al., 1992). The release of PDGF from aortic endothelial cells following mechanical challange may be involved in smooth muscle cell proliferation and intimal thickening in vessel walls following increased pressure and flow. Additionally, cyclic deformation of vascular smooth muscle cells in culture result in the increased release of PDGF which acts through an autocrine mechanism to stimulate mitogenic responses in these cells (Wilson et al., 1993). This effect is the result of increased expression of PDGF genes in these cells in response to mechanical challange (Wilson et al., 1993). Exposure of cultured aortic endothelial cells to shear or stretch also stimulates release of a number of vasoactive compound involved in acute regulation of flow, including endothelin-1 (Sadoshima, et al., 1993, Malek et al., 1993a) and nitric oxide (NO)(Dainty et al., 1990) as a result of increased expression of genes for endothelin-1 and NO synthase.

Substantial redistribution of the total cellular actin pool from the polymerized filamentous pool to the monomeric globular pool is observed rapidly in response to shear stress in cultured aortic endothelial cells (Morita et al., 1993). Increased expression and release of endothelin-1 is seen in the same time frame as actin pool redistribution. Endothelin-1 expression and release in response to shear can be mimicked by disruption of actin filaments with cytochalasin-B and is prevented by stabilization of cellular actin filaments by pretreatment with phalloidin (Morita, et al., 1993). These results implicate shear-induced actin cytoskeletal restructuring in the signaling pathway regulating shear-induced endothelin-1 production and release.

Expression of Adaptive Genes: As can be seen from the discussion above, the genes which are induced in the time immediately following mechanical challenge appear to either encode proteins whose function is related to acute regulatory events in which the challenged cell is involved or to encode proteins whose functions have been related to intracellular signaling and regulation. However, adaptive responses in cell structure and function also occur in response to mechanical challenge. Upregulation of genes involved in these adaptive responses appears to occur in a time frame which is much later relative to the onset of mechanical challenge. Speculation can be made as to whether the changes observed in signaling and regulatory genes early after mechanical challenge are involved in the upregulation of adaptive gene expression. Integration of multiple signaling events might be required which could also be responsible for the delayed increase in the messages for these adaptive proteins. Regardless, a second phase of increased gene expression is observed in cells with prolonged exposure to mechanical challenge, and the genes which constitute this program represent proteins whose function is involved in maintaining cell structure and function in the face of an altered mechanical environment.

In general, the heart responds to pressure overload (as well as a number of other hypertrophic stimuli) with a generalized increase in protein synthesis, characterized by an upregulation of expression of genes for most cardiac proteins (Boheler and Dillman, 1988). Pressure overload also stimulates shifts in the expression of genes for specific isoforms of contractile proteins as well as stimulating expression of genes which are dormant in the adult heart. Mechanical challenge and subsequent hypertrophic responses in cardiac myocytes, both in culture and *in vivo*, are accompanied by an upregulation of a program of contractile protein genes which resembles patterns expressed in the fetal and neonatal stages of cardiac development (Sadoshima et al., 1992, Komuro et al, 1991). This program of genes includes an actin isoform found in adult skeletal muscle as well as shifts in myosin isoform gene expression. Additionally, adrenergic agonists which stimulate increases in ribosome formation and protein synthesis in cultured myocytes also stimulate increases in tubulin mRNA content (Rothblum and Hannan, personal communication). Preliminary results indicate that mechanical challenge also results in the accumulation of tubulin mRNA in cultured neonatal cardiac myocytes in a time frame consistent with the secondary adaptive phase of mechanoresponsive gene expression (Hannan, Rothblum, and Watson; unpublished observations). It is interesting to note that alterations in the tubulin cytoarchitecture have been linked to impairment of contractile function in myocytes isolated from hypertrophic hearts (Tsutsui et al, 1993).

The endothelial cells which line the arteries and arterioles of the vasculature are involved in the acute regulation of blood flow. These cells are acutely sensitive to changes in pressure and flow and, as described above, release a number of agents which influence cells throughout the vasculature. These cells also demonstrate significant structural remodeling in the face of altered flow involving cellular alignment with the flow (Langille and Adamson, 1981, Reneman, 1993, Levesque and Nerem, 1985). This remodeling appears to be the result of restructuring of the cellular cytoarchitecture (Kim et al., 1989) as well as the focal adhesions through which these cells attach to their substratum and to which the supporting cytoarchitecture is anchored (Davies et al., 1993). While rapid alterations in the relative distribution of cytoskeletal components is seen immediately following exposure to shear stress in cultured endothelial cells (Morita et al., 1993), long term reorientation of cells and morphological alterations accompanying this reorientation appears to be somewhat delayed. These delayed morphological responses to shear stress may involve both reorganization of focal adhesions as well as the synthesis of new cytoarchitectural components. It is possible that the rapidly induced signals involved in the cytoarchitecturally-regulated expression of endothelin-1 may also initiate a cascade of signaling events involved in the delayed morphological adaptations seen in endothelial cells exposed to shear for prolonged periods. The signaling trail which links mechanical challenge to both adaptive remodeling and endocrine function in aortic endothelial cells is, as yet, poorly defined.

Regulatory Signaling Responses to Mechanical Challange

Our knowledge of intracellular regulatory signal transduction stems from studies into endocrine regulation of cellular biochemistry and biology. Therefore, most of the studies examining signaling in cells following mechanical challenge have focused on changes in the activities of these "classic" hormonally-regulated signal transduction pathways. While a multitude of "classic" signaling events are triggered in cells exposed to mechanical challenges, it is possible that non-traditional signaling pathways may also act during cellular responses to mechanical challenges (Wang et al, 1993). For example, it is certainly the case that the interface between classic signal transduction pathways and their regulation of cytoskeletal structure following hormonal activation is reversed in a number of circumstances of mechanical challenge. Many shape-dependent or mechanically-stimulated phenotypes in a variety of cell types can either be mimicked or reversed by disruption of the actin cytoskeleton with a variety of antagonists. In this section I will examine our current understanding of the activation of classic signal transduction pathways by mechanical challenge as well as the interplay between these signaling events and dynamic remodeling of the cytoarchitecture in response to mechanical challenge.

Adenylyl cyclase-cyclic AMP signaling pathway. A number of cell types and organ systems respond to mechanical challenge with activation of the cyclic AMP signal transduction pathway. Pressure overload in the isolated perfused rat heart stimulates anabolic responses including accelerated rates of ribosome formation and protein synthesis, which are accompanied by increases in cyclic AMP and cyclic AMP-dependent protein kinase activity (Watson et al., 1989, Xenophontos et al, 1989). Volume overload of the rat heart *in vivo* results in hypertrophic responses which are also accompanied by increased activity of the cyclic AMP-dependent protein kinase (Lavandero et al., 1993). The anabolic responses observed in the perfused heart with pressure overload can be mimicked by agonists which increase cyclic AMP in hearts and are prevented by muscarinic-cholinergic antagonism of cyclic AMP accumulation in response to pressure overload (Watson et al., 1989, Xenophontos et al, 1989). As these anabolic responses are not seen in cultured neonatal cardiac myocytes in response to mechanical challenge (Sadoshima et al. 1992b), this response may be a function of the nonmyocyte population of cells in the heart. Consistent with this hypothesis is the recent observation that release of nonmyocyte growth factor from cardiac nonmyocytes is increased following beta-adrenergic stimulation of this cell population in culture (Long et al, 1993). This factor has been shown to exert anabolic influences upon cardiac myocytes (Long et al. 1993). Increased fibroblast cyclic AMP and subsequent release of such a factor in response to overload could account for the cyclic AMP dependent anabolic responses seen with pressure in the perfused whole heart.

Other cell types and tissues respond to a variety of mechanical challenges with increases in cyclic AMP or the activity of adenylyl cyclase. Increased ventilation of the rat lung following partial pneumonectomy leads to anabolic responses in the remaining tissue (Rannels, 1989) as well as accumulation of cyclic AMP (Russo et al., 1989). Exposure of cultured aortic endothelial cells to cyclic patterns of mechanical deformation also influences adenylyl cyclase activity (Letsou et al., 1990). Osmotic challenges to a number of cell types including turkey erythrocytes (Morgan et al., 1989), an osteosarcoma cell line used as a model of bone remodeling (UMR-106 cells)(Sandy et al., 1989), isolated rat hepatocytes (Baquet et al., 1991), XC rat sarcoma cells (Watson, unpublished observations), and S49 mouse lymphoma cells (Watson 1989, Watson 1990) results in accumulation of cyclic AMP.

Studies have been initiated to investigate the molecular requirements for cyclic AMP accumulation following mechanical challenge. Results of these studies indicate that accelerated adenylyl cyclase activity is responsible for increases in cyclic AMP in response to osmotic challenge in S49 mouse lymphoma cells (Watson 1990). Adenylyl cyclase, like many components of signal transduction pathways, exists as a family of related proteins in mammalian cells with eight unique isoforms having been identified (Krupinski et al., 1992), of which six have been cloned, sequenced, and characterized (Krupinski et al. 1989, Krupinski et al. 1992, Feinstein et al. 1991, Bakalyar et al. 1990, Gao and Gilman, 1991, Yoshimura and Cooper, 1992, Ishikawa et al, 1992, Premont et al., 1992, Katsushika et al., 1992). The remaining two isoforms, one of which is the predominant isoform expressed in S49 cells (Type VII), have recently been cloned and characterized (Cali et al., Watson et al., in press). Experiments were performed to ascertain if the ability to respond to osmotic challenge is a

Table 1. Influence of Gs-alpha expression on adenylyl cyclase activity in cyc- mutant S49 mouse lymphoma cells

Gs-alpha gene expressed	cAMP (pmol/mg cell protein)			
	Basal (NormOsm)	HypOsm	Forskolin	Isoproterenol
Mouse (S49 WT) long and short	7.8 ± 0.4	17.2 ± 0.9	1092 ± 110	934 ± 7
None (S49 cyc-)	1.13 ± 0.05	1.37 ± 0.12	25.2 ± 0.01	1.12 ± 0.02
Rat Gs - alpha① long (Gs_α L1.1)	5.71 ± 0.2*	10.0 ± 0.4*	1144 ± 16*	981 ± 3*
Rat Gs - alpha① long (Gs_α L2)	4.95 ± 0.08*	7.39 ± 0.15*	653 ± 24*	2314 ± 73*
Rat Gs - alpha① short Gs_α S1	4.03 ± 0.11*	5.68 ± 0.28*	868 ± 21*	3727 ± 117*

①Transformed clonal S49 cyc- cell line provided by Dr. Jeff Stadel, SmithKline Beecham Pharmaceuticals.
HypOsm indicates reduction of medium osmolarity by 33%.
Forskolin and isoproterenol were at final concentrations of 10 uM and 1 uM, respectively.
* $p < 0.05$ vs. S49 cyc- response

conserved property of the adenylyl cyclase family of proteins. Results indicate that when adenylyl cyclase isoforms are overexpressed in 293 embryonal kidney cells, only expression of Types I, VIII, and VII impart osmotic responsive changes in cyclic AMP content (Watson, et al., unpublished observations). The cyclic AMP responses to osmotic challenge in 293 cells overexpressing Type I and Type VIII adenylyl cyclases, both calcium-stimulated isoforms of the enzyme, are dependent on a calcium-stimulated influx of extracellular calcium. The response of Type VII adenylyl cyclase appears to occur independent of increases in intracellular calcium.

While the mechanism of activation of Type VII following osmotic challenge is not clearly defined, evidence indicates that the potential exists for a role for the stimulatory guanine-nucleotide binding protein, Gs, which is involved in hormonal activation of adenylyl cyclase. The response to osmotic challenge in a mutant clonal line of S49 cells (cyc^-) which lacks Gs alpha-subunit expression is severely attenuated (Watson 1990). Three potential explanations exist for this observation, the presence of a second-osmotic sensitive mutation in these cells, a critical role for Gs in mediating the majority of the osmotic stimulation of the type VII enzyme, or a role for Gs in amplifying Type VII responsiveness to osmotic challenge. A precedent exist for Gs amplification of Type VII adenylyl cyclase activation. Direct stimulation of S49 adenylyl cyclase by forskolin is also attenuated when comparing activity in wild type and cyc^- S49 cell lines (Watson 1990, Katada et al. 1986), an effect which can be reversed in membranes prepared from this cell line by inclusion of Gs-alpha (Katada et al. 1986). Experiments were performed to determine if transformation and expression of Gs-alpha in S49 cyc^- cells could reconstitute wild type osmotic responsiveness to these cells. Exogenous expression of either the long or short forms of Gs-alpha in cyc^- cells restored beta-adrenergic responsiveness, potentiated forskolin stimulation, and wild type responsiveness to osmotic challenge (Table 1). It was concluded that (1) a secondary mutation in S49 cyc^- cells does not account for attenuated responsiveness to osmotic challenge, and (2) Gs-alpha plays a significant role in full responsiveness of the S49 adenylyl cyclase to osmotic challenge either through a direct role in the signaling cascade or by amplifying the direct activation of the S49 adenylyl cyclase by osmotic challenge. Further characterization of the responsiveness of the Type VII adenylyl cyclase isoform is currently being performed in 293 cells transformed to express the cDNA for this enzyme.

Additional levels of control of adenylyl cyclase activity during osmotic challenge are implied by the results of experiments involving actin cytoskeletal antagonists in cells overexpressing specific isoforms of the enzyme. The ability of a specific adenylyl cyclase isoform to respond to osmotic challenge cosegregates with the ability of treatment with cytochalasin B to stimulate cyclic AMP accumulation and ablate further increases in response to osmotic challenge (Table 2). The implication derived from these observations is that integrity of the actin cytoarchitecture can influence adenylyl cyclase activity, with reorganization resulting from either cytochalasin treatment or dynamic volume changes with osmotic challenge resulting in activation of specific adenylyl cyclase isoforms (Types I and VII). This observation is consistent with the observations that actin cytoskeletal disruption can influence or modify a variety of cellular responses to mechanical challenges (see below).

Phospholipase C-generated second messenger signaling pathways. The hydrolysis of phosphatidylinositol 4,5 bisphophate (PIP_2) by phospholipase C (PLC) results in the generation of several second messengers which function through distinct signaling pathways. Inositol trisphosphate (IP_3) is a water soluble product of PIP_2 hydrolysis by PLC which binds to specific receptors on the endoplasmic reticulum resulting in the release of intracellular calcium stores. 1,2 diacylglycerol (DAG) is also generated in the PLC-mediated hydrolysis of PIP_2 which activates the lipid-dependent, calcium-dependent protein kinase C, which exist as a family of differentially expressed isoforms in mammalian cells.

Table 2. Sensitivity of Adenylyl Cyclase Isoforms to Cytochalsin B Disruption of Actin Cytoskeleton

Predominant① Adenylyl cyclase Isoform	Cell Model	cAMP (pmol/mg cell protein)	
		(--) Cyto B	(+) Cyto B
Type I Type III Type VI Type VII	293 Embryonal Kidney Cells	3.9 ± 0.1	4.2 ± 0.2
Type I	Transformed 293 cells	10.4 ± 1.7	21.5 ± 1.2 ®
Type III	Transformed 293 cells	5.3 ± 0.5	5.4 ± 0.7
Type VI	Transformed 293 cells	6.1 ± 0.2	6.9 ± 0.3
Type VII	XC rat sarcoma cells	6.9 ± 0.2	27.7 ± 1.6 ®
Type VII Type VI	S49 mouse lymphoma cells	3.2 ± 0.3	13.4 ± 0.7 ®

Cytochalasin treatment is 50 ug/ml for 40 min prior to experiment and during the experiment.

® $p < 0.05$ vs. (-) Cytochalasin B

①Predominant isoform is that introduced by transfection with a cDNA in transformed cell lines, but reflects endogenous expression in the other entries.

These bifurcating signaling pathways are activated in cells by a variety of growth factors and hormones, and have also been shown to be activated following mechanical challenge in a number of cells types and tissues .

Mechanical challenge of neonatal cardiac myocytes in culture has been shown to increase generation of IP_3 and activate PKC (Yazaki, et al. 1993, Sadoshima and Izumo, 1993a). Activation of PKC following mechanical challenge has been linked to the increased expression of several IEGs including *c-fos* and *ras* (Komuro, et al., 1990, Komuro, et al, 1991, Sadoshima, et al. 1992b). Studies by Izumo et al. indicate that the activation of PKC is required for angiotensin II-initiated autocrine/paracrine activation of *c-fos* upregulation (Sadoshima and Izumo, 1993a, 1993b, Sadoshima et al., 1993). These signaling events also appear to be involved in the upregulation of "late" markers for cardiac hypertrophy in stretched cardiac myocytes in culture (Sadoshima et al., 1992b, Komuro et al, 1991). These include genes of the aforementioned fetal program, including genes encoding skeletal alpha-actin, beta-myosin heavy chain, and atrial natriuretic factor (ANF).

Mechanical challenge of cultured aortic endothelial cells by shear stress as well as cyclic stretch has been shown to increase PKC activity and IP_3 generation (Nollert et al., 1990, Bhagyalakshmi, et al., 1992, Rosales and Sumpio, 1992, Shen et al., 1992, Dassouli et al., 1993). The response in PLC-generated signals to mechanical challenge in these cells appears to be related less to the absolute strain imposed upon the cells than to changes in the status of strain (Rosales and Sumpio, 1992, Brophy et al., 1993, Winston et al., 1993). In other words, an increase or decrease in the strain load on these cells stimulates an acute increase in IP_3 concentration, which with prolonged exposure to the altered level of strain decays to steady state levels approaching values found in cells prior to the alteration in strain. The downstream ramifications of alterations in strain and generation of PLC-generated signals in cultured aortic endothelial cells includes activation of *c-fos* expression (Ranjan and Diamond, 1993).

Osmotic challenge also increases the generation of IP_3 in a number of cell types including isolated rat hepatocytes (Baquet et al., 1991), skate erythrocytes (Musch and Goldstein, 1990), and UMR-106 osteosarcoma cells (Sandy et al., 1989). The relative physiological importance of such responses in these cells is, for the most part, unknown. In the case of osmotic challange to isolated rat hepatocytes, downstream targets for the increased activity of the cAMP and PLC singaling pathways involves upregulation of actin gene expression (Theodoropoulos et al., 1992). It is possible that these osmotically-stimulated signaling events may be related to regulatory volume decreases following swelling in other cells.

Stretch-activated ion channels (SACs). The first identified mechanotransducing activity in cells was a stretch-activated ion conductance. Due to its lack of resemblance in pharmacology and electrophysiology to known voltage and agonist regulated channels, this activity was ascribed to a novel class of previously unidentified channels called SACs (for Review, see Sachs 1987, Morris 1990). Although the first mechanotransducing function to be identified, this particular class of channels remains to this date unidentified at the protein or mRNA level in mammalian cells. Thus, characterization of these channels has been limited to channel properties in isolated patch clamp experiments and identification of downstream effects regulated by these channels limited to correlative biological responses. These channel activities have been identified in almost every cell in which they have been sought (Morris 1990). Potential targets for SAC signals include stress-induced responses in excitable tissues such as contractility in cardiac myocytes (Sigurdson et al., 1992), volume regulatory responses in a wide variety of cell types, and regulatory events in cultured endothelial cells in response to shear stress (Naruse and Sokabe, 1993, Lansman et al., 1987). Exact identification of functions for SAC activities and characterization of chemical properties to this class of channels will require either purification of a SAC protein or molecular cloning of a cDNA encoding one of the members of this family of mechanotransducers.

Mitogen-activated protein (MAP) kinases / tyrosine kinases. Recent work in a number of systems has indicated that activation of members of the MAP kinase family and as yet unidentified tyrosine kinases is involved in the signaling cascade connecting mechanical challenge and biological response. These events appear to lie downstream of the initial mechanotransducing events but appear to be intimately involved in the events which directly modulate expression of specific deformation-responsive genes.

Responses to mechanical challenge of cardiac myocytes appear to involve activation of MAP kinase kinase (or MEF) and potential downstream activation of S6 kinase activities (Yamazaki et al. 1993). These responses appear to lie downstream from increases in PKC activity seen following mechanical challenge of cultured myocytes, as inhibition by staurosporine or PKC-down regulation by prolonged phorbol ester treatment attenuate these responses (Yamazaki et al. 1993). Incomplete reversal of mechanically-induced MAP kinase activation implies that a second non-PKC dependent mechanism for MAP kinase activation with mechanical challenge may exist in cultured cardiac myocytes. These mechanoresponsive events also appear to be activated by the autocrine/paracrine actions of angiotensin II release from the local cardiac cell renin-angiotensin system (Sadoshima et al., 1993). Treatment of stretched myocyte-enriched cultures with angiotensin-converting enzyme (ACE) inhibitors or AT_1 angiotensin receptor antagonists (Sadoshima et al, 1993, Sadoshima and Izumo, 1993b) prevented stretch-induced activation of IEGs and "late" response genes linked to activation of MAP kinase in these cells. Cyclic stretching of rat glomerular mesangial cells in culture also results in activation of S6 kinase activity (Homma et al., 1992). Once again, this activation appears to lie downstream of increased PKC activity resulting from mechanical challange.

Osmotic challenge of certain cells may also activate MAP kinase and tyrosine kinase activities. Volume regulatory responses in the human intestine 407 cell line (Tilly et al., 1993) were potentiated following inhibition of protein tyrosine phosphatases with sodium orthovanadate and were reduced following inhibition of tyrosine kinase activity with herbimycin A or genistein. Osmotic challenge resulted in the tyrosine phosphorylation of several proteins as well as phosphorylation of MAP kinase. The potential exists for a growth factor mediated role in responses of cells to osmotic challenge. Once again, it appears that activation of these particular signaling pathways lies downstream of other initial mechanotransducing events.

Regulatory interactions at the cytoskeletal-signal transduction interface. As stated in the "Introduction" of this chapter and mentioned in various sections regarding gene expression and signaling, cytoskeletal structure appears to play an integral role in the appearance and maintenance of shape-dependent and mechanically-induced phenotypes. Activation of a number of signaling pathways and expression of both "immediate" and "late" responses which are related to mechanical challenge can be mimicked and or prevented by disruption of the actin cytoskeleton. Activation of adenylyl cyclase in a number of cell types by osmotic challenge can be mimicked by cytochalasin treatment, and further activation by simultaneous osmotic challenge prevented (Table 2). Phorbol ester- induction *c-fos* and *c-myc* gene expression in 3T3-L1 cells through activation of PKC is enhanced following treatment of cells with antagonists to either microtubules (vinblastine sulfate) or actin microfilaments (cytochalasin B) (Pavlath et al., 1993). Shear-induced upregulation of endothelin-1 gene expression in cultured aortic endothelial cells can be prevented by actin cytoskeletal disruption and mimicked by phalloidin stabilization of the actin cytoarchitecture in the absence of mechanical challange (Morita et al., 1993). Expression of the *c-fos* gene in WI-38 and HeLa cells can be induced by treatment with the actin cytoskeletal antagonist cytochalasin D (Zambetti et al., 1991). Expression of differentiated phenotypes, seen *in vivo* and produced by culture of cells in 3-dimensional gels can be induced in monolayer cultures of rat hepatocytes (Caron 1990), rabbit articular chondrocytes (Benya et al., 1988), and rabbit synovial fibroblasts (Unemmori and Werb, 1986) by treatment with cytochalasins at doses which alter actin cytoskeletal structure but do not grossly change cell morphology. Many other examples exist in the literature, all implying that a bidirectional transfer of information exists between the cytoarchitecture and signaling pathways which are involved in the determination of cell structure and function.

Information regarding crosstalk between the signaling and cytoskeletal components of cells is limited. The identities of several proteins whose functions include both regulation of signaling and regulation or interaction with the cytoskeleton have been determined. However, their roles in communication from cellular structure to signal generation is not known. Several actin binding proteins have been shown to modulate hydrolysis of PIP_2 by PLC-gamma, limiting access to this substrate to forms of the enzyme phosphorylated following exposure to a number of growth factors (Goldschmidt-Clermont et al., 1991, Jamney and Stossel 1987, Lassing and Lindberg 1985). PIP_2 hydrolysis releases these proteins from PIP_2 sequestration and allows them to interact and modify the actin cytoarchitecture. Function of these proteins in the opposite direction during mechanical challenge can be speculated, but this form of regulation has not been demonstrated. An actin binding protein has been identified in yeast which interacts with the *ras*-mediated activation of adenylyl cyclase and sits at a similar cytoskeletal-signal transduction interface (Cyclase-Associated Protein, CAP)(Vojtek et al., 1991). Although a homolog of this protein has been identified in mammalian cells (Matviw et al., 1992), the function of this protein remains to be determined.

Several tyrosine kinase activities have been found associated with focal adhesions in cultured mammalian cells. These tyrosine kinase activities are accentuated following adhesion and spreading of cells on extracellular matrix substrata (Burridge et al., 1992, Bockholt and Burrgidge, 1993, Zachary and Rozengurt, 1992). One of these tyrosine kinases has been identified as a novel focal adhesion-associated tyrosine kinase and designated $pp125^{FAK}$. Activation of $pp125^{FAK}$ in adhesive cells requires an intact actin cytoskeleton, as disruption of this structure with cytochalasins results in loss of tyrosine kinase activity (Burridge et al., 1992, Lipfert et al., 1992, Bockholt and Burridge, 1993). As focal adhesions are intimately involved in the organization and stability of the actin cytoarchitecture in adherent cells, the location of $pp125^{FAK}$ at a critical site of architectural regulation and its function as a tyrosine kinase signaling protein make it a potential candidate as an interpreter of cell shape and strain information for biological response and adaptation. Once again, no information regarding involvement in this process has been obtained.

While a great deal of descriptive work has been published regarding signaling events activated by mechanical challenge in cells and tissues, these reports have done little to define the regulatory pathways responsible for alterations in cellular structure and function in response to such challenges. This may be the result of the great number of signaling events induced concurrently in cells by mechanical challenge. With the immense potential for crosstalk between these signaling pathways, discrimination between initial signaling events and coincident activation of signal transduction pathways through crosstalk will be difficult. Identification of the program of genes activated in a specific cell type in response to mechanical challenge and identification of the signaling pathways which can directly activate expression of the members of this program will be required to sort true signals from crosstalk noise.

Acknowledgements

I wish to thank Dr. Jeffrey Stadel (SmithKline Beecham) for provision of stably transformed S49 *cyc-* cells expressing Gs-alpha, Dr. John Krupinski (Geisinger Clinic) and Dr. Randall Reed (Johns Hopkins University) for provision of cDNAs for adenylyl cyclase isoforms, and Carole Frankenfield and Kathryn Giger in my laboratory for technical assistance.

References:

Bakalyar, H.A., and Reed, R.R. Identification of a specialized adenylyl cyclase that may mediate oderant detection. Science 250:1403-1406, 1990.

Baquet, A., Meijer, A.J., and Hue, L. Hepatocyte swelling increases inositol 1,4,5-risphosphate, calcium, and cyclic AMP concentration but antagonizes phosphorylase activation by Ca^{2+}-dependent hormones. FEBS Lett. 278:103-106, 1991.

Benya, P.D., Brown, P.D., and Padilla, S.R. Microfilament modification by dihydrocytochalasin B causes retinoic acid-modulated chondrocytes to reexpress the differentiated collagen phenotype without a change in shape. J. Cell Biol. 106:161-170, 1988.

Bhagyalakshmi, A., Berthiaume, F., Reich, K.M., and Frangos, J.A. Fluid shear stress stimulates membrane phospholipid metabolism in cultured human endothelial cells. J. Vasc. Res. 29:443-449, 1992.

Bockholt, S.M., and Burridge, K. Cell spreading on extracellular matrix proteins induces tyrosine phosphorylation of tensin. J. Biol. Chem. 268:14565-14567, 1993.

Boheler, K.R., and Dillman, W.H. Cardiac response to pressure overload in the rat: the selective alteration of *in vitro* directed RNA translation products. Circ. Res. 63:448-456, 1988.

Brophy, C.M., Mills, I, Rosales, O. Isales, C., and Sumpio, B.E. Phospholipase C: A putative mechanotransducer for endothelial cell response to acute hemodynamic changes. Biochem. Biophys. Res. Comm. 190:576-581, 1993.

Burridge, K., Turner, C.E., and Romer, L.H. Tyrosine phosphorylation of paxillin and $pp125^{FAK}$ accompanies cell adhesion to extracellular matrix: a role in cytoskeletal assembly. J. Cell Biol. 119:893-903, 1992.

Cali, J.J., Zwaagstra, J.C., Mons, N., Cooper, D.M.F., and Krupinski, J. Type VIII Adenylyl cyclase: A Calcium/calmodulin-stimulated enzyme with a potential role in long-term potentiation. J. Biol. Chem, in press.

Caron, J.M. Induction of albumin gene transcription in hepatocytes by extracellular matrix proteins. Mol. Cell. Biol. 10:1239-1243, 1990.

Dainty, I.A., McGrath, J.C., Spedding, M., and Templeton, A.G.B. The influence of the initial stretch and the agonist-induced tone on the effect of basal and stimulated release of EDRF. Br. J. Pharmacol. 100:767-773, 1990.

Dassouli, A., Sulpice, J-C., Roux, S., and Crozatier, B. Stretch-induced inositol trisphosphate and tetrakisphosphate production in rat cardiomyocytes. J. Mol. Cell. Cardiol. 25:973-982, 1993.

Davies, P.F., Robotewskyj, A., and Griem, M.L. Endothelial cell adhesion in real time. Measurements *in vitro* by tandem scanning confocal image analysis. J. Clin. Invest. 91:2640-2652, 1993.

Diamond, S.L., Sharefkin, J.B., Dieffenbach, C., Frasier-Scott, K., McIntire, L.V., and Eskin, S.G. Tissue plasminogen activator messenger RNA levels increase in cultured human endothelial cells exposed to laminar shear stress. J. Cell. Physiol. 143:364-371, 1990.

Feinstein, P.G., Schrader, K.A., Bakalyer, H.A., Tang, W-J., Krupinski, J., Gilman, A.G., and Reed, R.R. Molecular cloning and characterization of a Ca^{2+}/calmodulin-insensitive adenylyl cyclase from rat brain. Proc. Nat. Acad. Sci. 88:10173-10177, 1991.

Gao, B., and Gilman, A.G. Cloning and expression of a widely distributed (type IV) adenylyl cyclase. Proc. Nat. Acad. Sci. 88:10178-10182, 1991.

Goldschmidt-Clermont, P., Kim, J., Machesky, L., Rhee, S., and Pollard, T. Regulation of phospholipase C-gamma 1 by profilin and tyrosine phosphorylation. Science 251:1231-1233, 1991.

Homma, T., Akai, Y., Burns, K.D., and Harris, R.C. Activation of S6 kinase by repeated cycles of stretching and relaxation in rat glomerular mesangial cells. J. Biol. Chem. 267:23129-23135, 1992.

Hsieh, H-J., Li, N-Q., and Frangos, J.A. Shear-induced platelet-derived growth factor gene expession in human endothelial cells is mediated by protein kinase C. J. Cell. Physiol. 150:552-558, 1992.

Iba, T., Shin, T., Sonodo, T., Rosales, O, and Sumpio, B.E. Stimulation of endothelial secretion of tissue-type plasminogen activator by repetitive stretch. J. Surg. Res. 50:457-460, 1991.

Ishikawa, Y., Katsushika, S., Chen, L., Halnon, N.J., Kawabe, J-I., and Homcy, C.J. Isolation and characterization of a novel cardiac adenylylcyclase cDNA. J. Biol. Chem. 267:13553-13557, 1992.

Jamney, P., and Stossel, T. Modulation of gelsolin function by phosphatidylinositol 4,5 bisphosphate. Nature 325:362-364, 1987.

Katada, T., Oinuma, M., and Ui, M. Mechanisms for inhibition of the catalytic activity of adenylate cyclase by the guanine nucleotide-binding proteins serving as the substrate of islet-activating protein, pertussis toxin. J. Biol. Chem. 261:5215-5221, 1986.

Katsushika, S., Chen, L., Kawabe, J, Nilakantan, R., Halton, N.J., Homcy, C.J., and Ishikawa, Y. Cloning and characterization of a sixth adenylyl cyclase isoform: Types V and VI constitute a subgroup within the mammalian adenylyl cyclase family. Proc. Nat. Acad. Sci. U.S.A. 89:8774-8778, 1992.

Kim, D.W., Gotlieb, A.I., and Langille, B.L. *In vivo* modulation of endothelial F-actin microfilaments by experimental alterations in shear stress. Arteriosclerosis 9:439-445, 1989.

Komuro, I., Kaida, T., Shibasaki, Y., Kurabayashi, M., Katoh, Y., Hoh, E., Takaku, F., and Yazaki, Y. Stretching cardiac myocytes stimulates protooncogene expression. J. Biol. Chem. 265:3595-3598, 1990.

Komuro, I., Katoh, Y., Kaida, T., Shibasaki, Y., Kurabayashi, M., Takaku, F., and Yazaki, Y. Mechanical loading stimulates cell hypertrophy and specific gene expression in cultured rat cardiac myocytes. J. Biol. Chem. 266:1265-1268, 1991.

Krupinski, J., Lehman, T.C., Frankenfield, C.D., Zwaagstra, J.C., and Watson, P.A. Molecular diversity in the adenylylcyclase family: evidence for eight forms of the enzyme and cloning of Type VI. J. Biol. Chem. 267:24858-24862, 1992.

Krupinski, J., Coussen, F., Bakalyar, H.A., Tang, W.J., Feinstein, P.G., Orth,K., Slaughter, C., Reed, R.R., and Gilman, A.G. Adenylyl cyclase amino acid sequence: Possible channel- or transporter-like structure. Science 244:1558-1564, 1989.

Langille, B.L., and Adamson, S.L. Relationship between blood flow direction and endothelial cells orientation at branch sites in rabbits and mice. Circ. Res. 48:481-488, 1981.

Lansman, J.B., Hallam, T.J., and Rink, T.J. Single stretch-activated ion channels in vascular endothelial cells as mechanotransducers? Nature 325:811-812, 1987.

Lassing, I., and Lindberg, U. Specific interactions between phosphatidylinoitol 4,5 bisphosphate and profilactin. Nature 314:472-474, 1985.

Lavandero, S., Cartagena, G., Guarda, E., Corbalan, R., Godoy, I., Sapag-Hagar, M., and Jalil, J.E. Changes in cyclic AMP dependent protein kinase and active stiffness in the rat volume overload model of heart hypertrophy. Cardiovasc. Res. 27:1634-1638, 1993.

Letsou, G.V., Rosales, O., Maitz, S., Vogt, A., and Sumpio, B.E. Stimulation of adenylate cyclase activity in cultured endothelial cells subjected to cyclic stretch. J. Cardiovasc. Surg. 31:634-639, 1990.

Levesque, M.J., and Nerem, R.M. The elongation and orientation of cultured endothelial cells in response to shear stress. J. Biomech. Eng. 107:342-347, 1985.

Lipfert, L., Haimovich, B., Schaller, M.D., Cobb, B.S., Parsons, J.T., and Brugge, J.S. Integrin-dependent phosphorylation and activation of the protein tyrosine kinase pp125^{FAK} in platelets. J. Cell Biol. 119:905-912, 1992.

Long, C.S., Hartogensis, W.E., and Simpson, P.C. Beta-Adrenergic stimulation of cardiac non-myocytes augments the growth-promoting activity of non-myocyte conditioned medium. J. Mol. Cell. Cardiol. 25:915-925, 1993.

Malek, A.M., Gibbons, G.H., Dzau, V.J., and Izumo, S. Fluid shear stress differentially modulates expression of genes encoding basic fibroblast growth factor and platelet derived growth factor B chain in vascular endothelium. J. Clin. Invest. 92:2013-2021, 1993b.

Malek, A.M., Greene, A.L. and Izumo, S. Regulation of endothelin 1 gene by fluid shear stress is transcriptionally mediated and independent of protein kinase C and cAMP. Proc. Nat. Acad. Sci. U.S.A. 90:5999-6003, 1993a.

Matviw, H., Yu, G., and Young, D. Identification of a human cDNA encoding a protein that is structurally and functionally related to the yeast adenylyl cyclase-associated CAP proteins. Mol. Cell. Biol. 12:5033-5040, 1992.

Morgan, H.E., Xenophontos, X.P., Haneda, T., McGlaughlin, S., amd Watson, P.A. Stretch-Anabolism Transduction. J. Appl. Cardiol. 4:415-422, 1989.

Morris, C.E. Mechanosensitive ion channels. J. Membr. Biol. 113:93-107, 1990.

Morita, T., Kurihara, H., Maemura, K., Yoshizumi, M., and Yazaki, Y. Disruption of cytoskeletal structures mediates shear stress-induced endothelin-1 gene expression in cultured porcine aortic endothelial cells. J. Clin. Invest. 92:1706-1712, 1993.

Musch, M.W., and Goldstein, L. Hypotonicity stimulates phosphatidylcholine hydrolysis and generates diacylglyserol in erythrocytes. J. Biol. Chem. 265:13055-13059, 1990.

Naruse, K., and Sokabe, M. Involvement of stretch-activated ion channels in calcium mobilization to mechanical stretch in endothelial cells. Am. J. Physiol. 264:C1037-C1044, 1993.

Nollert, M.U., Eskin, S.G., and McIntire, L.V. Shear stress increases inositol trisphosphate levels in human endothelial cells. Biochem. Biophys. Res. Comm.170:281-287, 1990.

Pavlath, G.K., Shimizu, Y., and Shimizu, N. Cytoskeletal active drugs modulate signal transduction in the protein kinase C pathway. Cell Struct. Funct. 18:151-160, 1993.

Premont, R.T., Chen, J., Ma, H-W., Pommapalli, M., and Iyengar, R. Two members of a widely expressed subfamily of hormone-stimulated adenylyl cyclases. Proc. Nat. Acad.Sci. 89:9809-9813, 1992.

Ranjan, V., and Diamond, S.L. Fluid shear stress induces synthesis and nuclear localization of *c-fos* in cultured human endothelial cells. Biochem. Biophys. Res Comm. 196:79-84, 1993.

Rannels, D.E. Role of mechanical forces in compensatory growth of the lung. Am. J. Physiol. 257:L179-189, 1989.

Reneman, R.S. Endothelial cells as mechanoreceptors. News Physiol. Sci. 8:55-56, 1993.

Russo, L.A., Rannels, S.R., Laslow, K.S., and Rannels, D.E. Stretch-related changes in lung cAMP after partial pneumonectomy. Am. J. Phyiol. 257:E261-E268, 1989.

Rosales, O.R., and Sumpio, B.E. Changes in cyclic strain increase inositol trisphosphate and diacylglycerol in endothelial cells. Am.J. Physiol. 262:C956-C962, 1992.

Sachs, F. Baroreceptor mechanisms at the cellular level. Fed. Proc. 46:12-16, 1987.

Sadoshima, J., and Izumo, S. Signal transduction pathways of angiotensin II-induced *c-fos* gene expression in cardiac myocytes *in vitro*. Roles of phospholipid-derived second messengers. Circ. Res. 73:424-438, 1993.

Sadoshima, J., and Izumo, S. Molecular characterization of angiotensin II - induced hypertrophy of cardiac myocytes and hyperplasia of cardiac fibroblasts. Critical role of the AT_1 receptor subtype. Circ. Res. 413-423, 1993.

Sadoshima, J., Jahn, L., Takahashi, T., Kulik, T.J., and Izumo, S. Molecular characterization of the stretch-induced adaptation of cultured cardiac cells. J. Biol. Chem. 267:10551-10560, 1992a.

Sadoshima, J., Takahashi, T., Jahn, L., and Izumo, S. Role of mechanosensitive ion channels, cytoskeleton, and contractile activity in stretch-induced immediate-early gene expression and hypertrophy of cardiac myocytes. Proc. Nat. Acad. Sci. U.S.A. 89:9905-9909, 1992b.

Sadoshima, J., Xu, Y., Slayter, H.S., and Izumo, S. Autocrine release if angiotensin II mediates stretch-induced hypertrophy of cardiac myocytes *in vitro*. Cell 75:977-984, 1993.

Sandy, J.R., Meghji, S., Farndale, R.W., and Meikle, M.C. Dual elevation of cyclic AMP and inositol phosphates in response to mechanical deformation of murine osteoblasts. Biochim. Biophys. Acta 1010:265-269, 1989.

Shen, J., Luscinskas, F.W., Connolly, A., Dewey, C.F. Jr., and Gimbrone, M.A. Jr. Fluid shear stress modulates cytosolic free calcium in vascular endothelial cells. Am. J. Physiol. 262:C384-C390, 1992.

Sigurdson, W., Ruknudin, A., and Sachs, F. Calcium imaging of mechanically induced fluxes in tissue-cultured chick heart: role of stretch-activated ion channels. Am. J. Physiol. 262:H1110-H1115, 1992.

Theodoropoulos, P.A., Stournaras, C., Stoll, B., Markogiannakis, E., Lang, F., Gravanis, A., and Haussinger, D. Hepatocyte swelling leads to rapid decrease of the G-/total actin ratio and increases actin mRNA levels. FEBS Lett. 311:241-245, 1992.

Tilly, B.C., van den Berghe, N., Tertoolen, L.G.H., Edixhoven, M.J., and de Jonge, H.R. Protein tyrosine phosphorylation is involved in osmoregulation of ionic conductances. J. Biol. Chem. 268:19919-19922, 1993.

Tsutsui, H., Ishihara, K., Cooper, G. 4th, Cooper, G. Cytoskeletal role in the contractile dysfunction of hypertrophied myocardium. Science 260:682-687, 1993.

Unemmori, E.N., and Werb, Z. Reorganization of polymerized actin: a possible trigger for induction of procollagenase in fibroblasts cultured in and on collagen gels. J. Cell Biol. 103:1021-1031, 1986.

Vojtek, A., Haarer, B., Field, J., Gerst, J., Pollard, T.D., Brown, S., and Wigler, M. Evidence for a functional link between profilin and CAP in the yeast *S. cerevisiae*. Cell 66:497-505, 1991.

Wang, N., Butler, J.P., and Ingber, D.E. Mechanotransduction across the cell surface and through the cytoskeleton. Science 260:1124-1127, 1993.

Watson, P.A. Accumulation of cAMP and calcium in S49 mouse lymphoma cells following hyposmotic swelling. J. Biol. Chem. 264:14735-14740, 1989.

Watson, P.A. Direct stimulation of adenylate cyclase by mechanical forces in S49 mouse lymphoma cells during hyposmotic swelling. J. Biol. Chem. 265:6569-6575, 1990.

Watson, P.A., Haneda, T., and Morgan, H.E. Effect of higher aortic pressure on ribosome formation and cAMP content in rat heart. Am. J. Physiol. 256:C1257- C1261, 1989.

Wilson, E., Mai, Q., Sudhir, K., Weiss, R.H., and Ives, H.E. Mechanical stimulation induces growth of vascular smooth muscle cells via autocrine action of PDGF. J. Cell Biol. 123:741-747, 1993.

Winston, F.K., Thibault, L.E., and Macarak, E.J. An analysis of the time-dependent changes in intracellular calcium concentration in endothelial cells in culture induced by mechanical stimulation. J. Biomech. Eng. 115:160168, 1993.

Xenophontos, X.P., Watson, P.A., Chua, B.H.L., Haneda, T., and Morgan H.E. Increased cyclic AMP content accelerates protein synthesis in rat heart. Circ. Res. 65:647-656, 1989.

Yamazaki, T., Tobe, K., Hoh, E., Maemura, K., Kaida, T., Komuro, I., Taemoto, H., Kadowaki, T., Nagai, R., and Yazaki, Y. Mechanical loading activates mitogen-activated protein kinase and S6 peptide kinase in cultured rat cardiac myocytes. J. Biol. Chem. 268:12069-12076, 1993.

Yazaki, Y., Komura, I., Yamazaki, T., Tobe, K., Maemura, K., Kadowaki, T., and Nagai, R. Role of protein kinase system in the signal transduction of stretch-mediated protooncogene expression and hypertrophy of cardiac myocytes. Mol. Cell Biochem. 119:11-16, 1993.

Yoshimura, M, and Cooper, D.M.F. Cloning and expression of a Ca^{2+}-inhibitable adenylyl cyclase from NCB-20 cells. Proc. Nat. Acad. Sci. 89 :6716-6720, 1992.

Zachary, I., and Rozengurt, E. Focal adhesion kinase ($p125^{FAK}$): a point of convergence in the action of neuropeptides, integrins, and oncogenes. Cell 71:891- 894, 1992.

Zambetti, G., Ramsey-Ewing, A., Bortell, R., Stein, G., and Stein J. Disruption of the cytoskeleton with cytochalasin D induced *c-fos* gene expression. Exper. Cell Res.192:93-101, 1991.

16
Cytoskeletal Plaque Proteins as Regulators of Cell Motility, and Tumor Suppressors

A. Ben-Ze'ev, J.L. Rodríguez Fernández, B. Geiger,
M. Zöller, U. Glück

Introduction

Malignant transformation is accompanied by diverse cellular manifestations including alterations in cell growth rate, and major changes in cell structure affecting cell adhesion and shape, motile activity, and cytoskeletal organization. Morphological changes are perhaps among the most conspicuous features of the transformed phenotype in culture, characterized by rounded cell shape with poorly organized microfilament bundles (Weber et al 1974; Pollack et al 1975), and aberrant adhesions (Ben-Ze'ev 1985; Raz and Ben-Ze'ev 1987). These multiple phenotypic changes between normal and tumor cells are consistent with the multistep theory of tumorigenesis, implying defects in numerous genes, including genes for molecules mediating cell adhesion (Ben-Ze'ev 1992; Hedrick et al 1993; Tsukita et al 1993). Cell adhesion to neighboring cells and to the extracellular matrix is mediated by transmembrane receptors of the cadherin and integrin families of receptors (Hynes, 1987; Takeichi 1991, Figures 1,2). These receptor molecules are associated with cytoskeletal plaque proteins in the cytoplasmic face of the membrane to form different cellular junctions (Burridge et al 1988; Tsukita et al 1993). Recent studies suggest that these junctional proteins, in addition to their structural role, are also involved in signal transduction (Schwartz 1992; Juliano and Haskill 1993). This

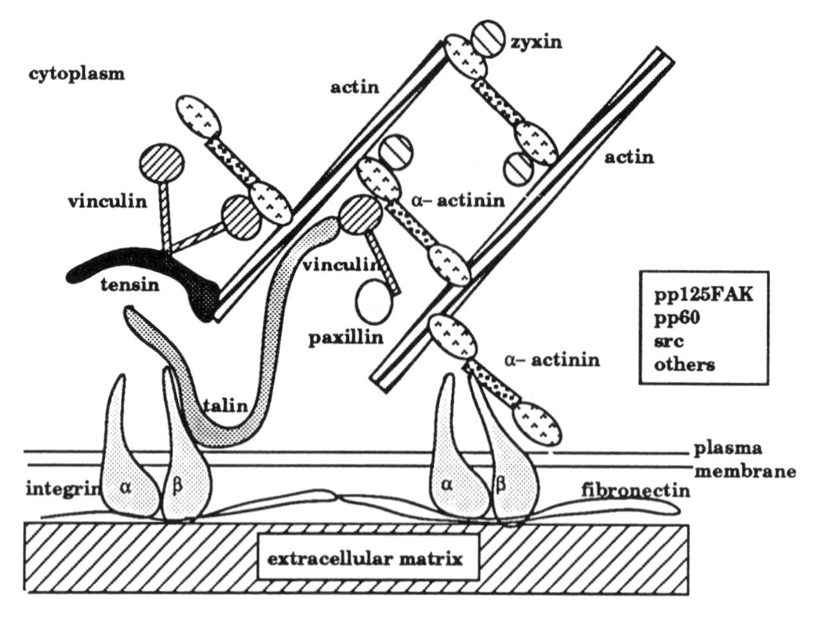

Figure 1. Schematic representation of the molecular interactions at cell-ECM contact sites involving the microfilaments. Note that the actin filaments can be linked to the transmembrane integrin receptors by α-actinin via an interaction involving three molecules (α-actinin, vinculin and talin), or directly by an interaction between α-actinin and integrin (For details, see Otey et al 1993).

interesting idea is based on findings demonstrating the localization of a large number of oncogene products and regulatory molecules, known to play a role in signal transduction, in the junctional plaque (Geiger and Ginsberg 1991; Burridge et al 1992; Geiger et al 1992). The current view of cell adhesion-mediated signal transduction implies the involvement of phosphorylation-dephosphorylation mechanisms, mainly on tyrosine (Volberg et al 1992; Guan and Shalloway 1992). Thus, a constitutive activation of this pathway in tumor cells, may explain the ability of cancer cells to grow without adhesion in soft agar. In addition, changes in the availability of the transmembrane receptors and the cytoskeletal plaque proteins may also be involved in controlling the tumorigenic ability of cells (Takeichi 1993; Juliano and Varner 1993; Ben-Ze'ev, 1991,1992; Tsukita et al 1993). Here we review recent findings from our work in support of a possible role for changes in the expression of junctional plaque proteins, which link the microfilaments to the membrane. We show that by regulating the level of these proteins it is possible to affect cell motility, and that junctional plaque proteins can act as effective tumor suppressors.

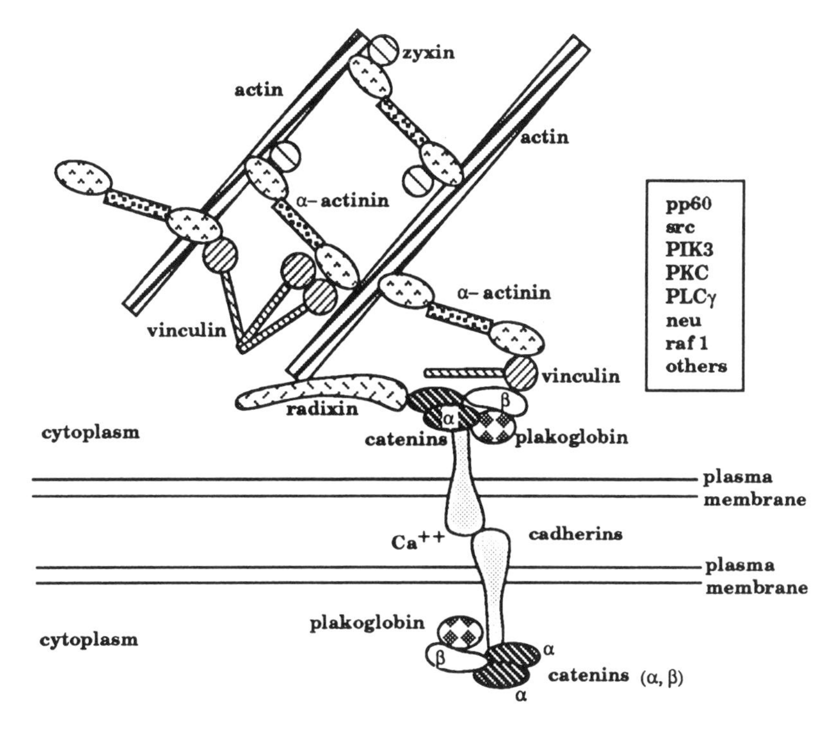

Figure 2. Schematic representation of the molecular interactions at cadherin-mediated cell-cell junctions involving the microfilaments. Note the presence of vinculin, α-actinin and of regulatory molecules and oncongene products in both cell-ECM (Figure 1) and cell-cell junctions.

Regulation of plaque proteins during growth activation and differentiation

Cytoskeletal proteins are among the most abundant and stable cellular proteins. Their expression in many cell types is, however, not constitutive, but is regulated in response to changes in the organization of the respective cytoskeletal filamentous structure (Ben-Ze'ev 1986a,b). For example, the synthesis of tubulin, the major component of microtubules, is tightly coupled to the state of polymerization of microtubules (Ben-Ze'ev et al 1979; Cleveland et al 1981), and the expression of different types of intermediate filament proteins responds to changes in cell-cell and cell-substrate contacts (Ben-Ze'ev 1983, 1984). The expression of actin and the adhesion plaque component vinculin (Figure 1), are also regulated in cells establishing contacts with different substrates and upon cell spreading (Farmer et al 1983; Ungar et al 1986; Bendori et al 1987).

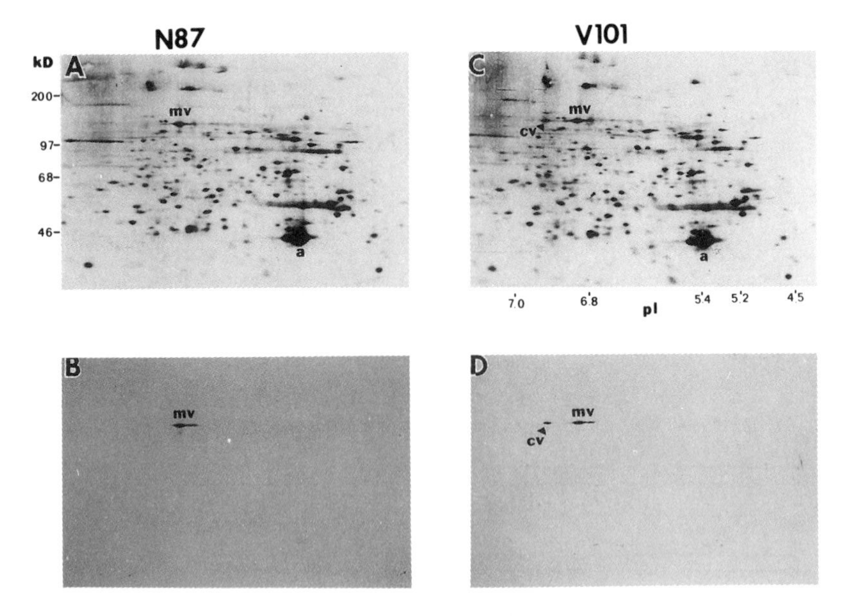

Figure 3. Identification and levels of chicken vinculin expressed in mouse 3T3 cells determined by 2-D gel electrophoresis. Cells of a neo^{r} clone, N87 (A, B), and of a chicken vinculin-transfected clone V101 (C, D) were labeled with ^{35}S-methionine and analyzed by 2-D gel electrophoresis. The separated proteins were transferred to nitrocellulose and the protein pattern obtained by autoradiography (A,C). The same blots were used to identify vinculin by immunodetection (B, D). mv, mouse vinculin; cv, chicken vinculin; a, actin. The level of cv is 20% of mv (From Rodríguez Fernández et al 1992a).

Changes in cell shape and microfilament organization are part of the rapid responses of quiescent cells to stimulation by growth factors (Schlessinger and Geiger 1981). The disruption of actin stress fibers by PDGF in quiescent 3T3 cells results apparently from release of vinculin from adhesion plaques (Herman and Pledger 1985). Moreover, vinculin gene expression is rapidly stimulated after activation of cells into the growth cycle (Ben-Ze'ev et al 1990; Bellas et al 1991). The expression of other components of the ECM-microfilament structural complex is part of the "immediate-early" response of quiescent cells to growth factor stimulation (Rysek et al 1989; Wades et al 1992; Glück et al 1992). This suggestion is supported by the transient increase in the expression of these genes in regenerating liver after partial hepatectomy (Glück et al 1992).

The expression of differentiated cell functions are are also dependent on the establishment of appropriate cell contacts and cytoskeletal organization in vivo and in cultured cells. In many cell types, isolation from the tissue of origin results in the loss of

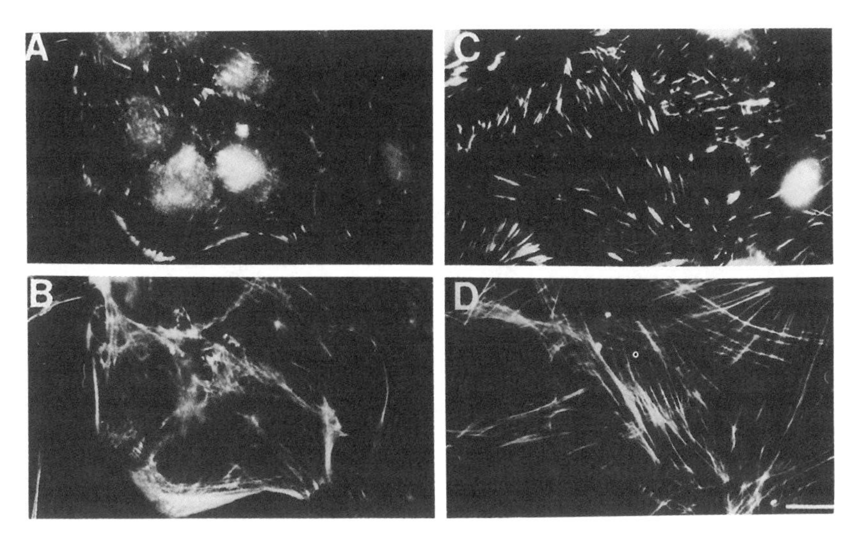

Figure 4. The organization of actin and vinculin in 3T3 cells overexpressing vinculin. Control 3T3 cells (A,B) and clone V101 (C,D see Figure 3) were double stained for vinculin (A,C) and actin (B,D). (From Geiger et al 1992).

differentiation specified gene expression. Restoration of "proper" junctional assembly with the ECM and/or neighboring cells, and the subsequent cell shape changes can restore cell type specific gene expression (Ben-Ze'ev 1992; Ben-Ze'ev et al 1990). The regulation of microfilament and junctional plaque protein expression is tightly associated with the changes in cell contacts and shape which characterize the establishment of the differentiated phenotype (Rodríguez Fernández and Ben-Ze'ev 1989; Ben-Ze'ev and Amsterdam 1986, 1987; Ben-Ze'ev et al 1988; Ben-Ze'ev 1989).

Modulation of junctional plaque protein expression affects cell motility

The abundance and stability of cytoskeletal proteins raised the question of the significance of these modulations in their expression in cell physiology. To address this question more directly, we prepared cells with a stable, targeted increase or decrease of vinculin or α-actinin expression. 3T3 cells overexpressing a transfected vinculin at 20% of the endogenous vinculin level (Figure 3), showed a more elaborate adhesion plaque and stress fiber system when compared to control cells (Figure 4). Moreover, these cells had a significantly decreased ability to migrate and close an artificial wound introduced in a confluent monolayer (Figure 5), and formed shorter

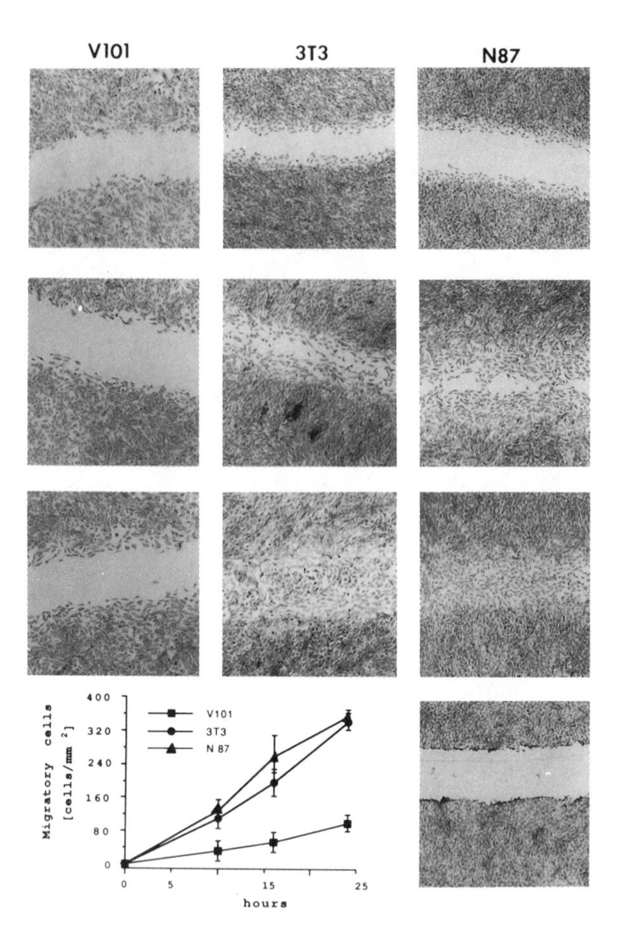

Figure 5. Decreased ability to migrate into a "wound" of vinculin-transfected cells. Confluent cultures of 3T3 cells, a neo^r control (N87), and cells transfected with chicken vinculin (V101) were "wounded", and at various times after the "wounding", the cultures were fixed and stained with Giemsa. The number of cells migrating into an area of 1 mm^2 was determined randomly in 10 different areas along the wound for each cell type at each time point. (From Rodríguez Fernández et al 1992a).

phagokinetic tracks (Figure 6B,C). Studies with 3T3 clones overexpressing a transfected α-actinin at levels between 40% and 60% of the endogenous protein level similarly displayed a reduced cell motility (Glück and Ben-Ze'ev 1994). These results strongly suggest that even modest increases in the expression of adherens junction plaque proteins can dramatically effect cell locomotion. The mechanisms underlying this effect of increased expression of vinculin and α-actinin on cell motility are presently unknown. As vinculin and α-actinin are involved in linking actin to the membrane in adhesion plaques, it is conceivable that the modulations in their

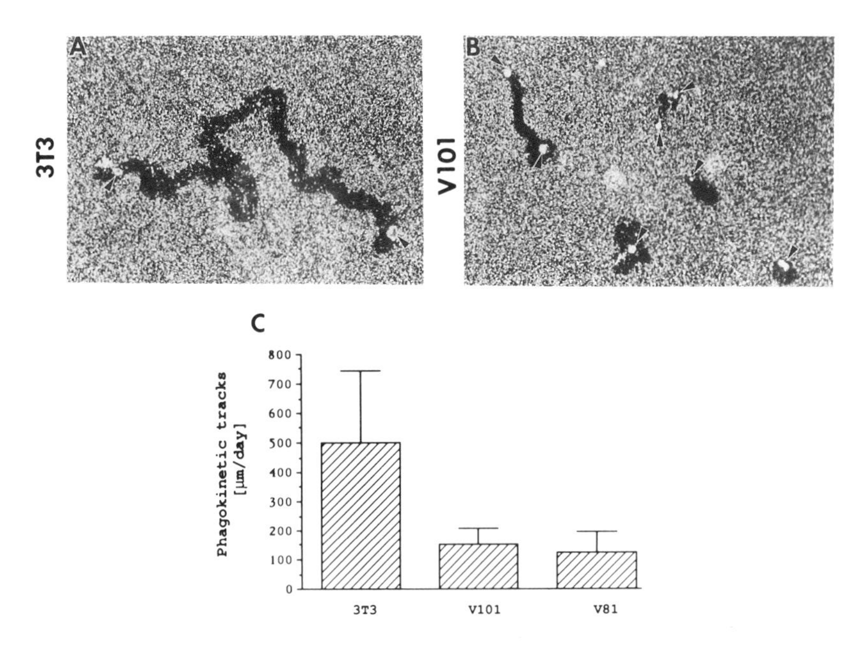

Figure 6. The migratory ability of individual vinculin transfected 3T3 cells is reduced. Cells were seeded sparsely on coverslip precoated with colloidal gold. The phagokinetic tracks produced by 3T3 cells (A) were compared to those of vinculin-transfected cells (clones V81 and V101) by darkfield microscopy (B). The differences in the length of tracks produced by individual cells were determined by projecting the tracks on a screen and measuring randomly the length of about 30 tracks from each cell type (C). (From Rodríguez Fernández et al 1992a).

expression can affect cells by changing cell adhesion and/or microfilament structure. The cooperative model for the assembly of adhesion plaques from soluble components (Geiger et al 1992; Figure 7) would predict that overexpression of α-actinin or vinculin may enhance a cooperative and more efficient assembly of adhesion plaques, which will lead to increased adhesion and decreased motility.

According to this model, reduced expression of either vinculin or α-actinin should result in an inefficient assembly of adhesion plaques, thus leading to decreased adhesion and increased motility. This hypothesis was tested by a targeted suppression of vinculin in 3T3 cells by transfection with antisense vinculin. Clones expressing vinculin at about 10% to 30% of control levels (Figure 8, Table 1) showed a round phenotype, with fewer vinculin positive plaques at the cell periphery (Figure 9). Moreover, suppression of vinculin expression also conferred an increase in 3T3 cell motility and the acquisition of anchorage independent growth in soft agar (Figure 10, Table 1). Furthermore, the targeted inactivation of vinculin in F9

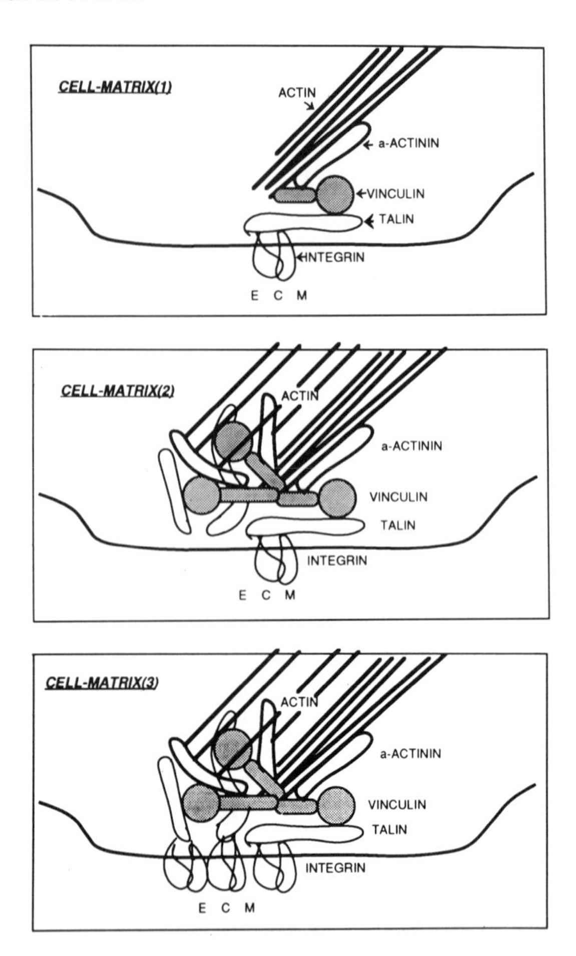

Figure 7. Schematic models describing the co-operative molecular assembly of adhesion plaques. Cell-Matrix (1) shows a linear array of molecular interactions. Cell-Matrix (2) shows the cooperative effect resulting from vinculin oligomerization. Cell-Matrix (3) shows inside-out immobilization of transmembrane receptors to the newly formed contact sites.

teratocarcinoma cells, by homologous recombination, produced vinculin deficient cells which displayed a decreased adhesion and spreading on the substrate, and an increase in cell motility (Coll et al 1994). In addition, 3T3 cells in which α-actinin expression was reduced to 10% of control levels, by antisense transfection, became tumorigenic in nude mice (Glück and Ben-Ze'ev 1994). Taken together, these results strongly suggest that regulation of junctional plaque protein expression can control cell motility and the tumorigenic ability of cells.

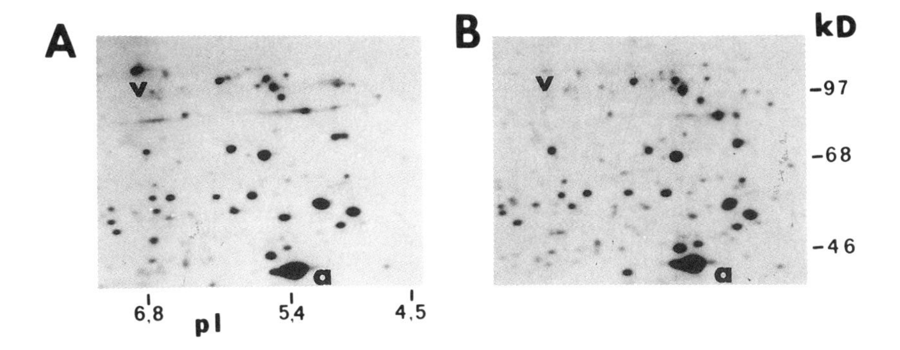

Figure 8. Suppression of vinculin synthesis in antisense vinculin-transfected cells. 3T3 cells were transfected with antisense vinculin and the neo^r gene, and neo^r colonies expressing decreased vinculin by Western blotting were isolated. Such clones synthesize 10 fold less vinculin (B) than control 3T3 (A), as determined by 2-D gel electrophoresis of ^{35}S-methionine labelled cells. v, vinculin; a, actin.

Table 1. Morphology, motility, and growth properties of antisense-vinculin transfected clones

Cell line	Vinculin levels % of 3T3[1]	Cell shape	Colonies in agar[2]	Migration (cell number)[3]
3T3	100	flat	30	180±20
D31	108.3	flat	ND	ND
A41	106.7	flat	4	210±60
C11	102.7	flat	ND	150±60
C41	98.8	flat	20	ND
D51	78.9	flat	ND	ND
A11	69.6	flat	ND	ND
D41	58.4	flat	6	ND
C51	43.4	flat	ND	ND
A51	31.6	round	188	400±80
B51	16.6	round	575	460±100
B61	11.0	round	143	360±50
C61	10.3	round	113	ND

[1] The levels of vinculin were determined by quantitative computerized scanning of immunoblots. The level of vinculin in 3T3 cells was taken as 100%.

[2] The number of colonies containing >50 cells was determined 2 weeks after seeding 10^4 cells in agarose-containing medium.

[3] Cell motility is the number of cells which migrated into an area of 1 mm^2 16 hours after introducing a "wound" into a confluent monolayer (From Rodríguez Fernández et al 1993).

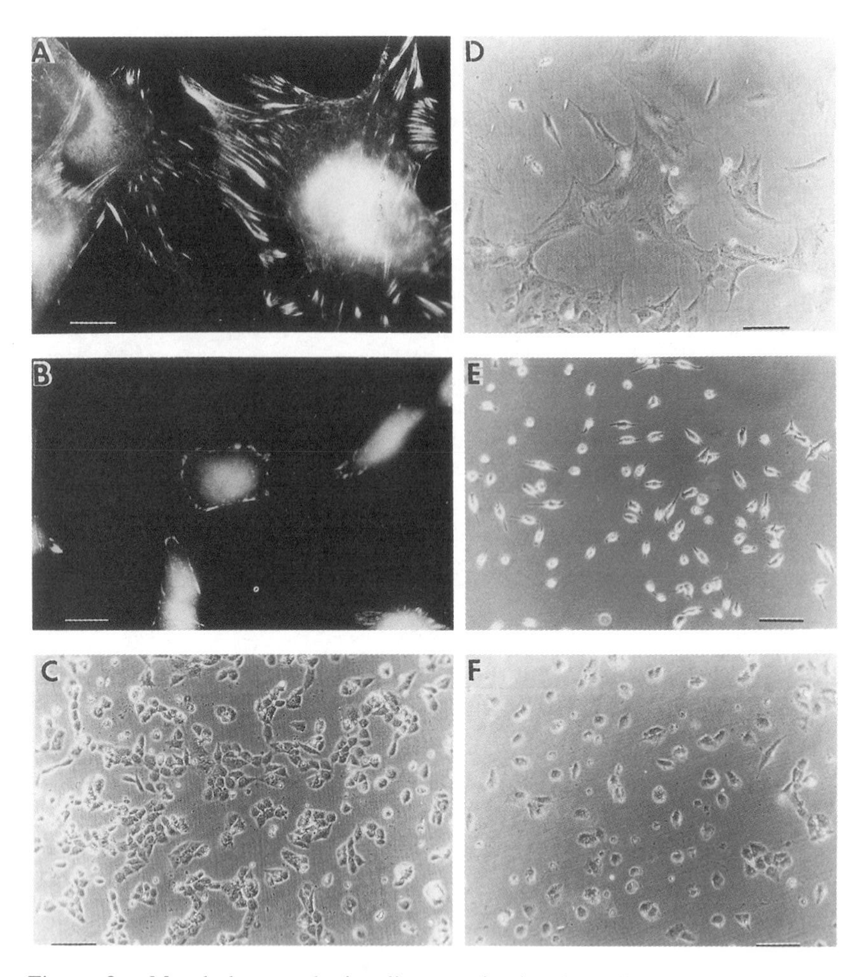

Figure 9. Morphology and vinculin organization in cells expressing antisense vinculin. A B, cells grown on coverslips were fixed, permeabilized, and immunostained with anti vinculin antibody followed by rhodamine-labeled anti mouse IgG. A,D, neo^r control; B-F, individual clones expressing decreased levels of vinculin. The bars in A and B indicate 10 μm, in C-F the bar indicates 50 μm (From Rodríguez Fernández et al 1993).

Suppression of tumorigenicity in transformed cells after transfection with vinculin and α-actinin

Cytoarchitectural changes and altered adhesive properties are characteristic to many transformed cells in culture (Ben-Ze'ev 1985; Raz and Ben-Ze'ev 1987). These include a decrease in, or even lack of, microfilament bundles (Figure 11) and a reduced expression of microfilament proteins such as actin, vinculin and α-actinin (Figure

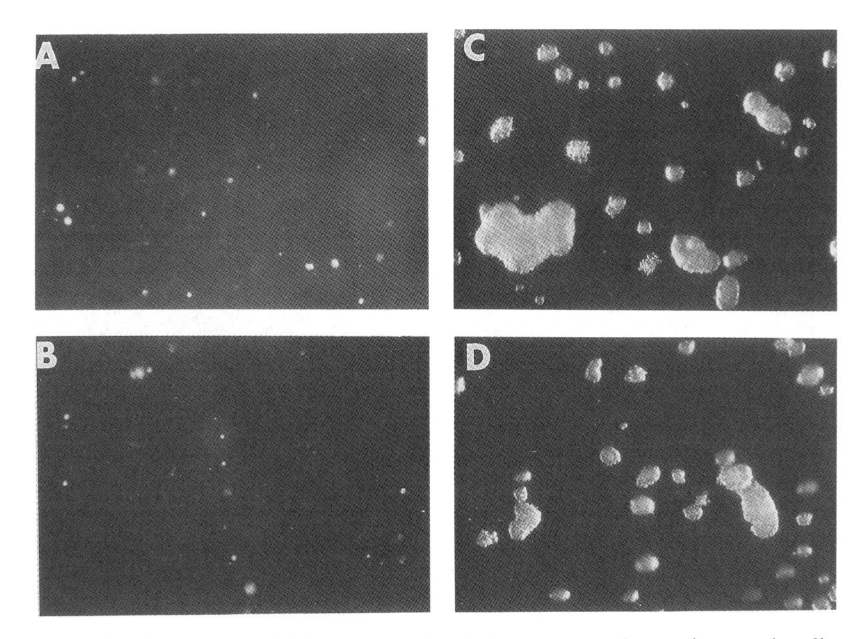

Figure 10. Anchorage independent growth of clones expressing antisense vinculin. Cells were seeded in 0.3% agarose containing medium (10^4 cells per 35 mm diameter dish). Pictures were taken after 2 weeks of incubation at 37°C. A,B, neo^r clones; C,D, antisense vinculin transfected clones (From Rodríguez Fernández et al 1993).

Table 2. Tumorigenicity of SVT2 clones expressing different vinculin levels

Clone	Vinculin Level[1] Chicken	Mouse	Tumor Incidence (%)	Growth in Agar[2] (%)
3T3	0	750	0/6 (0)	< 0.01
neo 1	0	ND	6/6 (100)	11.0±0.8
neo 3	0	140	6/6 (100)	13.3±1.0
SVT2	0	135	6/6 (100)	12.3±0.5
D 34	344	170	1/6 (16)	5.7±1.0
D 43	135	ND	5/6 (83)	ND
D 41	80	ND	5/6 (83)	ND
D 44	670	160	0/6 (0)	3.6±1.5

Balb/c mice were injected with $5x10^6$ cells per animal.

[1]Arbitrary O.D. units of the chicken vinculin levels were obtained by densitometer scanning of cell lysates separated by 2-D gel electrophoresis.

[2]Cells were seeded in duplicates at 250 and 10^3 cells per 35 mm dish in soft agar and the number of colonies (per cent) formed per plate was determined. ND, not done. (From Rodríguez Fernández et al 1992b).

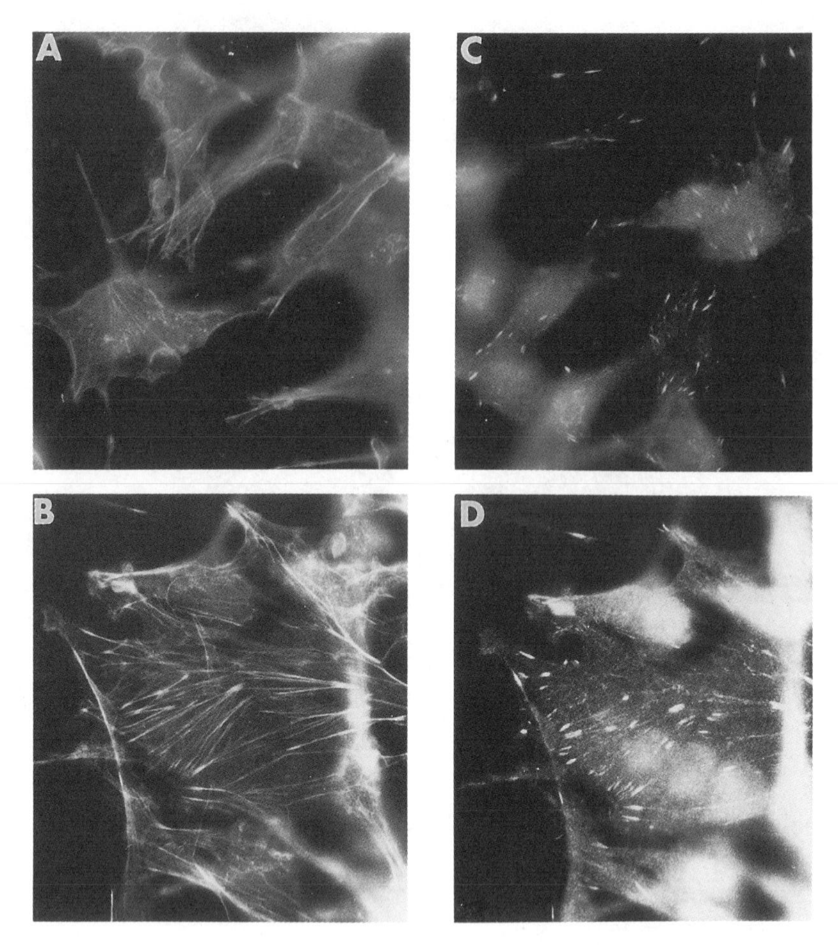

Figure 11. Organization of actin and vinculin in SV40-transformed 3T3 cells (SVT2) and in SVT2 cells transfected with vinculin. SVT2 (A,C) and vinculin-transfected SVT2 were double labeled for actin (A,B) and vinculin (C,D). Note the more elaborate microfilament bundles and vinculin-plaques in the vinculin-transfected SVT2.

12). The restoration of vinculin (Rodríguez Fernández et al 1992b) or α-actinin expression (Figure 13) in SV40-transformed 3T3 cells to that found in nontransformed 3T3 resulted in cells with an apparent increase in cell substrate adhesiveness and stress fiber formation (Figure 11), and a decreased ability to grow in soft agar (Tables 2,3). Moreover, the ability of these cells to develop tumors was completely suppressed in syngeneic animals, and was markedly reduced in athymic nude mice (Tables 2,3).

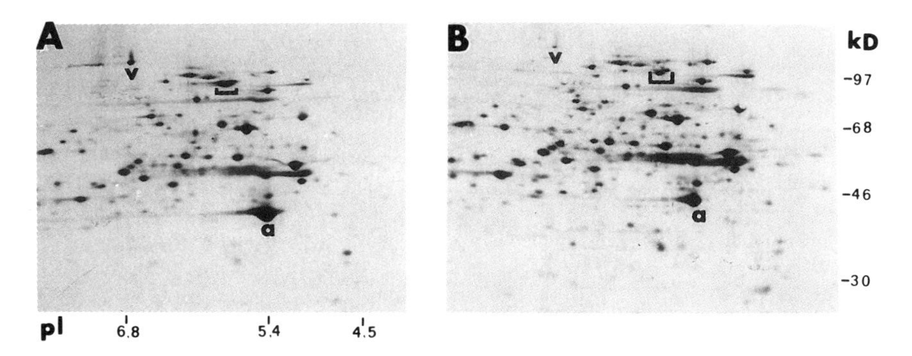

Figure 12. Decreased synthesis of microfilament proteins in SVT2 cells. 3T3 (A) and SVT2 (B) cells were pulse labeled with ^{35}S-methionine and analyzed by 2-D gel electrophoresis. a, actin; v, vinculin; bracket marks the position of α-actinin. Note the decrease in the synthesis of actin, vinculin and α-actinin in SVT2 cells.

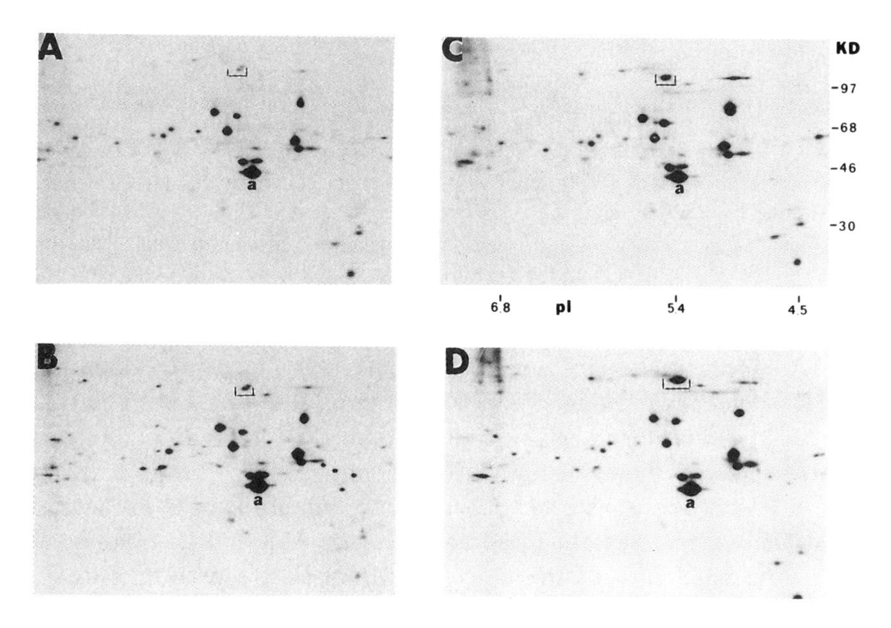

Figure 13. Levels of α-actinin expressed by different SVT2 clones transfected with human α-actinin cDNA. SVT2 cells were co-transfected with the neo^r gene and with a full length human α-actinin cDNA. Neo^r control and clones expressing high levels of α-actinin were labeled with ^{35}S-methionine and analyzed by 2-D gel electrophoresis. A, SVT2 neo^r control; B-D, α-actinin transfected clones: B, Sα1; C, Sα8; D, Sα29. For levels of α-actinin in these clones see Table 3. (From Glück et al 1993).

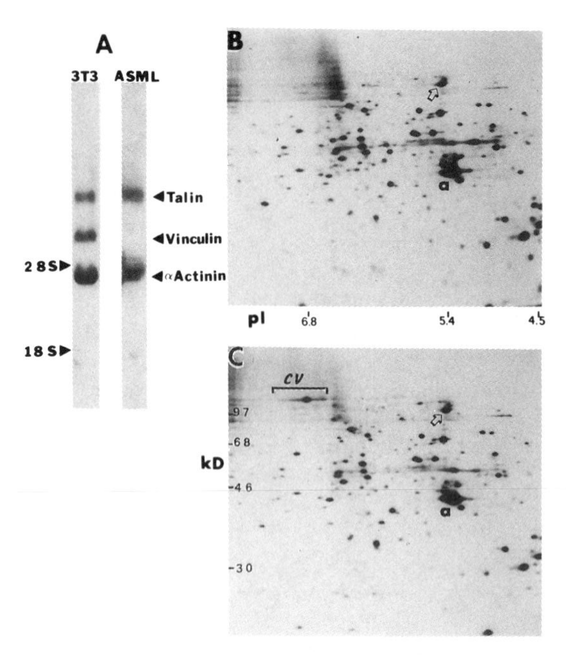

Figure 14. Lack of vinculin expression by the ASML adenocarcinoma cell line. A, Northern blot hybridization of 3T3 and ASML cell RNA preparations with cDNA probes to talin, vinculin and α-actinin. B, 2-D gel electrophoretic analysis of ^{35}S-methionine-labelled proteins of ASML cells. C, Analysis of an ASML clone transfected with vinculin cDNA. cv, vinculin; a, actin; arrow, α-actinin. Note the undetectable level of vinculin in ASML cells, and the high level of vinculin expression in the transfected ASML cells. (From Rodríguez Fernández et al 1992b).

A highly malignant adenocarcinoma cell line which does not express detectable levels of vinculin (Raz et al 1986), but expresses normal levels of the adhesion plaque proteins α-actinin and talin, was also transfected with vinculin cDNA (Figure 14). Clones expressing high levels of the transfected vinculin showed an increase in the cell substrate contact areas of these cells (Figure 15), and the malignant metastatic capability of the cells was dramatically reduced (Figure 16). Detailed analysis of clones expressing different levels of the transgene (Tables 2,3 and Rodríguez Fernández et al 1992b; Glück et al 1993), revealed that there is a direct correlation between the level of expression of the transgene (either vinculin or α-actinin), and the suppression of the tumorigenic phenotype. Taken together, these results demonstrate that cytoskeletal plaque proteins of adherens junctions can act as determinants of cell shape and motility, and are also effective suppressors of the tumorigenic phenotype.

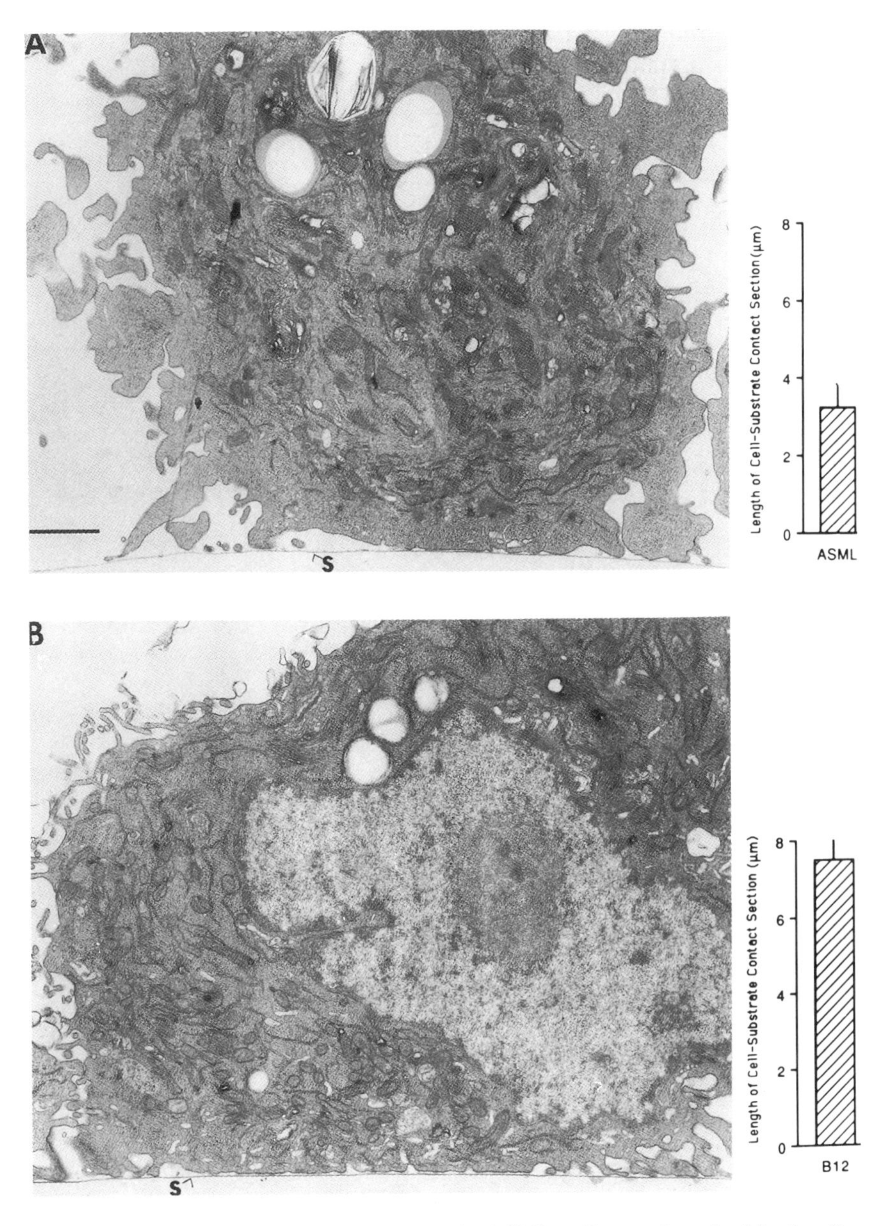

Figure 15. Increased cell-substrate contact in ASML cells transfected with vinculin. Transmission electronmicrographs of sections perpendicular to the substrate at areas of cell-substrate contact in ASML (A) and vinculin-transfected ASML clone B 12 (B). S, substrate. Note the 2 fold increase in the length of cell-substrate contacts in vinculin-transfected cells.

Table 3. Tumorigenicity of SVT2 clones expressing different α-actinin levels

Cell line	α-Actinin level[1]	Tumor incidence in Balb/C mice (%)	Growth in agar (%)[2]
3T3	658.0	0/6 (0)	<0.01
SVT2	108.8	6/6 (100)	12.3±0.9
Sneo	135.9	4/4 (100)	13.3±1.0
Sα16	250.1	4/5 (80)	11.5±0.8
Sα1	497.3	2/4 (50)	ND
Sα8	693.0	1/5 (20)	5.1±0.7
Sα29	1297.6	0/5 (0)	1.8±0.1

[1] Arbitrary O.D. units obtained by computerized densitometer scanning of 2-D gels as shown in Fig. 4.
[2] % number of colonies formed in agar compared to the number of colonies formed on plastic per same number of cells seeded (From Glück et al 1993).

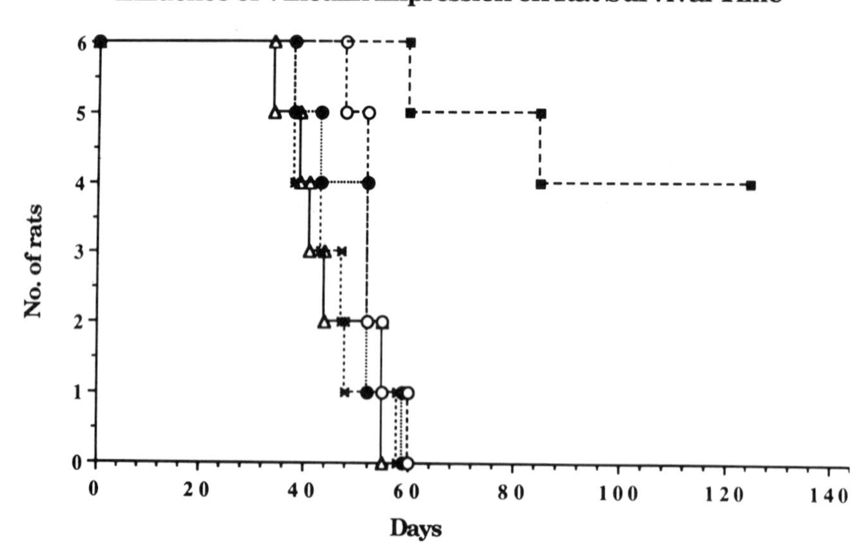

Figure 16. Rat survival after injection with ASML cells expressing different chicken-vinculin levels. Groups of 6 rats were injected in the footpad with $5x10^5$ cells of ASML clones expressing different levels of chicken vinculin. The survival time of individual rats, in days, was determined. Δ----Δ, untransfected ASML cells; ■----■, clone B 12; X----X, clone B 12 r, a revertant of clone B12. Note the increase in the survival of animals injected with clone B 12 which expresses the highest level of vinculin, and the reversion to the highly malignant phenotype in the revertant clone B 12 r, which has lost the vinculin gene (From Rodríguez Fernández et al 1992b).

Discussion

The studies presented here demonstrated that structural components of adherens junctions which link the microfilaments to areas of cell-ECM and cell-cell contact sites are involved in regulating cell motility and the tumorigenic ability of cells. The results are consistent with the view that these molecules not only determine cell shape and adhesion, but are governing major cellular processes such as growth, differentiation and transformation. In addition to the adherens junction proteins vinculin and α-actinin described above, the regulation of other microfilament proteins were shown to alter the motile and tumorigenic properties of cells. The overexpression of gelsolin, by 25%, which is a microfilament severing protein, lead to increased motility in 3T3 cells (Cunningham et al 1991), and the expression of actin binding protein in human melanoma cells which lack this protein, reduced membrane blebbing and increased cell motility (Cunningham et al 1992). Tropomyosin 1, an actin crosslinker whose level is reduced in tumor cells of various origins (Matsumura and Yamashiro-Matsumura 1986), suppressed the tumorigenicity of *ras*-transformed 3T3 cells after its level was increased by cDNA transfection (Prasad et al 1993).

Recently, α-catenin, a vinculin-like molecule (Nagafuchi et al 1991; Herrenknecht et al 1991), which is localized in adherens type cell-cell junctions (Figure 2), was found to extensively modulate cell aggregation upon transfection into α-catenin-deficient cells (Hirano et al 1992). Moreover, several types of the more invasive (the scattered phenotype) human adenocarcinomas were shown to have lost α-catenin expression (Kadowaki et al 1994; Morton et al 1993).

Other cytoplasmic plaque components of cell-cell adherens junctions include the ERM family which displays an 84% identity at the amino-terminal half of its members (ezrin, radixin, moesin, Figure 2). These molecules which also function in linking actin filaments to the membrane, were recently shown to be closely related to the tumor suppressor neurofibromatosis 2 (NF2), a major human genetic disorder involved in the formation of brain tumors (Trofatter et al 1993; Rouleau et al 1993). The candidate gene (for NF2) codes for merlin which is 62% identical in its amino-terminal half to the ERM family members.

Taken together, these results favor the concept that junctional plaque proteins which are closely localized to oncogene products and signal transduction molecules (Figures 1,2), can act as tumor suppressors. Whether this new function for cytoskeletal plaque proteins is exerted by their structural role (regulation of cell shape, adhesion and motility), as shown it the studies presented above, or by their action to physically bring together signal transducing molecules specialized in negative regulation of growth, remains to be determined. Future studies based on information obtained from the human genome project, and targeting experiments of these genes in transgenic mice will provide important information on the relationship between junctional plaque molecules and the malignant properties of cells.

Acknowledgments

This study was supported by grants from The Council For Tobacco Research-USA, from the Minerva Fund, from the Israel Cancer Research Fund (ICRF), from the USA-Israel Binational Fund, and by a grant from the Leo and Julia Forchheimer Center for Molecular Genetics at The Weizmann Institute of Science. A. B-Z is the Lunenfeld-Kunin Professor for Genetics and Cell Biology.

References

Ben-Ze'ev A. Mechanisms of regulating tubulin synthesis in cultured mammalian cells. Cell 17:319-325; 1979.

Ben-Ze'ev A. Cell configuration-related control of vimentin biosynthesis and phosphorylation in cultured mammalian cell. J. Cell Biol. 97:858-865; 1983.

Ben-Ze'ev A. Differential control of cytokeratins and vimentin synthesis by cell-cell contact and cell spreading in cultured epithelial cells. J. Cell Biol. 99:1424-1433; 1984.

Ben-Ze'ev A. The cytoskeleton of cancer cells. Biochim. Biophys. Acta 780:197-212; 1985.

Ben-Ze'ev A. Regulation of cytoskeletal protein synthesis in normal and cancer cells. Cancer Rev. 4:91-116; 1986a.

Ben-Ze'ev A. The relationship between cytoplasmic organization, gene expression and morphogenesis. Trends Biochem. Sci. 11:478-481; 1986b.

Ben-Ze'ev A. Cell shape and cell contacts: molecular approaches to cytoskeleton expression. In: Stein W. D.; Bronner F., ed. Cell Shape: determinants, regulation, and regulatory role. Academic press; 1989:p. 95-119.

Ben-Ze'ev A. Animal cell shape changes and gene expression. BioEssays 13:207-212; 1991.

Ben-Ze'ev A.; Amsterdam A. Regulation of cytoskeletal proteins involved in cell contact formation during differentiation of granulosa cells on extracellular matrix. Proc. Natl. Acad. Sci. USA 83:2894-2898; 1986.

Ben-Ze'ev A.; Amsterdam A. In vitro regulation of granulosa cell differentiation: Involvement of cytoskeletal protein expression. J. Biol. Chem. 262:5366-5376; 1987.

Ben-Ze'ev A.; Reiss R.; Bendori R.; Gorodecki. B. Transient induction of vinculin gene expression in 3T3 fibroblasts stimulated by serum-growth factors. Cell Regulation 1:621-636; 1990.

Ben-Ze'ev A.; Robinson G. S.; Bucher N. L. R.; Farmer S. R. Cell-cell and cell-matrix interactions differentially regulate the expression of hepatic and cytoskeletal genes in primary cultures of rat hepatocytes. Proc. Natl. Acad. Sci. USA 85:2161-2165; 1988.

Ben-Ze'ev A.; Rodríguez Fernández J.-L.; Baum G.; Gorodecki B. Regulation of cell contacts, cell configuration and cytoskeletal gene expression in differentiating cells. In: Fisher P. B., ed. Mechanisms of Differentiation. Boca Raton, FL: CRC Press; 1990:p. 143-173.

Bellas R. E.; Bendori R.; Farmer S. R. Epidermal growth factor activation of vinculin and $beta_1$-integrin gene transcription in quiescent Swiss 3T3 cells. J. Biol. Chem.` 266:12008-12014; 1991.

Bendori R.; Salomon D.; Geiger B. Contact-dependent regulation of vinculin expression in cultured fibroblasts: a study with vinculin specific cDNA probes. EMBO J. 6:2897-2905; 1987.

Burridge K.; Fath K.; Kelly T.; Nuckolls G.; Turner C. Focal adhesions: Transmembrane junctions between the extracellular matrix and the cytoskeleton. Annu. Rev. Cell Biol. 4:487-525; 1988.

Burridge K.; Turner C. E.; Romer L. H. Tyrosine phosphorylation of paxillin and $pp125^{FAK}$ accompanies cell adhesion to extracellular matrix: a role in cytoskeletal assembly. J. Cell Biol. 119:893-903; 1992.

Cleveland D. W.; Lopata M. A.; Sherline P.; Kirschner M. W. Unpolymerized tubulin modulates the level of tubulin mRNAs. Cell 25:537-546; 1981.

Coll J.-L.; Ben-Ze'ev A.; Rodríguez Fernández J. L.; Ezzell R. M.; Barribault H.; Oshima R. G.; Adamson E. Targeted disruption of the vinculin gene in F9 cells changes cell morphology, adhesion and locomotion. Submitted.

Cunningham C. C.; Garlin J. B.; Kwiatkowski D. J.; Hartwig J. H.; Janmey P. A.; Byers R. H.; Stossell T. P. Actin-binding protein requirement for cortical stability and efficient locomotion. Science 255:325-327; 1992.

Cunningham C. C.; Stossel T. P.; Kwiatkowsk D. J. Enhanced motility in NIH 3T3 fibroblasts that overexpress gelsolin. Science 251:1233-1236; 1991.

Farmer S. R.; Wan K. M.; Ben-Ze'ev A.; Penman S. Regulation of actin mRNA levels and translation responds to changes in cell configuration. Mol. Cell. Biol. 3:182-189; 1983.

Geiger B.; Ayalon O.; Ginsberg D.; Volberg T.; Rodríguez Fernández J. L.; Yarden Y.; Ben-Ze'ev A. Cytoplasmic control of cell-adhesion. Cold Spring Harb. Symp. Quant. Biol. 57:631-642; 1992.

Geiger B.; Ginsberg D. The cytoplasmic domain of adherens-type junctions. Cell Motil. Cytoskel. 20:1-6; 1991.

Glück U.; Ben-Ze'ev A. Transfection of 3T3 cells with α-actinin and antisense α-actinin affects cell motility and tumorigenicity. Submitted.

Glück U.; Kwiatkowski D. J.; Ben-Ze'ev A. Suppression of tumorigenicity in simian virus 40-transformed 3T3 cells transfected with α-actinin cDNA. Proc. Natl. Acad. Sci. USA 90:383-387; 1993.

Glück U.; Rodríguez Fernández J. L.; Pankov R.; Ben-Ze'ev A. Regulation of adherens junction protein expression in growth-activated 3T3 cells and in regenerating liver. Exp. Cell Res. 202:477-486; 1992.

Guan J.-L.; Shalloway D. Regulation of focal adhesion associated protein tyrosine kinase by both cellular adhesion and oncogenic transformation. Nature 358:690-692; 1992.

Hedrick L.; Cho K. R.; Vogelstein B. Cell adhesion molecules as tumour suppressors. Trends Cell Biol. 3:36-39; 1993.

Herman B.; Pledger J. Platelet-derived growth factor-induced alterations in vinculin and actin distribution in Balb/C-3T3 cells. J. Cell Biol. 100:1031-1040; 1985.

Herrenknecht K.; Ozawa M.; Eckerskorn C.; Lottspeich F.; Lenter M.; Kemler R. The uvomorulin-anchorage protein α catenin is a vinculin homologue. Proc. Natl. Acad. Sci. USA 88:9156-9160; 1991.

Hirano S.; Kimoto N.; Shimoyama Y.; Hirohashi S.; Takeichi M. Identification of a neural α-catenin as a key regulator of cadherin function and multicellular organization. Cell 70:293-301; 1992.

Hynes R. O. Integrins: a family of cell surface receptors. Cell 48:549-554; 1987.

Juliano R. L.; Haskill S. Signal transduction from the extracellular matrix. J. Cell Biol. 120:577-585; 1993.

Juliano R. L.; Varner J. A. Adhesion molecules in cancer: the role of integrins. Curr. Opin. Cell Biol. 5:812-818; 1993.

Kadowaki T.; Shiozaki H.; Inoue M.; Tamura S.; Oka H.; Doki Y.; Iihara K.; Matsui S.; Iwazawa A.; Nagafuchi A.; Tsukita S.; Mori T. E-cadherin and α-catenin expression in human esophageal cancer. Cancer Res. 1994. In press.

Matsumura F.; Yamashiro-Matsumura S. Tropomyosin in cell transformation. Cancer Rev. 6:21-39; 1986.

Morton R. A.; Ewing C. M.; Nagafuchi A.; Tsukita S.; Isaacs W. B. Reduction of E-cadherin levels and deletion of the α-catenin gene in human prostate cancer cells. Cancer Res. 53:3585-3590; 1993.

Nagafuchi A.; Takeichi M.; Tsukita S. The 102 kd cadherin-associated protein: similarity to vinculin and posttranscriptional regulation of expression. Cell 65:849-857; 1991.

Otey C. A.; Vasquez G. B.; Burridge K.; Erickson B. W. Mapping of the α-actinin binding site within the β_1-integrin cytoplasmic domain. J. Biol. Chem. 268:21193-21197; 1993.

Pollack R.; Osborn M.; Weber K. Patterns of organization of actin and myosin in normal and transformed cells. Proc. Natl. Acad. Sci. USA 72:944-998; 1975.

Prasad G. L.; Fulder R. A.; Cooper H. L. Expression of transduced tropomyosin 1 cDNA supresses neoplastic growth of cells transformed by the *ras* oncogene. Proc. Natl. Acad. Sci. USA 90:7039-7043; 1993.

Raz A.; Ben-Ze'ev A. Cell-contact and architecture of malignant cells and their relationship to metastasis. Canc. Metastas. Rev. 6:3-21; 1987.

Raz A.; Zöller M.; Ben-Ze'ev A. Cell configuration and adhesive properties of metastasizing and non-metastasizing Bsp73 rat adenocarcinoma cells. Exp. Cell Res. 162:127-141; 1986.

Rodríguez Fernández J. L.; Ben-Ze'ev A. Regulation of fibronectin, integrin and cytoskeleton expression in differentiating adipocytes: inhibition by extracellular matrix and polylysine. Differentiation 42:65-74; 1989.

Rodríguez Fernández J. L.; Geiger B.; Salomon D.; Ben-Ze'ev A. Overexpression of vinculin suppresses cell motility in Balb/c 3T3 cells. Cell Motil. Cytoskel. 22:127-134; 1992a.

Rodríguez Fernández J. L.; Geiger B.; Salomon D.; Ben-Ze'ev A. Suppression of vinculin expression by antisense transfection confers changes in cell morphology, motility and anchorage-dependent growth of 3T3 cells. J. Cell Biol. 122:1285-1294; 1993.

Rodríguez Fernández J. L.; Geiger B.; Salomon D.; Sabanay I.; Zöller M.; Ben-Ze'ev A. Suppression of tumorigenicity in transformed cells after transfection with vinculin cDNA. J. Cell Biol. 119:427-438; 1992b.

Rouleau G. A.; Merel P.; Lutchman M.; Sanson M.; Zucman J.; Marineau C.; Hoang-Xuan K.; Demczuk S.; Desmaze C.; Plougastel B.; Pulst S. M.; Lenoir G.; Bijlsma E.; Fashold R.; Dumanski J.; de Jong P.; Parry D.; Eldrige R.; Aurias A.; Delattre O.; Thomas G. Alteration in a new gene encoding a putative membrane-organizing protein causes neuro-fibromatosis type 2. Nature (Lond.) 363:515-521; 1993.

Rysek R.-P.; MacDonald-Bravo H.; Zerial M.; Bravo R. Coordinate induction of fibronectin, fibronectin receptor, tropomyosin, and actin genes in serum-stimulated fibroblasts. Exp. Cell Res. 180:537-545; 1989.

Schlessinger J.; Geiger B. Epidermal growth factor induces redistribution of actin and α-actinin in human epidermal carcinoma cells. Exp. Cell Res. 134:273-279; 1981.

Schwartz M. A. Transmembrane signaling by integrins. Trends Cell Biol. 2:304-308; 1992.

Takeichi M. Cadherin cell adhesion receptors as a morphogenetic regulator. Science (Wash. DC) 251:1451-1455; 1991.

Takeichi M. Cadherins in cancer: implications for invasion and metastasis. Curr. Opin. Cell Biol. 5:806-811; 1993.

Trofatter J. A.; MacCollin M. M.; Rutter J. L.; Murrell J. R.; Duyao M. P.; Parry D. M.; Eldridge R.; Kley N.; Menon A. G.; Pulaski K.; Haase V. H.; Ambrose C. M.; Munroe D.; Bove C.; Haines J. L.; Martuza R. L.; MacDonald M. E.; Seizinger B. R.; Short M. P.; Buckler A. J.; Gusella J. F. A novel moein-, ezrin-, radixin-like gene is a candidate for the neurofibromatosis 2 tumor suppressor. Cell 72:791-800; 1993.

Tsukita S.; Itoh M.; Nagafuchi A.; Yonemura S.; Tsukita S. Submembranous junctional plaque proteins include potential tumor suppressor molecules. J. Cell Biol. 123:1049-1053; 1993.

Ungar F.; Geiger B.; Ben-Ze'ev A. Cell contact- and shape-dependent regulation of vinculin synthesis in cultured fibroblasts. Nature (London) 319:787-791; 1986.

Volberg T.; Zick Y.; Dror R.; Sabanay I.; Gilon C.; Levitzki A.; Geiger B. The effect of tyrosine-specific protein phosphorylation on the assembly of adherens type junctions. EMBO J. 11:1733-1742; 1992.

Waites G. T.; Graham I. R.; Jackson P.; Millake D. B.; Patel B.; Blanchard A. D.; Weller P. A.; Eperon I. C.; Critchley D. R. Mutually exclusive splicing of calcium-binding domain exons in chick α-actinin. J. Biol. Chem. 267:6263-6271; 1992.

Weber K.; Lazarides E.; Goldman R. D.; Vogel A.; Pollack R. Localization and distribution of actin fibers in normal, transformed and revertant cell lines. Cold Spring Harb. Symp. Quant. Biol. 39:363-369; 1974.

17

Mechanical Signal Transduction and G proteins

S.R.P.Gudi and J.A.Frangos

Introduction

Mechanical stress can alter the structural and functional properties of cells at the cellular, molecular and genetic levels resulting in both rapid responses and slower adaptive changes to a sustained mechanical environment. The later often results in intracellular alterations to counter the external mechanical stress. Cellular responses to direct mechanical stress appear to involve an interplay between structural elements and the biochemical secondary messengers.

There is an increasing awareness of the signal transduction mechanisms by which mechanical signals can be converted into electrophysiological and biochemical responses in the cells, and their adaptation to the environment by altered gene expression. With increased interest in the molecular changes, that accompany shear–induced cellular adaptation, scientists and engineers have turned to in vitro methods to investigate mechanical stress effects on individual cells. Although many cellular functions are effected by mechanical stress, the mechanism through which the cells transduce physical forces remains unknown. Little is known on the early mechanisms involved in the transduction of mechanical signal across the cell membrane.

In this chapter we focus on possible pathways by which a mechanical signal may be transduced into a biochemical signal inside the cell, with a special emphasis on G-proteins as candidates for the transducer, and to give a survey of the current hypotheses to describe this phenomenon.

Mechanical Stimulus

Tissues and cells in vivo experience more than one form of mechanical stress. This may be a complex combination of tension, compression and shear stress, (having a biaxial or a simple uniaxial field) represents a complex rather than uniform gradient, and finally, may be cyclic with intermixed static periods or periods of high activity followed by lower activity and rest. An example of this biologic variation in strain fields is the aorta, which is subjected to flow, inducing largely unidirectional shear stress on endothelial cells on the luminal side, as well as tension on endothelial cells and the underlying smooth–muscle cells.

Striated skeletal muscle is subject to tension as sarcomeres contract. The central bulk of muscle experiences compression as actin and myosin filaments slide over each other, creating shear stress. The force of muscle contraction applies tension to attached tendons. Likewise bone is subjected to cyclic loading during standing, walking or running. Torque is applied to long bones and joints, while compression and shear stress to joints.

How is this cascade of mechanical signals delivered to cells and tissues? How these signals are interpreted, and what are the key cellular early and late responses?

Biochemical Responses of Cells Subjected to Mechanical Stimuli

During the last two decades, researchers began to focus on the effects of mechanical stress on isolated cells in culture. These included endothelial cells, osteoblasts and muscle cells. The field of application of mechanical activity to cultured cells to simulate dynamic environments has grown now to include tissue studies by researchers in a wide spectrum of disciplines and has encompassed cells from the aorta, skin, lung, brain, uterus and other sources. Devices to apply compression, tension, and shear have been developed to study cells from tissues that are derived from mechanically active environments in vivo.

When endothelial cells are subjected to shear stress or mechanical stretching, a diverse set of responses are generated; the fastest response occurring within milliseconds and the slowest response taking several days. Mechanisms linking early changes of intracellular calcium are measured by the activation nitric oxide synthase, a calcium dependent enzyme which is responsible for the synthesis of the nitrovasodilator nitric oxide from arginine. Preceding the mobilization of intracellular calcium is an elevation of inositol triphosphate whose mobilization in turn is often preceeded by activation of a membrane receptor. A similar cascade applies to the synthesis and release of prostacyclin (PGI_2), detectable approximately 1 minute after stimulus. IP_3 elevation and calcium elevation are detectable within several seconds after mechanical stimulation. Long term changes of endothelial structure and morphology occur over a time course of several hours. Alterations in the location and size of stress fibers is discernible after several hours of stimulation and the morphological changes of cell shape and alignment that accompany such changes are not surprising.

ENDOTHELIAL CELLS

The arterial endothelium plays a key role in regulating vascular tone. It is located at the interface between blood flow and the vessel wall and is exposed to forces that incorporate diverse mechanical characteristics. They produce a variety of substances involved in the control of vascular tone, platelet aggregation, and cell adhesion and proliferation. Most flow studies have been carried out using human umbilical-vein endothelial cells (HUVECs), bovine aortic endothelial cells (BAECs), and pig aortic endothelial cells (PAECs). It has been demonstrated that fluid flow modulates the release of several vasoactive compounds such as prostacyclin, endothelium-derived relaxing factor (NO) and endothelin, and increases transendothelial permeability. The release of several other compounds, such as tissue-type plasminogen activator, platelet-derived growth factor, fibronectin, interleukin 6 and proteoglycans, is affected too.

Endothelial cells have hormone receptors linked to guanine nucleotide-binding proteins (G proteins) which modulate the function of enzymes that produce intracellular messengers such as adenosine 3':5'-cyclic monophosphate (cAMP), inositol 1,4,5-triphosphate(IP_3), and 12-diacyl glycerol (Flavahan and Vanhoutte, 1990). These intracellular agonists trigger a cascade of reactions inside the cell eventually leading to a physiological response. One of the most important signal transduction pathway involves the activation of phospholipase C, which cleaves membrane-bound phosphotidylinositol 4,5-diphosphate into IP_3 and diacylglycerol. This mechanism has been shown to mediate several agonist-induced endothelial cell responses (Newby and Henderson, 1990).

IP_3 : Inositol 1,4,5-triphosphate (IP_3) causes the release of calcium into the cytosol from intracellular stores and diacylglycerol activates protein kinase C in the presence of elevated levels of calcium in the cytoplasm (Berridge and Irvine, 1989). IP_3 levels in HUVEC monolayers subjected to shear stress were increased, with a half-maximal response below 1.4 dyn/cm2 (Bhagyalakshmi and Frangos, 1989). The implications of the increase in IP_3 levels by shear is very important, considering the central role played by this second messenger in the activation of the calcium dependent pathways in the endothelial cell.

cAMP is generated by the action of adenylate cyclase on adenosine triphosphate(ATP), and the activity of adenylate cyclase is modulated by stimulatory and inhibitory guanine nucleotide-binding proteins (G_s and G_i). In HUVECs subjected to a shear stress of 4.3 dyn/cm^2 for 15 min, the levels of cAMP increased at least four times in sheared cells (Reich et al.1990). The physiological implications of the flow-induced cAMP increase are unclear, but may be involved in the control of trans-endothelial permeability (Oliver 1990).

Resting cytosolic calcium levels in endothelial cells are normally less than 100 nM, but in BAECs subjected to shear stresses ranging between 3.7 to 10.3 dyn/cm^2, a sharp increase in intracellular calcium was observed at the onset of flow (Ando et al. 1988). Calcium released from intracellular stores may trigger an influx of potassium ions from calcium activated channels, which then hyperpolarizes the membrane, thereby facilitating the entry of calcium through ion channels in BAECs (Luckhoff and Busse, 1990). Direct activation of G proteins by aluminum fluoride in PAECs causes a calcium influx without stimulating the phospholipase C pathway, suggesting that calcium channels may also be linked to G proteins (Graier et al. 1990). It is possible that flow could activate these G proteins or their parent hormone receptors. Fluid flow might also increase the passive permeability of the plasma membrane to calcium ions or stretch activated ion channels (Lansman et al. 1987).

PGI_2 : Endothelial cells produce prostacyclin (PGI_2), a prostanoid derived from arachidonic acid. This vasodilator inhibits platelet aggregation and is thought to play a major role in keeping a nonthrombogenic surface on the vascular endothelium. The rate of PGI_2 synthesis is increased by fluid flow in HUVECs (Frangos et al. 1985) and BAECs (Grabowski et al. 1985). Shear induced PGI_2

production in HUVECs is inhibited by Quin 2/AM, an intracellular calcium chelator, and in the presence of ethylene glycol bis (β-aminoethyl ether) N,N,N',N'-tetraacetic acid (EGTA), which chelates the calcium present in the external medium, suggesting that both intra- and extracellular calcium are essential (Bhagyalakshmi and Frangos, 1989). Fluid flow may cause a rise in intracellular calcium levels in BAECs, and the ions are apparently released from intracellular stores. The later is generally mediated by IP_3, suggesting that phospholipase C activation would occur before the increase in intracellular calcium levels in the pathway leading to PGI_2 production by flow. In fact, the PLC pathway seems to be the main regulator of arachidonic release and PGI_2 synthesis in HUVECs treated with fluoride, a general activator of G proteins (Magnusson et al.1989).

To explain the effects of shear stress on phospholipid metabolism, the most plausible mechanism would first involve the activation of PLC with concomitant phosphotidylinositol (PI) turnover. The second messenger IP_3 then presumably triggers the release of calcium from intracellular stores and the resulting increased cytosolic calcium levels would act synergistically with diacylglycerol to activate protein kinase C. Both the elevated intracellular calcium and active protein kinase C would stimulate phospholipase A_2 (PLA_2) and/or phospholipase D. PLA_2 may be independently activated by a G protein or via the entry of calcium through receptor-operated channels. Arachidonic acid could be liberated from phosphotidylinositol by the actions of phospholipase C and diacylglycerol lipase. Diacylglycerol may also inhibit lysophosphatide acyltransferase, thereby reducing the rate of reincorporation of arachidonic acid into the phospholipids and increasing the availability of free arachidonic acid for for PGI_2 synthesis.

ET-1 : Endothelin is a potent vasoconstrictor produced by endothelial cells,that stimulates smooth muscle cell growth (Komuro et al. 1989). Endothelin–1, –2, and –3 have been identified. ET–1 is one of the most potent vasoconstrictors, causes contractions that develop slowly which are powerful and long lasting (Yanagisawa et al.1988). Its potent mitogenic properties in vascular smooth muscle cells, together with vasoconstrictive properties, imply an important role in the chronic and acute control of local blood flow. The influence of shear stress on the release of ET–1 from cultured endothelial cells is controversial. Yoshizumi et al. (1989) observed an elevated cumulative production of ET–1 after 24 and 48 hr. of exposure to shear. In contrast, Sharefkin et al. (1991) have observed a dramatic shear stress mediated depression in ET–1 release using higher levels of shear. This discrepancy has recently been clarified (Kuchan and Frangos, 1993). Agonist mediated stimulation of ET–1 release has been shown to be mediated via the phospholipase C–protein kinase C (PKC)pathway (Emori et al.1991), and NO mediated changes in cGMP have been implicated as an inhibitor of thrombin-mediated ET–1 production. Kuchan and Frangos (1993) demonstrated that an elevation in ET–1 release may be mediated by PKC followed by a sustained dramatic depression caused by shear–mediated production of cGMP. Both the duration of the elevation and the magnitude of the suppression are dependent on the level of shear stress and resulting accumulations of cGMP.

EDRF : Endothelium–derived relaxing factors (EDRFs) are usually described as short half–life vasodilatory compounds released by endothelial cells stimulated by a variety of agonists such as calcium ionophore, bradykinin, acetylcholine and substance P (Cocks et al. 1985). EDRFs are thought to be derived from the enzymatic oxidation of L–arginine followed by cleavage of the molecule nitric oxide (NO). The enzymes responsible for NO production appear to be mostly membrane– associated and calcium dependent (Forstermann et al. 1991; Mitchell et al. 1991). The NO produced can diffuse out and cause smooth muscle relaxation or can be converted into S–nitrosothiols, which are more stable.
PDGF : Platelet–derived growth factor (PDGF) is a potent stimulator of smooth muscle cell growth and a vasoconstrictor produced by endothelial cells. It consists of dimers consisting of A and B chains, each expressed by different genes (Betsholtz et al. 1986). It has been demonstrated that steady shear causes a transient increase of PDGF A and B chain mRNA levels in HUVECs (Hsieh et al. 1991) and also shear–induced PDGF gene expression is mainly mediated by protein kinase C activation and requires intracellular calcium. Furthermore, G proteins seem to be involved in this process (Hsieh et al. 1992). Shear stress, either directly or indirectly (G protein–mediated), enhances the membrane phosphoinositide turnover via phospholipase C, producing diacylglycerol, an activator of protein kinase C. The activated protein kinase C then triggers the subsequent PDGF gene expression. Resnick et al. (1993) reported that a cis–acting element in the PDGF–B gene is required for shear stress responsiveness. A core binding sequence within this shear stress responsive element (GAGACC) appears to bind to transcription factors present in the nuclear extracts of shear stressed BAECs. This sequence is also present in the promoters of certain other endothelial genes that have been found to be up–regulated by shear stress, including fos, jun, tPA, intercellular adhesion molecule 1 (ICAM–1), and TGF–β1, thus suggesting a general mechanism for induction of gene transcription by this mechanical stimulus.

OSTEOBLASTS

Osteoblasts are bone producing cells that line the surfaces in trabecular bone. When bones are mechanically loaded, flow of the extra-capillary fluid filling the trabecular meshwork is induced. It has been hypothesized that, this flow stimulates bone producing cells such as osteoblasts and thereby could induce bone remodeling.

Reich et al (1990) demonstrated that the levels of cAMP were significantly increased when cultured osteoblasts were subjected to shear stress. This stimulation of cAMP levels by flow was inhibited by the cyclooxygenase inhibitor ibuprofen, indicating that prostaglandins mediate the cAMP response caused by shear stress in osteoblasts. PGE_2 synthesis in osteoblasts was also shown to be stimulated by flow in a shear dependent manner (Reich and Frangos 1991). The effect of flow on inositol 1,4,5-triphosphate (IP_3) levels was also investigated. High shear stress (24 dyn/cm2) caused a transient increase in IP_3 levels. Inhibition of this flow stimulated IP_3 by a cyclooxygenase inhibitor, suggests that prostaglandin E_2 mediates the IP_3 as well as the cAMP responses. These observations show that osteoblasts are sensitive to flow and support the hypothesis that extracapillary fluid flow, induced by mechanical loading, stimulates bone metabolism.

MUSCLE CELLS

All cells are endowed to some extent with cytoplasmic filaments capable of providing contraction to the cell. Muscle cells are particularly well endowed with actin and myosin contractile elements. It has been known that pressure overload in cardiac muscle for a period of weeks can increase ventricle weight and enzyme content (Dowell et al. 1983). Komuro et al (1990; 1991) reported that the protooncogene c-fos was up-regulated in statically stretched cardiac muscle cells. A passive stretch of 10% for 24h increased rat myocyte total RNA by 45% and increased skeletal actin mRNA by 16% and that c-fos mRNA was increased 15-fold within 30 min in a protein kinase C dependent manner. It is believed that mechanical strain might directly stimulate protein kinaseC activity via phospho-lipase C activation to induce early and late genes yielding c-fos, c-myc, and β-type myosin heavy-chain and skeletal α-actin, respectively. Recently Yamazaki et al (1993) demonstrated that myocyte stretching increases the activities of MAP kinase and S6 peptide kinase which may play an important role in the induction of the specific genes and the increase in the protein synthesis.

The Mechanochemical Transducers

Although many cellular functions are affected by stress, the mechanism through which the cells transduce physical forces remain largely in debate. Lansman et al. (1987) showed that stretch, exerted by applying suction to a microelectrode at the cell surface, activated ion channels permeable to calcium. Olesen et al. (1988) showed that shear stress activates endothelial cell potassium channels. Activation of ion channels may represent a direct mechanism by which endothelial cells respond to shear, but it is also possible that ion channel activation is downstream of a more fundamental transduction process. Wang et al. (1993) hypothesize that transmission of physical forces to cytoskeletal elements represents the primary means of mechanotransduction. They also suggest that integrins act as mechanoreceptors and transmit mechanical signals to the cytoskeleton. Mechano-trasduction, in turn, may be mediated simultaneously at the multiple locations inside the cell through force–induced rearrangements within a tensionally integrated cytoskeleton. But this hypothesis lacks proper experimental support to explain the mechanism of mechano-chemical transduction and also not account for the activation of membrane processes known to mediate many cellular responses. It is also possible that membrane–associated structures are redistributed by shear in a phenomenon similar to the "capping" of receptors that is driven by ligand binding in some cell types, and that this process elicits intracellular responses. Studies with red cells also clearly indicate the capacity of shear stress to mechanically redistribute cell–membrane components (Schmid-Schoenbein, 1981). A more basic process like G protein-activation might be involved in shear induced conformational changes of transmembrane proteins that drive intracellular responses in a fashion analogous to many receptor systems.

ION CHANNELS

Many cells, of mammalian and primitive origin, express stretch–activated ion channels (Guhary and Sachs, 1984). Both stretch–activated and shear stress activated ion channels have been identified in cultured endothelial cells. In the intact cell, the net effect of activation of a stretch–sensitive channel would be an influx of ions resulting in depolarization of the cell. Recently, stretch–inactivated channels have been identified in primitive (snail) cells (Morris and Sigurdron, 1989). They occur in the same patches as stretch–activated K^+ channels. As tension increased, the stretch–inactivated channels closed as the stretch–activated channels opened. Similar channels have not been reported in endothelial cells, though they have been reported in dystrophic muscle cells.

The intensity of forces in a local region of the membrane rarely reach a level which will activate the full complement of stretch–activated channels distributed throughout the cell and when a cell is subjected to external forces there is a distribution of force throughout the cell for alterations of cell tension (Morris and Horn, 1991). The local forces are reduced below critical threshold levels for activation of channels by dissipation throughout the cell. So, the activation of stretch channels may be quite modest even at sites of disturbed flow. Even then, activation of a small fraction of the available stretch-activated channels on the endothelial cells may leak significant amounts of the important second messenger calcium, with consequences for activation of relaxing factor synthesis and calcium / calmodulin dependent gene regulation.

An important consequence of activation of K^+ channels by flow is depolarization of the cell, a phenomenon that has been confirmed using membrane potential–sensitive dyes (Nakache and Gaub, 1988) as well as patch clamping measurements (Oleson et al.1988). The mechanisms by which shear stress elicits channel activation is not clear. It is possible that the ion channel may be linked in some way to a discrete protein mechanosensor in the membrane. K^+ channel activation by receptor ligand coupling is usually mediated through G proteins. Recently, Ohno et al. (1993) reported that shear stress elevates endothelial cGMP via an NO–dependent mechanism. They suggest, that the effect of shear stress is mediated by a unique signal transduction pathway that is coupled to a pertussis toxin sensitive G protein and requires the activity of an endothelial K^+ channel. However Kuchan et al. (1993) postulated that a PTX-refractory G protein, G_q, coupled to PLC, is activated by shear leading to secondary signaling events and ultimately to NO production.

CYTOSKELETON

Endothelial cells (anchorage dependent), maintain their shape on a substratum by inherent cell tension (Ingber, 1991). Such tension is generated when the cytoskeleton interacts with other regions of the cell, particularly at sites of adhesion to the subendothelial matrix. The cytoskeleton may play a key role in changing the internal cellular tension to counter the external forces.

The prominent response of F–actin stress fibers to external forces implicates them as a principal force transmission structure in endothelial cells. Disruption of stress fibers inhibits a number of primary and secondary responses in the cell. There is reduced stretch–activated ion channel activity in response to membrane deformation if actin filaments are disrupted (Guhary and Sachs, 1984), and endothelial cell shape change, realignment to flow, and focal adhesion remodelling are all inhibited by drugs that interfere with microfilament turnover (Wechezak et al. 1989).

Cytoskeleton interacts with components associated with integral membrane proteins in the plasma membrane (Burridge et al. 1988) and that the stress fibers in turn are connected elsewhere in the focal adhesion sites to intercellular proteins, that link plasma membranes of adjacent cells, and to the nuclear membrane. The focal adhesion sites are enriched in tyrosine kinase $pp60^{src}$, FAK as well as threonine and serine kinases whose activities have been demonstrated to change upon mitogen stimulation. The integrin family of heterodimeric transmembrane glycoproteins act as cell surface receptors for extracellular matrix (ECM) molecules (Bras et al. 1993), and are associated with the cytoskeletal proteins inside the cell. Recent evidence suggests that the integrins may elicit biochemical signals by tyrosine kinase phosphorylation as well as by interaction with closely associated intracellular proteins, Endothelial integrins primarily consists of the subunits α and β. The β subunit binds α–actinin, which in turn binds to F–actin. Tyrosine phosphorylated proteins (including integrins, talin, vinculin, α–actinin, and paxillin) and the cytoplasmic tyrosine kinase $pp60^{src}$ have been localized to focal adhesions. The induction of phosphorylation by mechanical stimulation at focal adhesion sites in endothelial cells would demonstrate a convergence of mechanotransduction and hormone–receptor pathways. Cellular mechanical force transduction mechanisms are probably a combination of force transmission via cytoskeletal elements and force transduction of the transmitted mechanical stress to biochemical signals at mechano-transducer sites somewhere in the cell.

Cells Transduce Mechanical Signals via G proteins

Cells have the ability to process and respond to enormous amounts of information. Much of this information is provided to individual cells in the form of physical or chemical signals. These ligands interact with transmembrane receptors and transduced to an intracellular signal. The heterotrimeric guanine nucleotide binding proteins (G proteins) are signal trasnsducing proteins that couple a large number of membrane bound receptors to a variety of intracellular effector systems.

G proteins are made up of α ,β and γ subunits. Interaction of the G protein with the activated receptor promotes the exchange of guanosine diphosphate (GDP), bound to the α subunit, for guanosine triphosphate (GTP) and the subsequent dissociation of the α-GTP complex from the βγ heterodimer. A single receptor can activate multiple G protein molecules thus amplifying the ligand binding event. The α subunit with GTP bound and the free βγ subunit may interact with effectors like ion channels and enzymes that generate regulatory molecules or second messengers. Second messengers such as cyclic

AMP or inositol triphosphate, in turn, generate dramatic intracellular changes including selective protein phosphorylation, gene transcription, cytoskeleton reorganization, secretion, and membrane depolarization. Induced signal is terminated with the hydrolysis of GTP to GDP.

Agonist-induced phospholipase activation is mediated by G proteins (Tkachuk and Voyno-Yasanetskaya, 1991). Direct stimulation of G proteins by AlF_4^- and GTPγS lead to PGI_2 production in HUVECs. In BAECs, pertussis toxin (PTX) inhibits the stimulation of PGI_2 release by leukotrienes C4 and D4, but potentiates that induced by ATP. Fluid flow induced a burst in PGI_2 release in HUVECs (Frangos et al. 1985). Flow stimulated endothelial phospholipid turnover (Bhagyalakshmi and Frangos, 1989), suggesting that flow triggers a pathway leading to PGI_2 synthesis which probably involves phospholipase activation. However, the earlier events in this signal transduction are not clearly known. Berthiaume and Frangos (1992) demonstrated that a PTX-sensitive G protein mediates the flow induced PGI_2 synthesis in HUVECs. G protein inhibitors, GDPβS and PTX inhibited both the initial burst, and the later steady phase of PGI_2 synthesis stimulated by flow. The production of PGI_2 by stationary cells was unaffcted by GDPβS or PTX. They also suggested, that flow synergizes with histamine in stimulating the producion of PGI_2 explaining that flow can modulate the stimulus-response coupling of other agonists in HUVECs via an intracellular mechanism.

Steady shear stress induced a transient increase in PDGF A and B mRNA levels in HUVECs (Hsieh et al. 1991), but the mechanism for shear induced PDGF gene expression is not well understood. It has been shown that shear stress can stimulate membrane phospholipid turnover (perhaps via phospholipase C) in endothelial cells, producing the second messengers inositol triphosphate (IP_3) and diacyl glycerol (DAG) (Bhagyalakshmi and Frangos, 1989; Nollert et al.,1990; Bhagyalakshmi et al. 1992). IP_3 triggers the release of calcium from intracellular pools, and DAG is an activator of protein kinase C a major kinase responsible for many cellular responses, e.g.,gene expression, cell proliferation, etc.(Nishizuka, 1986). Hsieh et al. (1992) demonstrated that shear-induced PDGF gene expression in HUVECs was mediated by protein kinase C activation, and not by cAMP- or cGMP-dependent kinases. They also showed that inhibition of G proteins by GDPβS caused a reduction in the shear-induced PDGF gene expression, indicating that G proteins were involved in this process.

Exposure of endothelial cells to fluid flow stimulates the production of NO (Kuchan and Frangos, 1993). The release of NO caused by acute increases in flow is regulated differently from that resulting from a sustained exposure to steady flow (Kuchan and Frangos, 1993). While production stimulated by acute increases in flow were calcium and calmodulin dependent, that stimulated by sustained exposure to flow was dependent on neither. It has also been shown that NO release caused by acute changes in flow, requires stimulation of a calcium activated potassium channel (Cooke et al., 1991). NO production stimulated by acute changes in flow is similar to that mediated by exposure to agonists which also require elevation of intracellular calcium levels. In addition Graier et al. (1990) observed elevated calcium levels in endothelial cells treated with AlF_4^-. Together these observations suggest that exposure to flow may lead to activation of a G protein mediated production of NO. PTX, which inhibits G_i and G_o proteins via ADP-ribosylation is routinely used to identify the class of G proteins involved in a biological response. Members of G_i family G_{i1} and G_{i3} are coupled to phospholipase C in HUVECs (Voyno-Yasenetskaya et al., 1989). Kuchan et al. (1993) demonsrated the lack of PTX-sensitive G proteins in NO production. Stimulation of NO production by AlF_4^- and GTPγS was only partally inhibited by PTX at a dose which completely blocked stimulation by an α_2-adrenergic agonist (Shibano et al. 1992). In the case of shear stress, the G protein(s) required is PTX-refractory. It appears that more than one class of G proteins can activate transduction pathways leading to NO production. While α_2-adrenergic and 5-HT_1 serotonergic receptors are coupled to PTX sensitive G proteins, shear and bradykinin stimulated NO production are PTX-refractory.

Members of PTX-refractory family, G_q, are coupled to PLC (Simon et al., 1991) are candidates for mediators of flow induced NO production. Recently, Liao and Homcy (1993) identified and characterized G protein components of both the pertussis toxin-sensitive and -insensitive bradykinin signaling pathways for EDRF release. The bradykinin receptor is predominantly coupled to pertussis toxin-insensitive $G_{\alpha q}$, although pertussis toxin-sensitive $G_{\alpha i2}$ and $G_{\alpha i3}$ can also contribute to this signaling pathway. It is clear that activation of G proteins play an important role in the effects of shear stress on numerous vasoactive substances of potential physiological significance.

Reich and Frangos (1991) demonstrated that prostaglandin E_2 levels in osteoblasts increase in a shear dependent manner. PGE_2 mediates the signal transduction of mechanical stretch and pressure on osteoblasts (Somjen et al. 1980 ; Imamura et al. 1990). G_{i2}, a pertussis toxin sensitive G protein, is responsible for mediating the PGE_2 and $PGF_{2\alpha}$ induced increases in inositol phosphates in osteoblasts (Miwa et al. 1990; Tokuda et al. 1991). Thus, a PTX sensitive G protein links the prostaglandin receptor to phospholipase C. Reich and Frangos (1991) showed that IP_3 levels are increased by fluid flow, indicating that phospholipase C is activated by flow. Reich et al. (1994) proposed a mechanism for flow-induced PGE_2 production. Fluid flow activates (directly or indirectly) G proteins and causes changes in intracellular calcium levels either by release from intracellular stores or by calcium influx. Arachidonic acid which is liberated via phospholipase A_2 and phospholipase C / diacylglycerol lipase is then metabolized to PGE_2.

When human blood platelets are exposed to hypotonic medium, they swell first, but shortly revert to their original volume, by a process called regulatory volume decrease(RVD). Platelet RVD is mediated by enhanced independent K^+ and Cl^- effluxes and associated water (Livne et al.1987). This RVD response is controlled by a lipoxygenase-derived product, hepoxilin A_3. It can also be detected after subjecting the platelets to mechanical forces like centrifugation and laminar flow, which indicates that hepoxilin A_3 metabolism in human platelets is initiated by a common mechanical-biochemical transduction mechanism. Margalit et al. (1993) demonstrated that the arachidonic acid metabolism which leads to hepoxilin A_3 formation is initiated by Ca^{2+} insensitive PLA_2 and this phospholipase is activated by PTX sensitive G proteins. They also showed that

RVD is inhibited by PTX. Margalit et al. (1993) suggested that mechanical activation causes conformational changes in G protein coupled receptors and these mechanoreceptors transform a mechanical stimulus to a biochemical response i.e. hepoxilin A_3 formation in human platelets. Therefore, it seems that G proteins are involved in the initial step of the mechanical-biochemical transduction leading to hepoxilin A_3 formation in human platelets.

Conclusions

It is long known that cells have highly specialized force sensing mechanisms such as those involved in hearing and touch, which can convert a mechanical stimulus into electrical and biochemical responses. There is strong evidence that flow by itself is another stimulus,exists in the cell environment. Flow regulation of vascular tone via the endothelium and smooth muscle cells involves mechanical signal transduction via G proteins. The experimental support is limited,however recent. More studies are needed to understand a well defined transduction mechanism by which flow signal is conveyed across the cell membrane.

References

Ando J, Komatsuda T,Kaamiya A: Cytoplasmic calcium response to fluid shear stress in cultured vascular endothelial cells. In Vitro Cell.Devel.Biol. 24:871-877;1988

Berridge MJ, Irvine RF: Inositol phosphates and cell signalling. Nature. 341:197-205;1989

Berthiaume F, Frangos JA: Flow-induced prostacyclin production is mediated by a pertussis toxin-sensitive G protein. FEBS lett. 308:277-279;1992

Betsholtz C, Johnsson A, Heldeir C-H, Westermark B, Lind P, Urdea Ms, Eddy R, Shows TB, Philpott K, Mellor AL, Knott TJ, Scott J: cDNA sequence and chromosomal localization of human platelet derived growth factor A-chain and its expression in tumor cell lines. Nature. 320:695-699;1986

Bhagyalakshmi A, Frangos JA: Mechanism of shear-induced prostacyclin production in endothelial cells. Biochem.Biophys.Res.Commun. 158:33-37;1989

Bhagyalakshmi A, Berthaume F, Reich KM, Frangos JA: Fluid shear stress stimulates membrane phospholipid metabolism in cultured human endothelial cells. J.Vasc.Res.29:443-449;1992

Bras CK,Gibbon LF, Jeevaratnam P, Wilkins J, Dedhar S: Stimulation of tyrosine phosphorylation and accumulation of GTP-bound $p21^{ras}$ oupon antibody-mediated $\alpha_2\beta_1$ integrin activation in T-lymphoblastic cells. J.Biol.Chem. 268:20701-20704;1993

Burridge K, Fath K,Kelly T, Nuckells G, Turner C: Focal adhesion: transmembrane junctions between the extracellular matrix and the cytoskeleton. Annu. Rev. CellBiol. 4:487-525;1988

Cocks TM, Angus JA, Campbell JH, Campbell GR: Release and properties of endothelium-derived rlaxing factor(EDRF) from endothelial cells in culture. J.Cell.Physiol. 123:310-320;1985

Cooke JP, Rossitch E.Jr., Andon NA, Loscalzo L, Dzan VT: Flow activity on endothelial potassium channel to release an endogenous nitrovasodilator. J.Clin. Invest. 88:1663-1671;1991

Dowell RT, Haithcoat JL, Hasser EM: Metabolic enzyme response in the pressure-overloaded heart of weanling and adult rats. Proc.Soc. Exp.Biol. Med. 174: 368-376;1983

Emori T,Hirata Y, Ohta K, Kaumo K, Eguchi S, Imai T, Shichiri M, Marumo F: Cellular mechanism of endothelin-1 release by angiotensin and vasopressin. Hypertension. Dallas. 18:165-170;1991

Flavahan NA, Vanhoutte PM: G proteins and endothelial responses. Blood vessels. 27:218-229;1990

Forstermann U, Pollock JS, Schmidt HHHW, Heller M, Murad F: Calmodulin dependent endothelium derived relaxing factor/nitric oxide synthase activity is present in the particulate and cytosolic fractions of bovine aortic endothelial cells. Proc.Natl.Acad.Sci.USA. 88:1788-1792;1991

Frangos JA, Eskin SG, McIntire LV, Ives CL:Flow effects on prostacyclin production in cultured human endothelial cells. Science. 227:1477-1479;1985

Grabowski EF, Jaffe EA, Weksler BB:Prostacyclin production by cultured endothelial cell monolayers exposed to step increases in shear stress. J.Lab.Clin..Med. 103:36-43;1985

Graier WF, Schmidt K, Kukowetz WR: Effect of sodium fluoride on cytosolic free Ca^{2+} concentrations and cGMP levels in endothelial cells. Cell.Signal. 2:369-375;1990

Guharay F, Sachs F: Stretch activated single ion channel currents in tissue cultured embryonic chick skeletal muscle cells. J.Physiol. Lond. 352:685-701;1984

Hsieh H-J, Li N-Q, Frangos JA: Shear stress increases endothelial platelet derived growth factor mRNA leverls. Am.J.Physiol. 260:H642-H646;1991

Hsieh H-J, Li N-Q, Frangos JA: Shear-induced platelet derived growth factor gene expression in human endothelial cells is mediated by protein kinase C. J.Cell.Physiol. 150:552-558;1992

Imamura K, Ozawa H, Hiraidi T, Takahasi N, Shibasaki Y, Fukuhara T, Suda T: Continuously applied compressive pressure induces bone resorption by a mechanism involving prostaglandin E_2 synthesis. J.cell.Physiol. 144:222-228;1990

Ingber D: Integrins as mechanochemical transducers. Curr.Opin.Cell Biol. 3:841-848;1991

Komuro I, Kaida T, Shibazaki Y, Kurabayashi M,Katoh Y, Hoh E, Takaku F, Yazaki Y: Stretching cardiac myocytes stimulates protooncogene expression. J.Biol.Chem. 265: 3595-3598;1990

Komuro I, Katoh Y, Kasida T, Shibazaki Y, Kurabayashi EH, Takaku F, Yazaki Y:Mechanical loading stimulates cell hypertrophy and specific gene expression in cultured rat cardiac myocytes. J.Biol.Chem. 266:1265-1268 ; 1991

Komuro I, Kurihara H, Sugiyama T, Yoshizumi M, Takaku F, Yazaki Y: Endothelin stimulates c-fos and c-myc expression and proliferation of vascular smooth muscle cells. FEBS lett. 238:249-252;1988

Kuchan MJ, Frangos JA: Role of calcium and calmodulin in flow-induced nitric oxide production in endothelial cells. 1993. (in press)

Kuchan MJ, Frangos JA: Shear stress regulates endothelin-1 release via protein kinase and cGMP in cultured endothelial cells. Am.J.Physiol. 264: 150-156;1993

Kuchan MJ, Jo H, Frangos JA:Role of G-proteins in shear-mediated nitric oxide production. 1993. (in press)

Lansman JB, Hallam TJ, Rink TJ: Single stretch activation channels in vascular endothelial cells as mechanotransducers? Nature. 294:667-668;1987

Lansman JB, Hallam TJ, Rink TJ: Single stretch-activated ion channels in vascular endothelial cells as mechanotrasducers. Nature. Lond. 325: 811-813;1987

Liao JK, Homcy CJ: The G protein of $G_{\alpha i}$ and $G_{\alpha q}$ family couple the bradykinin receptor to the release of the endothelium-derived relaxing factor. J.Clin. Invest. 92:2168-2172;1993

Livne A, Grinstein S, Rothstein A: Volume-regulating behavior of human platelets. J.Cell.Physiol. 131:354-363;1987

Luckhoff A, Busse R: Calcium influx into endothelial cells and formation of endothelium- derived relaxing factor is controlled by the membrane potential. Pflugers Arch. 416:305-311;1990

Magnusson MK, Halldorssn H, Kjeld M,Thorgeirssn G: Endothelial inositol phosphate generation and prostacyclin production in response to G-protein activation by AlF_4 . Biochem.J. 264:703-711;1989

Margalit A, Livne AA, Funder J, Granot Y: Initiation of RVD response in human platelets:Mechanical-biochemical transduction involves pertussis toxin-sensitive G protein and phospholipase A_2. J.Memb.Biol. 136: 303-311;1993

Mitchell JA, Fostermann U, Warner TD, Pollock JS, Schmidt HHHW, Heller M, MuradF: Endothelial cells have a particulate enzyme system responsible for EDRF fomation: measurement by vascular relaxation. Biochem.Biophys. res.Commun. 176: 1417- 1423 ;1991

Miwa M, Tokuda H, Tsushita K, Kotoyori J, Takahasi Y, Ozaki N, Kozawa O,Oiso Y: Involvement of prtussis toxin-sensitive GTP binding protein prostaglandin $F_{2\alpha}$ induced phosphoinositide hydrolysis in osteoblast-like cells. Biochem. Biophys. Res.Commun. 171:1229-1235;1990

Morris CE, Horn R: Failure to elicit neuronal macroscopic mechanosensitive currents anticipated by single channel studies. Science. 251:1246-1249;1991

Morris CE, Sigurdron WS: stretch-inactivated ion channels coexist with stretch-activated ion channels. Science. 243:807-809;1989

Nakache M, Gaub WE: Hydrodynamic hyperpolarization of endothelial cells. Proc.Natl. Acad.Sci.USA. 85:1841-1843;1988

Newby AC, Henderson,AH: Stimulus-secretion coupling in vascular endothelial cells. Annu.Rev.Physiol. 52:661-674;1990

Nishizuka Y: Studies and perspectives of protein kinase C. Science. 233:305-312; 1986

Nollert MV, Eskin SG, McIntire LV: Shear stress increases inositol triphosphate levels in human endothelial cells. Biochem.Biophys.Res. Commun. 170:281-287;1990

Ohno M, Gibbons GH, Dzan VJ, Cooke JP:Shear stress elevates endothelial cGMP. Role of potassium channel and G protein coupling. Circulation. 88: 193-197;1993

Olesen SP, Clapham DE, Davies PF: Hemodynamic shear stress iactivates a K^+ current in vascular endothelial cells. Nature. Lond. 331:168-170;1988

Oliver JA: Adenylate cyclase and protein kinase C mediate opposite actions on endothelial junctions. J.Cell.Physiol. 145:536-542 ;1990

Reich KM, Frangos JA: Effect of flow on prostaglandin E_2 and inositol triphosphate levels in osteoblasts. Am.J.Physiol. 261:C428-C432;1991

Reich KM, Gay CV, Frangos JA: Fluid shear stress as a mediator of osteoblast cyclic adenosine monophosphate production. J.Cell.Physiol. 143: 100-104 ;1990

Reich KM, Gudi SRP, Frangos JA: Activation of G protein mediates flow-induced prostaglandin E_2 production in osteoblasts.1994.(communicated)

Resnick N, Collins T, Atkins W, Bonthorn DT, Dewey CF.Jr.,Gimborne MA.Jr.: Platelet derived growth factor B chain promoter contains a cis-acting fluid shear-sress-responsive element. Proc.Natl.Acad.Sci.USA. 90:4591-4595; 1993

Schmid-Schoenbein H: factors promoting and preventing the fluidity of blood. In microcirculation..pp.249-266.Effros RM,Schmid-schoebein H,Ditzel J. eds. Academic press,New York.1981

Shibano T, Codina J, Birnbaumer L, Vanhoutte PM: Guanosine 5'-o-(thio triphosphate) causes endothelium dependent, pertussis toxin-sensitive relaxation in porcine coronary arteries. Biochem.Biophys. Res. Commun. 189:324-329;1992

Sharefkin JB,Diamond SL,Eskin SG, McIntire LV, Dieffenbach CW: Fluid flow decreases prepro endothelin mRNA levels and suppress endothelin-1 release in cultured human endothelial cells. J.Vasc.Surg. 14:1-9; 1991

Simon MI,Strathman MP, Gautam N:Diversitt of G proteins in signal transduction. Science. 252:802-808;1991

Somjen D, Binderman I, Berger E, Harell A: Bone remodeling induced by physical stress is prostaglandin E_2 mediated. Biochim.Biophys. Acta. 627:91-100; 1980

Tkachuk VA, Voyno-Yasanetskaya TA: Two types of G proteins involved in regulation of phosphoinositide turnover in pulmonary endothellial cells. Am.J.Physiol.suppl. 261:118-122;1991

Tokuda H, KOzawa O, Yoneda M, Oiso Y, Takatsuki K, Asano T,Kato K: Possible coupling of prostaglandin E_2 receptor with petussis toxin sensitive guanine nucleotide binding protein in osteoblast like cells. J.Biochem. 109:229-233; 1991

Voyno-yasenetskaya TA, Panchenko MP, Nupenko EV, Rybin VO, Tkachunk VA: Histamine and bradykinin stimulate the phosphoinositide turnover in human umbilical vein endothelial cells via different G proteins. FEBS lett. 259:67-70;1989

Wang N, Butler JP, Ingber DE: Mechano transduction across the cell surface and through the cytoskeleton. Science. 260:1124-1127;1993

Wechezak AR, Wight TN, Viggers RF, Sauvage LR: Endothelial adherence under shear stress is dependent upon microfilament reorganizaqtion. J.Cell. Physiol. 139: 136-146;1989

Yamazaki T, Tobe K, Hoh E, Macmura K, Kaida T, Komuro I, Tammoto H, Kadowaki T, Nagai R, Yazaki Y: Mechanical loading activates mitogen -activated protein kinase and S6 peptide kinase in cultured rat cardiac myocytes. J.biol.Chem. 268:12069-12076;1993

Yanagisawa M, Kurihara H, Kumura S, Tomobe Y, Kobayashi M, Masaki T: A novel potent vasconstrictor peptide produced by vascular endothelial cells. Nature Lond. 332:411-415;1988

Yoshizumi M, Kurihara H, Sugiyama T,Takaku F, Yanagisawa M, Masaki T, Yazaki Y: Hemodynamic shear stress stimulates endothelin production by cultured endothelial cells. Biochem.Biophys.Res.Commun. 161:859-864;1989

18
Modeling Mechanical-Electrical Transduction in the Heart

F. Sachs

Introduction

The rate and rhythm of the heart is sensitive to mechanical deformation. As early as 1915, Bainbridge reported that distention of the atria produced an increase in heart rate (Bainbridge, 1915). Later studies showed that stretching alters action potential configuration (Dudel and Trautwein, 1954; Rosen et al. 1981; Dean and Lab, 1989; Lab, 1980; Taggart et al. 1992c; Taggart et al. 1992a; Taggart et al. 1992b; White et al. 1993), cause quiescent tissue to become spontaneously active and cause the generation of extra systoles (Hansen et al. 1990; Rajala et al. 1990; Stacy, Jr. et al. 1992; Sideris et al. 1990; Sideris et al. 1989; Hansen et al. 1991).

Aside from its intrinsic scientific interest, mechanical-electrical transduction in the heart may have important clinical implications. The origin of the extra systoles that cause fatal arrhythmias following myocardial ischemia and cardiac failure are not known (Francis, 1986; Meinertz et al. 1984). However, these arrhythmias tend to be correlated with the degree of mechanical myocardial dysfunction rather than electrophysiological status (Shultz et al. 1977). The damaged myocardium, weakened by ischemia, may be stretched during contraction of surrounding healthy tissue and generate excitatory currents leading to extra beats and arrhythmias. The mechanism by which stretch produces changes in excitability is postulated to occur via mechanosensitive ion channels, particularly those that activate with stretch (SACs) (Sigurdson et al. 1987; Craelius et al. 1988; Ruknudin et al. 1993; Sipido and Marban, 1991; Kim, 1992) .These channels are found throughout the phylogenetic ladder from *E. Coli* to *Homo Sapiens.* These channels are a family in the same sense as ligand gated and voltage gated channels. They may have ionic selectivities that vary from highly K^+ selective to cation selective to weakly anion selective. No Ca^{+2} selective or Na^+ selective channels have been observed to date although recently a stretch sensitive *l* type Ca^{2+} current has been reported (Langton, 1993). The defining property of SACs is that they respond to membrane stress with a large increase in the probability of being open. For those interested in reading further about these channels, recent reviews are available (Lecar and Morris, 1993; Sachs, 1992; Sokabe and Sachs, 1992; Sachs, 1990; Sachs, 1991; Martinac, 1991).

A variety of SACs have been observed in cardiac tissues using single channel patch clamp recording. The first reports were from Morris and co-workers who reported K^+ selective SACs in molluscan heart cells (Sigurdson et al. 1987). Later, Craelius et al (1988) reported that embryonic rat ventricular cells had a cation selective SAC. In rat atrial cells, Van Wagoner (1993) has shown that the K-ATP channels are activated by stretch. In rat heart, Kim has shown an arachidonic acid sensitive, K^+ selective, SAC (Kim, 1992) as well as a multistate cation selective SAC (Kim, 1993). In tissue cultured embryonic chick heart cells, Ruknudin et al (1993) showed five kinds of SACs distinguished on the basis of selectivity and conductance: two K^+ selective and three cation selective.

Although the single channel studies establish a substrate for defining the physical mechanism of mechano-electric transduction, there is little data available to substantiate the role that these channels may play in generating the observed effects. In part, this gap in the data comes from a lack of pharmacological agents specific for the different kinds of SACs. We should not expect to find "a blocker" for SACs any more than we possess "a blocker" for voltage or ligand gated ion channels. A number of agents have been shown to block different SACs — Gd^{+3} (Ruknudin et al, 1993; Yang and Sachs, 1989), amiloride (Lane et al. 1991; Hamill et al. 1992), quinidine (Sigurdson et al. 1987), tolbutamide and intracellular ATP (Van

Wagoner, 1993), TTX and diltiazem (Ruknudin et al. 1993) — but none are specific for SACs. It seems unlikely that a solid demonstration of the role of SACs will be made until a specific blocker is available.

In this paper I have asked the question of whether the presence of SACs is sufficient to explain the observed effects of stretch on cardiac action potentials. I have added to the "Hodgkin-Huxley" equations for different cardiac tissues, equations describing a cation selective SAC and, with minimal assumptions, tried to model experimental data available in the literature.

Methods

In the absence of detailed information on SACs found in the tissues for which cellular models exist, I have chosen to use a simple SAC model based upon results in the literature and recent results obtained in my laboratory. The choice of SAC channel parameters is inevitably somewhat arbitrary. The mean channel current, the variable that affects the action potential, depends upon the product of the channel conductance, density, and the probability of being open, and these cannot be separated by any mean current simulation. The goal here is to choose parameters that match data from single channel and whole cell recordings, leaving as a free parameter the sensitivity of the channel to whole cell stretch. Some choices in gating parameters were made on the basis of consistency with observed data and in some cases a few cellular model parameters had to be altered to match experimental data.

Reversal potential, V_r

Since stretch invariably causes depolarization of the heart cells, I selected a cation selective SAC. Cation selective SACs are common in many cells. In chick heart cells, although the cells have both cation and K^+ selective SACs, we have found that whole-cell, perforated-patch, recordings of the mechanosensitive current show a reversal potential in the range of -20 mV (Hu and Sachs, in preparation). This suggests that cation selective channels dominate the *in vivo* response and I have chosen a SAC model with a reversal potential of -20 mV. The results were not sensitive to variations between -20 and -40 mV.

Density, ρ.

Based upon the observed frequency of recording the different channel types and the patch area measured using electron microscopy, a reasonable estimate for the channel density was taken as $0.3/\mu m^2$ (Ruknudin et al. 1993).

Conductance, γ

Many cation selective SACs have a conductance in the range of 25-40 pS (Sachs, 1992; Sachs, 1991; Sachs, 1990a), and based on the work on the chick heart cells, I have chosen 25 pS as the channel conductance.

Gating

The probability of being open was modeled using the steady state response, i.e. the SAC channel kinetics are considered infinitely fast. The kinetics of the channels have been solved for a few cases (Yang and Sachs, 1990; Guharay and Sachs, 1985; Guharay and Sachs, 1984) and these suggest that the relaxation times of the channels themselves are probably fast compared with the rise time of realistic stimuli. The role of possible adaptation (Hamill and McBride, 1992) and viscoelastic coupling of forces to the channel was ignored in the absence of any critical data on the issue. Since whole hearts show sustained effects of stretch (Franz et al. 1989; Franz et al. 1992), and rate dependent effects of the stimulus are relatively small (Franz et al. 1992) , adaptation is probably not a major factor. The SAC model was necessarily simplified and will eventually need a number of modifications to better reflect the tissue behavior. The mechanosensitive conductance may actually represent the combined activity of several channels with varying stretch and voltage sensitivity and differing selectivities (Ruknudin et al. 1993). I have not included the specific effects of Ca^{2+} flux through SACs (Sigurdson et al. 1992; Yang and Sachs, 1990; Cooper et al. 1986) which is capable of modifying Ca^{2+} release (Sigurdson et al. 1992) and exchanger currents. I have also left out specific Na^+ and K^+ fluxes and possible coupling between active tension and channel activation. There is no data available on the latter topic.

Parameterization

Using a simple Boltzmann relationship, the SAC current, I_{sac}, was taken to be activated by deviations in sarcomere length from the minimal tension length (L_o=1 μm). The current equation is:

$$I_{sac} = -(V - V_{rev})\gamma\rho A / (1 + K \exp(-\alpha(L - L_0))) \qquad (1)$$

where V is the membrane potential, A is the cell area, K is an equilibrium constant controlling the amount of current at $L_{O,}$ L is the sarcomere length, α is a sensitivity parameter and the others symbols have been previously defined. Clearly there are correlations between L_O α and K. Although it is possible to separate them, the necessary data doesn't exist in the literature.

Integration

The equations governing the action potentials, and in some cases the tension, were integrated on a 486PC using the program HEART (Oxsoft Ltd., Oxford, England) with modifications to include I_{sac}. This program uses a variable step size Adams integrator. Unless noted, no attempt was made to change other aspects of the cell models included in the program. Since many of the literature experiments were made with rabbit and dog ventricular models for which no adequate set of parameters exists, I simulated guinea pig and rat ventricular models as analogs.

Results

Purkinje fibers

The only experimental data come from a paper by Rosen et al (1981) on the dog Purkinje fiber (the paper actually only contains data from one fiber, and no raw data was shown!). Figure 1 shows simulated action potentials at different degrees of stretch. Figure 2A shows a summary of the effects of stretch on maximum diastolic potential, MDP, action potential amplitude, AMP, and the maximum rate of rise of the action potential, $\dot{V}_{max}$. The experimental data (Fig. 2A) was obtained on the ascending limb of the Starling curve and the main effects of stretch were to decrease the diastolic potentials, $\dot{V}_{max}$, and the action potential amplitude. The simulated values are shown in Fig. 2B. There is generally good agreement except that the values of $\dot{V}_{max}$ are several times lower than those of Rosen et al (1981). This discrepancy is a result of the model not accounting for the impedance of intercellular clefts that reduce the effective capacitance. The

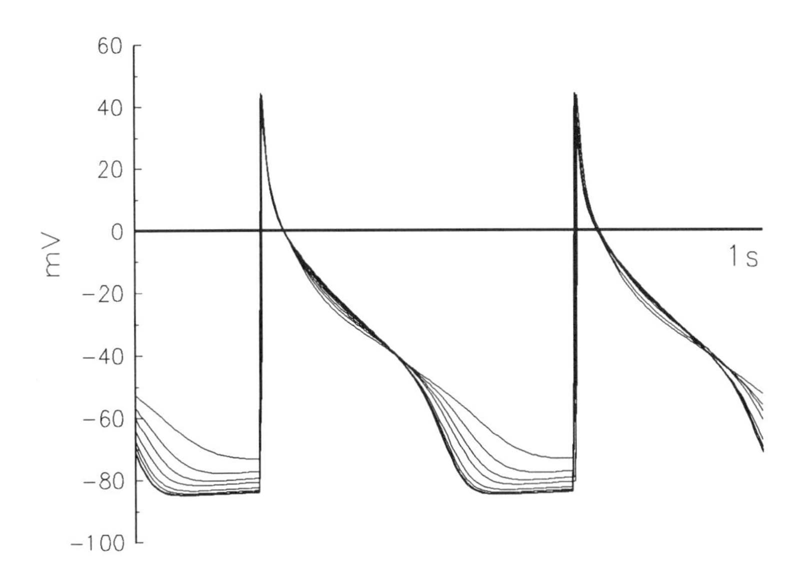

Figure 1. The simulation of the effect of stretch upon Purkinje fiber action potentials. The figure shows superimposed action potentials at different degrees of stretch. The Purkinje model was used with the following modifications to make the simulation better correspond to the data of Rosen et al (their Fig. 4): $[K_i]$ was decreased to 120mM to depolarize the MDP to -85 mV at rest. Pacing was at 2/s. The parameters of I_{sac} were as stated in methods above with K=3000 and α=3. Simulations were run for 3 seconds and the data shown is taken from the fifth and sixth action potentials in the sequence to allow time for the initial conditions to relax. The simulation was done using constants from DiFrancesco and Noble (1985) with corrections as listed in the manual to HEART (Anonymous, 1992).

simulation results (Fig. 2A) quantitatively mimic much of the experimental behavior with the notable exception of a peak in the $\dot{V}_{max}$ curve.

The peak was obtained with different values for the SAC reversal potential. I don't know whether the peak represents a prediction of real behavior or is a result of errors in the model equations. (Errors in integration were checked by limiting the step size, HIACC=1 in HEART).

The peak doesn't appear in the experimental data of Fig 2A. In the simulation, the peak appears to result from increased Ca^{2+}current activation at depolarized levels. The amplitude of the peak in $\dot{V}_{max}$ can be re-

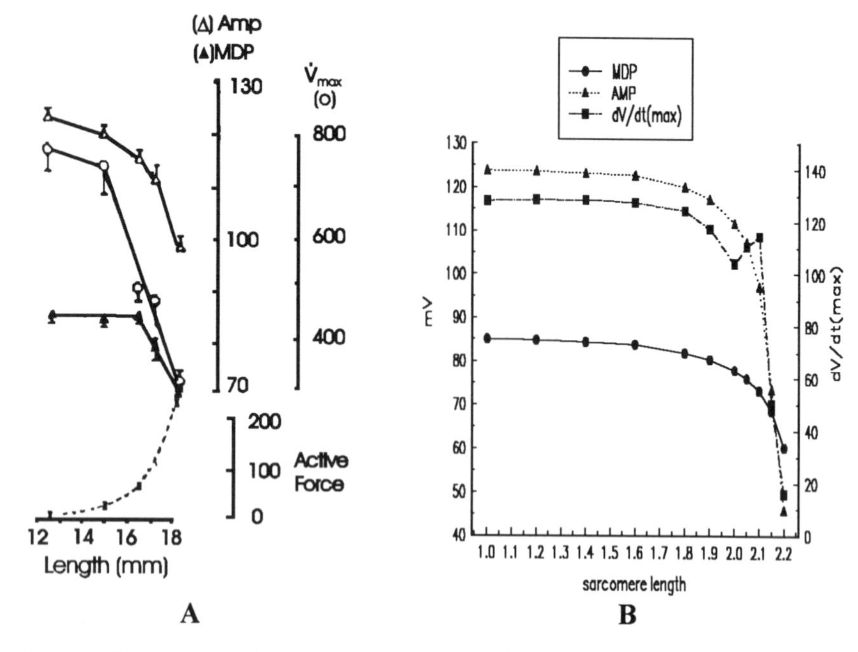

Figure 2. **A** Experimental results of stretch on Purkinje fibers of the dog (after Rosen et al (1981). **B**. Simulated results of stretching Purkinje fibers.. MDP = maximum diastolic potential, AMP = amplitude of action potential, dV/dt(max) = $\dot{V}_{max}$ = maximum rate of rise of the action potential, sarcomere length in μm.

duced by shifting h_∞ to more negative potentials so that the main inward current carried by Na^+ is more easily reduced by stretch induced depolarization. For example, in Figure 2B, the simulation was done using a 6 mV negative shift in h_∞. Without such a shift, $\dot{V}_{max}$ was constant at ≈ 160 V/s between 1 and 2.1 μm. At 2.2 μm it jumped to ≈ 340V/s, and at 2.3 μm it fell to ≈ 60 V/s. It is not uncommon for isolated cardiac cells and tissues to show a leftward shift in h_∞, and the data of Rosen et al may have come from such a fiber.

Simulations using the older parameters of McAllister, Noble and Tsien (1975) (MNT) showed no peak – only a monotonic decrease of $\dot{V}_{max}$ with stretch (Sachs and Jing, unpublished). The Na^+ current gating models in HEART have been extensively modified from MNT to incorporate newer data. Regardless of the finer details of the $\dot{V}_{max}$ curve, it is clear that very little I_{sac} is necessary to mimic the main features of the experimental data. At a sarcomere length of 2.2 μm (just past the peak of the active tension curve), the maximum current is about 130 pA in a cell of 75 nF. The kinds of changes in action potential shape seen for Purkinje fibers,

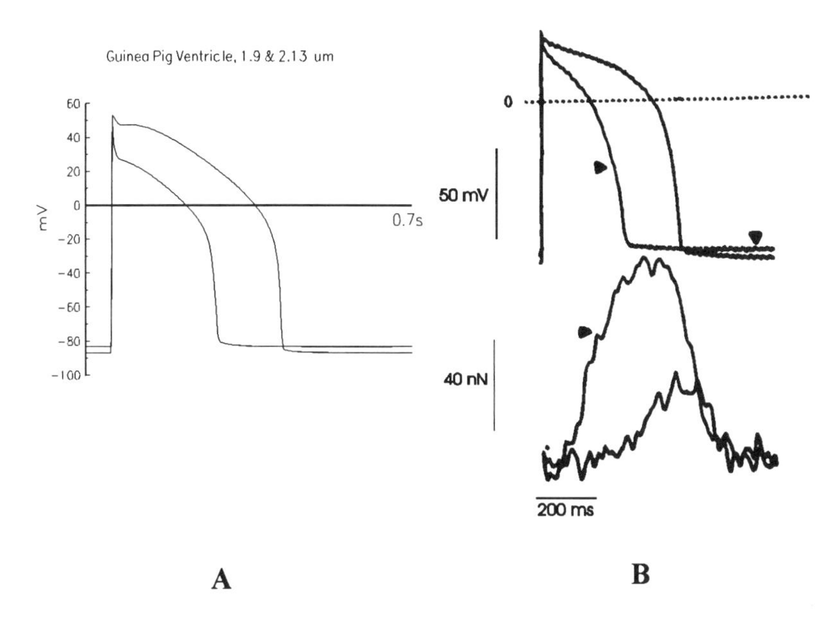

Figure 3.**A**. Simulation of stretch upon a guinea pig ventricular cell. The shorter action potential is from the cell stretched to 2.13 μm. To make the action potential sufficiently long to mimic the data in Fig. 3B, the Ca^{2+} and delayed rectifier, i_K, were slowed by a factor of 5 relative to the 37 ^{O}C version. Also, $[K_i]$ was set to 110 mM to lower the resting potential to the observed -86 mV; $\alpha = 10.7$, $K= 1{,}345{,}000$, $V_r = -50$ mV.
B. Experimental data on the effects of stretch on action potentials and force of a guinea pig ventricular cell. Stretch from 1.9 - 2.13 μm. Arrows point to responses from the stretched cell. (after White et al., 1993).

have earlier been reported for a wide variety of cardiac tissues by Penefsky and Hoffman (1963).

It is difficult to establish the correlation of preparation length and sarcomere length. Although Rosen et al (1981) include electron micrographs of stretched fibers, they do not include a scale bar, only a magnification. This scaling apparently does not allow for scaling during publication since sarcomere lengths calculated from the figures are in error, e.g. their Fig. 8B, shows a fiber at twice rest length with a sarcomere length of only 1.4 μm. The micrographs show a significant variation in sarcomere length that presumably will add dispersion to the action potential properties if SAC activation is actually coupled to sarcomere length.

Ventricular Cells

Data on stretch induced effects at the single cell level are rare. The only results I am aware of come from a wonderful paper by White et al (1993) in which they looked at the effects of stretch on intracellular Ca^{2+}, auxotonic force, sarcomere length and action potential configuration in isolated guinea pig cells. The main effect of stretch on the action potential, illustrated in Figure 3A, was to decrease the duration of the action potentials with a slight tendency to depolarize with stretch. Since White et al's (1993) experiments were conducted at room temperature, the simulation (whose rates were set for 37 ^{o}C), had to be slowed down. Using the guinea pig ventricular model in HEART, which is taken largely from (DiFrancesco and Noble, 1985), I was readily able to simulate the observed shortening of the action potential and slight depolarization of the maximum diastolic potential.

To duplicate the data in Fig. 3B, the reversal potential for the SACs had to be shifted to -50 mV (Fig. 3A). With V_r = -20 mV, the shortening was pronounced at short times, but the duration near completion of repolarization was nearly that of controls. The simulation of Fig. 3B required extreme values of the activation parameters: $\alpha = 10.7$, K= 1,345,000. These values were necessary to produce an order of magnitude change in conductance (≈. 0.002 - 0.02 μS between 1.9 and 2.13 μm. This sensitivity is probably an artifact of non-uniform stresses in the experiment. White et al (1993) applied force through carbon filaments stuck to the lateral surface of the cells. This would lead to high local membrane stresses near the attachment points and higher than average SAC activation. *In vivo,* stresses are probably applied more uniformly.

For the guinea pig ventricular model, the program HEART includes calculation of cell tension. Comparison of the experimental data with the simulated data on tension indicates that the passive compliance of the real preparation is much larger than the model predicts and that in the experiment, active tension increases much faster with stretch than the model predicts. These differences could arise from relatively high compliance attachments of the carbon fibers to the cell membrane and also high local stress in the membrane that might cause pronounced Ca^{2+} influx through SACs (Sigurdson et al. 1992;1993). It is interesting that White et al (1993) observed a slight stretch dependent increase in Ca^{2+} with stretch, an effect expected from work on chick heart cells (Sigurdson et al., 1992). They also observed that responses in some cells were observed on the first stretch of a cell, while in others they appeared only after repeated stretches.

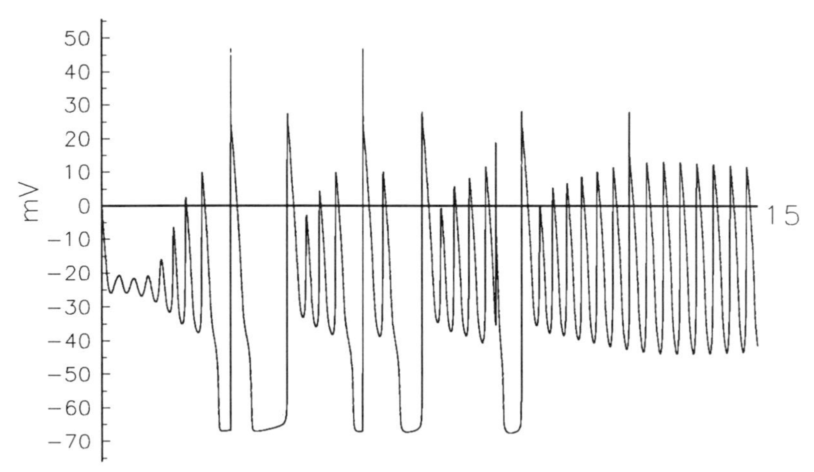

Figure 4. Simulation of activity at large, constant stretch, 2.58 μm. Constants: V_r = -20, α= 3, K=2000. Fluctuations in Ca^{2+} mimic the voltage fluctuations (data not shown).Abscissa is in seconds.

This is analogous to results seen with isolated patches in which repeated flexing of the patch may lead to pronounced positive or negative changes in SAC sensitivity (Hamill and McBride, 1992).

White et al (1993) also observed that prolonged stretching could sometimes lead to sustained depolarization with sub threshold oscillations and an elevation in intracellular Ca^{2+} levels. The simulation shows similar behavior with increasing SAC current as shown in Fig. 4. A puzzling feature of White et al's (1993) data is that the increase in Ca^{2+} levels can be sustained without a measurable increase in cell force (their Fig. 5). This suggests that some of the Ca^{2+}-indo fluorescence arose from pools not available to the contractile apparatus.

Whole Hearts

It is clearly of physiological and clinical relevance to understand how an intact heart responds to stretch. Unfortunately, it is difficult to obtain an intracellular recording for an extended period from a beating heart, but a number of investigators have used monopolar recordings from the endo or epicardial surfaces (Franz et al. 1992; Stacy,Jr. et al. 1992; Taggart et al. 1992c; Taggart et al. 1992a; Taggart et al. 1992b; Hansen et al. 1991; Lab and Dean, 1991; Hansen et al., 1990). These experiments have typically been made with dog or rabbit hearts that are large enough to insert balloons for stretching the ventricles by inflation. The electrophysiological models have usually been constructed using data from small hearts such

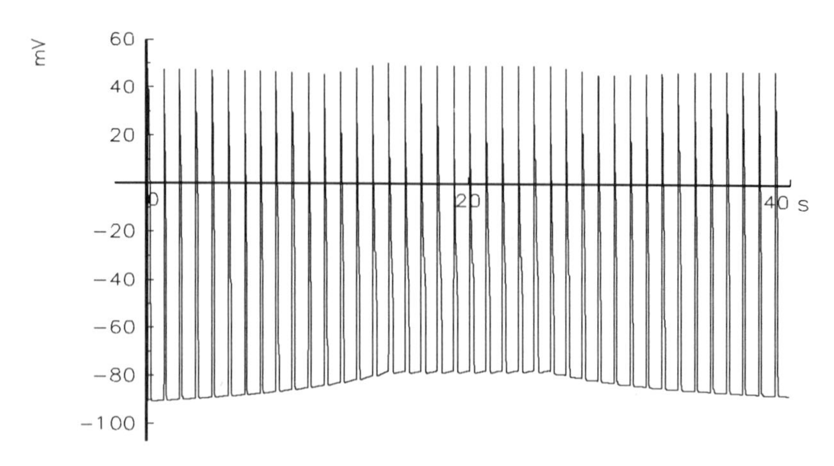

Figure 5. Simulation of the trapezoidal stretch experiment of Franz et al (1992) (their Fig. 4, similar to Figure 7, below). A trapezoidal stretch was applied at 0.3 s beginning at a sarcomere length of 1.5 μm. Sarcomere length was ramped reaching 2 μm at 15 s. It was sustained there for 15 s and then returned in a symmetrical manner to the rest length. There were no significant changes in action potential shape during the periods of constant stress, i.e. the kinetics relaxed rapidly compared to the stimulus interval of 1s. The apparent increase in the overshoot is an artifact of aliasing in the plotting program. Fig. 6 shows an expanded view of the action potential shape. Constants: $V_r = -20$, $K = 100$, $\alpha = 3$.

as rat or guinea pig. In an attempt to test for generality of the conclusions I have simulated the experiments using two preparations, guinea pig and rat, although only results from guinea pig are shown. When comparing whole heart experiments to individual cell experiments, the degree of stretch is difficult to determine. In whole heart experiments the common variable is a volume change but the reference volume was not always reported. Making the biophysicist's tried and true assumption that the ventricle is a thin walled sphere, I estimate the maximal linear strains to be in the range of 20-40 %.

Franz et al (1992) and others (Hansen et al. 1990; Hansen et al. 1991) have found that stretching the ventricle can elicit premature ventricular responses with a probability approaching unity. Figure 8 shows the experimental data (Figure 5 from the Franz et al., 1992). Figure 9 shows a simulation of this experiment. The basic response is readily reproduced.

Working with rabbit heart and feedback control of ventricular volume, Franz et al (1992) showed that slowly applied stresses caused a depolari-

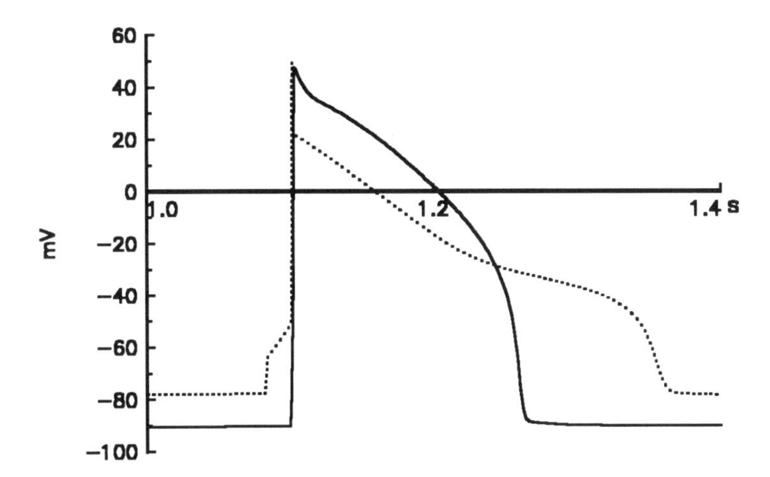

Figure 6. Expanded view of action potentials from Figure 5. Solid line is data at the rest length, the dotted line is during stretch. The action potentials have been aligned on the rising phase of the action potential, not the stimulus, which can be seen to precede the stretched action potential. Note that the maximal overshoot is unaffected by stretching, but the early plateau is shortened while the later plateau is lengthened. The original data of Franz et al (1992), their Fig. 4, shown below, shows an apparent drop in action potential amplitude with stretch. Given the limited bandwidth of the recording and spatial averaging of the monopolar recording, the drop in amplitude probably represents a drop in the average plateau potential rather than a drop in the peak overshoot.

zation of the resting potential and reduction of the action potential amplitude, similar to the results of Rosen et al (1981) on Purkinje fibers. In the case of the guinea pig model, similar results are seen as shown in Fig. 5. The action potentials shown in Fig. 6, with a shortening of the early repolarization and a lengthening of later repolarization are almost identical to the monophasic action potentials (MAPs) recorded in canine hearts under stretch (Franz et al., their Fig. 5).

Franz et al (1992) observed that the likelihood of evoking a premature excitation was dependent not only on the amplitude of the stimulus, but its rise time. Faster rising stimuli were more likely to evoke a response. Their stimuli were trapezoidal ramps with varying rise time, a short (50-100 ms) plateau, and a 200 ms fall time. They found that the threshold for a premature ventricular excitation obeyed the relationship: $v = -0.02\dot{v} + 500$ μl where v is the trigger volume and $\dot{v}$ is the rate of rise of volume. Assuming a starting volume of 500 μl, this would correspond in linear strain, ε (spherical shell model), to $\varepsilon = .02\dot{\varepsilon} + 1.74$ as shown in Figure 10.

I simulated this experiment for guinea pig ventricle and found an extremely weak dependence of action potential generation on strain rate. critical sarcomere length, SL, was given by: $SL = -5.7x10^{-5}\dot{SL} + 1.01$. For Purkinje fibers the rate dependency was also small but of the opposite sign: faster rates of rise required larger amplitudes, $SL = 3.4x10^{-4}\dot{SL} + 0.95$! [Simulation constants: Purkinje, K = 3000, α = 3, V_r = -20, i_f was reduced by 1/3 to decrease spontaneous beating rate. Stretch was applied following the action potential beginning at various times to vary the slope, but always ending at 1s. Guinea pig, K=100, α = 3.13, V_r = -20. Stimulus was maximum at 0.4s after the pacing stimulus which was 0.4/s].

One of the consistent findings from the work of Franz et al (1992) and Hansen's group (Stacy, Jr. et al. 1992), is that the reliable generation of

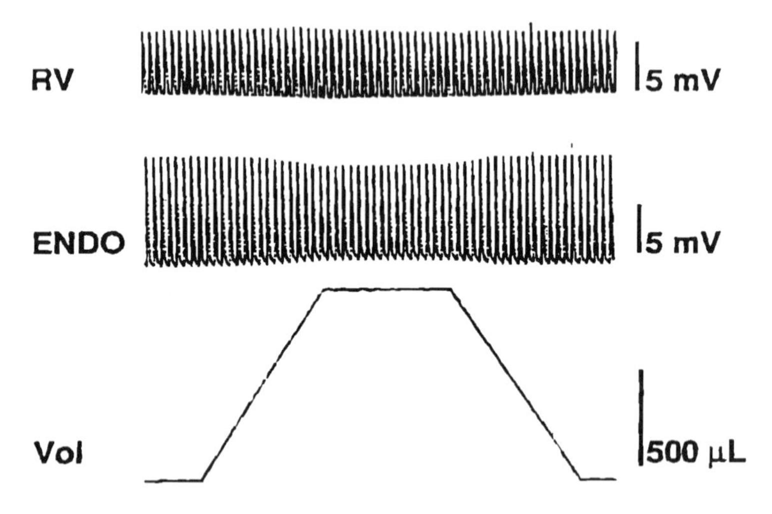

Figure 7. Monopolar action potentials in response to a trapezoidal change in volume (after Franz et al (1992)). ENDO and RV refer to the endocardium of the left ventricle and the epicardium of the right ventricle, respectively. The ventricle inflating balloon was placed in the left ventricle. Compare to the simulation of Figure 5.

PVEs is dependent upon relatively rapid changes in pressure. Slow stretching produces changes in action potential configuration, but not extra systoles. The simulations above suggest that the reason phasic responses generate PVEs lies beyond standard cellular electrophysiology. It is possible that rapid changes in volume or pressure cause increased viscoelastic stress in the wall of the ventricle, increasing activation of SACs. Alternatively, SACs themselves may show temporal inactivation *in vivo* as some do *in vitro* (Hamill and McBride, 1992).

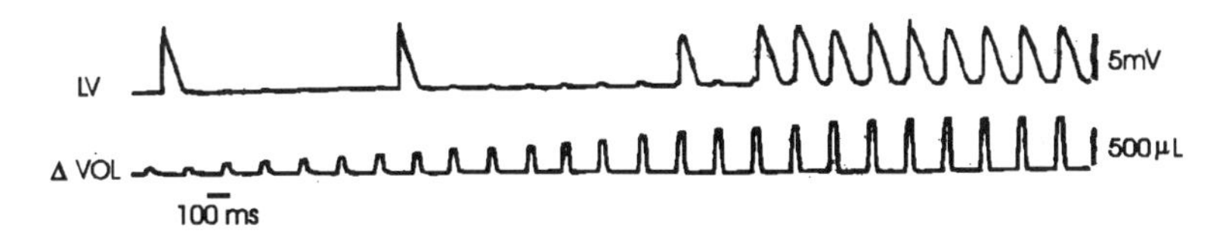

Figure 8. Increasing steps of volume increase the probability of observing a premature ventricular excitation in rabbit ventricle. (After Franz et al ,1992).

Franz et al (1992) measured the refractory period for volume induced PVEs and found that the threshold increased markedly as the volume steps encroached on repolarization. Figure 11 shows the simulated behavior of a guinea pig myocyte and a Purkinje fiber. Because the Purkinje fiber is spontaneously active, the time dependent effects are larger. During repolarization, stretch produced a sub-threshold depolarization, which in an intact heart might have lead to a propagated response. The results for the guinea pig shown in Figure 11 are generally similar to the data shown by Franz et al (1992, their Figure 6). This suggests that stretch induced PVEs do not arise from the Purkinje system. I have not made an exhaustive study of different cellular models, but the results on guinea pig ventricle are similar to the results on rat ventricle.

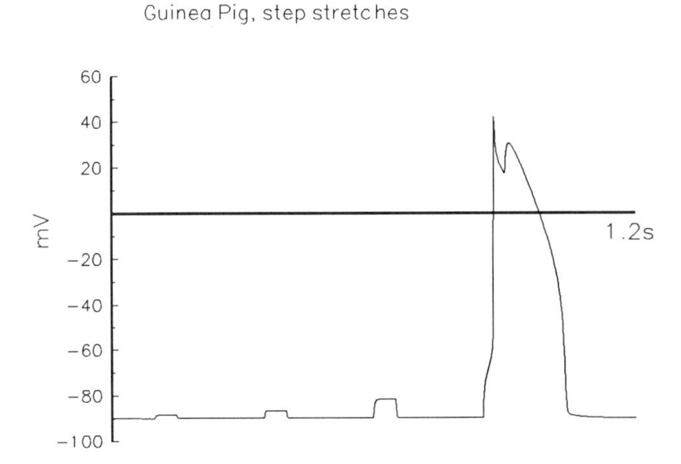

Figure 9. Simulation of experiment by Franz et al (1992) (their Fig. 5 reproduced above as Fig. 8). Four increasing steps of stretch were applied producing depolarizations, and in the last instance, an action potential. Note that when stretch was released during the plateau, the action potential rebounded, an effect previously noted in the literature (Lab, 1982; Kaufmann et al. 1971). Stimuli were 1.6,1.7,1.85 and 2 μm (6-22% strain) lasting 50 ms. Constants: V_r = -20, α= 3.3, K=100.

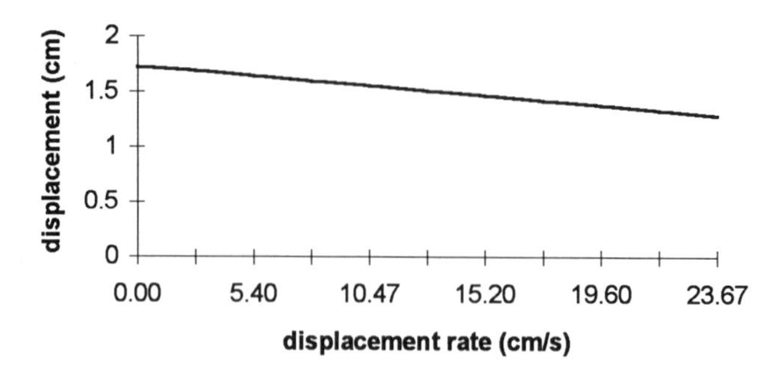

Figure 10. Displacement required to evoke a PVE as a function of the rate (adapted from Figure 9 of Franz et al (1992).

Discussion

This paper is a first attempt to model the electrophysiological consequences of the presence of SACs in heart cells. The main results suggest that the observed data can be explained by the presence of a conductance < 75 pS/pF with a reversal potential in the range of -20 mV, similar to the conductance of many cation selective SACs (Yang and Sachs, 1993; Morris, 1990). Although K^+ selective SACs have been reported, such negative reversal potentials cannot explain the published cellular or tissue data. Since there are a variety of SACs even in a single cell (Ruknudin et

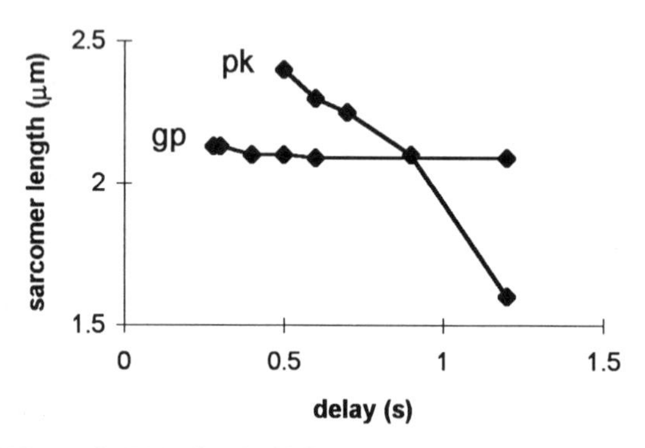

Figure 11. Change in PVE threshold for a 50 ms stretch as function of the time from the stimulus. The simulations were done using the constants of Figures 9 and 11, with the exception that i_f for Purkinje fibers was restored to the normal 3μS. Threshold was taken to be the stretch that produced an action potential with overshoot. Notation: pk refers to Purkinje fibers and gp to guinea pig ventricular cells.

al. 1993), it is possible, however, that K^+ selective SACs make a partial contribution Specific Ca^{2+} linked or other mechanisms do not appear necessary to explain the basic observations. The role of rate dependent effects requires further work. Viscoelastic elements at some level seem necessary. In the whole heart, after depolarizations result from rapid cross clamp of the aorta (Franz et al.1989) but not from steady afterload (Hansen, 1993). The coupling of contraction to electrical activity also needs study in order to permit better modeling. Kaufmann et al's (1971) observation on cat papillary muscle that action potential duration decreases with force development suggests some form of coupling does exist, but the variation of free Ca^{2+} during these processes cannot be ignored as a possible influence. It is possible that stretch of sympathetic nerves within the heart may play a role in mechano-electric responses. On the basis of indirect evidence of stretch induced effects in elevated Ca^{2+}, Lerman et al (1985) suggest that β-adrenergic activity is not involved, but evidence for mechanically induced release of neurotransmitters goes back at least to Fatt and Katz (1952).

The gating model (Eq. 1) is a sigmoidal, Boltzmann relationship. It is equivalent to a three state kinetic model and chosen for its physical simplicity, but it is probably incorrect in detail. Hansen's group (Stacy, Jr. et al. 1992) indicates that in the whole heart, stretch-sensitive changes in diastolic potential are linear, not exponential, in strain. If the above model applies, this suggests that the exponential term is small and that in the physiological case, the channels do not approach saturation. In the simulations shown above, the exponential behavior was not always small and curvature can sometimes be seen in the single cell response, e.g. MDP of Fig. 5.

The modelling is limited by an appalling lack of knowledge about the responses of isolated heart cells to stretch. There is no published voltage clamp data and only a few papers on the effects of stretch upon the action potential. If we are to understand how stretch causes arrhythmias we will need at minimum to address the following issues.

Whole heart

1. What is the reversal potential of stretch induced changes? To what ions is it sensitive?
2. Is there a component of nervous system involvement, i.e. does stretch affect the release of sympathetic transmitters?
3. What are the wall stresses and strains involved, particularly during transient changes in volume?
4. What pharmacological agents affect the mechanical response?

Whole Cell

1. What is the ionic basis of the whole cell mechanical response?
2. What is the steady state relationship between strain and conductance?
3. What is the transient response of the mechanosensitive current?
4. What is the temperature dependence of the mechanosensitive response?
5. What is the voltage dependence of the mechanical response?
6. What pharmacological agents affect the mechanical response?
7. How is the mechanosensitive conductance modulated by contraction?
8. How do mechanical responses vary among cells from different parts of the heart?

Single channel

1. What mechanosensitive channels are present in cells isolated from tissues whose responses to stretch have been studied?
2. What is the transient response of SACs? Is there adaptation, and if so, does it arise from the channels themselves or from associated cytoskeletal structures?
3. What pharmacological agents affect SACs?
4. Do all SACs in a patch feel the same forces?

Modeling

1. What happens if tight coupling is included between stretch induced Ca^{2+} influx and Ca^{2+} release (c.f. Sigurdson et al. 1993)?
2. What are the effects of including K^+, Na^+ and Ca^{2+} permeabilities explicitly in SAC activation?
3. If SAC gating is affected by active tension, what are the predicted effects?
4. What are the predicted effects of SAC activation upon tension and Ca^{2+} levels?
5. If stretch induced nerve stimulation is present, what are its predicted effects?
6. What are the predicted effects of SACs upon tension?
7. What are the predicted effects of extracellular ions and drugs on mechanosensitive responses?
8. What are the mechanosensitive properties of coupled heart cells (one, two and three dimensional arrays)?

Clinical Implications

A number of labs have emphasized the fact that the occurrence of arrhythmias following an infarct is more closely correlated with the mechanical status of the heart rather than electrical activity (Lab and Dean, 1991). It is observed clinically that following an infarct passive stretch of the damaged wall may occur, particularly during systole. It is not difficult to imagine that during systole, the periphery of the infarct will be stretched and may cause the generation of excitatory currents, leading to the generation of arrhythmias. The data from intact hearts shows that PVEs can be produced with absolute reliability by a rapid stretch (Franz et al. 1992; Stacy, Jr. et al. 1992). It has been shown by Hansen and co workers (Hansen et al. 1991) that stretch induced PVEs can be blocked by Gd^{+3}, a good, but non-specific, blocker of many SACs (Yang and Sachs, 1989; Ruknudin et al. 1993), but not by traditional organic Ca^{2+} channel blockers. This raises the possibility of finding prophylactic agents of a new class that could be used in the critical few days following an infarct.

Acknowledgments

This work was supported by the American Heart Association, NY Chapter, and the USARO, grant DAAL0392G0014. I would like to thank Hu Hai, Max Lab, David Hansen and Michael Franz for helpful discussions.

References

Anonymous HEART PROGRAM MANUAL, V3.8. Oxford, England: Oxsoft Ltd., 1992.

Bainbridge, F. A. The influence of venous filling upon the rate of the heart. *J. Physiol.* 50:65-84, 1915.

Cooper, K. E., Tang, J. M., Rae, J. L., and Eisenberg, R. S. A cation channel in frog lens epithelia responsive to pressure and calcium.. *J. Membrane Biol.* 93:259-269, 1986.

Craelius, W., Chen, V., and El-Sherif, N. Stretch activated ion channels in ventricular myocytes. *Bioscience Reports* 8:407-414, 1988.

Dean, J. W. and Lab, M. J. Arrhythmia in heart failure: Role of mechanically induced changes in electrophysiology. *The Lancet* 1:1309-1312, 1989.

DiFrancesco, D. and Noble, D. A model of cardiac electrical activity incorporating ionic pumps and concentration changes. *Phil. Trans. Roy. Soc.* B307:353-398, 1985.

Dudel, J. and Trautwein, W. Das Aktionspotential und Mechanogramm des Herzmuskels unter dem Einfluss der Dehnung.. *Cardiologia* 25:334-362, 1954.

Fatt, P. and Katz, B. Spontaneous subthreshold activity at motor nerve endings. *J. Physiol.* 117:109-128, 1952.

Francis, G. S. Development of arrhythmias in the patient with congestive heart failure: pathophysiology, prevalence and prognosis. *Am.J.Cardiology* 57:3B-7B, 1986.

Franz, M. R., Burkhoff, D., Yue, D. T., and Sagawa, K. Mechanically induced action potential changes and arrhythmia in isolated and in situ canine hearts. *Cardiovasc. es.* 23:213-223, 1989.

Franz, M. R., Cima, R., Wang, D., Profitt, D., and Kurz, R. Electrophysiological effects of myocardial stretch and mechanical determinants of stretch-activated arrhythmias. *Circulation* 86:968-978, 1992.

Guharay, F. and Sachs, F. Stretch-activated single ion channel currents in tissue-cultured embryonic chick skeletal muscle. *J. Physiol. (Lond)* 352:685-701, 1984.

Guharay, F. and Sachs, F. Mechanotransducer ion channels in chick skeletal muscle: the effects of extracellular pH. *J. Physiol. (Lond)* 353:119-134, 1985.

Hamill, O. P., Lane, J. W., and McBride, D. W. Amiloride: a molecular probe for mechanosensitive ion channels. *Trends Pharmacol Sci* 13:373-376, 1992.

Hamill, O. P. and McBride, D. W. Rapid adaptation of single mechanosensitive channels in Xenopus oocytes. *Proc Natl Acad Sci U S A* 89:7462-7466, 1992.

Hansen, D. E., Craig, C. S., and Hondeghem, L. M. Stretch-induced arrhythmias in the isolated canine ventricle. Evidence for the importance of mechanoelectrical feedback. *Circulation* 81:1094-1105, 1990.

Hansen, D. E., Borganelli, M., Stacy, G. P.,Jr., and Taylor, L. K. Dose-dependent inhibition of stretch-induced arrhythmias by gadolinium in isolated canine ventricles. Evidence for a unique mode of antiarrhythmic action. *Circ.Res.* 69:820-831, 1991.

Hansen, D. E. Mechanoelectrical feedback effects of altering preload, afterload and ventricular shortening. *Am. J. Physiol.* 264:H423-H432, 1993.

Kaufmann, R. L., Lab, M. J., Hennekes, R., and Krause, H. Feedback interaction of mechanical and electrical events in the isolated mammalian ventricular myocardium (cat papillary muscle). *Pflugers Arch.* 324:100-123, 1971.

Kim, D. A mechanosensitive K+ channel in heart cells - activation by arachidonic acid. *J. Gen. Physiol.* 100(6):1021-1040, 1992.

Kim, D. Novel cation-selective mechanosensitive ion channel in the atrial cell membrane. *Circ.Res.* 72:225-231, 1993.

Lab, M. and Dean, J. Myocardial mechanics and arrhythmias. *J. Cardiovasc. Pharm.* 18(S2):S72-S79, 1991.

Lab, M. J. Transient depolarization and action potential alterations following mechanical changes in isolated myocardium. *Cardiovas. Res.* 14:624-637, 1980.

Lab, M. J. Contraction-excitation feedback in myocardium. *Circ.Res.* 50:757-765, 1982.

Lane, J. W., McBride, D., and Hamill, O. P. Amiloride blocks the mechanosensitive cation channel in Xenopus oocytes. *J Physiol (Lond)* 441:347-366, 1991.

Langton, P. D. Calcium channels currents recorded from isolated myocytes of rat basilar artery are stretch sensitive. *J Physiol (Lond)* 471:1-11, 1993.

Lecar, H. and Morris, C. Biophysics of mechanotransduction. In: *Mechanoreception by the Vascular Wall*, edited by G. M. Rubanyi. Mount Kisco, NY: Futura Publishing, 1993, p. 1-11.

Lerman, B. B., Burkhoff, D., Yue, D. T., and Sagawa, K. Mechanoelectrical Feedback: Independent Role of Preload and Contractility in Modulation of Canine Ventricular Excitability. *J. Clin. Invest.* 76:1843-1850, 1985.

Martinac, B. Mechanosensitive ion channels: biophysics and physiology. In: *Thermodynamics of cell surface receptors*, Ed. M.B.Jackson. CRC Press, 327-352, 1993.

McAllister, R. E., Noble, D., and Tsien, R. W. Reconstruction of the electrical activity of cardiac purkinje fibres. *J. Physiol. (Lond.)* 231:1-59, 1975.

Meinertz, T., Hoffman, T., Kasper, W., and et al, Significance of ventricular arrhythmias in idiopathic dilated cardiomyopathy. *Am.J.Cardiology* 53:902-907, 1984.

Morris, C. E. Mechanosensitive ion channels. *J. Membrane Biol.* 113:93-107, 1990.

Penefsky, Z. A. and Hoffman, B. F. Effects of stretch on mechanical and electrical properties of cardiac muscle. *Am. J. Physiology* 204:433-438, 1963.

Rajala, G. M., Pinter, M. J., and Kaplan, S. Response of the quiescent heart tube to mechanical stretch in the intact chick embryo. *Developmental Biology* 61:330-337, 1990.

Rosen, M. R., Legato, M. J., and Weiss, R. M. Developmental changes in impulse conduction in the canine heart. *Am. J. Physiol.* 240:H546-H554, 1981.

Ruknudin, A., Sachs, F., and Bustamante, J.O. Stretch-activated ion channels in tissue-cultured chick heart. *Am. J. Physiol.* 264:H960-H972, 1993.

Sachs, F. Stretch-sensitive Ion Channels. *The Neurosciences* 2:49-57, 1990.

Sachs, F. Mechanical transduction by membrane ion channels: a mini review. *Mol. Cell. Biochem.* 104:57-60, 1991.

Sachs, F. Stretch sensitive ion channels: an update. In: *Sensory Transduction*, edited by D. P. Corey and S. D. Roper. NY: Rockefeller Univ. Press, Soc. Gen. Physiol., 1992, p. 241-260.

Shultz, R. A., Strauss, H. W., and Pitt, B. Sudden death following myocardial infarction: relation to ventricular premature contractions in the late hosptal phase and left ventricular ejection fraction. *Am. J. Med. Sci.* 62:1921977.

Sideris, D. A., Toumanidis, S. T., Kostis, E. B., Diakos, A., and Moulopoulos, S. D. Arrhythmogenic effect of high blood pressure: some observations on its mechanism. *Cardiovasc. Res.* 23:983-992, 1989.

Sideris, D. A., Toumanidis, S. T., Kostis, E. B., Stagiannis, K., Spyropoulos, G., and Moulopoulos, S. D. Response of tertiary centres to pressure changes. Is there a mechano-electrical association?. *Cardiovasc. Res.* 24:13-18, 1990.

Sigurdson, W. J., Morris, C. E., Brezden, B. L., and Gardner, D. R. Stretch activation of a K+ channel in molluscan heart cells. *J. Exp. Biol.* 127:191-209, 1987.

Sigurdson, W. J., Ruknudin, A., and Sachs, F. Calcium imaging of mechanically induced fluxes in tissue-cultured chick heart: role of stretch-activated ion channels. *Am. J. Physiol.* 262:H1110-H1115, 1992.

Sigurdson, W. J., Sachs, F., and Diamond, S. L. Mechanical perturbation of cultured human endothelial cells causes rapid increases of intracellular calcium. *Am. J. Physiol.* 264:H1745-H1752, 1993.

Sipido, K. R. and Marban, E. L-type calcium channels, potassium channels, and novel nonspecific cation channels in a clonal muscle cell line derived from embryonic rat ventricle. *Circ.Res.* 69:1487-1499, 1991.

Sokabe, M. and Sachs, F. Towards a molecular mechanism of activation in mechanosensitive ion channels. In: *Advances in Comparative and Environmental Physiology, v10*, edited by F. Ito. Berlin: Springer-Verlag, 1992, p. 55-77.

Stacy, G. P.,Jr., Jobe, R. L., Taylor, L. K., and Hansen, D. E. Stretch-induced depolarizations as a trigger of arrhythmias in isolated canine left ventricles. *Am. J. Physiol.* 263:H613-H621, 1992.

Taggart, P., Sutton, P., John, R., Lab, M., and Swanton, H. Monophasic action potential recordings during acute changes in ventricular loading induced by the Valsalva manoeuvre. *Br. Heart J.* 67:221-229, 1992a.

Taggart, P., Sutton, P., and Lab, M. Interaction between ventricular loading and repolarization: relevance to arrhythmogenesis. *Br. Heart J.* 67:213-215, 1992b.

Taggart, P., Sutton, P., Lab, M., Runnalls, M., O'Brien, W., and Treasure, T. Effect of abrupt changes in ventricular loading on repolarization induced by transient aortic occlusion in humans. *Am. J. Physiol.* 263:H816-H823, 1992c.

Van Wagoner, D. R. Mechanosensitive gating of atrial ATP-sensitive potassium channels. *Circ.Res.* 72:973-983, 1993.

White, E., Le Guennec, J. Y., Nigretto, J. M., Gannier, F., Argibay, J. A., and Garnier, D. The effects of increasing cell length on auxotonic contractions: membrane potential and intracellular calcium transients in single guinea-pig ventricular myocytes. *Exp. Physiol.* 78:65-78, 1993.

Yang, X. and Sachs, F. Block of stretch-activated ion channels in Xenopus oocytes by gadolinium and calcium Ions. *Science* 243:1068-1071, 1989.

Yang, X. C. and Sachs, F. Characterization of stretch-activated ion channels in *Xenopus* oocytes. *J. Physiol. (Lond.)* 431:103-122, 1990.

Yang, X. C. and Sachs, F. Mechanically sensitive, non-selective, cation channels. In: *Non-selective ion channels*, edited by D. Siemen and J. Hescheler. Heidelberg: Springer-Verlag 79-92, 1994.

19

Cellular Tensegrity and Mechanochemical Transduction

D.E. Ingber

Introduction

To explain how biological tissues form and function, we must first understand how different types of regulatory signals, both chemical and mechanical, integrate inside the cell. A clue to the mechanism of signal integration comes from recognition that the action of a force on any mass, regardless of scale, will result in a change in three dimensional structure. This is critical because recent studies reveal that many of the molecules that mediate signal transduction and stimulus-response coupling are physically bound to insoluble structural scaffoldings within the cytoskeleton and nucleus (Ingber 1993a). In this type of "solid-state" regulatory system, mechanically-induced structural arrangements could provide a mechanism for regulating cellular biochemistry and hence, efficiently integrating structure and function. However, this is a difficult question to address using conventional molecular biological approaches because this problem is not based on changes in chemical composition or local binding interactions. Rather, it is a question of architecture. As a result of this challenge, a new scientific discipline of "Molecular Cell Engineering" is beginning to emerge which combines elements of molecular cell biology, bioengineering, architecture, and biomechanics. The purpose of this chapter is not to present a detailed analysis of the work from different laboratories that has contributed to this field since this has been recently published (Ingber 1993a,b). Instead, I hope to provide a brief overview of recent work from my own laboratory which attempts to place biochemical mechanisms of regulation within the context of a specific architectural paradigm that is known as "tensegrity". I also will explore the implications of this approach for cell biology as well as the future of engineering.

Control of Morphogenesis

My laboratory is interested in the mechanism by which three dimensional tissue form is generated or what is known as morphogenesis. Our work has shown that the development of functional tissues, such as branching capillary networks, requires both soluble growth factors and insoluble cell anchoring proteins that are known as extracellular matrix (ECM) molecules. Interestingly, we found that ECM molecules are the dominant regulators since they dictate whether individual cells will either proliferate, differentiate, or involute in response to soluble stimuli (Ingber et al. 1987; Ingber and Folkman 1988; Ingber and Folkman 1989a,b; Ingber 1990; Mooney et al. 1992a,b).

Analysis of the molecular basis of these effects revealed that ECM molecules alter cell growth via both biochemical and biomechanical signaling mechanisms (Ingber 1991). ECM molecules cluster specific integrin receptors on the cell surface and thereby activate intracellular chemical signaling pathways (e.g., Na^+/H^+ antiporter, inositol lipid turnover, protein tyrosine phosphorylation), stimulate expression of early growth response genes (e.g., c-fos, jun-B), and induce quiescent cells to pass through the G_0/G_1 transition (Ingber et al. 1990; Schwartz et al. 1991; Plopper et al. 1991; McNamee et al. 1993; Dike and Ingber 1993; Hansen et al. 1993). However, while activation of these chemical signaling pathways is necessary for growth, it is not sufficient. In addition, immobilized ECM components must physically resist cell tension and promote changes of cell and nuclear shape (Sims et al. 1992; Ingber 1993b) in order to promote entry into S phase (Ingber et al. 1987; Ingber and Folkman 1989a; Ingber 1990; Mooney et al. 1992a,b; Singhvi et al. 1994).

Cellular Tensegrity

But how could cell shape changes alter cell function? Our working hypothesis has been that cell, cytoskeletal, and nuclear form alterations result from changing the balance of mechanical forces that are distributed across transmembrane ECM receptors on the cell surface and that it is this change in force distributions that provides regulatory information to the cell (Ingber 1991). This concept is based on studies we carried out with "stick and string"models of cells that are built using tensegrity architecture (Ingber 1993b; Ingber et

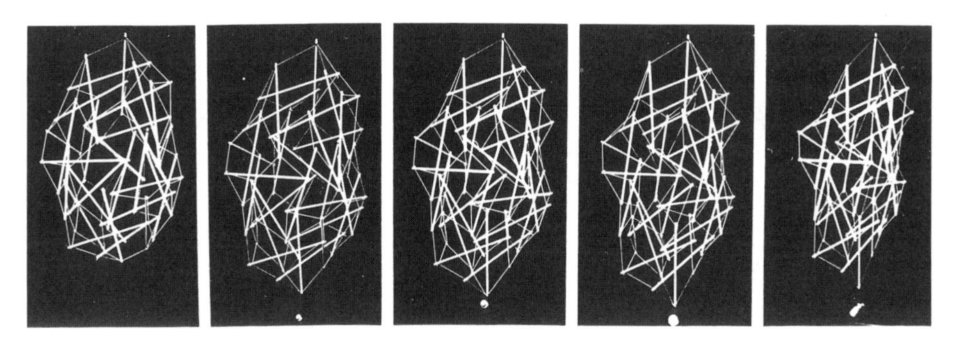

Figure 1. Tensegrity models constructed from wooden applicator sticks and elastic string and loaded with metal weights of increasing mass (from left to right). Note that the structure exhibits a global response to applied stress in that all of its mechanically-interdependent elements rearrange rather than deform locally (reprinted with permission from Wang et al. 1993).

al. 1994). Tensegrity structures depend on tensional integrity rather than compressional continuity for their stability. This novel form of structural stabilization is often visualized using a number of different compression-resistant struts that do not physically touch, but are pulled up and open through interconnection with a continuous series of tension elements (Fig. 1). This form of architectural stabilization, which is unusual in man-made structures, provides the structural basis for the incredible efficiency and stability of geodesic domes. In living cells, microtubules and cross-linked actin filaments may resist compression whereas actin-containing contractile microfilaments and intermediate filaments appear to serve as tension elements (Ingber 1993b).

Most importantly, tensegrity cell models predict many behaviors that are exhibited by living cells including the dependence of cell shape on the ECM mechanics (Ingber 1991; Ingber 1993b), the coordination between cell and nuclear shape changes that is observed in spreading cells (Ingber et al. 1987; Ingber 1990), and the polarization of nuclei to the cell base that occurs when cells adhere to ECM (Ingber et al. 1986). Tensegrity models also explain how specific molecular patterns (e.g., stress fibers, polygonal nets, geodomes) develop within actin cytoskeleton in response to mechanical stress (Ingber 1993b; Ingber et al. 1994).

Studies with tensegrity models suggest that transmembrane receptors that link support elements within the contractile cytoskeleton to resistance sites within the ECM anchoring foundation should play a pivotal role in mechanotransduction (Ingber 1991). The most common and best characterized type of cell surface ECM receptors are known as "integrins". Integrins are members of a superfamily of transmembrane receptors that were first identified based on their ability to bind to a specific three amino acid sequence (RGD) that is found within many ECM proteins (Hynes, 1992;Schwartz 1992). The intracellular portion of certain integrin subunits (e.g., ß1) binds to actin-associated proteins (e.g., talin, vinculin, paxillin, α-actinin) and thereby, physically interconnects ECM with the cytoskeleton within a region of the cell that is known as the "focal adhesion" (Burridge et al., 1988). Integrins also have been found to activate many of the same intracellular chemical signaling pathways that mediate the growth-promoting effects of soluble mitogens (Ingber et al. 1990; Schwartz et al. 1991;Schwartz 1992; Hynes 1992; McNamee et al. 1993). For these reasons, integrins are perfectly poised to integrate both types of signals, chemical and mechanical, at the cell surface.

Experimental Measurements

Recently, in experiments with membrane-permeabilized cells, we confirmed that the structural stability of the cell and nucleus depends on a dynamic balance of tensile and compressive forces that are distributed across cell surface integrin receptors (Sims et al. 1992). We found that mechanical tension is generated within contractile microfilaments via an actomyosin filament sliding mechanism similar to that found in muscle, transmitted across transmembrane integrin receptors, and resisted by ECM anchoring points. When cytoskeletal tension overcame the mechanical resistance of the substratum in a spread cell, rapid and coordinated retraction of the cell, cytoskeleton, and nucleus resulted. These findings confirmed that all mechanical loads are imposed on a pre-existing cellular force balance and are consistent with our concept that transmembrane integrin receptors act as mechanochemical transducers (Ingber 1991).

To analyze the molecular basis of mechanotransduction, we developed a magnetic twisting device in which controlled mechanical stresses can be applied directly to cell surface integrin receptors. The stresses are applied by twisting surface-bound ferromagnetic microbeads (5.5 μm diameter) that are coated with integrin ligands (e.g., fibronectin, antibodies, synthetic RGD-peptides) and the cellular response is measured simultaneously using an in-line magnetometer (Wang et al.1993). In these studies which were carried out by Ning Wang and myself in collaboration with Jim Butler at the Harvard School of Public Health, we found that ß1 integrin effectively transferred mechanical loads across the cell surface and supported a force-dependent cytoskeletal stiffening response whereas non-adhesion receptors (e.g., acetylated-LDL receptor) did not. Force transfer correlated with recruitment of focal adhesion proteins and linkage of integrins to the actin cytoskeleton. Yet the cytoskeletal response to stress involved higher order structural interactions between all three different types of cytoskeletal filaments: microfilaments, microtubules and intermediate filaments. This was demonstrated by demonstrating that individually disrupting these filament systems using cytochalasin D, nocodazole, and acrylamide, respectively, prevented the cytoskeletal stiffening in response to stress (Fig. 2).

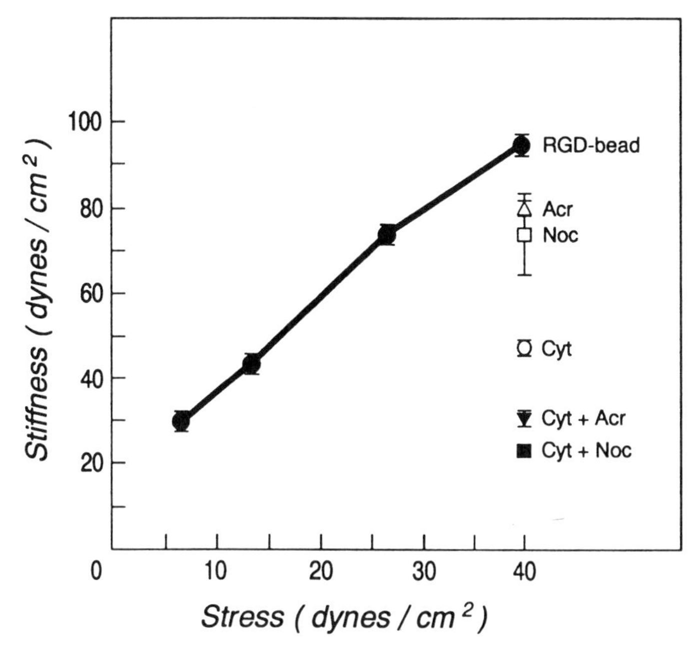

Figure 2. Continuum mechanics analysis of living endothelial cells. Stiffness (ratio of stress to strain) of the cytoskeleton was determined using the magnetic twisting device. This response depended on the presence of intact microtubules, intermediate filaments, and microfilaments. Noc, nocodazole (10 μg/ml); Acr, acrylamide (4 mM); Cyt, cytochalasin D (0.1μg/ml) (reprinted with permission from Wang et al. 1993).

Interestingly, the intact cytoskeleton responded to mechanical stress as if it were a continuous, pre-stressed structural lattice in that it exhibited a linear stiffening response (Fig. 2). Stiffness increased in direct proportion as the level of applied stress was raised. While this type of mechanical behavior is observed in many biological tissues, it can not be explained by conventional biomechanical theories (McMahon, 1984; Stamenovic, 1990). However, this linear stiffening behavior was mimicked precisely using tensegrity cell models (Fig. 3). In other words, this linear stiffening response can be explained if the cytoskeleton is a tensegrity structure, that is, a molecular continuum of mechanically-interdependent struts and tensile elements that rearrange globally, rather than deform locally, in response to stress (Fig. 1). If the cytoskeleton functions as a tensegrity structure, then its ability to resist angular deformation and hence its "stiffness" would not be a function of the deformability of individual filaments, rather it would be a property of the integrated cytoskeletal lattice. This is exactly what we observed in living cells.

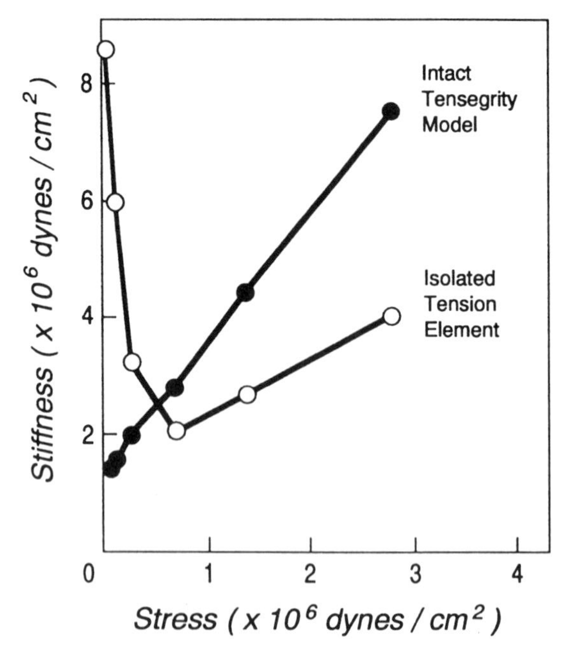

Fig. 3. Continuum mechanics analysis of a three-dimensional tensegrity model. The model consisted of a geodesic spherical array of wood dowels (0.3 cm by 15 cm) and thin elastic threads (0.06 cm by 6 cm). The model was suspended from above and loaded with metal weights at its lower end as shown in Fig. 1. Similar measurements were carried out with a single tension element (elastic thread) isolated from the model. Only the intact tensegrity structure exhibited linear stiffening (reprinted with permission from Wang et al. 1993).

More recently, we carried out additional studies with the magnetic probe to analyze the biophysical mechanism by which ECM alters the structure of the cytoskeleton and controls cell form (Wang and Ingber, 1994). Formation of basal ECM adhesive contacts was independently varied by altering the density of fibronectin immobilized on plastic dishes. Raising the ECM coating density promoted cell spreading and increased cytoskeletal stiffness, permanent deformation, and apparent viscosity. When the applied stress was independently varied, the stiffness and apparent viscosity of the cytoskeleton increased in parallel although neither cell shape, ECM contacts, nor permanent deformation was altered. Application of the same stresses over a lower number ECM contacts using smaller beads (1.4 μm versus 5.5 μm diameter) resulted in decreased cytoskeletal stiffness and apparent viscosity, again without altering permanent deformation. The finding that use of smaller beads resulted in a decreased stress for a given strain, rather than an increase, also confirmed that this technique probes into the depth of the cytoskeleton and not just the cortical membrane.

To discriminate between osmotic and cytoskeletal effects on cell mechanics as well as active (ATP-dependent) versus passive properties of the cytoskeleton, cells were permeabilized with saponin. In the absence of ATP, cytoskeletal stiffness and apparent viscosity were significantly higher than in controls whereas permanent deformation decreased by more than half. Addition of ATP under conditions that promote cytoskeletal tension generation (Sims et al. 1992) resulted in decreases in cytoskeletal stiffness and apparent viscosity which could be detected within 2 min following ATP addition, prior to any measurable change in cell size. No change in permanent deformation was seen at early times, however, a significant decrease was observed once the cytoskeletal lattice had physically contracted. Importantly, regardless of cell shape or membrane continuity, cytoskeletal stiffness increased in direct proportion to the applied stress. In other words, both intact living cells and permeabilized cells that lack membrane integrity behaved as if they were tensegrity structures.

These results suggest that the effects of ECM on cytoskeletal mechanics and cell shape are not due to changes in osmotic or hydrodynamic pressures. Rather, ECM alters cytoskeletal stiffness and apparent viscosity by binding integrins, promoting formation of molecular links with the cytoskeleton, transmitting mechanical stresses across these linkages, and inducing structural rearrangements within a continuous, tensionally-integrated cytoskeletal lattice. These effects are observed when cells spread on ECM, however, they can be induced independently of cell shape. In contrast, the ability of the cytoskeleton to sustain stress-induced deformations is more tightly coupled to cell extension and depends on both passive cytoskeletal plasticity and dynamic molecular remodeling events.

Mechanochemical Transduction

These experiments using the magnetic probe clearly show that integrins act as cell surface mechanoreceptors in that they transmit mechanical signals to the cytoskeleton via a specific molecular pathway. They also show that application of a local mechanical stress results in global structural rearrangements throughout the interconnected cytoskeleton. Mechanochemical transduction, in turn, may be mediated simultaneously at multiple locations by force-induced cytoskeletal rearrangements that result in redistribution of associated elements of the cell's metabolic machinery (Ingber 1993a). This is because much of cellular biochemistry functions in a "solid-state" in living cells. As an example: George Plopper and Helen McNamee in my laboratory have found that many of the chemical signaling molecules that are sensitive to ECM binding (e.g., protein kinases, lipid kinases, phospholipase C, Na+/H+ antiporter) are immobilized on cytoskeletal filaments that form the backbone of the cell's focal adhesion complex (Plopper et al. 1991; Plopper and

Ingber 1993; McNamee and Ingber, 1993). Metabolic enzymes similarly associate with cytoskeletal filaments in the cytoplasm and regulatory molecules that are responsible for control of DNA replication and transcription are often immobilized along scaffolds within the nucleus (rev. in Ingber 1993a).

If cells use a tensegrity mechanism to stabilize their structure, then local distortion of integrin receptors on the cell surface should produce coordinated structural changes throughout the cell (Fig. 1). Deformation of intracellular scaffolds due to mechanical force transfer across integrins may regulate cellular biochemistry and function by changing local proximity between immobilized regulatory enzymes, substrates, and regulatory molecules along cytoskeletal or nuclear scaffolds (Fig. 4). Nuclear distortion also might promote expansion of nuclear pores and therefore enhance nuclear transport of molecules that are required for cell cycle progression (Hansen and Ingber 1992). Interestingly, an example of geometric contol of biochemical reactions already exists: actin preferentially polymerizes from vertices of actin "geodomes" within the cytoskeleton of migrating cells (rev. in Ingber 1993b).

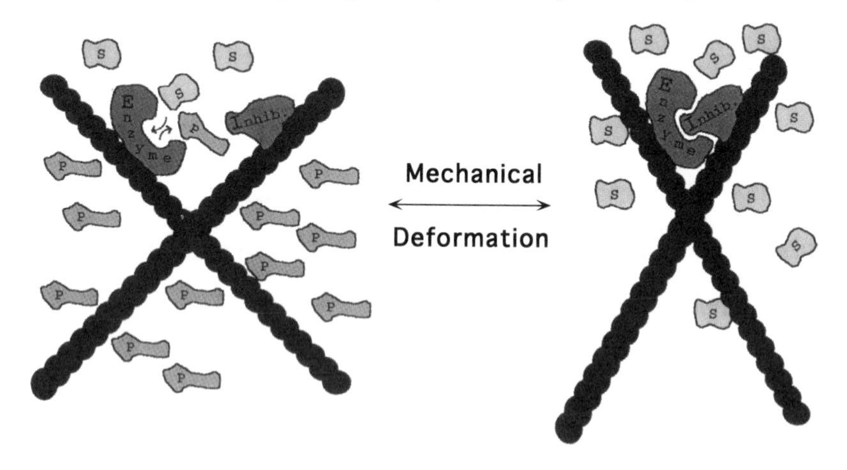

Figure 4. Mechanoregulation of Solid-State Biochemistry. Diagram showing that mechanical deformation of the cytoskeleton may alter cellular biochemistry by altering proximities between immobilized enzymes, substrates, and regulatory molecules. S, substrate; P, product; Inhib., enzyme inhibitor.

Thermodynamic Mechanisms

The tensegrity paradigm also provides a thermodynamic mechanism by which changes in the cellular force balance may feed back to regulate the assembly of the cytoskeletal filaments that comprise the orienting lattice of this solid-state regulatory system. For example, a thermodynamic model of microtubule polymerization which incorporates complementary force interactions between microtubules and the ECM according to the tensegrity paradigm has been published (Buxbaum and Heideman, 1988). In this model, microtubules stabilize specialized cell extensions and support cell spreading by physically resisting mechanical loads that are generated within the surrounding actin cytoskeleton. Direct evidence that MTs are under compression has been obtained in studies with neurites (Joshi et al. 1985). Changes in the force applied to microtubules also have been shown to regulate microtubule polymerization in these cells (Dennerll et al., 1988).

Interestingly, cells have evolved a mechanism to maintain a constant steady-state mass or "set-point" of monomeric tubulin; this process is known as tubulin autoregulation. Increases in the free tubulin monomer are compensated for by a concomitant decrease in tubulin protein synthesis which, in turn, results from selective destabilization of tubulin mRNA (Ben-Ze'ev et al., 1979; Cleveland et al., 1981). Maintaining a fixed set-point for tubulin monomer in this manner is consistent with the observation that cells also tend to contain a constant steady-state mass of tubulin polymer (Kirschner and Mitchison, 1986).

However, thermodynamic analysis of tubulin polymerization predicts that increasing the mechanical load on a microtubule will result in microtubule disassembly if the tubulin monomer set-point is held constant (Hill, 1981; Buxbaum and Heideman, 1988). This occurs because compressing a microtubule increases the critical concentration of tubulin that is required to maintain tubulin in a polymerized form. Tubulin subunits within the compressed polymer that were previously in equilibrium with free tubulin monomers would be released until the tubulin monomer concentration rises sufficiently to re-establish equilibrium, or the pre-existing force balance is regained. Thus, continued force application would be expected to promote rampant microtubule depolymerization if the concentration of tubulin monomer could not increase.

If this is true, then cells would lose the mechanical support provided by microtubules precisely when it is needed most. This paradox suggests that either our understanding of cellular tubulin regulation is incomplete, the tensegrity model is wrong, or a simple thermodynamic analysis of microtubule polymerization in cells that autoregulate tubulin synthesis is incorrect.

In an effort to reconcile these two contradictory cellular mechanisms for regulating tubulin mass, David Mooney, a graduate student with Bob Langer (MIT) who was working in my laboratory, measured changes in tubulin monomer and polymer levels that resulted from altering cell-ECM contact formation (Mooney et al. 1991). Primary rat hepatocytes were cultured on bacteriological petri dishes that were pre-coated with different densities of purified ECM molecules. We have previously shown that altering cell-ECM contacts in this manner promotes different degrees of cell extension and switches hepatocytes between growth and differentiation (Mooney et al., 1992a). He found that decreasing the cell's ECM adhesive contacts increased the steady-state mass of tubulin monomer, while the steady-state mass of microtubule polymer remained relatively constant. Importantly, tubulin autoregulation functioned normally in these cells, resulting in a decreased rate of tubulin synthesis in cells which contain a higher mass of tubulin monomer. This effect was offset, however, by a retardation of tubulin degradation, allowing different tubulin monomer masses to be maintained in cells with varying ECM contacts.

These results indicate that the set-point for the tubulin monomer mass in hepatocytes is regulated by the density of ECM contacts, and are consistent with a mechanism of microtubule regulation in which the ECM stabilizes microtubules by both accepting transfer of mechanical loads and altering tubulin degradation in cells that continue to autoregulate tubulin synthesis. Here is an elegant example of how chemical and mechanical signaling systems integrate inside living cells. This biomechanical perspective also resulted in identification in an entirely new form of biochemical regulation: control of tubulin protein half-life by mechanical signaling across integrins.

The Future

Implications for Cell Biology and Medicine

Much of our work is now focused on the molecular basis of mechanochemical transduction. We believe that this approach should lead to important insights into fundamental biological processes, including cell growth and migration as well as specialized forms of mechanoreception, such as gravity sensation. Specifically, we are using our magnetic twisting device to quantitate changes in intracellular biochemistry that result when controlled mechanical stresses are applied to cell surface integrin receptors. Biochemical responses to be measured include release of chemical second messengers, ion fluxes, cytoskeletal filament polymerization, nuclear transport, and gene expression. We are also trying to identify the molecular elements that mediate force transfer between integrins, the cytoskeleton and the nucleus.

In this manner, we hope to begin to understand how chemical and mechanical signals interplay to regulate cell form and function as well as gene expression. Because the cytoskeleton and nuclear matrix differ considerably between specialized cells, we believe that the molecular cell engineering approach also may lead to the identification of new molecular targets for therapeutic intervention in a wide variety of diseases.

Implications for Bioengineering

The combined architectural and molecular biological approach we have used opens up entirely new avenues of research in biomedical engineering. For example, based on the concept that alterations in cell mechanics that are associated with shape changes may serve as a target for cell regulation, we have begun to develop new methods for controlling cell growth and function for biotechnology and engineering applications. Joyce Wong, a graduate student with Bob Langer (MIT), and I have found that electrically conducting polymers may be used as a new type of tissue culture substrate. Most importantly in the present context, application of an electrical potential to these polymers provides a non-invasive means to control the shape and hence, function of adherent cells, independent of any medium alteration (Wong et al. 1994).

In studies carried out by Rahul Singhvi, a graduate student in the laboratory of Dan Wang (MIT) and Amit Kumar, a fellow in the laboratory of George Whitesides (Harvard), we have fabricated an elastomeric stamp, containing defined features on the micrometer scale, using photolithographic etching procedures. The stamp was then used to imprint gold surfaces with specific patterns of self-assembled monolayers of alkanethiols, and thereby, to create islands of defined shape and size that support ECM protein adsorption and cell attachment. Using this technique, it was possible to place cells in predetermined locations and arrays, separated by defined distances, and to dictate their shape. Literally, "square" cells with corners approximating 90° could be created. Most importantly, limiting the degree of cell extension provided control over both cell growth and secretion of specialized cell products (Singhvi et al. 1994). This approach may be very useful for automating drug screening and procedures in toxicology that utilize single-cell functional microassays; for applications in genetic engineering that require micromanipulation or microinjection, such as generation of transgenic animals; and for fabricating sheets of cells in defined patterns and shapes. Combination of this micropatterning approach with a method for fabricating synthetic bioerodible "ECM" scaffolds (Cima et al. 1991a,b), may provide a new way to engineer "artificial tissues" for use in organ transplantation and surgical reconstruction (Ingber 1993c).

Finally, based on the tensegrity paradigm, we are beginning to use continuum mechanics analysis and computerized architectural design techniques to characterize and model biological structures at the subcellular and molecular levels. Our long term goal is to develop a mathematical basis for the tensegrity model which would facilitate structural analysis of cells as well as the development of more rigorous design criteria for future engineering applications.

Acknowledgments

This work was supported by grants from N.I.H., American Cancer Society, and N.A.S.A.. Dr. Ingber is a recipient of a Faculty Research Award from American Cancer Society.

References

Ben-Ze'ev, A.; Farmer, S.R.; Penman, S. Mechanisms of regulating tubulin synthesis in cultured mammalian cells. Cell 17:319-325; 1979.

Burridge, K.; Fath, K.; Kelly, T.; Nucko, G.; Turner, C. Focal adhesions: transmembrane junctions between the extracellular matrix and cytoskeleton. Ann. Rev. Cell Biol. 4, 487-525; 1988.

Buxbaum, R.E.; Heideman, S.R. A thermodynamic model for force integration and microtubule assembly during axonal elongation. J. Theor. Biol. 134:379-390; 1988.

Cima, L.; Ingber, D.; Vacanti, J.; Langer, R. Hepatocyte culture on biodegradable substrates. Biotech. Bioengineer. 38:145-158; 1991a.

Cima, L.; Vacanti, J.; Vacanti, C.; Mooney, D.; Ingber, D.; Langer, R. Tissue Engineering by Cell Transplantation using degradable polymer substrates. J. Biomech. Engin. 113:143-151; 1991b.

Cleveland, D.W.; Lopata, M.A.; Sherline, P.; Kirschner, M.W. Unpolymerized tubulin modulates the level of tubulin mRNAs. Cell 25:537-546; 1981.

Dennerll, T.J.; Joshi, H.C.; Steel, V.L.; Buxbaum, R.E.; Heideman, S.R. Tension and compression in the cytoskeleton of PC-12 neurites II: quantitative measurements. J. Cell Biol. 107:665-674; 1988.

Dike, L.; Ingber, D.E.; Fibronectin activates growth-associated genes in endothelial cells. Mol. Biol. Cell 4: 299a; 1993.

Hansen, L.K.; Ingber, D.E.; Regulation of nucleocytoplasmic transport by mechanical forces transmitted through the cytoskeleton. In: Feldherr, C., ed. Nuclear Trafficking. San Diego: Academic Press; 1992:p. 71-86.

Hansen, L.K.,; Mooney, D.J.; Ingber, D.E. Regulation of cell cycle progression in hepatocytes by extracellular matrix and cell spreading. Mol. Biol. Cell 4: 348a; 1993.

Hill, T.L. Microfilament and microtubule assembly or disassociation against a force. Proc. Nat. Acad. Sci. U.S.A. 78:5613-5617; 1981.

Hynes, R.O. Integrins: versatility, modulation, and signaling in cell adhesion. Cell 69:11-25; 1992.

Ingber, D.E. Fibronectin controls capillary endothelial cell growth by modulating cell shape. Proc. Natl. Acad. Sci. U.S.A. 87:3579-3583; 1990.

Ingber, D.E. Integrins as mechanochemical transducers. Curr. Opin. Cell Biol. 3:841-848; 1991.

Ingber, D.E. The riddle of morphogenesis: a question of solution chemistry or molecular cell engineering? Cell 75:1249-1252; 1993a.

Ingber, D.E. Cellular Tensegrity: defining new rules of biological design that govern the cytoskeleton. J. Cell Sci. 104:613-627; 1993b.

Ingber, D.E. Extracellular matrix, cellular mechanics, and tissue engineering. In: Bell, E., ed. Tissue Engineering: Current Perspectives. Boston: Birkhauser; 1993c:p. 69-82.

Ingber, D.E.; Dike, L.; Hansen, L.; Karp, S.; Liley, H.; Maniotis, A.; McNamee, H.; Mooney, D.; Plopper, G.; Sims, J.; Wang, N. Cellular tensegrity: exploring how mechanical changes in the cytoskeleton regulate cell growth, migration, and tissue pattern during morphogenesis. Int. Rev. Cytol. 1994 - in press.

Ingber, D.E.; Folkman, J. Inhibition of angiogenesis through inhibition of collagen metabolism. Lab. Invest. 59:44-51; 1988.

Ingber, D.E.; Folkman, J. Mechanochemical switching between growth and differentiation during fibroblast growth factor-stimulated angiogenesis in vitro: role of extracellular matrix. J. Cell Biol. 109: 317-330; 1989a.

Ingber, D.E.; Folkman, J. How does extracellular matrix control capillary morphogenesis? Cell 58:803-807; 1989b.

Ingber, D.E.; Madri, J.A.; Folkman, J. Extracellular matrix regulates endothelial growth factor action through modulation of cell and nuclear expansion. In Vitro Cell Dev. Biol. 23:387-394; 1987.

Ingber, D.E.; Madri, J.A.; Jamieson, J.D. Basement membrane as a spatial organizer of polarized epithelia: exogenous basement membrane reorients pancreatic epithelial tumor cells in vitro. Am. J. Pathol. 122:129-139;1986

Ingber, D.E.; Prusty, D.; Frangione, J.; Cragoe, E.J., Jr.; Lechene, C.; Schwartz, M. Control of intracellular pH and growth by fibronectin in capillary endothelial cells. J. Cell Biol. 110:1803-1812; 1990.

Joshi, H.C.; Chu, D.; Buxbaum, R.E.; Heideman, S.R. Tension and compression in the cytoskeleton of PC 12 neurites. J. Cell Biol. 101:697-705; 1985.

Kirschner, M.; Mitchison, T. Beyond self-assembly: from microtubules to morphogenesis. Cell 45:329-342; 1986.

McMahon, T.A. In: Muscles, Reflexes, and Locomotion. Princeton Univ. Press, Princeton, N.J.; 1984.

McNamee, H.; Ingber, D.; Schwartz, M. Adhesion to fibronectin stimulates inositol lipid synthesis and enhances PDGF-induced inositol lipid breakdown. J. Cell Biol. 121:673-678; 1993.

McNamee, H.P.; Ingber, D.E. Integrin-dependent activation of phosphoinositide kinases occurs within the focal adhesion complex. Mol. Biol. Cell 4:362a;1993.

Mooney, D.; Hansen, L.; Farmer, S.; Vacanti, J.; Langer, R.; Ingber, D. Switching from differentiation to growth in hepatocytes: control by extracellular matrix. J. Cell Physiol. 151:497-505; 1992a.

Mooney, D.J.; Langer, R.; Hansen, L.K.; Vacanti, J.P.; Ingber, D.E. Induction of hepatocyte differentiation by the extracellular matrix and an RGD-containing synthetic peptide. Proc. Mat. Res. Soc. 252:199-204; 1992.

Mooney, D.J.; Langer, R.; Ingber, D.E. Intracellular tubulin monomer levels are controlled by varying cell-extracellular matrix contacts. J. Cell Biol. 115:38a; 1991.

Plopper, G.; Ingber, D.E. Rapid induction and isolation of focal adhesion complexes. Biochem. Biophys. Res. Commun. 193:571-578; 1993.

Plopper, G.; Schwartz, M.A.; Chen, L.B.; Lechene, C.; Ingber, D.E. Binding of fibronectin induces assembly of a chemical signaling complex on the cell surface. J.Cell Biol. 115:130a; 1991.

Schwartz, M. Transmembrane signaling by integrins. Trends in Biol. Sci. 2:304-308; 1992.

Schwartz, M.A.; Lechene, C.; Ingber, D.E. Insoluble fibronectin activates the Na+/H+ antiporter by clustering and immobilizing integrin $\alpha 5 \beta 1$, independent of cell shape. Proc. Natl. Acad. Sci. U.S.A. 88:7849-7853; 1991.

Sims, J.; Karp, S.; Ingber, D.E. Altering the cellular mechanical force balance results in integrated changes in cell, cytoskeletal, and nuclear shape. J. Cell Sci. 103:1215-1222; 1992.

Singhvi, R.; Kumar, A.; Lopez, G.; Stephanopoulos, G.N.; Wang, D.I.C.; Whitesides, G.M., Ingber, D.E. Engineering cell shape and function. Science 1994 - revised ms in review.

Stamenovic, D. Micromechanical foundations of pulmonary elasticity. Physiol. Rev. 70:1117-1140; 1990.

Wang, N.; Butler, J.P.; Ingber, D.E. Mechanotransduction across the cell surface and through the cytoskeleton. Science 260: 1124-1127; 1993.

Wang, N.; Ingber, D.E. Control of cytoskeletal mechanics by extracellular matrix, cell shape, and mechanical tension. Biophys. J. 1994 - revised ms in review.

Wong, J.Y.; Langer, R.S.; Ingber, D.E. Electrically conducting polymers can non-invasively control the shape and growth of mammalian cells. Proc. Natl. Acad. Sci. U.S.A. 1994 - in press.

Part VI

Physical Regulation of Tissue Metabolic Activity

20

Stress, Strain, Pressure and Flow Fields in Articular Cartilage and Chondrocytes

V.C. Mow, N.M. Bachrach, L.A. Setton, F. Guilak

Introduction

Articular cartilage serves as the load-bearing material of joints, with excellent friction, lubrication and wear characteristics (Mow *et al.* 1992a). Under normal physiological conditions, these essential biomechanical functions are provided with little or no degenerative changes over the lifetime of a human joint. However, biomechanical factors such as excessively high impact loads, repetitively applied loads, joint immobilization and instability, and abnormal range of motion can alter the composition, structure and material properties of articular cartilage (e.g., Armstrong *et al.* 1985; Caterson and Lowther 1978; Donohue *et al.* 1983; Helminen *et al.* 1987; Palmoski *et al.* 1979-1981; Vener *et al.* 1992). These changes in the biochemical composition and material properties are due, in part, to the alterations in the metabolic activities of the chondrocytes attempting to remodel articular cartilage in an effort to adapt to their new biomechanical environment (e.g., Helminen *et al.* 1987; Howell *et al.* 1992; Stockwell 1979).

Many *in vivo* and *in vitro* studies have demonstrated that mechanical loading is required for the normal maintenance of articular cartilage (e.g., Helminen *et al.* 1987 and 1992; Jones *et al.* 1982; Jurvelin *et al.* 1989; Kiviranta *et al.* 1987; Palmoski *et al.* 1979-1984; Sah *et al.* 1989 and 1991; Setton *et al.* 1992a; Tammi *et al.* 1987). The ability of the chondrocytes to modulate their metabolic activity in response to their mechanical environment provides a means by which articular cartilage can alter its structure and composition, and hence regulate its material properties and biomechanical function in response to the physiological demands of the body. Thus, the mechanical environment of the chondrocytes is an important factor affecting the health of articular cartilage and consequently the function of the diarthrodial joint, and the progression of joint degeneration (Howell *et al.* 1992; Mankin and Brandt 1992; Mow *et al.* 1992a).

Structure and Composition of Articular Cartilage

Articular cartilage consists of a solid matrix saturated with water and dissolved salts. The solid matrix contains mostly type II collagen and aggregating proteoglycans, and a number of other functionally important, though quantitatively minor collagen types (e.g., type IX) and glycoproteins (e.g., biglycan and decorin). The assemblage of these biomacromolecules comprises the solid matrix and contributes to ~ 20% of the tissue by wet weight (Buckwalter and Rosenberg 1982; Eyre 1990; Muir 1977 and 1983; Wight *et al.* 1991). The fluid phase represents the remaining ~ 80% of the tissue wet weight, and consists of water and dissolved salts (e.g., Na^+, K^+, Ca^{++}, SO_3^{--}, PHO_4^{--}; Linn and Sokoloff, 1965). In purely mechanical terms, articular cartilage can be considered as a mixture of two phases: a porous-permeable solid phase and a freely flowing fluid phase (e.g., Edwards 1967; Mow *et al.* 1980-1992b). The solid matrix of articular cartilage has a highly organized distribution of collagen and proteoglycans through the tissue depth (e.g., Broom and Poole 1983; Bullough and Goodfellow 1968; Clark 1985; Clarke 1971; Hunziker and Schenk 1987; Muir *et al.* 1970). The ultrastructural organization of the tissue provides cartilage with its characteristic heterogeneous and anisotropic mechanical behaviors (e.g., Akizuki *et al.* 1986; Broom and Myers 1980; Kempson *et al.* 1973 and 1977; Myers *et al.* 1984; Roth and Mow 1980; Woo *et al.* 1976).

Mechanical and Swelling Behavior of Articular Cartilage

Numerous interdisciplinary studies of composition, structure and material properties have been performed to determine how the macromolecular components of the extracellular matrix (ECM) provide articular cartilage with its mechanical properties in tension (e.g., Akizuki *et al.* 1986; Kempson *et al.* 1973; Roth and Mow 1980; Schmidt *et al.* 1990; Woo *et al.* 1976), compression (Armstrong and Mow, 1982; Athanasiou *et al.* 1991; Hayes and Mockros 1971; Kempson *et al.* 1971; Mow *et al.* 1980), and shear (Hayes and Bodine 1978; Spirt *et al.* 1989; Woo *et al.* 1987; Zhu *et al.* 1993). In addition, the composition and structure of articular cartilage also provides the physicochemical mechanisms which regulate tissue swelling, and fluid and ion transport properties (Edwards 1967; Eisenberg and Grodzinsky 1985; Gu *et al.* 1992 and 1993; Lai *et al.* 1991; Mansour and Mow 1976; Maroudas 1968-1979; Mow *et al.* 1980-1992b; Myers *et al.* 1984, Setton *et al.* 1993a and 1993b). These swelling, fluid and ion transport properties, together with the intrinsic mechanical properties of the collagen-proteoglycan matrix, ultimately govern the complete deformational behavior of cartilage. In general, the manner by which forces or deformations are applied to the tissue determines the characteristics of the stress, strain and fluid flow fields within the tissue as well as the magnitude and distribution of these quantities (see Mow *et al.* 1992a and 1992b for detailed reviews of the mechanical behavior of articular cartilage and load support in joints).

Chondrocytes of Articular Cartilage

During skeletal development articular cartilage forms from densely packed mesenchymal cells that differentiate into chondrocytes, proliferate rapidly and synthesize large volumes of ECM. As the tissue matures during skeletal growth, the cells become separated from each other and encased within a dense interterritorial matrix (Ogden 1979). Eventually, chondrocytes within adult articular cartilage only occupy 1-10% of the tissue by volume (Figure 1), and exhibit a heterogeneous variation in volume and morphology with depth through the cartilage layer (Stockwell, 1971-1979). Since adult articular cartilage is avascular, the viability and metabolic activity of the chondrocytes depend on the diffusion and convective transport of nutrients to, as well as the disposal of waste products from, the cells through the porous-permeable solid matrix (Gu *et al.* 1992 and 1993; Maroudas 1968 and 1979; Mansour and Mow 1976; Mow *et al.* 1980-1992b).

Despite their relative sparsity and limited nutrient supply, the chondrocytes are capable of successfully maintaining the articular cartilage throughout life in the intensely loaded mechanical environment found in diarthrodial joints (Mow *et al.* 1992a). Indeed, the entire organization and structure of the ECM and hence the material properties of the tissue are maintained and regulated by the anabolic and catabolic activity of the chondrocyte population. In normal cartilage, chondrocyte

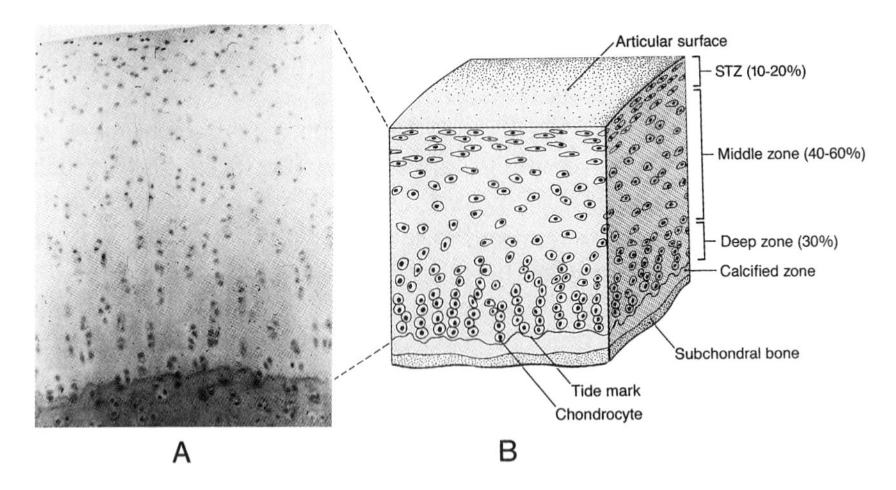

Figure 1. Photomicrograph (A) and schematic diagram (B) of chondrocytes in articular cartilage, showing the zonal variation in cell morphology and distribution between the surface tangential zone (STZ), middle zone and deep zone. Note that chondrocytes in articular cartilage occupy less than 10% of the solid matrix volume. (Reprinted with permission from Mow VC *et al.*: *Basic biomechanics of the musculoskeletal system* Lea and Febiger, 2nd edition, 1989, p.32).

metabolic activity is regulated by both genetic and environmental factors which include soluble mediators (cytokines, growth factors and pharmaceutical agents), matrix composition (Howell *et al.* 1992; Mankin and Brandt 1992; Morales 1992; Mow *et al.* 1992a; Muir 1977 and 1983), as well as mechanical factors associated with physiological joint loading, such as applied tractions and hydrostatic pressure (Gray *et al.* 1988 and 1989; Guilak *et al.* 1994c; Hall *et al.* 1989 and 1991; Jones *et al.* 1982; Palmoski *et al.* 1979-1984; Sah *et al.* 1989 and 1991; Schneiderman *et al.* 1986; Tammi *et al.* 1987 and 1991). The ability of chondrocytes to perceive stimuli from their mechanical environment provides a means for the tissue to respond to changes in the physiological loading demands of the joint.

Chondrocytes as Mechanical Inclusions

Since the chondrocytes are embedded and attached to the extracellular (solid) matrix, the states of stress and strain they experience will be directly influenced by the deformation within the surrounding tissue. In mechanics this situation (i.e., a body of one material embedded within another material) describes an "inclusion" problem for which mathematical analyses are quite difficult. For articular cartilage, an

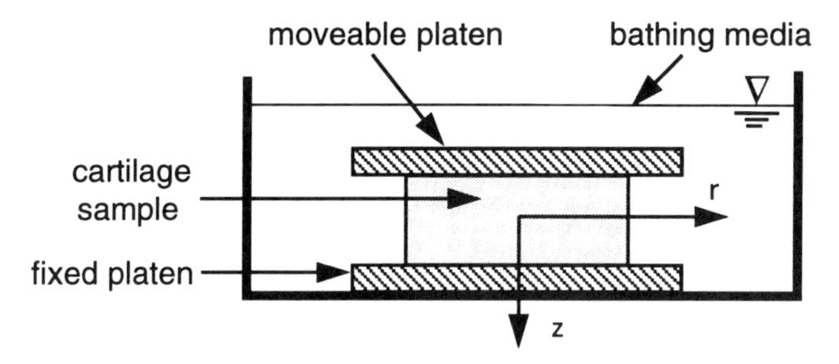

Figure 2. Schematic diagram of the unconfined compression testing configuration. In a creep experiment, a constant compressive load will be applied to the top platen and the resulting translation of the platen will be monitored. The direction of load application corresponds to the z-direction shown. Note that the platens are rigid and may be permeable or impermeable.

inclusion problem is even more formidable due not only to the intricate 3D morphology of the chondrocytes and their distributions within the tissue, but also to the largely unknown mechanical properties of the cells and the complex nature of the ECM. In order to take a first step towards analyzing the complex problem of chondrocyte deformation within the ECM, a number of simplifying assumptions may be made both experimentally and theoretically. In the studies presented in this chapter, the cartilage explant system will be analyzed as a two-dimensional, axisymmetric testing configuration (Figure 2) in order to facilitate theoretical modeling. The material properties of the ECM will be considered to be homogeneous, with the exception of the cellular inclusions. Finally, the chondrocyte shape will be assumed to be spherical and loaded in such a manner as to produce an axisymmetric deformational state. This set of assumptions provides a tractable mathematical theory which may be solved analytically or by finite element analysis. The goal of using a mathematical analysis of the cartilage explant and chondrocyte is to develop a theory which can be used to predict the states of stress, strain, pressure (hydrostatic and osmotic), interstitial flow and electric fields within the ECM and in the environment of the cell during *any* prescribed experimental loading. This type of precise knowledge of the mechano-electrochemical events in cartilage is required in order to provide insight into possible signal transduction mechanisms and to understand the response of the chondrocytes to distinct mechano-electrochemical stimuli.

In this chapter, we begin with a review of what is known of the relationships between mechanical loading and chondrocytic response which have been determined from a number of *in vitro* cartilage explant studies. We then present a detailed analysis of the states of stress, strain, interstitial fluid-flow and pressure fields in articular cartilage in several often used testing configurations. These theoretical predictions are based on the biphasic theory for articular cartilage developed by Mow *et al.* (1980) to describe hydrostatic loading and static and cyclical loading in an unconfined compression experiment (Armstrong *et al.* 1984; Guilak *et al.* 1988 and 1990b; Spilker *et al.* 1990). The resulting temporal and spatial variations in the magnitudes of these fields are then used to calculate the surface tractions acting on chondrocytes *in situ* within the ECM. In addition, some results of a finite element analysis of the chondrocyte as a biphasic inclusion are presented in order to investigate more precisely a potential mechanical role for the chondrocyte deformation *in situ* (Guilak *et al.* 1990a). Finally, the results of experimental measurements of the deformation of live chondrocytes in mechanically-loaded explant culture systems will be presented, based on a confocal microscopy technique for quantitative measurement of the morphology of the chondrocyte within the ECM (Guilak *et al.* 1992 and 1994a,e). This body of knowledge will serve as the foundation for interpreting the results of *in vitro* studies on the modulation of chondrocytic metabolic activity in loaded explants.

Mechanical Regulation of Chondrocyte Activity

In vivo and *in vitro* studies have been reported on the effects of mechanical loads on the metabolic activity of the chondrocyte. Investigations of cartilage response to altered patterns of joint loading *in vivo* have included models of surgically-induced instability and altered physical activity ranging from immobilization to strenuous physical exercise (Caterson and Lowther 1978; Guilak *et al.* 1994d; Helminen *et al.* 1992; Jurvelin *et al.* 1989; Moskowitz 1992; Palmoski *et al.* 1979-1982; Ratcliffe *et al.* 1993 and 1994; Setton *et al.* 1992a and 1994; Tammi *et al.* 1987). These *in vivo* studies have provided strong evidence that mechanical factors can induce changes in the composition, morphology, metabolic activity and mechanical properties of articular cartilage and other joint tissues. In general, however, these studies are limited by an inability to precisely determine the magnitude and distributions of the loading that cartilage experiences *in vivo*, so that a determination of the relevant mechanical signals to the chondrocyte *in vivo* is similarly limited.

Other *in vitro* studies have been designed to examine the metabolic response of cartilage to load in explant culture systems (Gray *et al.* 1988 and 1989; Guilak *et al.* 1994c; Hall *et al.* 1991; Jones *et al.* 1982; Sah *et al.* 1989 and 1991; Schneiderman *et al.* 1986). These studies have the advantage of providing for a precise control of the biomechanical and biochemical environments within the explants during testing. These studies have been used to isolate particular mechanical stimuli and their effects on chondrocytic activity. For example, experiments have been designed to study the effects of hydrostatic pressure (e.g., Hall *et al.* 1991; Parkkinen *et al.* 1993a) in which matrix deformations are negligible (Guilak *et al.* 1988). Others have studied static or cyclic loading regimes in order to compare the metabolic response of cartilage in an equilibrium condition where no fluid-flow or fluid pressure is expected (i.e., static) to that in which interstitial fluid-flow and fluid pressure may be an important factor (i.e., cyclic). In this section we present a short review of some of the major experimental findings of these *in vitro* culture studies.

Hydrostatic Pressure

It has been shown that the high loads transmitted across conforming joints (e.g., ankle, hip) during weight bearing are predominantly supported via the hydrostatic pressure developed within the interstitial fluid (Mow *et al.* 1992b; Ateshian *et al.* 1994). This load carriage mechanism may be important for regulating the mechanical signaling and metabolic response of chondrocytes in the articular layer. Several researchers have studied changes in the biochemical composition of cartilage explants subjected to hydrostatic pressure by pressurizing the fluid bath surrounding the tissue. The synthesis of proteoglycan, protein and RNA have been shown to increase at intermediate levels of hydrostatic pressure (2.6-5 MPa) (Kaye *et al.*1980, Lippiello *et al.* 1985, Parkkinen *et al.* 1993a). In contrast, lower pressures (0.35-2.1 MPa) have been found to decrease, or produce no change in chondrocytic activity (Kimura *et al.* 1985; Lippiello *et al.* 1985). At higher and *non-physiologic* levels of pressure (20 or 50 MPa), a decrease in proteoglycan synthesis has been observed which did not revert to control levels after removal of the hydrostatic pressure (Hall *et al.* 1991). However, in this study short durations of high pressure loading (20s) did stimulate elevated proteoglycan synthesis. These studies demonstrate that the chondrocyte is capable of perceiving hydrostatic pressure changes, although neither the mechanism of signal perception, nor the precise threshold of signal magnitude required for stimulation are known.

Static Compression

An often used protocol in cartilage explant studies is the application of compressive loads to a radially unconfined cylindrical specimen of articular cartilage (Figure 2). Static compression under loads corresponding to 0.001-3 MPa of compressive stress has been shown to cause a decrease in proteoglycan synthesis and release from the ECM (Gray *et al.* 1988; Guilak *et al.* 1994c; Jones *et al.* 1982; Palmoski and Brandt 1984), as well as a decrease in protein synthesis (Gray *et al.* 1988). In some cases, a "rebound" phenomenon has been reported, with elevated levels of biosynthetic activity after removal of the compressive load (Burton *et al.* 1993; Guilak *et al.* 1994c). Studies investigating the rate of release of proteoglycans from cartilage explants have shown that the magnitude of compressive load can similarly influence the catabolic activity of proteoglycans (Guilak *et al.* 1994c; Jones *et al.* 1982).

Dynamic Compression

In order to determine if cyclic loading conditions, corresponding to physiological loading, are themselves stimulatory in cartilage, a number of investigators have studied the effects of dynamic compression on unconfined cartilage explants. These studies indicate that application of compressive loads in a sinusoidal or intermittent manner can stimulate biosynthetic activity (Burton *et al.* 1993; Parkkinen *et al.* 1992; Palmoski and Brandt 1984; Sah *et al.* 1989 and 1991; Tammi *et al.* 1991). These effects show a strong dependence on the magnitude and frequency of the applied stresses. Palmoski and Brandt (1984), using an intermittent, unconfined loading configuration with impermeable loading platens, observed a significant suppression of proteoglycan synthesis rates at a duty cycle of 60s on/ 60s off (0.0083 Hz) and for compressive stresses of 0.001-0.011 MPa. At a duty cycle of 4s on/ 11s off (0.0667 Hz), stresses of 0.001-0.006 MPa resulted in a net stimulation of proteoglycan synthesis, leading to the conclusion that slow duty cycles cause a reduction in proteoglycan synthesis while more rapid duty cycles promote stimulation. This conclusion was supported by later studies of proteoglycan and protein synthesis at a higher frequency range of 0.25-0.50 Hz (Larsson *et al.* 1991; Tammi *et al.* 1991). A number of studies have applied compressive deformations in a cyclic manner to cartilage explants (i.e., 1-7% @ 0.0001-0.5 Hz) with similar findings for the altered biosynthesis (Copray *et al.* 1985; Sah *et al.* 1989). In addition, Copray *et al.* (1985) found a decrease in cell proliferation at 10 and 20%

compression at 0.7Hz. Parkkinen *et al.* (1992) examined proteoglycan synthesis rates in cartilage explants which were compressed at 0.5-1.0 MPa using an indenting head with a diameter of half that of the explant. These studies showed that loads applied for 50ms at intervals of 2s (~0.5 Hz) and 4s (~0.25 Hz) stimulated proteoglycan synthesis in the loaded region, while loading intervals of 20s (0.05 Hz) or 60s (0.0167 Hz) did not have a significant effect.

All these results indicate that distinct external mechanical stimuli can produce changes in the metabolic events within the tissue. However, it is essential to recognize that the results of these experiments, as with all explant studies, represent the collective response of all chondrocytes within the tissue. The heterogeneities in the composition, molecular ultrastructure and chondrocyte morphology within articular cartilage suggest that the mechanical and physicochemical changes that the chondrocytes experience during tissue loading will significantly differ with depth from the articular surface (Aydelotte *et al.* 1988; Gore *et al.* 1983; Guilak *et al.* 1992 and 1994e; Zanetti *et al.* 1985). In addition, theoretical analyses of the internal pressure, deformation and fluid-flow fields in articular cartilage based on the biphasic theory indicate that the mechanical environment of the chondrocyte will vary both with time and with position in the cartilage explant (Armstrong *et al.* 1984; Guilak *et al.* 1990b and 1994e; Mow *et al.* 1980-1992b; Suh and Li 1993). These variations are expected to differ depending on the loading configuration as well as the mechanical properties of the tissue. All of these considerations complicate the determination of a precise relationship between mechanical signal and chondrocytic response.

Chondrocytes and Mechanical Signal Transduction

It is evident from the wealth of *in vivo* and *in vitro* experiments that the mechanical environment in articular cartilage plays a critical role in regulating the metabolic activity of the chondrocytes. However, the mechanisms by which these cells control rates of synthesis and catabolism (e.g., regulation of specific gene transcription, the maintenance of a level of steady state mRNA, post-translational effects, activation and deactivation of the enzymes responsible for matrix breakdown) remain to be determined. One major difficulty encountered in isolating the mechanisms of mechanical signal transduction in articular cartilage at present is the inability to precisely characterize the micro-mechanical environment around chondrocytes, as explained above. In addition, due to the charged, hydrated nature of the ECM, other mechanical, physicochemical and electrical effects may result from mechanically loading tissue. These phenomena include spatially varying changes in matrix deformation, interstitial fluid flow, hydrostatic and osmotic pressures, and electrokinetic effects such as streaming potentials and streaming currents (Broom and Myers 1980; Eisenberg and Grodzinsky 1985; Frank and Grodzinsky 1987; Gu *et al.* 1992 and 1993; Guilak and Mow 1992; Lai *et al.* 1991, Maroudas 1979; Mow *et al.* 1980-1992b). Associated with the deformation of the ECM are other effects such as deformation of the chondrocytes, an increase in the pericellular concentrations of proteoglycan and fixed charge density, as well as changes in the interstitial ion concentrations and pH (Gray *et al.* 1988; Guilak 1990a, 1992 and 1994e; Guilak and Mow 1992; Lai *et al.* 1991; Maroudas 1979; Warden *et al.* 1994). Currently, there is evidence that any (or all) of these phenomena are involved in the signal transduction mechanisms of the chondrocyte.

In our proposed model of chondrocytes as inclusions in the ECM, some possible mechanisms of signal transduction associated with the stress-strain environment around the cell are of particular significance. Several proposed signal transduction pathways in the literature are consistent with the predictions and observations that cells deform when the ECM is deformed (Ben Ze'ev 1991; Getzenberg *et al.* 1990; Guilak *et al.* 1994b; Ingber 1991; Watson 1991). One suggested pathway implicates the role of the cytoskeleton in transmitting extracellular deformation to intracellular organelles (Getzenberg *et al.* 1990; Ingber 1991). This paradigm suggests that mechanical and physical changes in the ECM or stretching of the cell membrane may be transduced into metabolic events via integrins and microfilament-associated proteins which are physically linked to the ECM. Conceivably, these mechanical signals could regulate gene expression through a direct physical connection from the ECM to the genome via nuclear matrix proteins. Alternatively, changes in cell shape or size could initiate intracellular signals through stretch-activated and stretch-inactivated ion channels on the plasma membrane (Sachs 1991). Cellular deformation, and associated membrane deformation, may therefore provide a means of directly regulating second messenger activity (Guilak *et al.* 1994b; Horoyan *et al.* 1990; Sachs 1991; Watson 1991). The transport of even a relatively small number of ions may have a substantial biologic effect through second messengers such as cyclic adenosine monophosphate, inositol triphosphate, or calcium ion (Ca^{++}). All these studies point to the need for precisely quantifying and isolating the various specific mechanical effects in the ECM of articular cartilage which may influence chondrocyte synthetic and catabolic activities.

Mechanical Environment within the Extracellular Matrix

When a load is applied to the articular surface of a joint, complex distributions of tensile, shear and compressive stresses, and flow and pressure fields are generated within the cartilaginous layer. The analyses for these fields are complex, but similar to those found in contact mechanics problems. However, in diarthrodial joints, these analyses are between two biphasic cartilage layers of differing properties (Ateshian *et al.* 1994; Mow *et al.* 1992a). As a mixture of solid and fluid phases, cartilage responds to the applied load through two primary mechanisms: 1) *deformation* of the porous-permeable solid matrix which leads to increased contact areas and decreased contact stresses (Mow *et al.* 1992a and 1992b); and 2) interstitial fluid flow and exudation thus enhancing lubrication at the articulating surface (Dowson *et al.* 1981, Hou *et al.* 1990; Jin *et al.* 1992). These deformation and recovery mechanisms are time-dependent (or viscoelastic) since they depend on the time required for interstitial fluid movements. Movement of the interstitial fluid, as well as relative movements between macromolecules of the solid matrix, give rise to dissipative and viscoelastic effects such as creep, stress-relaxation and hysteresis (Armstrong *et al.* 1984; Hayes and Mockros 1971; Hayes and Bodine 1978; Holmes *et al.* 1984; Mow *et al.* 1980-1992b; Setton *et al.* 1993b; Zhu *et al.* 1993), and significant electrokinetic effects such as streaming potential and streaming current, and electroosmosis (Frank and Grodzinsky 1987; Gu *et al.* 1992 and 1993). In carefully crafted *in vitro* experiments, it is possible to create simplified states of stress or deformation which lead to a precise characterization of the mechanical environment within the cartilage explant. The most important goal of this study is to develop a detailed understanding of the mechanical signals transmitted to the cells in articular cartilage during the various loading conditions commonly used in explant studies. In the sections below we present a detailed analysis of the internal stress, strain, pressure, and flow fields that exist when the tissue is subjected to hydrostatic loading, uniaxial static compression and uniaxial dynamic compressive loading.

Constitutive Modeling of Articular Cartilage

In order to understand the fluid-solid interactions occurring within cartilage, a binary mixture theory, with accompanying experiments, were developed by Mow and co-workers (1980) to describe articular cartilage. The basis for this binary theory is the well known mixture theory developed in rational continuum mechanics (Bowen 1976; Crane *et al.* 1970). These theories have significant similarities with the poroelasticity theory of Biot (1941), although they are not the same. In the biphasic formulation for articular cartilage, the tissue is modeled as a mixture of two incompressible phases: a solid phase, representing collagen and proteoglycan, and a fluid phase, representing the interstitial fluid and dissolved electrolytes. Tissue compressibility occurs through volume gain or loss due to fluid exudation or imbibition. In the most general case, the theory includes a solid phase which is nonlinearly permeable, anisotropic, inhomogeneous and viscoelastic undergoing finite deformation, and a fluid phase which is dissipative. A hyperelastic formulation for the solid matrix has also been derived (Holmes and Mow 1990). Diffusional couples may also exist in the general formulation. Major advances were made in the 1980s and early 1990s toward the development of nonlinear finite deformation biphasic theories to describe articular cartilage. These nonlinear theories included: 1) the nonlinear strain dependent permeability effect (Holmes *et al.* 1985; Lai and Mow 1980; Lai *et al.* 1981); and 2) finite-deformation effects (Mow *et al.* 1986; Holmes and Mow, 1990). These are effects of great importance since cartilage is likely to undergo large deformations *in situ.*

As a first approximation of the mechanical environment within the ECM, one may consider the special case of a biphasic medium with a solid matrix which is isotropic, homogeneous and linearly elastic. Furthermore, if the solid phase is subject to infinitesimal deformations, the constitutive equations for the solid matrix stress $\mathbf{T}^s$, fluid stress $\mathbf{T}^f$ and the interaction force between the two phases π are given by:

$$\mathbf{T}^s = -\phi^s p\mathbf{I} + \lambda_s \mathrm{tr}(\mathbf{E})\mathbf{I} + 2\mu_s\mathbf{E} \tag{1}$$

$$\mathbf{T}^f = -\phi^f p\mathbf{I} - \frac{2}{3}\mu_a(\mathrm{div}\ \mathbf{v}^f)\mathbf{I} + 2\mu_s\mathbf{D} \tag{2}$$

$$\pi = p\nabla\phi^f \mathbf{I} + K(\mathbf{v}^f - \mathbf{v}^s) \tag{3}$$

where ϕ^s and ϕ^f are solidity and porosity, λ_s and μ_s are Lame constants for the solid phase, μ_a is the apparent viscosity of the fluid phase in cartilage, **E** is the infinitesimal strain tensor for the solid matrix, **D** is the rate of deformation tensor for the interstitial fluid, $\mathbf{v}^s$ and $\mathbf{v}^f$ are velocity of solid and fluid phases respectively, and K is the coefficient of the diffusive drag caused by the relative motion between the two phases. This latter coefficient is related to the permeability coefficient k by the equation $k = (\phi^f)^2/K$ (Lai and Mow 1980). In addition, the incompressibility of the phases is given by the equation:

$$\text{div}\,(\phi^s\mathbf{v}^s + \phi^f\mathbf{v}^f) = 0. \tag{4}$$

For the special case where the fluid is assumed to be inviscid, these equations, together with the momentum equations have been used to study the biphasic viscoelastic behaviors of articular cartilage under mechanical loading conditions (Mow *et al.* 1980-1992b). Indeed mathematical solutions for several commonly used experimental configurations have been obtained and used to determine the material properties. These include the uniaxial permeation problem (Lai and Mow 1980), the confined compression creep and stress relaxation problem (Mow *et al.* 1980), the unconfined compression creep and stress relaxation (Armstrong *et al.* 1984) and the biphasic indentation creep (Mak *et al.* 1987; Mow *et al.* 1989). The solution for the indentation creep test has been successfully used to determine simultaneously the two elastic constants and the permeability k of articular cartilage from a single experiment (Athanasiou *et al.* 1991). Using such an algorithm for tissue samples taken from the bovine patellar groove samples, Mow and co-workers (1989) obtained the following material parameters: Poisson's ratio, $\nu = 0.25 \pm 0.07$, aggregate modulus, $H_A = 0.47 \pm 0.15$ MPa and permeability, $k = 1.42 \pm 0.58 \times 10^{-15}$ m^4/Ns.

The viscoelastic behavior of normal articular cartilage in compression is due predominantly to the frictional drag associated with interstitial fluid flow. By using the biphasic theory, the mechanism through which such viscoelasticity manifests itself can be easily understood. For example in the confined compression stress-relaxation test, results of the biphasic theoretical analysis demonstrate how the movement of interstitial fluid controls the stress history. During the ramp compression phase of a stress-relaxation test, fluid exudation gives rise to a peak stress due to the diffusive resistance of the solid matrix to fluid flow. During the relaxation phase, fluid redistribution gives rise to a "relief" of the compacted region at the surface and hence, stress relaxation. For other tissues, e.g., the cartilaginous end-plate of the intervertebral disc, or degenerated articular cartilage, the permeability may be naturally high. This high permeability may negate the large drag force due to interstitial fluid flow, and thus diminish the biphasic viscoelastic effect. Under this circumstance, the viscoelasticity of the collagen-proteoglycan solid matrix begin to play an important role in governing tissue behavior (Setton *et al.* 1993b; Zhu *et al.* 1993).

Hydrostatic Pressure

Both solid and fluid phases of the tissue have been shown to be effectively incompressible (Guilak *et al.* 1988). A direct consequence of this result is that even at high pressures (Hall and Urban 1989; Hall *et al.* 1991), a state of zero deformation and strain will exist throughout a biphasic material, such as the ECM of articular cartilage. However, the state of stress in the tissue is non-zero, isotropic and uniform throughout the matrix, with the hydrostatic pressure being supported by both solid and fluid phases. The surface tractions (loads) experienced by a spherical cell in this environment are isotropic, i.e., they act normal to the cell membrane. Such a hydrostatic pressure loading mechanism has been shown to affect chondrocyte metabolism, as described above. Suggested pathways for this metabolic effect are through alterations of the Na^+/K^+ membrane pump (Hall *et al.* 1991) or through a cytoskeletally-mediated pathway (Parkkinen *et al.* 1993b, Wright *et al.* 1992).

Static Compression

In the "static loading" explant experiment, a compressive load or displacement is applied to a cylindrical explant of articular cartilage via a rigid platen, which may be permeable or impermeable (Figure 2). This corresponds to the unconfined compression experiment which has been analyzed using the biphasic theory for either permeable or impermeable platens (Armstrong *et al.* 1984, Spilker *et al.* 1990). Using this biphasic analysis and characteristic material properties for articular cartilage (i.e., compressive modulus H_A = 0.45 MPa, Poisson's ratio ν_s = 0.1 and hydraulic permeability $k = 1 \times 10^{-14} m^4/N \cdot s$), the internal fluid-pressure fields and interstitial fluid-flow fields were calculated for a cartilage explant subjected to step application of a compressive load (-0.022 MPa). The mechanical behavior in response to this load can be divided into three phases: 1) an instantaneous response at $t=0^+$ involving a change in shape of the tissue without a volumetric change; 2) a transient period during which the solid matrix deformation increases (i.e., creep) as the interstitial fluid exudes from the free surfaces of the explant; and 3) an equilibrium state as $t \rightarrow \infty$ when there is no more fluid movement within or across the tissue. The volumetric change of the tissue under these compressive loading conditions equals the fluid lost from the tissue.

Internal Flow and Pressure Fields

The biphasic theory provides precise descriptions of all mechanical quantities at any time and any position within the sample. To illustrate the necessity for such an analysis and demonstrate its application we discuss here only the interstitial fluid flow and pressure. As shown in Figure 3, the maximum interstitial fluid pressure is attained for short times after loading in the center of the explant. The gradient in fluid pressure across the cartilage explant will drive the process of fluid exudation and movement across the sample in the radial direction. This fluid flux is initially very rapid and localized at the radial edge (r=a) of

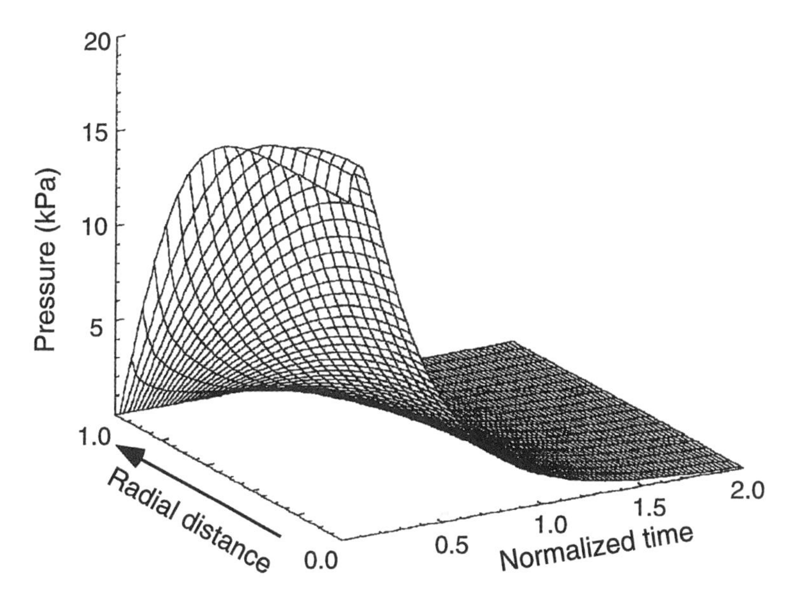

Figure 3. Variation in the interstitial fluid pressure with both radial position and time after application of a compressive load (f_0=-0.022 MPa) in an unconfined compressive creep experiment. The radial distance has been normalized by the explant radius so that values for r=0 and 1 correspond to the center and free surface of the explant, respectively. Values for time have been normalized by the gel diffusion time (t_g).

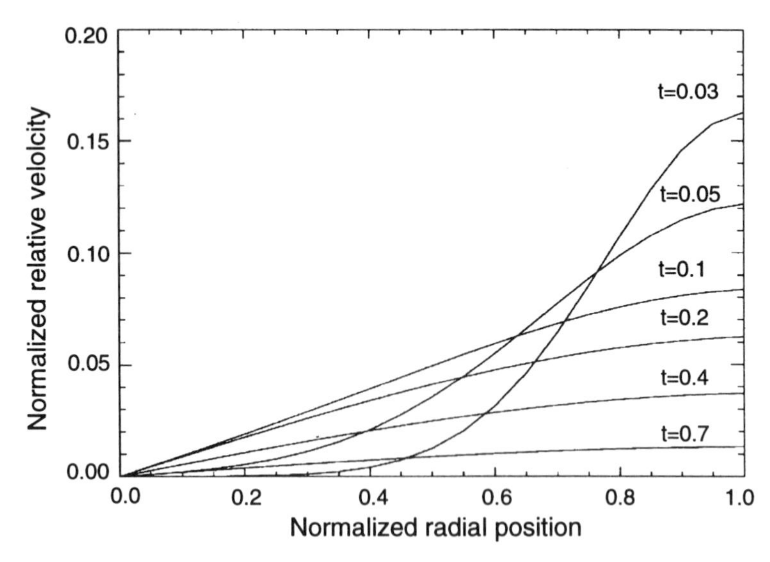

Figure 4. Variation of the relative velocity (fluid velocity relative to solid phase) with radial position in an unconfined compressive creep configuration, shown for select times after application of the compressive load (f_0=-0.022 MPa). The relative velocity is normalized by a characteristic parameter ($a/H_A k$), and values for time are normalized by the gel diffusion time ($t_g=a^2/H_A k$).

the specimen (Figure 4), and gradually decreases in magnitude to zero and propagates toward the center of the disc as the tissue equilibrates in creep. Fluid flow in this configuration is associated with transient changes in the matrix strain in both the axial (compressive) and radial (tensile) directions. The time to equilibrium is seen to be $t \sim 1.5\ t_g$, where t_g is the "gel diffusion time" (Mow *et al.* 1980), a parameter which depends on the material properties and geometry of the cartilage explant ($t_g = a^2/H_A k$, for a = sample radius). As an example, for the material parameters used in this study and an explant of 6mm diameter, the time to equilibrium can be larger than 3000s. This biphasic analysis demonstrates that *spatial* variations, as well as *temporal* variations exist throughout the tissue, and thus the mechanical environment around a chondrocyte within the ECM must vary, even during static compression.

Cell Tractions

To assess how the mechanical environment may affect the chondrocyte, the components of surface traction acting on the cell membrane may be calculated from the stress tensor **T** acting within the tissue ($\mathbf{T} = -p\mathbf{I} + \sigma$, where p is the hydrostatic pressure and σ is the effective elastic stress in the biphasic theory) and the unit outward normal and unit tangent vectors on the cell boundary. In Figure 5, the normal surface tractions arising from the hydrostatic pressure alone are shown for a representative spherical cell within the tissue at different radial distances from the center of a cartilage explant, and at 30s and 300s after application of the compressive load. As expected, the hydrostatic pressure p gives rise to a state of isotropic traction acting on the spherical cell surface at all times after loading. The magnitude of this surface traction varies significantly with radial position, as can be seen for a chondrocyte near the center (r/a=0.1) to one at the radial periphery (r/a=0.9). With increasing time (t=30s to t=300s), this surface traction for all cells will diminish.

In contrast, the tractions produced by the effective stress σ vary more significantly across the cell boundary than with radial position in the cartilage explant. In Figures 6 and 7 the components of the effective elastic stress normal and tangent (respectively) to the cell boundary are plotted against angular position on the cell surface. Note that the component of normal stress varies from tensile in the radial direction,

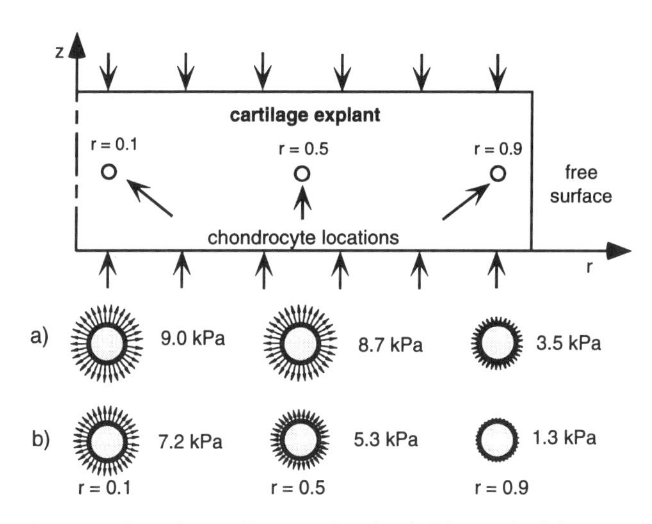

Figure 5. Cross-section of a cartilage explant loaded in a uniaxial compressive creep experiment, where the arrows on the explant denote the direction of the applied compressive load. The resulting tractions acting on a spherical chondrocyte due to hydrostatic pressurization of the extracellular matrix were calculated for three positions in the explant as shown (note that r denotes radial position normalized by the explant radius). The variation in the hydrostatic pressure is shown at a) 30s and b) 300s after application of the compressive load. (note that the arrow convention is outward negative).

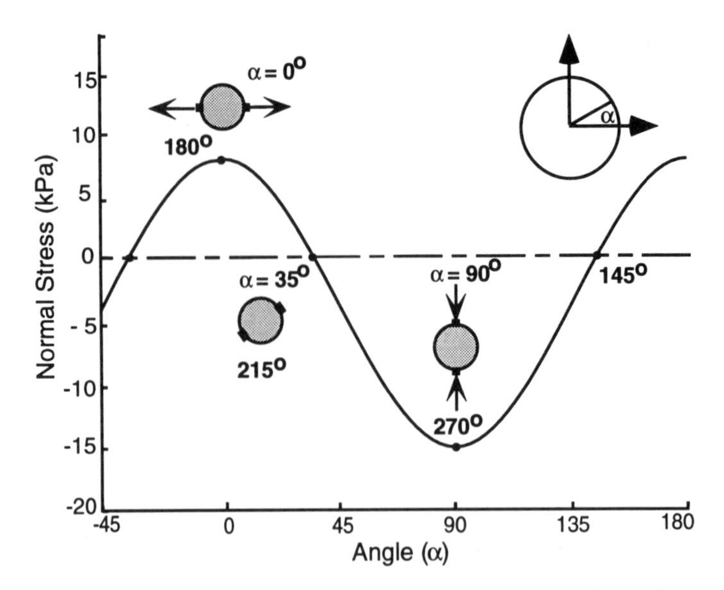

Figure 6. Variations in the normal stress acting on the surface of a spherical chondrocyte due to the existence of elastic stresses in the extracellular matrix generated in an unconfined compression creep test. The magnitude of normal stress is plotted as a function of position on the cell surface, and the direction of normal stress is shown schematically for 0°(180°), 35°(215°) and 90°(270°). Note that 0° corresponds to the r-direction, and 90° corresponds to the z-direction, the axis of applied loading. The normal stress is shown for a chondrocyte with a center at r=a/2 (a is the explant radius) at 30s after application of the compressive load.

(i.e., perpendicular to the axis of applied load $\alpha = 0°$) to compressive in the axial direction (i.e., the axis of the applied load $\alpha = 90°$) (Figure 6). At 35° and 215° there are no normal stresses acting on the cell membrane. In contrast, the hoop stress varies from tensile in the membrane at the surfaces facing the top and bottom edges of the explant i.e., α=90° and 270°, to compressive at the surfaces facing the center and radial edge i.e., α=0° and 180° (Figure 7). These results show that the pressures and stresses acting on a chondrocyte, even in this simple explant configuration is quite complex. Hence interpretation of biosynthetic response of chondrocytes to "compressive loading" must be exercised with care. This dependence of pressures and stresses on both time and space must be considered, since the paradigm is that chondrocytes can and do respond to their mechanical environment.

At equilibrium there will be no effects of interstitial fluid-flow or fluid pressurization so that matrix deformation may be considered to be the dominant mechanical factor (Armstrong *et al.* 1984; Mow *et al.* 1980-1992b). An explant study designed to investigate deformation as a

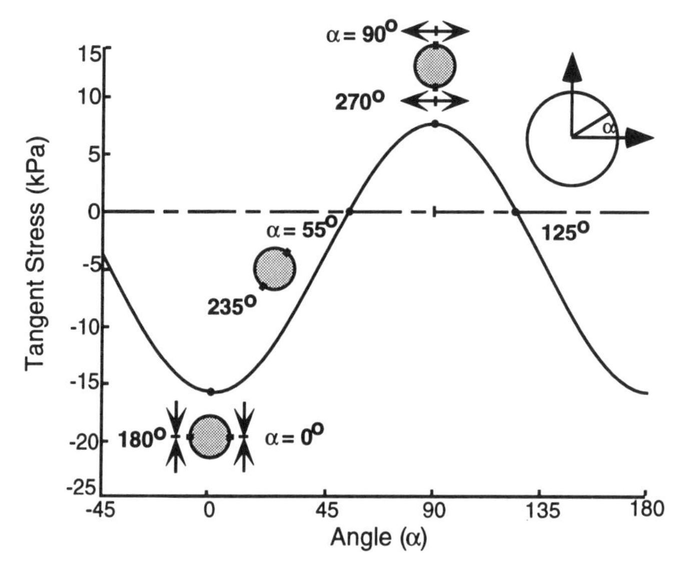

Figure 7. Variations in the tangent stress acting on the surface of a spherical chondrocyte due to the existence of elastic stresses in the extracellular matrix generated in an unconfined compression creep test. The magnitude of tangent stress is plotted as a function of position on the cell surface, and the direction of tangent stress is shown schematically for 0°(180°), 55°(235°) and 90°(270°). Note that 0° corresponds to the r-direction, and 90° corresponds to the z-direction, the axis of applied loading. The tangent stress is shown for a chondrocyte with a center at r=a/2 (a, explant radius) at 30s after application of the compressive load.

stimulus to chondrocyte metabolic activity and employing this configuration must be cognizant of the fact that the duration of the test must be sufficiently large so that the behavior (creep or stress-relaxation) does in fact reflect the equilibrium state of the tissue. This will be dependent on the time required by the tissue to reach compressive equilibrium which in turn depends on its gel diffusion time ($t_g = a^2/H_A k$).

Dynamic compression

In order to determine the mechanical environment within the ECM during cyclic compression, the biphasic theory and analysis was used as described by Guilak *et al.* (1990b) and Mow *et al.* (1990). The experimental protocol studied is that of cyclic load control with an oscillatory surface traction function $F(t) = -f_o \cdot (1-\cos\omega t)$ where f_o is the amplitude of the compressive stress and ω is the frequency of loading. The results shown are for the harmonic steady state response. The internal fluid pressure fields and interstitial fluid flow fields were calculated, and are shown in Figures 8 and 9 respectively for the material parameters given above, and a value for $f_0 = 0.022$ MPa and $\omega = 0.01$ Hz. Note that the fluid pressure is zero at the free surface of the cartilage explant (r/a=1), and attains a maximum magnitude at the center of the tissue (Figure 8). The large gradients in fluid pressure across the radial direction of the sample give rise to significant interstitial fluid flows (Figure 9), as well as matrix deformation and strain in the radial direction (not shown). The region of highest fluid velocities is found at the periphery of the cartilage explant, which is also the site of the lowest hydrostatic fluid pressure. Therefore, in cyclical loading, the central-most portions of the tissue are exposed to very small magnitudes of relative fluid motion and very large magnitudes of oscillatory fluid pressure. Increasing the frequency of the sinusoidal load only serves to increase the gradient in fluid pressure across the sample, so that larger magnitudes of interstitial fluid-flow will result.

Clearly the mechanical environment of the chondrocyte in this experiment will vary quite dramatically with radial position in the cartilage explant. In general, it can be shown that there will be a component of hydrostatic pressure traction acting on the cell surface which will be oscillatory, and vary from a maximum at the center to nearly zero at the periphery. The calculated effective elastic stress tractions have a pattern similar to that shown for the compressive creep case in that there is a large variation along the cell membrane (Figures 6 and 7). However, in this case, the cell tractions will be oscillatory and so expose the chondrocyte to a very different mechanical signal. The mechanical environment within the ECM produced by dynamic load control experiments (or displacement control) is dependent on the frequency and magnitude of compression, as well as tissue properties such as the permeability and elastic properties of the solid matrix, and the permeability condition of the rigid loading platen. Even at a relatively low frequency, e.g., ω=0.1Hz, in a displacement controlled experiment, the biphasic analysis predicts that the loading platen can lift-off from the surface (Mow *et al.* 1990; Sah *et al.* 1989). This occurs because the speed of movement of the loading platen exceeds the maximum speed at which surface of the specimen can rebound. When this occurs, the state of stress and strain within the explant becomes even more complex, so that any interpretation of the mechanical signal transduction mechanism in the explant experiment is necessarily uncertain.

The results presented here, both for the static and sinusoidal loading unconfined compression configurations, highlight the inherent spatial

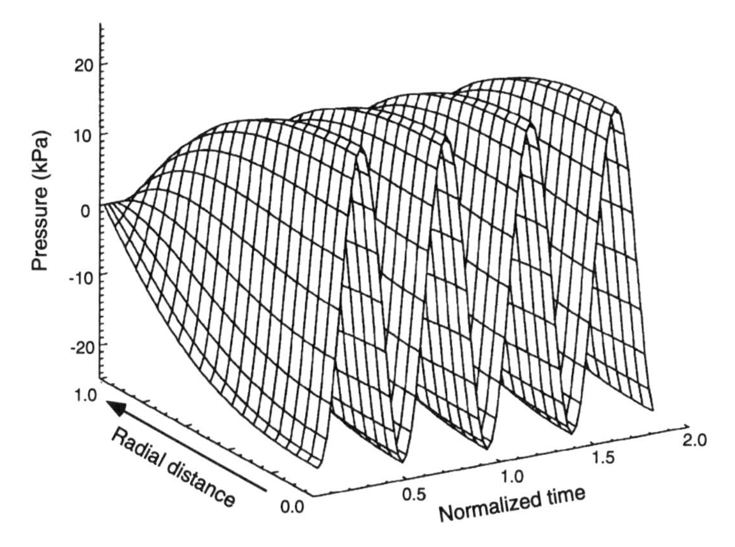

Figure 8. Variation in the interstitial fluid pressure with both radial position and time during an unconfined cyclic load control experiment: $F(t) = -f_o \cdot (1-\cos\omega t)$, where f_o=0.022 MPa and ω=0.01. The radial distance has been normalized by the explant radius so that values for r=0 and 1 correspond to the center and free surface of the explant, respectively. Values for time have been normalized by the characteristic gel diffusion time for the tissue ($t_g=a^2/H_A k$).

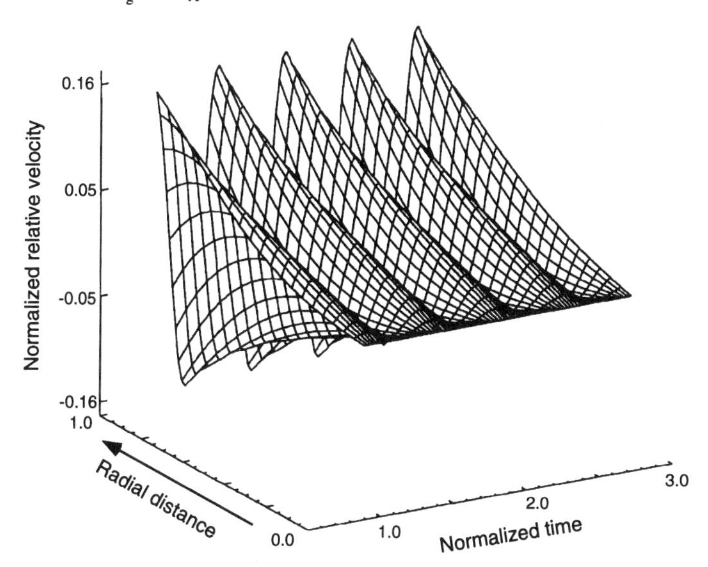

Figure 9. Variation of the relative velocity (fluid velocity relative to solid phase) with both time and radial position during an unconfined cyclic load control experiment: $F(t) = -f_o \cdot (1-\cos\omega t)$, where f_o=0.022 MPa and ω=0.01. Note that the relative velocity is normalized by a characteristic parameter for the tissue ($a/H_A k$), and values for time are normalized by the gel diffusion time ($t_g=a^2/H_A k$).

variations in a number of parameters which may be involved in the signaling of cellular response when the tissue is loaded. The results demonstrate the importance of a description of the stress, strain, pressure, fluid flow fields within the ECM of a loaded explant test. These results point to the fact that physiologically relevant models must be developed to analyze specific explant loading experiments in order to obtain the precise signal transduction mechanisms in cartilage.

Modeling of Chondrocyte-Matrix Interactions

The results of the previous section present the predicted cellular tractions assuming that the mechanical properties of the chondrocyte do not contribute significantly to the tissue as a whole, nor do they affect significantly the state of stress in the immediate surrounding matrix. If, however, the differences in the mechanical properties of the chondrocyte and surrounding ECM are significant, then the cells will affect the local or micro-mechanical environment. Since, at present, no information is available on the intrinsic mechanical properties of chondrocytes *in situ*, the only approach available is to assume a specific type of deformational behavior with the material properties used as parameters in the analysis. For example, such an approach has been used in assuming mammalian cells to be viscoelastic (Dong *et al.* 1988; Sung *et al.* 1988) where the cytoskeleton provides the main elastic framework of the cell, and that the fluid-solid interactions between the cytoskeleton and cytosol are responsible for the viscoelastic behavior of the cell. This type of modeling is similar to the fundamental ideas of the biphasic theory used for articular cartilage as a mixture of fluid and solid. Thus, we adopted the biphasic theory to model chondrocytes as a fluid-solid inclusion, embedded in and attached to the surrounding biphasic ECM (Guilak *et al.* 1990a). While the intrinsic properties (aggregate modulus, Poisson's ratio and permeability) defining the cell are assumed to be distinct from those of the ECM, it is assumed that, due to their small volume fraction in the tissue, the cells do not contribute to the overall mechanical behavior of the tissue as a whole. Due to the complex geometry of this problem, the biphasic finite element programs developed by Spilker *et al.* (1988) were used in the stress-strain analysis.

FEM Scaling Problem

The major obstacle in the analysis of all such problems is the large difference in the geometric scales (two orders of magnitude) between the macro-scale problem (i.e., mechanical environment of the explant) and the micro-scale problem (i.e., mechanical interactions between the cell and the ECM). By dividing the analysis into these two separate problems, a multiple scaling algorithm was developed to calculate the micro-scale response (Guilak and Mow 1992). In this algorithm, the FEM solution of the macro-scale problem was first used to determine the stresses, strains, fluid velocities and pressures at each node in a model representing the entire explant compressed between two porous, permeable loading platens (macro-scale). The results of this macro-scale FEM analysis were then used as the applied boundary conditions to a separate micro-scale FEM model, where a much finer mesh was constructed to represent the chondrocyte with its own distinct geometry and material properties. To match the micro-scale FEM problem to the macro-scale FEM problem, the time histories of the kinematic boundary conditions (fluid velocity and solid displacement) were calculated using the macro-scale model and these values were linearly interpolated across the faces of the micro-scale model. This technique has been used to examine the effects of several parameters, including relative material properties of the cell, pericellular and extracellular matrices, cell shape, and intercellular spacing (Guilak and Mow 1992).

Micro-Mechanical Environment

The results of this analysis indicate that under axial compression of the explant, the mechanical environment in the vicinity of the chondrocyte is highly non-uniform and dependent on the relative values of the elastic stiffness of the cytoskeleton, the extracellular and pericellular matrices. An important finding of this analysis was that the magnitudes of stresses and strains in the vicinity of the chondrocyte generally exceed the average magnitudes of stress and strain within the tissue. For example, changes in the relative Poisson's ratios of the cell (ν_c) and the ECM (ν_m) introduce significant changes in the radial stresses within the cell and the ECM. If $\nu_c > \nu_m$, and if ν_c is close to 0.5, the radial expansion of the cell under an axial compression of the ECM, produces a large radial compressive stress in the ECM, while the cell itself experiences a large tensile strain in the radial direction. Conversely, a cell with $\nu_c \sim 0$ will experience both radial and axial compressive strain and a large decrease in its volume. Cell shape also may play an important role in determining the stress-strain state of the chondrocyte (Mow and Guilak 1993). Larger principal stresses were predicted within and around the chondrocyte as the ellipticity (ratio of cell width to cell height) was increased. Slightly

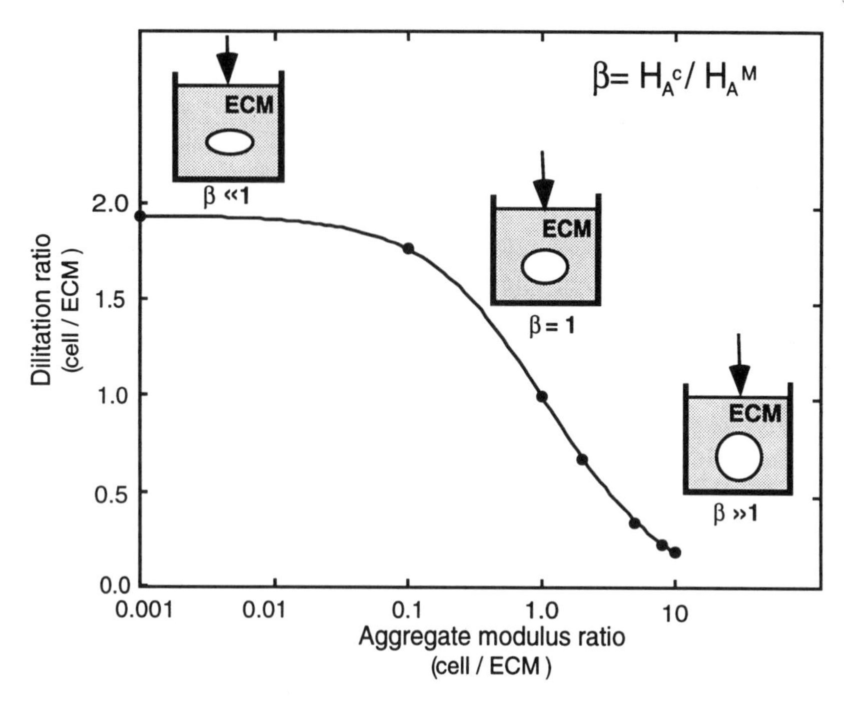

Figure 10. The biphasic theoretical model and finite element method were used to calculate the ratio of dilatation in the cell to that of the extracellular matrix (ECM) in the confined compression configuration at equilibrium. The predicted cellular dilatation relative to that of the ECM is plotted against the ratio of moduli of the cell (H_A^c) to that of the ECM (H_A^m). As H_A^c/H_A^m approaches zero, the intracellular dilatation increases to a maximum of almost two. Conversely, as H_A^c/H_A^m approaches ∞, the dilatation of the cell approaches zero. When cell and ECM have the same material properties (i.e., $H_A^c/H_A^m = 1$), they will experience the same dilatation.

prolate cells (ellipticity = 0.75) show even lower stress concentrations than spherical cells.

One parameter of interest that has been implicated in cell mechanotransduction and is known to be related to cell phenotype is change in cell volume. The FEM described above has been used to predict the intracellular dilatation induced by tissue compression. The same Poisson's ratio and permeability used above were chosen for both cell and matrix and, the ratio of the aggregate modulus of the cell (H_A^c) to that of the matrix (H_A^m) was varied from zero to ten. As H_A^c/H_A^m approaches zero, the predicted intracellular dilatation at equilibrium approaches twice that of the matrix (Figure 10). As the ratio approaches infinity, the cells see increasingly smaller values for the dilatation relative to that of the matrix.

Confocal Microscopy Studies of Chondrocyte Deformation

An important step towards understanding the mechanical environment of the chondrocyte is to determine the changes in chondrocyte shape and volume with applied compression of the tissue. This information is important not only for validation of FEM predictions of chondrocyte shape and volume changes under load, but also to better understand the role that chondrocyte deformation *per se* (e.g., shape and volume changes) plays in mechanical signal transduction. Currently, there is little quantitative information available on the deformation response of chondrocytes within the ECM (Guilak 1994a; Warden *et al.* 1994). While the mechanical properties of the cartilage matrix have been well characterized (Mow *et al.* 1980-1992b), there is little information available on the intrinsic mechanical properties of the chondrocytes *in situ*. Furthermore, chondrocyte shape and arrangement vary significantly with distance from the tissue surface *in situ* (Figure 1) (Stockwell 1971-1979). The shape and 3D organization of the chondrocytes are believed to be related to the local architecture of the collagen fibrils of the solid matrix (Stockwell 1979), and thus may possess a high degree of anisotropy. Because of these factors, as well as the heterogenous and anisotropic properties of the matrix, it is not clear how chondrocyte morphology may be affected by tissue deformation. To gain a better understanding of the role of chondrocyte deformation in regulating cartilage metabolic activity, an important first step is the accurate and precise quantification of the 3D changes in chondrocyte morphology during compression of the tissue.

The confocal scanning laser microscope (CSLM) provides a non-invasive means of examining viable cell morphology in relatively thick sections of tissue by forming serial optical sections through the depth of the specimen (Brakenhoff *et al.* 1989; Pawley 1990). Recently, new methods have been developed to obtain quantitative morphometric measurements on volume data recorded from the CSLM (Bachrach *et al.* 1993; Guilak 1994a; Guilak and Mow 1992). This method was used to quantify changes in chondrocyte morphology and local tissue deformation in the surface, middle and deep zones in explants of canine articular cartilage under physiological levels of matrix deformation (Guilak *et al.* 1994e). At 15% surface-to-surface compressive equilibrium strain in the tissue, decreases in cell height of 26%, 19% and 20%, and cell volume of 22%, 16% and 17% were observed in the surface, middle and deep zones respectively. Figure 11 shows one such cell *in situ*, represented by a shaded surface, before and after deformation

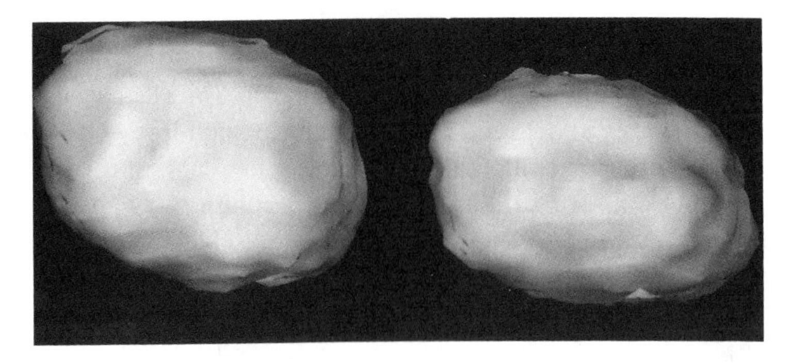

Figure 11. Laser scanning confocal microscopy was used to examine the morphology of live chondrocytes within the extracellular matrix of cartilage in an unconfined compression loading configuration. Geometric modeling algorithms were used to produce shaded surface solid models, as shown for a chondrocyte from the middle zone of articular cartilage a) before and b) after 20% axial compression of the tissue.

of the tissue. The deformation of surface zone chondrocytes was anisotropic, with significant lateral expansion of 5% occurring in the direction perpendicular to the local split-line direction. With removal of compression, complete recovery of cell morphology was observed in all cases. These observations indicate that chondrocyte deformation will occur during *in vivo* joint loading and therefore may have a role in mechanochemical coupling in cartilage.

The finding of a mechanically-induced decrease in chondrocyte volume is also intriguing considering the ability of mammalian cells to perceive and respond to volume changes. Various cells exhibit transport mechanisms activated by swelling and shrinkage (regulatory volume decrease and increase), and volume changes have been associated with mechanical transduction and signaling through mechano-sensitive channels involved in the transport of ions and organic substrates (Sarkadi and Parker 1991). Cellular volume changes have also been associated with rapid increases in intracellular second messengers such as Ca^{++} and cAMP (McCarty and O'Neil 1992; Pierce and Politis 1990).

Discussion and Conclusions

There are several major conclusions which may be derived from *in vivo* animal studies and *in vitro* explant studies on the effects of mechanical environment on cartilage metabolism. First, joint instability, immobilization and disuse do produce changes in cartilage anabolic and catabolic rates, and thus alter the gene expression, biochemical composition and mechanical properties of the cartilage in the affected joints *in vivo*. Second, mechanically loaded (including fluid pressurization) or osmotically loaded explants also produce altered chondrocyte metabolism *in vitro*. Static compression of explants seems to retard proteoglycan synthesis while certain frequencies of dynamic compression seem to enhance synthesis. The paradigm in all these studies seems to be that certain forms of mechanical or physicochemical stimulations *in vivo* or *in vitro* can modulate chondrocyte activities. Over the past decade or so, there has been an intensive search for the signal transduction mechanism(s) which might be responsible in transducing mechanical and physiochemical events into biologic events. At present, no hypothesis has surfaced as uniquely promising, partially due to the absence of attention paid to the complexity of the biomechanical problems in these studies. In this chapter, we have outlined some of the intrinsic difficulties in developing an understanding of the biomechanics of ECM and chondrocyte deformations, and suggested the biphasic model, with one set of assumptions, as a first attempt to quantify the mechanical information in more detail. The major advance from the results in the literature in these areas of investigation thus far has been the realization that the stress, strain, fluid flow and pressure (hydrostatic and osmotic) fields are highly non-uniform in the tissue, even when it is "statically loaded". Thus, in order to validate the hypothesis that chondrocytes in cartilage can respond to mechanical loading, and to ascertain specific mechanisms of signal transduction, further studies such as those outlined in this chapter must be pursued. The challenge, and therefore the opportunity, is to pursue these detailed studies which address many of these difficult biomechanics problems.

Acknowledgements

This study was supported by NIH Grant AR41913, a grant from the Whitaker Foundation and by the Frank E. Stinchfield Fellowship for Orthopaedic Research at the New York Hospital Orthopaedic Research Laboratory.

References

Akizuki, S.; Mow, V.C.; Muller, F.; Pita, J.C.; Howell, D.S.; Manicourt, D.H. Tensile properties of knee joint cartilage: I. Influence of ionic conditions, weight bearing, and fibrillation on the tensile modulus. J. Orthop. Res. 4:379-392; 1986.

Armstrong, C.G.; Lai, W.M.; Mow V.C. An analysis of the unconfined compression of articular cartilage. J. Biomech. Engng. 106:165-173; 1984.

Armstrong, C.G.; Mow, V.C. Variations in the intrinsic mechanical properties of human cartilage with age, degeneration and water content. J. Bone Joint Surg. 64A:88-94; 1982.

Armstrong, C.G.; Mow, V.C.; Wirth C.R. Biomechanics of impact-induced microdamage to articular cartilage--A possible genesis for chondromalacia patella. In: Finerman G. ed. AAOS Symposium on Sports Medicine: The Knee. St. Louis: CV Mosby Co.; 1985:p. 70-84.

Ateshian, G.A.; Lai, W.M.; Zhu, W.B.; Mow, V.C. An asymptotic solution for two contacting biphasic cartilage layers. J. Biomechanics, In Press, 1994.

Athanasiou, K.A.; Rosenwasser, M.P.; Buckwalter, J.A.; Malinin, T.I.; Mow, V.C. Interspecies comparison of *in situ* intrinsic mechanical properties of distal femoral cartilage. J. Orthop. Res. 9:330-340; 1991.

Aydelotte, M.B.; Greenhill, R.R.; Kuettner, K.E. Differences between sub-populations of chondrocytes. II. Proteoglycan metabolism. Conn. Tiss. Res. 18:223-234; 1988.

Bachrach, N.M.; Warden, W.H.; Chorney, G.S.; Ratcliffe, A.; Mow, V.C. A method for quantitative 3-D analysis of chondrocyte morphology using laser scanning confocal microscopy. In: Tarbell, J.M., ed. Advances in Bioengineering. New York: ASME; 1993: BED26:23-26.

Ben-Ze'ev, A. Animal cell shape changes and gene expression. BioEssays. 13:207-212; 1991.

Biot, M.A. General theory of three-dimensional consolidation. J. Appl. Physics.12:155-164; 1941.

Bowen, R.M. Theory of mixture. In: Eringen, A.E., ed. Continnum Physics. New York: Academic Press; 1976: V3:1-127.

Brakenhoff, G.A.; Van Spronsen, E.A.; Van Der Voort, H.T.M.; Nanninga, N. Three-dimensional confocal fluorescence microscopy. Methods Cell Biol. 30:379-98; 1989.

Broom, N.D.; Myers D.B. A study of the structural response of wet hyaline cartilage to various loading conditions. Conn. Tiss. Res. 7:227-237; 1980.

Broom, N.D.; Poole, A.C. Articular cartilage collagen and proteoglycans. Arthritis Rheum. 23:1111-1119; 1983.

Buckwalter, J.A.; Rosenberg, L.C. Electron microscopic studies of cartilage proteoglycans. J. Biol. Chem. 257:9830-9839; 1982.

Bullough, P.G.; Goodfellow, J. The significance of the fine structure of articular cartilage. J. Bone Joint Surg. 50B:852-857; 1968.

Burton-Wurster, N.; Vernier-Singer, M.; Farquhar, T.; Lust, G. Effect of compressive loading and unloading on the synthesis of total protein, proteoglycan, and fibronectin by canine cartilage explants. J. Orthop. Res. 11:717-729; 1993.

Caterson, B.; Lowther, D.A. Changes in the metabolism of the proteoglycans from sheep cartilage in response to mechanical stress. Biochim. Biophys. Acta. 540:412-422; 1978.

Clark, J.M. The organization of collagen in cryofractured rabbit articular cartilage: A scanning electron microscopic study. J. Orthop. Res. 3:17-29; 1985.

Clarke, I.C.; Articular cartilage: A review and scanning electron microscope study--1. The interterritorial fibrillar architecture. J. Bone Joint Surg. 53B:732-750; 1971.

Copray, J.V.C.M.; Jansen, H.W.B.; Duterloo, H.S. An *in vitro* system for studying the effect of variable compressive forces on the mandibular condylar cartilage of the rat. Arch. Oral Biol. 30(4):299-304; 1985.

Crane, R.E.; Green, A.E.; Naghdi, P.M. A mixture of viscous elastic materials with different constituent temperatures. Quart. J. Mech. Appl. Math. 23:171-184; 1970.

Dong, C.; Skalak, R.; Sung, K-L.P.; Schmid-Schonbein, G.W.; Chien, S. Passive deformation analysis of human leukocytes. J. Biomech. Engng. 110:27-36; 1988.

Donohue, J.M.; Buss, D.; Oegema, T.R.; Thompson, R.C. The effects of indirect blunt trauma on adult canine articular cartilage. J. Bone Joint Surg., 65A:948-957, 1983.

Dowson, D.; Unsworth, A.; Cooke, A.F.; Gvozdanovic, D. Lubrication of joints. In: Dowson, D; Wright, V., eds. An introduction to the biomechanics of joints and joint replacement. London: Inst. Mech. Eng.; 1981:p.120-145.

Edwards, J. Physical characteristics of articular cartilage. Proc. Inst. Mech. Eng. 181:16-24; 1967.

Eisenberg, S.R.; Grodzinsky, A.J. Swelling of articular cartilage and other connective tissues: Electromechanochemical forces. J. Orthop. Res. 3:148-159; 1985.

Eyre, D.R. Structure and function of the cartilage collagens: Role of type IX in articular cartilage. In: Brandt, K.D., ed. Cartilage Changes in Osteoarthritis. Indianapolis: Ciba-Geigy; 1990;p.12-16.

Frank, E.H.; Grodzinsky, A.J. Cartilage electromechanics - II. A continuum model of cartilage electrokinetics and correlation with experiments. J. Biomechanics. 20:629-639; 1987.

Getzenberg, R.H.; Pienta, K.J.; Coffey, D.S. The tissue matrix: Cell dynamics and hormone action. Endocrine Rev. 11(3):399-417; 1990.

Gore, D.M.; Higginson, G.R.; Minns, R.J. Compliance of articular cartilage and its variation through the thickness. Phys. Med. Biol. 28:233-247; 1983.

Gray, M.L.; Pizzanelli, A.M.; Grodzinsky, A.J.; Lee, R.C. Mechanical and physicochemical determinants of the chondrocyte biosynthetic response. J. Orthop. Res. 6:777-792; 1988.

Gray, M.L.; Pizzanelli, A.M.; Lee, R.C.; Grodzinsky, A.J.; Swann, D.A. Kinetics of the chondrocyte biosynthetic response to compressive load and release. Biochim. Biophys. Acta. 991:415-425; 1989.

Gu, W.Y.; Lai, W.M.; Mow, V.C. Analysis of fluid and ion transport through a porous charged-hydrated biological tissue during a permeation experiment. In: Salamon, N.J.; Sullivan, R.M., eds. Computational Mechanics of Porous Materials. New York: ASME; 1992:AMD136:29-41.

Gu, W.Y.; Lai, W.M.; Mow, V.C. Transport of fluid and ions through a porous-permeable charged-hydrated tissue, and streaming potential data on normal bovine articular cartilage. J. Biomechanics. 26:709-723; 1993.

Guilak, F. Volume and surface area measurement of viable chondrocytes in situations using geometric modeling of serial confocal sections. J. Microscopy; 1994a. In Press.

Guilak, F.; Donahue, H.J.; Zell, R.; Grande, D.; McLeod, K.J.; Rubin, C.T. Deformation-induced calcium signaling in articular chondrocytes. In: Mow, V.C.; Hochmuth, R.M.; Guilak, F.; Trans-Son-Tray, R., eds. Cell Mechanics and Cellular Engineering. New York: 1994b, In Press.

Guilak, F.; Hou, J.S.; Ratcliffe, A.; Mow, V.C. Articular cartilage under hydrostatic loading. In: Advances in Bioengineering. New York: ASME; 1988: BED15:183-186.

Guilak, F. Mow, V.C. Determination of the mechanical response of the chondrocyte in situ using confocal microscopy and finite element analysis. In: Advances in Bioengineering. New York: ASME; 1992: BED22:21-23.

Guilak, F.; Myers, B.C.; Ratcliffe, A.; Mow, V.C. The effect of static loading on proteoglycan metabolism in articular cartilage explants. Osteoarthritis and Cartilage, In Press, 1994c.

Guilak, F.; Ratcliffe, A.; Lane, N.; Rosenwasser, M.P.; Mow, V.C. Mechanical and biochemical changes in the superficial zone of articular cartilage in canine experimental osteoarthritis, J. Orthop. Res. In Press, 1994d.

Guilak, F.; Ratcliffe, A.; Mow V.C. The stress-strain environment around a chondrocyte: A finite element analysis of cell matrix interactions. In: Goldstein, S. A., ed. Advances in Bioengineering. New York: ASME; 1990a: BED17:395-398.

Guilak, F.; Ratcliffe, A.; Mow, V.C. Chondrocyte deformation and local tissue strain in articular cartilage: A confocal microscopy study. J. Orthop. Res. In Review, 1994e.

Guilak, F.; Spilker, R.L.; Mow V.C. A finite element model of cartilage extracellular matrix response to static and cyclic compressive loading. In: Goldstein, S. A., ed. Advances in Bioengineering. New York: ASME; 1990b: BED17:225-228.

Hall, A.C.; Urban, J.P.G. Responses of articular chondrocytes and cartilage to high hydrostatic pressure. Trans. Orthop. Res. Soc. 14:49; 1989.

Hall, A.C.; Urban, J.P.G.; Gehl, K.A. The effects of hydrostatic pressure on matrix synthesis in articular cartilage. J. Orthop. Res. 9:1-10; 1991.

Hayes, W.C.; Bodine, A.J. Flow-independent viscoelastic properties of articular cartilage matrix. J. Biomechanics. 11:407-420; 1978.

Hayes, W.C.; Mockros, L.F. Viscoelastic properties of human articular cartilage. J. Appl. Physiol. 31:562-568; 1971.

Helminen, H.J.; Jurvelin, J.; Kiviranta, I.; Paukkonen, K.; Saamanen, A-M.; Tammi, M. Joint loading effects on articular cartilage: A historical review. In: Joint loading. Helminen, H.J.; Kiviranta, I.; Tammi,M.; Saamanen, A-M.; Paukkonen, K.; Jurvelin, J., eds. Bristol, England: Wright; 1987:p.1-46.

Helminen, H.J.; Kiviranta, I.; Saamanen, A-M.; Jurvelin, J.S.; Arokoski, J.; Oettmeier, R.; Abendroth, K.; Roth, A.J.; Tammi, M. Effect of motion and load on articular cartilage in animal models. In: Kuettner, K.E.; Schleyerbach, R.; Peyron, J.C.; Hascall, V.C., eds. Articular cartilage and osteoarthritis. New York: Raven Press; 1992:p.501-510.

Holmes, H.; Hou J.S.; Mow, V.C. The energy dissipation mechanism in articular cartilage at low frequencies. In: Spilker, R.L., ed. Advances in Bioengineering. New York: ASME; 1984:41-42.

Holmes, M.H.; Mow, V.C. The nonlinear characteristics of soft gels and hydrated connective tissues in ultrafiltration. J. Biomechanics. 23:1145-1156; 1990.

Holmes, M.H.; Lai, W.M.; Mow V.C. Singular perturbation analysis of the nonlinear, flow-dependent, compressive stress-relaxation behavior of articular cartilage. J. Biomech. Engng. 107:206-218; 1985.

Horoyan, M.; Benoliel, A.M.; Capo, C.; Bongrand, P. Localization of calcium and microfilament changes in mechanically stressed cells. Cell Biophys. 17:243-256; 1990.

Hou, J.S.; Lai, W.M.; Holmes, M.H.; Mow, V.C. Squeeze film lubrication for articular cartilage with synovial fluid. In: Mow, V.C.; Ratcliffe, A.; Woo, S.L-Y., eds. Biomechanics of diarthrodial joints, Vol II. New York: Springer-Verlag; 1990:p.347-368.

Howell, D.S.; Treadwell, B.V.; Trippel, S.B. Etiopathogenesis of osteoarthritis. In: Moskowitz, R.W.; Howell, D.S.; Goldberg, V.M.; Mankin, H.J., eds. Osteoarthritis: Diagnosis and medical/surgical management. 2nd edition. Philadelphia: W.B.Saunders Co.; 1992:p.233-255.

Hunziker, E.B.; Schenk, R.K. Structural organization of proteoglycans. In: Wight, T.W.; Mecham, R.P., eds. Biology of Proteoglycans. Academic Press; 1987:p.155-183.

Ingber, D. Integrins as mechanochemical transducers. Curr. Opin. Cell. Biol. 3:841-848; 1991.

Jin, Z.M.; Dowson, D.; Fisher, J. The effect of porosity of articular cartilage on the lubrication of a normal human hip joint. J. Eng. Med. 206:117-124; 1992.

Jones, I.L.; Klamfeldt, A.; Sanstrom, T. The effect of continuous mechanical pressure upon the turnover of articular cartilage proteoglycans *in vitro*. Clin. Orthop. Rel. Res. 165:283-289; 1982.

Jurvelin, J.; Kiviranta, I.; Saammanen, A-M; Tammi, M.; Helminen, H.J. Partial restoration of immobilization-induced softening of canine articular cartilage after remobilization of the knee joint. J. Orthop. Res. 7:352-358; 1989.

Kaye, C.F.; Lippiello, L.; Mankin, H.; Numata, T. Evidence of pressure sensitive stimulus receptor system in articular cartilage. Trans. Orthop. Res. Soc. 5:1; 1980.

Kempson, G.E.; Muir, H.; Pollard, C.; Tuke, M. The tensile properties of the cartilage of human femoral condyles related to the content of collagen and glycosaminoglycans. Biochim. Biophys. Acta. 297:456-472; 1973.

Kempson, G.E.; Spivey, C.J.; Swanson, S.A.V.; Freeman, M.A.R. Patterns of cartilage stiffness on normal and degenerate femoral heads. J. Biomechanics. 4:597-609; 1971.

Kimura, J.H.; Schipplein, O.D.; Kuettner, K.E.; Andriacchi, T.P. Effects of hydrostatic loading on extracellular matrix formation. Trans. Orthop. Res. Soc. 16:53; 1991.

Kiviranta, I.; Jurvelin, J.; Tammi, M.; Saamanen, A-M; Helminen, H.J. Weight-bearing controls glycosaminoglycan concentration and articular cartilage thickness in the knee joint of young Beagle dogs. Arthritis Rheum. 30:801-809; 1987.

Lai, W.M.; Hou, J.S.; Mow, V.C. A triphasic theory for the swelling and deformational behaviors of articular cartilage. J. Biomech. Engng. 113:245-258; 1991

Lai, W.M.; Mow, V.C. Drag-induced compression of articular cartilage during a permeation experiment. Biorheology. 17:111-123; 1980.

Lai, W.M.; Mow, V.C.; Roth, V. Effects of a nonlinear strain-dependent permeability and rate of compression on the stress behavior of articular cartilage. J. Biomech. Engng. 103:61-66; 1981.

Larsson, T.; Aspden, R.M.; Heinegard. Effects of mechanical load on cartilage matrix biosynthesis *in vitro*. Matrix. 11:388-394; 1991.

Linn, F.C.; Sokoloff, L. Movement and composition of interstitial fluid of cartilage. Arthritis Rheum. 8:481-494; 1965.

Lippiello, L.; Kaye, C.F.; Neumata, T.; Mankin, H.J. *In vitro* metabolic response of articular cartilage segments to low levels of hydrostatic pressure. Connect. Tiss. Res. 13:99-107; 1985.

Mak, A.F.; Lai, W.M.; Mow, V.C. Biphasic indentation of articular cartilage--I: Theoretical analysis. J. Biomechanics. 20:703-714; 1987.

Mankin, H.J.; Brandt, K.D.; Biochemistry and metabolism of cartilage in osteoarthritis. In: Moskowitz, R.W.; Howell, D.S.; Goldberg, V.M.; Mankin, H.J., eds. Osteoarthritis: Diagnosis and medical/surgical management. 2nd edition. Philadelphia: W.B. Saunders Co.; 1992:p.109-154.

Maroudas, A. Balance between swelling pressure and collagen tension in normal and degenerate cartilage. Nature. 260:808-809; 1976.

Maroudas, A. Physicochemical properties of articular cartilage. In: Freeman, M.A.R., ed. Adult Articular Cartilage. Tunbridge Wells, England: Pitman Medical; 1979:p.215-290.

Maroudas, A. Physicochemical properties of cartilage in the light of ion-exchange theory. Biophys. J. 8:575-595; 1968.

Mansour, J.M.; Mow, V.C. The permeability of articular cartilage under compressive strain and at high pressures. J. Bone Joint Surg. 58A:509-516; 1976.

McCarty, N.A.; O'Neil, R.G. Calcium signaling in cell volume regulation. Physiol. Rev. 72:1037-1061; 1992.

Morales, T.I. Polypeptide regulators of matrix homeostasis in articular cartilage. In: Kuettner, K.E.; Schleyerbach, R.; Peyron, J.C.; Hascall, V.C., eds. Articular cartilage and osteoarthritis. New York: Raven Press; 1992:p.265-279.

Moskowitz, R.W.Experimental models of osteoarthritis. In: Moskowitz, R.W.; Howell, D.S.; Goldberg, V.M.; Mankin, H.J., eds. Osteoarthritis: diagnosis and medical/surgical management. 2nd edition. Philadelphia: W.B. Saunders Co.; 1992:p.213-232.

Mow, V.C.; Ateshian, G.A.; Ratcliffe, A. Anatomic form and biomechanical properties of articular cartilage of the knee joint. In: Finerman, G.A.M.; Noyes, F.R., eds. Biology and Biomechanics of the Traumatized Synovial Joint:The Knee as a Model. American Academy of Orthopaedic Surgeons Symposium. 1992b:p.55-81

Mow, V.C.; Gibbs, M.C.; Lai, W.M.; Zhu, W.; Athanasiou, K.A. Biphasic indentation of articular cartilage--II. A Numerical algorithm and an experimental study. J. Biomechanics. 22:853-861; 1989.

Mow, V.C.; Guilak, F.; Deformation of chondrocytes within the extracellular matrix of articular cartilage. In: Bell, E., ed. Tissue Engineering. Boston: Birkhauser; 1993:p.128-146.

Mow, V.C.; Holmes, M.H.; Lai, W.M. Fluid transport and mechanical properties of articular cartilage. J. Biomechanics. 17:377-394; 1984.

Mow V.C.; Hou, J.S.; Owens, J.M.; Ratcliffe, A. Biphasic and quasi-linear viscoelastic theories for hydrated soft tissues. In: Mow, V.C.; Ratcliffe, A.; Woo, S.L-Y., eds. Biomechanics of diarthrodial joints, I. New York: Springer-Verlag; 1990:p.215-260.

Mow, V.C.; Kuei, S.C.; Lai, W.M.; Armstrong, C.G. Biphasic creep and stress relaxation of articular cartilage in compression: Theory and experiments. J. Biomech. Engng. 102:73-84; 1980.

Mow, V.C.; Kwan, M.K.; Lai, W.M.; Holmes, M.H. A finite deformation theory for nonlinearly permeable soft hydrated biological tissues. In: Woo, S.L-Y.; Schmid-Schonbein, G.W.; Zweifach, B., eds. Frontiers in biomechanics. New York: Springer-Verlag; 1986:p.153-179.

Mow, V.C.; Ratcliffe, A.; Poole, A.R. Cartilage and diarthrodial joints as paradigms for hierarchical materials and structures. Biomaterials. 13:67-97; 1992a.

Muir, H. Molecular approach to understanding of osteoarthrosis. Ann. Rheum. Dis. 36:199-208; 1977.

Muir, H. Proteoglycans as organizers of the extracellular matrix. Biochem. Soc. Trans. 11:613-622; 1983.

Muir, H.; Bullough, P.; Maroudas, A. The distribution of collagen in human articular cartilage with some of its physiological implications. J. Bone Joint Surg. 52B:554-563; 1970.

Myers, E.R.; Lai, W.M.; Mow, V.C. A continuum theory and an experiment for the ion-induced swelling behavior of articular cartilage. J. Biomech. Engng. 106:151-158; 1984.

Ogden, J.A. The development and growth of the musculoskeletal system. In: Albright, J.A.; Brand, R.A. The scientific basis of orthopaedics. New York: Appleton-Century-Crofts; 1979; p.41-103.
Palmoski, M.J.; Perricone, E.; Brandt, K.D. Development and reversal of a proteoglycan aggregation defect in normal canine knee cartilage after immobilization. Arthritis Rheum. 22:508-517; 1979.
Palmoski, M.J.; Colver, R.A.; Brandt, K.D. Joint motion in the absence of normal loading does not maintain normal articular cartilage. Arthritis Rheum. 23:325-334; 1980.
Palmoski, M.J.; Brandt, K.D. Running inhibits reversal of atrophic changes in canine knee cartilage after removal of a leg cast. Arth Rheum 24:1329-1337; 1981.
Palmoski, M.J.; Brandt, K.D. Immobilization of the knee prevents osteoarthritis after anterior cruciate ligament section. Arthritis Rheum. 25:1201-1208; 1982.
Palmoski, M.J.; Brandt, K.D. Effect of static and cyclic compressive loading on articular cartilage plugs in vitro. Arthritis Rheum. 27:675-681; 1984.
Parkkinen, J.; Ikonen, J.; Lammi, M.J.; Laakkonen, J.; Tammi, M.; Helminen, H.J. Effects of cyclic hydrostatic pressure on proteoglycan synthesis in culture chondrocytes and articular cartilage explants. Arch. Biochem. Biophys. 300:458-465; 1993a.
Parkkinen, J.; Lammi, M.J.; Helminen, H.J.; Tammi, M. Local stimulation of proteoglycan synthesis in articular cartilage explants by dynamic compression *in vitro*. J. Orthop. Res. 10:610-620; 1992.
Parkkinen, J.; Lammi, M.J.; Pelttari, A.; Helminen, H.J.; Tammi. M.; Virtanen, I. Altered Golgi apparatus in hydrostatically loaded articular cartilage chondrocytes. Ann. Rheum. Dis. 52:192-198; 1993b.
Pawley, J. Fundamental limits in confocal microscopy. In: Pawley, J.B., ed. Handbook of Biological Confocal Microscopy. New York: Plenum Press; 1990.
Pierce, S.K.; Politis, A.D. Ca^{2+}-activated cell volume recovery mechanisms. Ann. Rev. Physiol. 52:27-42; 1990.
Ratcliffe, A.; Shurety, W.; Caterson, B. The quantitation of a native chondroitin sulfate epitope in synovial fluid and articular cartilage from canine experimental osteoarthritis and disuse atrophy. Arthritis Rheum. 36:543-551; 1993.
Ratcliffe, A.; Beauvais, P.J.; Saed-Nejad, F. Differential levels of aggrecan aggregate components in synovial fluids from canine knee joints with experimental osteoarthritis and disuse. J. Orthop. Res. In Press, 1994.
Roth, V.; Mow, V.C. The intrinsic tensile behavior of the matrix of bovine articular cartilage and its variation with age. J. Bone Joint Surg. 62A:1102-1117; 1980.
Sachs, F. Mechanical transduction by membrane ion channels: A mini review. Molecular Cell Biochem. 104:57-60; 1991.
Sah, R.L.Y.; Kim, Y.J.; Doong J-Y.H.; Grodzinsky, A.J.; Plaas, A.H.K.; Sandy, J.D. Biosynthetic response of cartilage explants to dynamic compression. J. Orthop. Res. 7:619-636; 1989.
Sah, R.L.Y; Doong J-Y.H.; Grodzinsky, A.J. Plaas, A.H.K; Sandy J.D. Effects of compression on the loss of newly synthesized proteoglycans and proteins from cartilage explants. Arch. Biochem. Biophys. 286: 20-29; 1991.
Sarkadi, B.; Parker, J.C. Activation of ion transport pathways by changes in cell volume. Biochim. Biophys. Acta. 1071:407-27; 1991.
Schneiderman, R.; Keret, D.; Maroudas, A. Effects of mechanical and osmotic pressure on the rate of GAG synthesis in the human adult femoral head cartilage: An *in vitro* study. J. Orthop. Res. 4:393-408; 1986.

Schmidt, M.; Mow, V.C.; Chun, L.E.; Eyre, D.R. Effects of proteoglycan extraction on the tensile behavior of articular cartilage. J. Orthop. Res. 8:353-363; 1990.

Setton, L.A.; Gu, W.Y.; Saed-Nejad, F.; Lai, W.M.; Mow, V.C. Swelling-induced pre-stress in articular cartilage and its physiological implications. Trans. Orthop. Res. Soc. 18:282; 1993a.

Setton, L.A.; Gu, W.Y.; Muller, F.J.; Pita, J.C.; Mow, V.C. Changes in the intrinsic shear behavior of articular cartilage with joint disuse. Trans. Orth. Res. Soc. 17:209; 1992a.

Setton, L.A.; Mow, V.C.; Muller, F.J.; Pita, J.C.; Howell, D.S. Mechanical properties of canine articular cartilage are significantly altered following transection of the anterior cruciate ligament. J. Orthop. Res. In Press, 1994.

Setton, L.A.; Zhu, W.B.; Mow, V.C. The biphasic poroviscoelastic behavior of articular cartilage in compression: Role of the surface zone. J. Biomechanics. 26:581-592; 1993b.

Spilker, R.L.; Suh, J-K.; Mow, V.C.; A finite element formulation for the nonlinear biphasic model for articular cartilage and hydrated soft tissues including strain-dependent permeability. In: Spilker, R.L., Simon, B.R. eds. Computational Methods in Bioengineering. New York: ASME; 1988; BED9:p.81-92.

Spilker, R.L.; Suh, J-K.; Mow, V.C.; Effects of friction on the unconfined compressive response of articular cartilage: A finite element analysis. J. Biomech. Engng. 112:138-146; 1990.

Spirt, A.A.; Mak, A.F.; Wassell, R.P. Nonlinear viscoelastic properties of articular cartilage in shear. J. Orthop. Res. 7:43-49; 1989.

Stockwell, R.A. The interrelationship of cell density and cartilage thickness in mammalian articular cartilage. J. Anatomy. 109:411-421; 1971.

Stockwell, R.A.; Meachim, G. The chondrocytes. In: Freeman, M.A.R., ed. Adult Articular Cartilage. London: Pitman Medical; 1973:p.51-99.

Stockwell, R.A. Biology of Cartilage Cells. Cambridge, England: Cambridge University Press; 1979.

Suh, J-K; Li, Z. Dynamic unconfined compression of articular cartilage under a cyclic compressive load. In: Langrana, N.A.; Friedman, M.H.; Grood, E.S. eds. 1993 Bioengineering Conference. New York: ASME; 1993:BED24;p.634-637.

Sung, K-L.P.; Dong, C.; Schmid-Schonbein, G.W.; Chien, S.; Skalak, R. Leukocyte relaxation properties. Biophys. J. 54:331-336; 1988.

Tammi, M.; Paukkonen, K.; Kiviranta, I.; Jurvelin, J.; Saamanen, A-M.; Helminen, H.J. Joint loading-induced alterations in articular cartilage. In Helminen H.J.; Kiviranta I.; Tammi, M.; Saamanen, A-M.; Paukkonen, K.; Jurvelin, J., eds. Joint Loading. Bristol, England: Wright; 1987:p.64-88.

Tammi, M.; Parkkinen, J.J.; Lammi, M.J.; Helminen, H.J. Pressure and frequency related $^{35}SO_4$ incorporation in cartilage explants by short term cyclic compression. Trans. Orthop Res Soc. 16:52; 1991.

Vener, J.M.; Thompson, R.C.; Lewis, J.L.; Oegema, T.R. Subchondral damage after acute transarticular loading: An in vitro model of joint injury. J. Orthop. Res. 10:759-765; 1992.

Warden, W.H.; Bachrach, N.M.; Chorney, G.S.; Ratcliffe, A.; Mow, V.C. Quantitation of anisotropic 3D deformation in different regions of the growth plate in compression. Trans. Orthop. Res. Soc. 19:144; 1994.

Watson, P.A. Function follows form: Generation of intracellular signals by cell deformation. Intracellular Signals. 5:2013-2019; 1991.

Wight, T.N.; Heinegard, D.K.; Hascall, V.C. Proteoglycans: structure and function. In: Hay, E.D. ed., Cell biology of extracellular matrix. 2nd edition. New York: Plenum Press; 1991:p.45-73.

Woo, S.L.-Y.; Akeson, W.H.; Jemmott, G.F. Measurements of nonhomogeneous directional mechanical properties of articular cartilage in tension. J. Biomechanics. 9:785-791; 1976.

Woo, SL-Y.; Kwan, M.K.; Lee, T.Q.; Field, F.P.; Kleiner, J.B.; Coutts, R.D. Perichondrial autograft for articular cartilage: Shear modulus of neocartilage studied in rabbits. Acta. Orthop. Scand. 58:510-515; 1987.

Wright, M.O.; Stockwell, R.A.; Nuki, G. Response of plasma membrane to applied hydrostatic pressure in chondrocytes and fibroblasts. Conn. Tiss. Res. 28:49-90. 1992.

Zanetti, M.; Ratcliffe, A.; Watt, F.M. Two subpopulations of differentiated chondrocytes identified with a monoclonal antibody to keratan sulfate. J. Cell. Biol. 101:53-59; 1985.

Zhu, W.; Mow, V.C.; Koob, T.J.; Eyre, D.R. Viscoelastic shear properties of articular cartilage and the effects of glycosidase treatments. J. Orthop Res. 11:771-781; 1993.

21

Deformation-Induced Calcium Signaling in Articular Chondrocytes

F. Guilak, H. J. Donahue, R. A. Zell, D. Grande,
K. J. McLeod, and C. T. Rubin

Introduction

Under normal physiological conditions, articular cartilage provides a nearly frictionless surface for the transmission and distribution of joint loads. The ultrastructure and composition of cartilage, which allow for this unique function, are maintained through a balance of the anabolic and catabolic activities of the chondrocyte cell population, which comprises a small fraction (1-10% by volume) of the tissue (Stockwell, 1979). Chondrocyte metabolic activity is regulated by both genetic and environmental factors, such as soluble mediators (e.g., cytokines, hormones) and physical stimuli (hydrostatic and osmotic pressures, mechanical load). This ability to regulate metabolic activity in response to the mechanical environment provides a means by which chondrocytes can alter the structure and composition, and hence the mechanical properties of the extracellular matrix, to the physical demands of the body (Gray *et al.,* 1988; Gray *et al.,* 1989; Guilak *et al.,* 1994b; Helminen *et al.*, 1987; Helminen *et al.*, 1992; Jones *et al.,* 1982; Sah *et al.,* 1989; Schneiderman *et al.,* 1986; Tammi *et al.,* 1987; also see review by Mow *et al.,* 1994, this volume). Under abnormal conditions, however, mechanical loads are believed to be an important factor in the initiation and progression of joint degeneration (Howell *et al.,* 1992; Moskowitz, 1992). Indeed, it is now believed that the mechanical environment of the chondrocytes is one of the most important factors affecting the health and function of the diarthrodial joint.

While there is considerable evidence indicating that chondrocytes can perceive and respond to their mechanical environment, the sequence of events in the transduction of mechanical signals to a biochemical response remains unclear. Because of the complex mechanical and physicochemical properties of articular cartilage, mechanical loading of the extracellular matrix exposes the chondrocytes to a diverse array of biophysical changes which are intrinsically coupled to one another and which could affect cell response (Table I, also see Mow *et al.,* 1994). Predominantly, these phenomena are related to the charged and hydrated nature of the cartilage matrix. For example, deformation of the solid matrix results in changes in tissue pH, fixed charge density, and pericellular concentrations of macromolecules such as proteoglycans (Gray *et al.,* 1988; Maroudas, 1979). Due to the hydrated nature of articular cartilage, matrix deformation also generates interstitial fluid pressurization and transport (Mow *et al.,* 1992), which in turn results in nonuniform electric currents within the charged solid matrix (Frank and Grodzinsky, 1987; Lai *et al.,* 1991). Further, deformation of the articular cartilage extracellular matrix results in shape and volume changes of the chondrocytes (Broom *et al.*, 1980; Guilak and Bachrach, 1993; Guilak *et al.*, 1994c). The sensitivity of chondrocyte response to the biophysical environment is emphasized by the findings that most of these phenomena have been shown to influence chondrocyte activity (Gray *et al.,* 1988; Gray *et al.,* 1989; Guilak *et al.,* 1994b; Hall *et al.;* 1991; Jones *et al.,* 1982; Parkkinen *et al.*, 1993; Sah *et al.,* 1989; Schneiderman *et al.,* 1986). Due to the intrinsic coupling of matrix and chondrocyte deformations, however, little information exists on the response of chondrocytes to deformation *per se* in the absence of the matrix-related effects described above.

Table I: BIOPHYSICAL EFFECTS OF CARTILAGE LOADING

Tissue Level Events	Cellular Level Events
Stress-strain fields	Changes in cell shape
Fluid pressure	Changes in cell volume
Interstitial fluid flow	Membrane deformation
Electrokinetic effects	Cytoskeletal deformation
Changes in fixed-charge density	Organelle (nucleus) deformation

In various cell types, it has been observed that deformation *per se* affects multiple cellular processes such as phenotypic expression, proliferation and metabolic activity (e.g., Banes *et al.*, 1994; Brighton *et al.,* 1992; Carosi *et al.,* 1992; Vandenburgh; 1992). Many of these responses are believed to be mediated through the activity of intracellular messengers such as cyclic adenosine monophosphate (cAMP), inositol phosphate (IP_3) and calcium ion (Ca^{2+}). Calcium ion, for example, is a ubiquitous intracellular second messenger which is involved in a large number of cellular processes. Hence a change in cytosolic concentration of Ca^{2+} ($[Ca^{2+}]_i$), which can be rapidly measured using fluorescent microscopy techniques, may be representative of other, longer-term effects on cell activity. Therefore, of particular interest are the findings in various cells that mechanical perturbation leads to transient increases in $[Ca^{2+}]_i$. For example, Goligorsky (1988) found that mechanical stimulation induced intracellular Ca^{2+} transients and membrane depolarization in cultured endothelial cells. Sanderson *et al.* (1990) observed similar Ca^{2+} responses in epithelial cells and reported intercellular propagation of Ca^{2+} waves which was mediated by inositol triphosphate (IP_3) (Boitano *et al.*, 1992). More recently, stretch- and deformation-induced Ca^{2+} transients have been reported in glial (Charles *et al.,* 1991a; Charles *et al.*, 1991b). osteoblastic (Jones and Bingmann, 1991; Xia and Ferrier, 1992), endothelial (Demer *et al.*, 1993; Sigurdson *et al.*, 1993) and fibroblastic cells (see Banes *et al.*, 1994, this volume). However, the mechanisms of Ca^{2+} mobilization in response to cellular deformation in these studies were not completely understood.

In this study, we examined the spatial and temporal characteristics of $[Ca^{2+}]_i$ wave propagation in response to mechanical deformation of the chondrocyte membrane. Experiments were performed to determine the source and pathway of increased Ca^{2+}, and the role that mechanosensitive ion channels may play in signal transduction. Further, the role of the cytoskeleton in deformation-induced calcium signalling was examined using agents which are known to inhibit cytoskeletal polymerization.

Materials and Methods

Chondrocyte Isolation and Culture

Bovine chondrocytes were isolated from the articular surface of the hock (ankle) joint of skeletally mature (3-5 year old) animals following the method of Burmeister *et al.* (1983). Cartilage plugs were removed and were enzymatically digested to isolate free chondrocytes using clostridial collagenase (0.37 g/l Type 1A; Sigma Chemical Corp., St. Louis, MO), testicular hyaluronidase (1 mg/l), and deoxyribonuclease (15 g/l; Worthington, Freehold, NJ). Digestion was carried out in a magnetic spinner flask at 37°C for 18h. The cell-enzyme mixture was passed through a sterile Nytex filter (120 μm) to remove the extracellular matrix debris. This was followed by centrifuging at 1500 rpm for 5 min. The cell pellet was resuspended in phosphate-buffered saline and recentrifuged. Cells were then counted and tested for viability by the trypan-blue-exclusion method. The freshly isolated cells were plated into 25 cm^2 flasks with RPMI 1640 medium supplemented with 10% fetal bovine serum, glutamine, and penicillin/streptomycin. All experiments were performed within the first two passages following isolation. Under these conditions, cells displayed characteristics typical of the chondrocyte phenotype, including expression of Type II collagen and sulfated proteoglycans, as well as the the ability to form a cartilaginous matrix.

Fluorescent Labeling and Confocal Microscopy

The confocal scanning laser microscope (CSLM) was used to examine the spatial and temporal distribution of intracellular fluorescence in response to cellular deformation at a single plane within the chondrocyte. Chondrocytes were plated at a density of 10,000/cm^2 on #0 glass coverslips in a viewing chamber (Biophysica Technologies, Sparks MD). Prior to testing, chondrocytes were incubated at 37°C for 45 min in 10 μM fluo-3-AM (Molecular Probes, Eugene, OR), a fluorescent indicator of $[Ca^{2+}]_i$. Fluorescent confocal microscopy was performed on a Nikon diaphot microscope (Nikon, Melville, NY) using a Noran/Odyssey real-time confocal system (Noran Instruments, Madison, WI) with a Nikon Plan Apo oil-immersion 40x, 1.3-NA objective. Argon laser excitation was set at 25% of full power at 488 nm with fluorescent emission recorded at 500 nm. A confocal slit of 25 μm was used with a gain setting of 1800 and contrast setting of 128.

Cell Deformation and Quantitative Fluorescence Measurements

Using a hydraulic micromanipulator (Narshige, Sea Cliff, NY), a single chondrocyte, in contact with several other chondrocytes, was deformed with the edge of a glass micropipette (o.d. 1 μm, Figure 1). The membrane of the chondrocyte was displaced by 1-2 μm, corresponding to maximum nominal deformations of ~25%. This magnitude of deformation was selected to correspond to chondrocyte deformation measured *in situ* during compression of the extracellular matrix (Guilak *et al.*, 1994c). Simultaneously, the average intracellular fluorescence intensity was recorded at 30 Hz within a rectangular 100 μm^2 area of the deformed chondrocyte and within all other chondrocytes within the field of view. Intracellular fluorescence intensities were monitored for 5 min following deformation. All experiments were performed at room temperature (21°C).

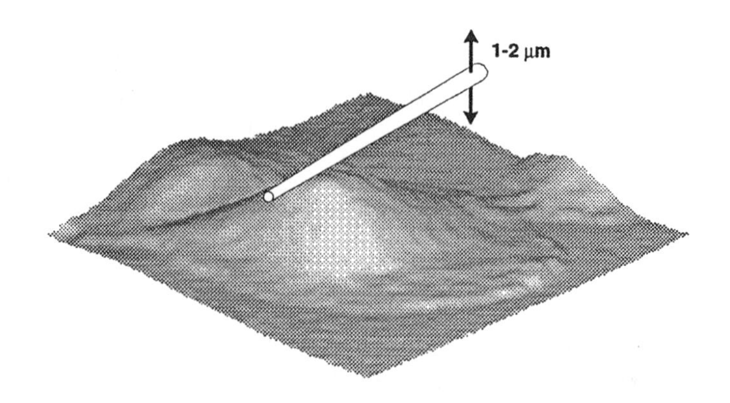

Figure 1: The chondrocyte membrane was displaced by 1-2 μm using the edge of glass micropipette. This configuration was used to apply physiologically relevant deformations to the chondrocyte while minimizing the possibility of damaging the membrane.

Mechanisms of Calcium Mobilization

To investigate the mechanisms involved in Ca^{2+} mobilization, various agents which are known to block specific Ca^{2+} pathways were utilized (Figure 2). To determine if the observed Ca^{2+} flux was due to transport from extracellular media experiments were performed in Ca^{2+}-free media with 10 mM EGTA.

To determine the role that mechanosensitive ion channels may play in deformation-induced Ca^{2+} mobilization, chondrocytes were treated with amiloride (1 μM) or gadolinium (10 mM) prior to testing. These agents have been reported as putative blockers of mechanosensitive ion channels (Andresen and Yang, 1992; Lorgensen and Ohmori, 1988; Lane *et al.,* 1992; Lane *et al.*, 1993; Yang and Sachs, 1989).

Calcium influx through voltage activated channels was studied using nifedipine (1 μM) and verapamil (20 μM). These agents have previously been utilized in studies to block L-type Ca^{2+} channels (Hille, 1992).

The role of the cytoskeleton in transducing membrane deformations into an intracellular biochemical signal was examined using cytochalasin D, an agent which inhibits polymerization of intracellular actin microfilaments. For these experiments, chondrocytes were incubated for 3 hrs at 37°C in cytochalasin D (20 μM) prior to testing (Schliwa, 1982; Yahara *et al.*, 1982).

Mechanisms of intercellular Ca^{2+} propagation were investigated using octanol, a blocker of functional gap junctions. For these experiments, chondrocytes were exposed to 3.8 mM octanol for 15-30 min prior to testing (Civitelli *et al.*, 1993).

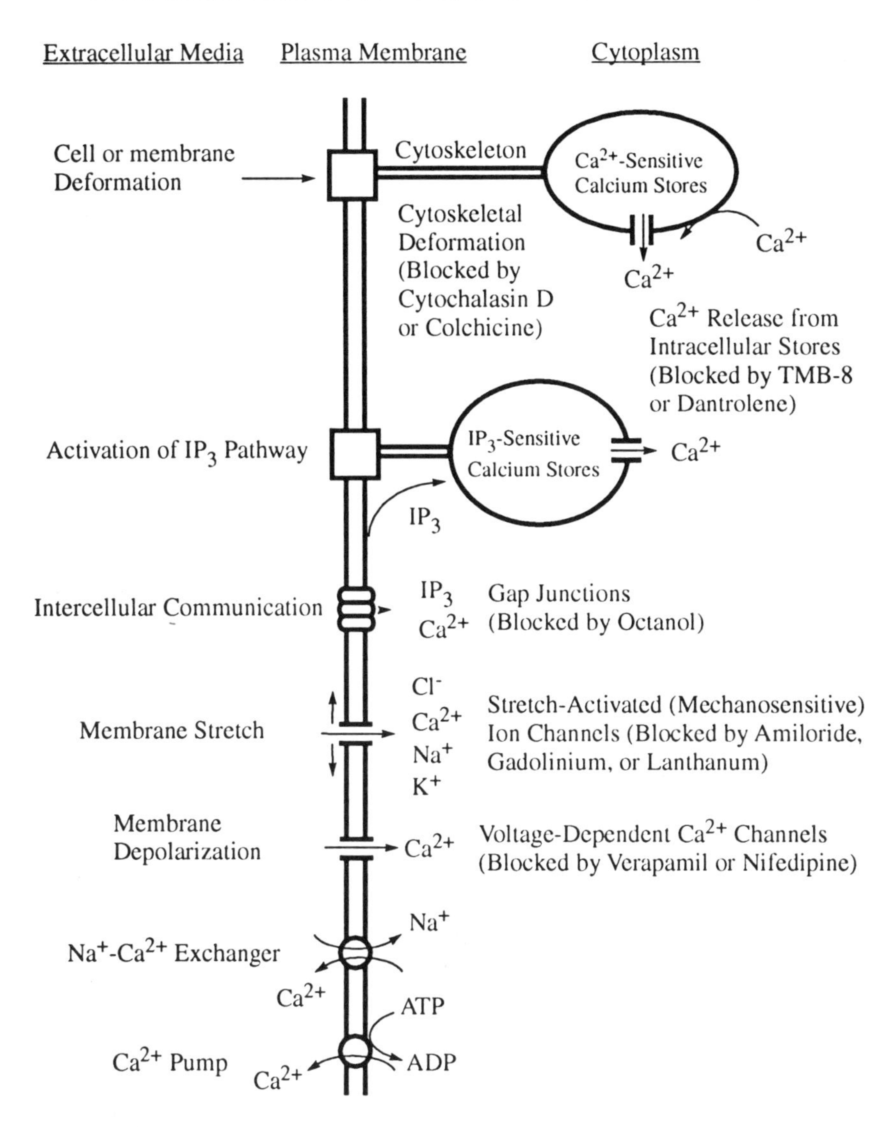

Figure 2: Mechanisms of intracellular Ca^{2+} homeostasis and mobilization. Increases in $[Ca^{2+}]_i$ could be initiated through stretch-activated, or mechanosensitive ion channels, release of Ca^{2+} from intracellular stores (Ca^{2+}-sensitive, or IP_3-sensitive), cytoskeletally-mediated effects, or by propagation of a signal from other cells via gap junctions.

Results

Cytosolic Calcium Response to Deformation

Small fluctuations in intracellular fluorescence with a peak range of ±15% around baseline were observed in all chondrocytes. An increase of greater than 15% over baseline was therefore regarded as a significant increase in $[Ca^{2+}]_i$ in response to deformation. In control media, (1.8 mM extracellular Ca^{2+}), mechanical deformation of the chondrocytes resulted in a wave of increased fluorescence which was initiated immediately at the site of contact with the micropipette (Figures 3 and 4). This wave propagated throughout the cell at a mean wave-front velocity of 10.8±5.6 µm/s (mean±s.d., n=6). Intracellular fluorescence intensities in the deformed cell peaked within 3-5 s following deformation and returned to baseline levels within 3-4 min (Figure 4).

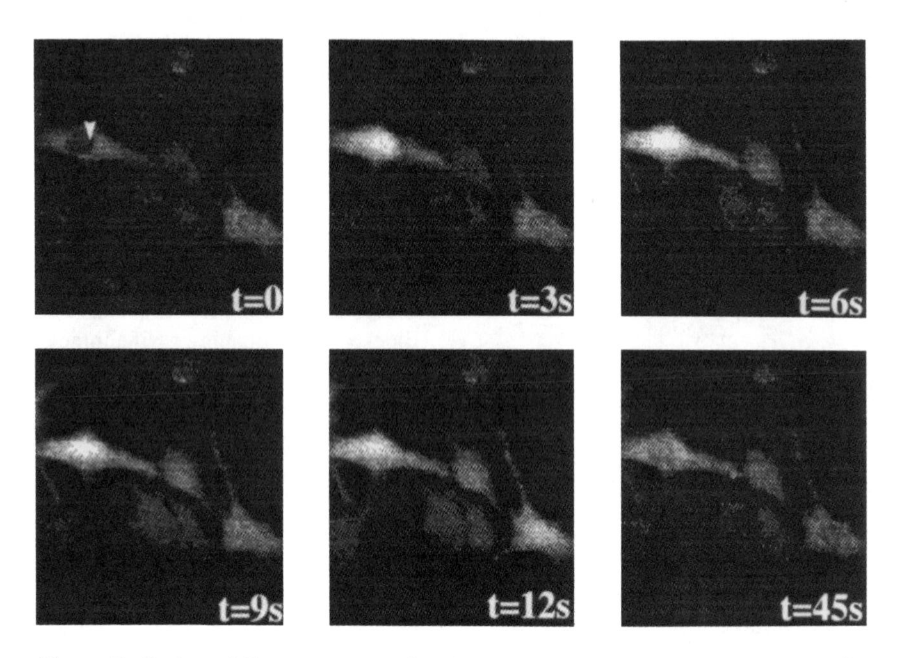

Figure 3: Series of fluorescent confocal images showing an increase in $[Ca^{2+}]_i$ which initiates at the site of applied deformation (arrow) and propagates to neighboring cells. Intracellular fluorescence levels peaked in the deformed cell within 3-5 s. By 30-45s following deformation, a decrease in fluorescence intensity was apparent in the deformed and neighboring cells.

Of all deformed chondrocytes bathed in control media, 95% (35 out of 37 tested) exhibited an immediate and transient increase in intracellular fluorescence. In these cells, peak fluorescence intensity was increased by 194±23% (mean±s.e., n=37) as compared to baseline in deformed cells and by 106±7% (n=247) in adjacent, undeformed cells. Removal of the deformation (raising of the micropipette) had no discernible effect on intracellular fluorescence.

Following a refractory period of approximately 5 min, this response could be initiated again, although peak fluorescence intensities were lower following subsequent applications of deformation as compared to the initial response (data not shown).

In several experiments, intracellular fluorescence levels were observed to decrease rapidly, presumably due to leakage of the dye through rupture of the chondrocyte membrane (Figure 5). This response was also accompanied by the formation of membrane blebs. In these cells, fluorescence levels did not return to baseline within 5 min, and mechanical stimulation did not result in any additional increases in fluorescence. Chondrocytes which exhibited this response were assumed to have been damaged and were excluded from the data analysis.

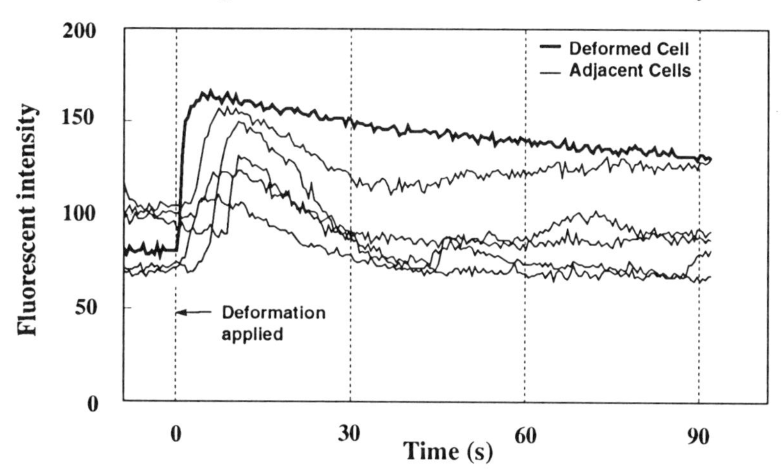

Figure 4: Intracellular fluorescence intensity versus time in deformed and neighboring chondrocytes in response to deformation of the chondrocyte membrane. An increase of intracellular fluorescence was observed immediately in the deformed cell and peaked within 3-5s. A delay was observed between the application of deformation and the rise of the fluorescence signal in neighboring cells, representing the time it took for the signal to propagate from cell to cell.

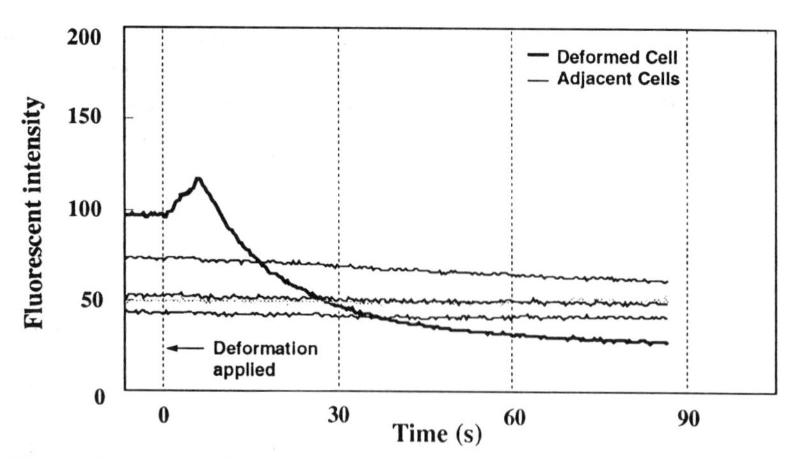

Figure 5: Intracellular fluorescence changes in response to apparent rupture of the chondrocyte membrane. Fluorescence levels did not return to baseline levels in these cells, suggesting that the intracellular dye had leaked out through a tear in the membrane. Chondrocytes exhibiting this response were not included in the data analysis.

Mechanisms of Calcium Mobilization

Mechanical deformation did not initiate a significant increase in cytosolic Ca^{2+} levels in chondrocytes bathed in Ca^{2+}-free media +EGTA. In these experiments, a mean fluorescence increase of 16±14% (n=14) was observed, with only 14% (2/14) of the cells responding to deformation with a positive increase in fluorescence. When an increase in intracellular fluorescence was observed in these experiments, the onset of the Ca^{2+} wave was significantly delayed as compared to control experiments. When the Ca^{2+}-free media in these experiments was replaced with control media, the characteristics of the deformation-induced Ca^{2+} transients were similar to those of the control chondrocytes.

Two of the channel-blocking agents utilized had a significant effect on the Ca^{2+} response of the chondrocytes. Treatment with amiloride also significantly decreased the percentage of cells responding to deformation (11 positive responses in 19 experiments, 58%, $p<0.05$ vs. control). Both amiloride and gadolinium significantly attenuated the peak increase in fluorescence to 84±12% (n=19, $p<0.05$) and 106±15% (n=15, $p<0.05$) over baseline, respectively.

Neither nifedipine nor verapamil affected the peak fluorescent intensity following deformation or the percentage of cells responding to deformation. Treatment with cytochalasin D resulted in significant changes in chondrocyte morphology to a more rounded shape, but did not affect the intracellular Ca^{2+} response.

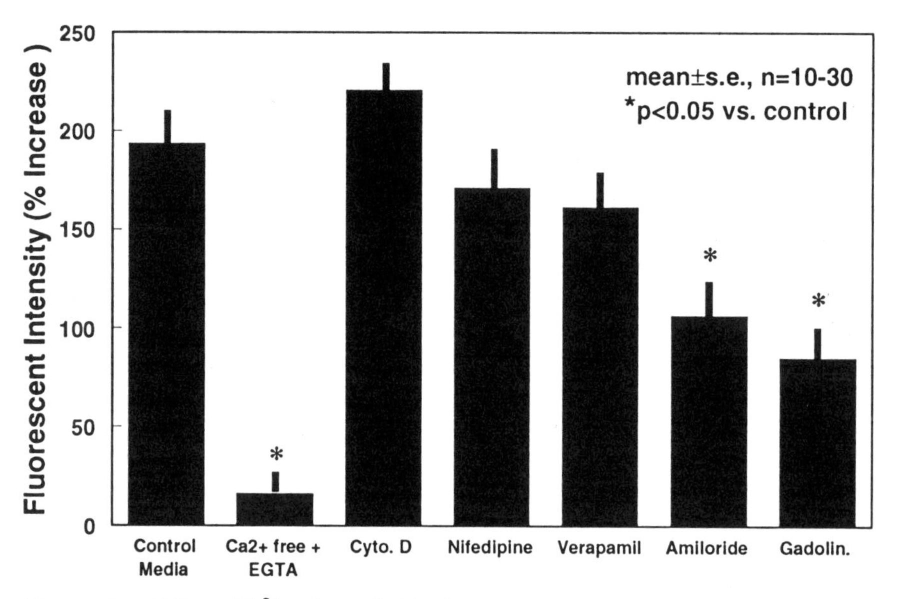

Figure 6: Effect Ca^{2+} channel blockers on deformation-induced calcium signalling in chondrocytes. The peak increase in fluorescence was significantly decreased in Ca^{2+}-free media or following incubation in gadolinium (mean±standard error of the mean, *$p<0.05$ vs. regular media, ANOVA with Student-Newman-Keuls).

Intercellular Propagation of Calcium Signals

Following deformation, the wave of increased fluorescence propagated to neighboring, undeformed chondrocytes at a velocity of 2.2±0.7 μm/s. On average, 64% (158/247) of the undeformed cells in the field of view demonstrated a significant increase in intracellular fluorescence. In these cells, peak fluorescence intensity was increased by 52±31%. In the presence of octanol, the percentage of undeformed cells responding was significantly decreased to 52% ($p<0.05$, comparison of proportions).

Discussion

The findings of this study support the hypothesis that chondrocyte deformation, in the absence of matrix-related effects, initiates an intracellular biochemical signal. In 95% of control cells exposed to mechanical stimulation, a significant increase in $[Ca^{2+}]_i$ was observed within milliseconds, peaking within 1-3 s. This Ca^{2+} pulse was propagated to neighboring, undeformed chondrocytes and in both deformed and neighboring cells $[Ca^{2+}]_i$ returned to baseline levels within 3-4 min following mechanical stimulation. The deformation-induced increase in $[Ca^{2+}]_i$ was attenuated by exposure to octanol, suggesting that Ca^{2+} signal propagation was via functional gap junctions. This is consistent with our previous studies showing that chondrocytes express connexin 43, a specific gap junction protein, and are functionally coupled *in vitro* (Donahue *et al.*, 1994)

Deformation-induced Ca^{2+} mobilization was abolished in the absence of extracellular Ca^{2+}. This finding, which is in agreement with observations on other cells (Charles *et al.,* 1991a; Charles *et al.,* 1991b; Sanderson *et al.,* 1990; Sanderson *et al.*, 1986), indicates that movement of Ca^{2+} across the cell membrane may well be required for initiation of a Ca^{2+} transient. In several of the experiments performed in Ca^{2+}-free media, increases in $[Ca^{2+}]_i$ were observed in neighboring cells in the absence of a Ca^{2+} response in the deformed cell. This finding is consistent with the study by Boitano *et al.* (1992) which suggested that cell-to-cell propagation of deformation-induced Ca^{2+} waves was mediated by IP_3.

An important new finding of this study was the observation that amiloride and gadolinium significantly attenuated the initiation and magnitude of deformation-induced Ca^{2+} waves, suggesting that the initial signal is transduced through mechanosensitive ion channels on the plasma membrane. A characteristic property of mechanosensitive ion channels is the relationship between their gating properties and mechanical tension of the plasma membrane (Guharay and Sachs, 1984; Sachs, 1988; Sachs, 1991). In preliminary finite element studies of micropipette indentation of the cell surface, our models predict significant levels of tensile strain in the vicinity of the micropipette. These results are consistent with the hypothesis that membrane stretch is the primary mechanical event which initiates intracellular Ca^{2+} wave propagation. The concept that mechanosensitive ion channels are indeed functional elements in the process of mechanical signal transduction is attractive since they have been identified in a number of different cells (Duncan and Misler, 1989; Guharay and Sachs, 1984; Sachs, 1988; Morris, 1990).

Another proposed pathway of cellular mechanical signal transduction is through the transmission of mechanical perturbations to intracellular organelles via the cytoskeleton (Ben-Ze'ev, 1991; Ingber, 1991). This paradigm suggests that mechanical deformation of the extracellular matrix may be transduced to the cytoskeleton through membrane-spanning integrins and linking proteins such as talin and vinculin. In our study, depolymerization of the actin microfilaments did not significantly affect the dynamics of Ca^{2+} mobilization in response to membrane deformation. These findings, consistent with those of Goligorsky (1988), imply that membrane-cytoskeleton or cytoskeleton-organelle interactions do not play a role in this signalling pathway and thus emphasize the contribution of mechanosensitive Ca^{2+} channels.

In previous studies, Sigurdson *et al.* (1993) predicted that a Ca^{2+} wave propagated by Ca^{2+} diffusion from the extracellular media could travel at a maximum velocity of 200-750 μm/s. Consistent with this prediction, a recent survey of Ca^{2+} waves recorded in various single cell preparations (Jaffe, 1993) showed that measured Ca^{2+} wave velocities can be grouped into three categories: the slow waves (0.1-1 μm/s associated with contractility; fast waves (5-30 μm/s) which move deep inside cells and are believed to be associated with intracellular signalling; and, the ultrafast waves (> 100 μm/s) which are surface waves and may well be important in initiating the fast waves. In our experiments, the observed Ca^{2+} waves propagated intracellularly with a mean wave speed of 10.8±5.6 μm/s placing these responses in the category of fast Ca^{2+} waves. These fast wave were first identified in fertilized eggs (Gilkey, *et al.*, 1978) and are believed to form a basis of intracellular Ca^{2+} signalling processes. That is, all intracellular Ca^{2+} transients may in fact be intracellular Ca^{2+} waves. Interestingly, fast Ca^{2+} waves within cells are found to occur with periodicities of 10-300s consistent with the refractory period of 180-300s seen in the chondrocytes.

The fact that the fast Ca^{2+} waves travel substantially below maximal velocities suggests two possible mechanisms of wave propagation in the cell. One is that the wave is simply dependent on free diffusion into the cell volume, in which case a wave velocity of 10 μm/s is consistent with Ca^{2+} diffusion coefficient on the order of 10^{-6} cm^2/s. This mechanism is referred to as calcium-induced calcium entry (CICE) a process arising from an initial increase in $[Ca^{2+}]_i$ stimulating an influx of extracellular Ca^{2+} through plasma membrane channels (Jaffe, 1993).

Alternatively, the Ca^{2+} wave may be surface waves within the cell which are rate-limited by mechanisms other than Ca^{2+} diffusion. Indeed, numerous reaction diffusion mechanisms have been proposed to account for the spatial propagation of Ca^{2+} through the cell. For example, even though the initial Ca^{2+} signal may originate from the extracellular media, a rate-limiting step may be the release of Ca^{2+} from intracellular stores. Endo *et al.* (1970) proposed a model for Ca^{2+} waves and oscillations based on the mechanism of calcium-induced calcium-release (CICR). This phenomenon was initially observed in muscle fibers, where Ca^{2+} itself caused increased release of Ca^{2+} from the sarcoplasmic reticulum (Fabiato, 1985). In this model, which has been used to describe Ca^{2+} spiking, it is postulated that cells contain a Ca^{2+} store within the endoplasmic reticulum (Meyer and Stryer, 1991). An initial increase $[Ca^{2+}]_i$, possibly through activation of mechanosensitive ion channels, could trigger an abrupt rise in $[Ca^{2+}]_i$ within the lumen of the endoplasmic reticulum which is then released into the cytosol.

Finally, the rate-limiting step in the propagation of Ca^{2+} waves may be the release of Ca^{2+} from an IP_3-sensitive pool. In non-excitable cells mechanical strains have been shown to activate phospholipase C levels and, subsequently, inositol phosphate metabolism (Brighton *et al.*, 1992;

Goldschmidt *et al.*, 1990). Formation of IP_3 could then release intracellular Ca^{2+} from an IP_3-sensitive pool, which in turn can trigger further release from the Ca^{2+}-sensitive pool. In recent studies, however, we have observed that blockers of intracellular Ca^{2+} release such as dantrolene or TMB-8 fail to attenuate Ca^{2+} wave propagation (Guilak *et al.*, 1994a), suggesting that these mechanisms may not be completely responsible for the observed transients.

In our experiments, the chondrocyte membrane was displaced to achieve maximum nominal strains of approximately 25% in order to correspond to representative chondrocyte deformations measured *in situ.* In these circumstances, local membrane strains in the region of contact with the micropipette may be hyperphysiologic in magnitude; hence it is important to note that these Ca^{2+} transients may represent a cellular response to wounding. However, our results suggest that any damage to the cells was not lethal. Cases characterized by a rapid decrease of intracellular fluorescence levels, presumably due to leakage of the dye through a tear in the membrane, were assumed to have damaged the chondrocytes through excessive deformation. Under these circumstances, membrane blebs appeared on the cell, and the Ca^{2+} response could not be reinitiated with further deformation, as reported in other cells (Sigurdson *et al.*, 1993). Importantly, the finding that amiloride and gadolinium attenuate $[Ca^{2+}]_i$ increase in response to deformation suggests that these transients are initiated by stretch and not by damage or tears in the membrane. While it is unlikely that cell damage occured in our experiments, it is possible that similar signal transduction mechanisms may signal cell damage and possibly wound healing in chondrocytes.

Our studies clearly show that mechanical deformation *per se* stimulates cytosolic Ca^{2+} mobilization in primary chondrocytes. However, the consequence of this increased $[Ca^{2+}]_i$ is unknown. Calcium is a nearly ubiquitous regulator of cell metabolism and is involved in the control of a large number of cellular functions (Carafoli, 1987; Rasmussen, 1986). The mechanism by which Ca^{2+} regulates cell function includes activation of calmodulin and protein kinases (Rasmussen, 1986), both of which can activate enzyme cascades resulting in hormone secretion, microtubule activation, exocytosis, protein synthesis and cell differentiation. In chondrocytes, it has been suggested that an increase in $[Ca^{2+}]_i$ may initiate matrix vesicle biogenesis (Iannotti *et al.*, 1989; Rosier, 1989), alter type I and type II collagen ratios (Deshmukh *et al.*, 1976; Deshmukh *et al.*, 1977), and affect proteoglycan synthesis rates (Eilam *et al.*, 1985, 1987; Iannotti *et al.*, 1990; Nevo *et al.*, 1988). The effect of $[Ca^{2+}]_i$ on proteoglycan synthesis is especially interesting in light of findings which suggest that static compression of articular cartilage inhibits proteoglycan synthesis (Gray *et al.*, 1988; Gray *et al.*, 1989; Guilak *et al.*, 1994b; Jones *et al.*, 1982; Sah *et al.*, 1989; Schneiderman *et al.*, 1986). Thus, one possible consequence of the deformation-induced changes in $[Ca^{2+}]_i$ could be modification of proteoglycan synthesis.

In summary, we have demonstrated that mechanical deformation of the chondrocyte membrane induces Ca^{2+} signal propagation within chondrocytes. Furthermore, the initiation of the Ca^{2+} signal is, at least in part, dependent on mechanosensitive channels in the chondrocyte membrane. These results suggest that in chondrocytes, physiologically relevant mechanical signals may be responsible for initiating a cytosolic Ca^{2+} signalling mechanism. Several important question still remain: 1) How do deformation-induced Ca^{2+} transients ultimately affect the phenotypic expression of the chondrocyte (e.g., gene expression, post-translational effects, biosynthetic and degradative activity)? 2) What are the thresholds of stress and strain which initiate Ca^{2+} signalling? 3) What other intracellular signalling pathways (e.g., IP_3, cyclic AMP) or ion transport mechanisms (e.g., Na^+/Ca^{2+}, Na^+/K^+) are activated by cellular deformation? 4) How are these signalling pathways affected by aging or degeneration of the articular cartilage?

Acknowledgements

The authors would like to thank Shirong Luo and Carl Paulino for their valuable assistance. We are also grateful to Dr. Peter Brink for use of his micromanipulator. This study was supported by a grant from the Whitaker Foundation and NIH AG10199.

References

Andresen, M.C.; Yang, M.Y. Gadolinium and mechanotransduction of rat aortic baroreceptors. Am J Physiol. 262:H1415-H1421; 1992.

Banes, A.J.; Sanderson, M.; Boitano, S.; Hu, P.; Baird, C.; Brigman, B.; Tsuzaki, M.; Fischer, T.; Lawrence, W.T. Mechanical load ± growth factors induce $[Ca^{2+}]_i$ release, cyclin D1 expression and DNA synthesis in avian tendon cells. In Mow, V.C.; Guilak, F.; Tran-Son-Tay, R.; Hochmuth, R.M. eds. Cell Mechanics and Cellular Engineering. New York: Springer Verlag, in press, 1994.

Ben-Ze'ev, A. Animal cell shape changes and gene expression. BioEssays 13:207-212; 1991.

Boitano, S.; Dirksen, E.R.; Sanderson, M.J. Intercellular propagation of calcium waves mediated by inositol triphosphate. Science Wash DC. 256:292-295; 1992.

Brighton, C.T.; Sennet, B.J.; Farmer, J.C.; Iannotti, J.P.; Hansen, C.A.; Williams, J.L.; Williamson, J. The inositol phosphate pathway as a mediator in the proliferative response of rat calvarial bone cells to cyclical biaxial mechanical strain. J Orthop Res, 10:385-393; 1992.

Broom, N.D.; Myers, D.B.: A study of the structural response of wet hyaline cartilage to various loading conditions. Conn Tiss Res. 7:227-237; 1980.

Burmeister G.R.; Menche, D.; Merryman, P.; Klein, M.; Winchester, R. Applications of monoclonal antibodies to the characterization of cells eluted from human articular cartilage. Arthr Rheum. 26:1187-1195; 1983.

Carafoli, E. Intracellular calcium homeostasis. Anu Rev Biochem. 56:395-404; 1987.

Carosi, J.A.; Eskin, S.G.; McIntire, L.V. Cyclical strain effects on production of vasoactive materials in cultured endothelial cells. J Cell Physiol. 151:29-36; 1992.

Charles; A.C.; Merrill, J.E.; Dirksen, E.R.; Sanderson, M.J. Intercellular signaling in glial cells: calcium waves and oscillations in response to mechanical stimulation and glutamate. Neuron. 6:983-992; 1991.

Charles; A.C.; Naus, C.; Zhu, D.; Dirksen, E.R.; Sanderson, M.J. Calcium waves propagate via gap junctions in glioma cells transfected with connexin 43. Soc Neurosci Abstr. 17:1147; 1991.

Civitelli, R.; Beyer, E.C.; Warlow, P.M.; Robertson, A.J.; Geist, S.T.; Steinberg, T.H. Connexin43 mediates direct intercellular communication in human osteoblastic cell networks. J Clin Invest. 91:1888-1896; 1993.

Deshmukh, K.; Kline, W.G.; Sawyer, B.D. Role of calcium in the phenotypic expression of rabbit articular chondrocytes in culture. FEBS Lett. 67:48-51; 1976.

Deshmukh, K.; Kline, W.G.; Sawyer, B.D. Effects of calcitonin and parathyroid hormone on the metabolism of chondrocytes in culture. Biochim Biophys Acta. 499:28-35; 1977.

Donahue, H.J.; Guilak, F.; Grande, D.; Bibb, M.; Porres, L.; McLeod, K.J.; Rubin, C.T.; Grine, E.; Hertzberg, E.; Brink, P. Intercellular communication via gap junctions in chondrocytes isolated from mature articular cartilage. Trans Orthop Res Soc. 19:375; 1994.

Duncan, R.; Misler, S. Voltage-activated and stretch-activated Ba^{2+} conducting channels in an osteoblast-like cell line (UMR 106). FEBS Lett. 251:17-21; 1989.

Eilam, Y.; Beit-Or, A.; Nevo, Z. Decrease in cytosolic Ca^{++} and enhanced proteoglycan derived growth factors in cultured chondrocytes. Biochem Biophys Res Comm, 132:770-779; 1985.

Eilam, Y.; Beit-Or, A.; Nevo, Z. Cytosolic free Ca^{++} as a signal for proteoglycan synthesis and cell proliferation in cultured chondrocytes. In Horowitz, S.; Sela, I., eds. Current Advances in Skeletogenesis III. Jerusalem, Israel: Heiliger, 1987:127-139.

Endo, M.; Tanaka, M.; Ogawa, Y. Calcium induced release of calcium from the sarcoplasmic reticulum of skinned skeletal muscle fibers. Nature. 228:34-36; 1970.

Fabiato, A. Simulated calcium current can both cause calcium loading in and trigger calcium release from the sarcoplasmic reticulum of a skinned canine cardiac Purkinje cell. J Gen Physiol. 85:291-320; 1985.

Frank, E.H.; Grodzinsky, A.J. Cartilage electromechanics - II. A continuum model of cartilage electrokinetics and correlation with experiments. J Biomechanics. 20:629-639; 1987.

Gilkey, J.C.; Jaffe, L.F.; Ridgeway, E.B.; Reynolds, G.T. A free calcium wave traverses the activating egg of the medaka, *Oryzias latipes*. J Cell Biol. 76:448-466; 1978.

Goldschmidt-Clermont, P.; Machesky, L.M.; Baldassare, J.J.; Pollard T.D.: The actin-binding protein profilin binds to PIP_2 and inhibits its hydrolysis by phospholipase C. Science Wash DC, 247:1575-1578; 1990.

Goligorsky, M.S. Mechanical stimulation induces $Ca^{2+}{}_i$ transients and membrane depolarization in cultured endothelial cells: Effects on $Ca^{2+}{}_i$ in co-perfused smooth muscle cells. FEBS Lett. 240:59-64; 1988.

Gray, M.L.; Pizzanelli, A.M.; Grodzinsky, A.J.; Lee R.C. Mechanical and physicochemical determinants of the chondrocyte biosynthetic response. J Orthop Res. 6:777-792; 1988.

Gray, M.L.; Pizzanelli, A.M.; Lee R.C.; Grodzinsky, A.J.; Swann, D.A. Kinetics of the chondrocyte biosynthetic response to compressive load and release. Biochim Biophys Acta. 991:415-425; 1989.

Guharay, F.; Sachs, F. Stretch-activated single ion channel currents in tissue-cultured embryonic chick skeletal muscle. J Physiol Lond. 352:685-701, 1984.

Guilak, F.; Bachrach, N.M. Compression-induced changes in chondrocyte shape and volume determined in situ using confocal microscopy. Trans Orthop Res Soc. 18:619; 1993.

Guilak, F.; Donahue, H.J.; Grande, D.A.; Zell, R.A.; McLeod, K.J.; Rubin, C.T. unpublished results, 1994a.

Guilak, F.; Meyer, B.C.; Ratcliffe, A.; Mow, V.C. Quantification of the effects of matrix compression on proteoglycan metabolism in articular cartilage explants. Osteoarthritis Cart, in press, 1994b.

Guilak, F.; Ratcliffe, A.; Mow, V.C. Chondrocyte deformation and local tissue strain in articular cartilage: A confocal microscopy study. J Orthop Res, submitted, 1994c.

Gupta, A.; Martin, K.J.; Miyauchi, A.; Hruska, K.A. Regulation of cytosolic calcium by parathyroid hormone and oscillations of cytosolic calcium in fibroblasts from normal and pseudohypoparathyroid patients. Endocrinology. 128:2825-2836; 1991.

Hall, A.C.; Urban, J.P.G.; Gehl, K.A. The effects of hydrostatic pressure on matrix biosynthesis in articular cartilage. J Orthop Res. 9:1-10; 1991.

Helminen, H.J.; Jurvelin, J.; Kiviranta, I.; Paukkonen, K.; Saamanen, A-M.; Tammi, M. Joint loading effects on articular cartilage: A historical review. In: Helminen H.J.; Kiviranta I.; Tammi, M.; Saamanen, A-M; Paukkonen, K.; Jurvelin, J. eds. Bristol, England: Wright and Sons, 1987:1-46.

Helminen, H.J.; Kiviranta, I.; Saamanen, A-M.; Jurvelin, J.S.; Arokoski, J.; Oettmeier, R.; Abendroth, K.; Roth, A.J.; Tammi, M. Effect of motion and load on articular cartilage in animal models. In: Kuettner, K.E.; Schleyerbach, R.; Peyron, J.C.; Hascall, V.C., eds. Articular Cartilage and Osteoarthritis. 1992:501-510.

Hille, B. Ionic channels of excitable membranes.Sunderland Massachusetts: Sinauer Associates, 1992.

Howell, D.S.; Treadwell, B.V.; Trippel, S.B.: Etiopathogenesis of osteoarthritis. In R.W. Moskowitz, R.W.; Howell, D.S.; Goldberg, V.M.; Mankin, H.J., eds. Osteoarthritis: diagnosis and medical/ surgical management, 2nd ed., Philadelphia PA: W.B. Saunders, 1992:233-252.

Iannotti, J.P.; Mechanism of action of parathyroid hormone-induced proteoglycan synthesis in the growth plate chondrocyte. J Orthop Res. 8:136-145; 1990.

Iannotti, J.P.; Naidu, S.; Noguchi, Y.; Hunt, R.M.; Brighton, C.T. Calcium induced matrix vesicle biogenesis. Trans Orthop Res Soc. 14:125; 1989.

Ingber, D. Integrins as mechanochemical transducers. Curr Opin Cell Biol. 3:841-848; 1991.

Jaffe, L.F. Classes and mechanisms of calcium waves. Cell Calcium. 14:736-745; 1993.

Jones, D.B.; Bingmann, D. How do osteoblasts respond to mechanical stimulation? Cells and Materials 1:329-340; 1991.

Jones, I.L.; Klamfeldt, A.; Sanstrom, T. The effect of continuous mechanical pressure upon the turnover of articular cartilage proteoglycans in vitro. Clin Orthop Rel Res. 165:283-289; 1982.

Jorgensen, F.; Ohmori, H. Amiloride blocks the mechano-electrical transduction channels of hair cells of the chick. J Physiol. 403:577-578; 1988.

Lai, W.M.; Hou, J.S.; Mow, V.C. A triphasic theory for the swelling and deformational behaviors of articular cartilage. J Biomech Engng. 113:187-197; 1991.

Lane, J.W.; McBride, D.W.; Hamill, O.P. Structure-activity relations of amiloride and its analogues in blocking the mechanosensitive channel in Xenopus oocytes. Br J Pharmacol. 106:283-286; 1992.

Lane, J.W.; McBride, D.W.; Hamill, O.P. Ionic effects on amiloride block of the mechanosensitive channel in Xenopus oocytes. Br J Pharmacol. 108:116-119; 1993.

Maroudas, A. Physicochemical properties of articular cartilage. In: Freeman, M.A.R., ed. Adult Articular Cartilage. Tunbridge Wells, England: Pitman Medical. p215-290; 1979.

Meyer, T.; Stryer, L. Calcium spiking. Annu Rev Biophys Chem. 20:153-174; 1991.

Morris, C.E. Mechanosensitive ion channels. J Membrane Biol. 113:93-107; 1990.

Moskowitz, R.W. Experimental models of osteoarthritis. In Moskowitz, R.W.; Howell, D.S.; Goldberg, V.M.; Mankin, H.J., eds. Osteoarthritis: diagnosis and medical/ surgical management, 2nd ed., Philadelphia PA: W.B. Saunders, 1992:213-232.

Mow, V.C.; Bachrach, N.M.; Setton, L.A.; Guilak, F. Stress, strain, pressure and flow fields in articular cartilage and chondrocytes. In Mow, V.C.; Guilak, F.; Tran-Son-Tay, R.; Hochmuth, R.M. eds. Cell Mechanics and Cellular Engineering. New York: Springer Verlag, in press, 1994.

Mow, V.C.; Ratcliffe, A.; Poole, A.R. Cartilage and diarthrodial joints as paradigms for heirarchial materials and structures. Biomaterials. 13:67-97; 1992.

Nevo, Z.; Beit-Or, A.; Eilam, Y. Slowing down aging of cultured embryonal chondrocytes by maintenance under lowered oxygen tension. Mech Aging Dev. 45:157-165; 1988.

Parkkinen, J.; Ikonen, J.; Lammi, M.J.; Laakkonen, J.; Tammi, M.; Helminen H.J. Effects of cyclic hydrostatic pressure on proteoglycan synthesis in culture chondrocytes and articular cartilage explants. Arch Biochem Biophys. 300:458-465; 1993.

Rasmussen, H. The calcium messenger system. New Eng J Med. 17:1094-1170; 1986.

Rosier, R.N. The role of intracellular calcium in matrix vesicle biogenesis. Orthop Trans. 8:238; 1984.

Sachs, F. Mechanical transduction in biological systems. CRC Crit Rev Biomed Eng, 16:141-169; 1988.

Sachs, F. Mechanical transduction by membrane ion channels: a mini review. Molecular Cell Biochem 104: 57-60; 1991.

Sah, R.L.Y.; Kim, Y.J.; Doong, J.Y.H.; Grodzinsky, A.J.; Plaas, A.H.K.; Sandy, J.D. Biosynthetic response of cartilage explants to dynamic compression. J Orthop Res, 7:619-636; 1989.

Sanderson, M.J.; Charles, A.C.; Dirksen, E.R. Mechanical stimulation and intercellular communication increases intracellular Ca^{2+} in epithelial cells. Cell Regulation 1:585-596; 1990.

Sanderson, M.J.; Chow, I.; Dirksen, E.R. Intercellular communication between ciliated cells in culture. Am J Physiol. 254:C63-C74; 1986.

Schliwa, M. Action of cytochalasin D on cytoskeletal networks. J Cell Biol. 92:79-91; 1982.

Schneiderman, R.; Keret, D.; Maroudas, A. Effects of mechanical and osmotic pressure on the fate of glycosaminoglycan synthesis in the human adult femoral head cartilage: an *in vitro* study. J Orthop Res, 4:393; 1986.

Sigurdson, W.J.; Sachs, F.; Diamond, S.L. Mechanical perturbation of cultured human endothelial cells causes rapid increases of intracellular calcium. Am J Physiol. 264:H1745-H1752; 1993.

Stockwell, R.A. Biology of cartilage cells. Cambridge, England: Cambridge University Press; 1979.

Tammi, M.; Paukkonen, K.; Kiviranta, I.; Jurvelin, J.; Saamanen, A-M.; Helminen, H.J. Joint induced alteration in articular cartilage. In Helminen H.J.; Kiviranta I.; Tammi, M.; Saamanen, A-M; Paukkonen, K.; Jurvelin, J. eds. Bristol, England: Wright and Sons, 1987:64-88.

Vandenburgh, H.H. Mechanical forces and their second messengers in stimulating cell growth in vitro. Am J Physiol. 262:R350-355; 1992.

Xia, S.L.; Ferrier J. Propagation of a calcium pulse between osteoblastic cells. Biochem Biophys Res Comm 186:1212-1219; 1992.

Yahara, I.; Harada, F.; Setsuko, S.; Yoshihira, K; Natori, S. Correlation between effects of 24 different cytochalasins on cellular structures and cellular events and those on actin in vitro. J Cell Biol. 92:69-78; 1982.

Yang, X.C.; Sachs, F. Block of stretch-activated ion channels in Xenopus oocytes by gadolinium and calcium ions. Science Wash DC. 243:1068-1071; 1989.

22
The Effects of Hydrostatic and Osmotic Pressures on Chondrocyte Metabolism

J.P.G. Urban and A.C. Hall

Introduction

Load-bearing cartilages are routinely exposed to significant mechanical forces. A wide range of studies *in vivo* have established that cartilage metabolism and ultimately the composition and mechanical properties of the tissue are influenced by the habitual load experienced. Mechanical forces can influence growth and development of cartilage, affecting the shape of the jaw for instance (Copray et al, 1985). In adults, cartilage is also influenced by mechanical stress. Loaded areas of cartilage are thicker and more resilient than unloaded regions. Removal of load causes cartilage to thin and become soft as proteoglycans are lost from the tissue while restoration of load can bring about recovery (reviewed by Tammi et al, 1987).

A range of *in vitro* studies have shown the complex responses of cartilage to load. Static loading invariably depresses synthesis of matrix in proportion to the applied load. For dynamic loading, the results are strongly dependent on frequency. Low frequencies can depress matrix synthesis rates, whereas high frequencies may stimulate them by 30-50% (Sah et al, 1989,1992; Parkkinen et al, 1993a).

Despite the marked effects of loading, and their apparent importance in the control of the composition and health of articular cartilage, the actual signals that the chondrocytes respond to are poorly understood. The signals to the chondrocytes arise from changes to the matrix during loading. The changes are complex and time-dependent and depend on the magnitude and duration of the load and on the material properties of the matrix. Table 1 describes the sequence following loading.

These changes occur to the physical environment of many cell types, although their magnitude differs, and therefore it is not surprising that cells may respond to some extent to any or all of these signals; changes to the ionic/osmotic environment for instance, can profoundly alter cell metabolism. Since cartilages, which are mainly subjected to compressive loading, experience particularly high hydrostatic pressures and alterations to their osmotic environment (Weightman and Kempson, 1979; Grodzinsky, 1983; Mow et al, 1984) it might be expected that chondrocytes can use these changes as signals to 'monitor' their unusual mechanical environment and respond accordingly to maintain appropriate mechanical properties. This chapter focuses on these two physical forces and their possible role in controlling matrix synthesis rates.

Table 1. The sequence of events following the application of load to articular cartilage. Events (1-3) are very rapid (msecs), and subsequent events have slower time courses (mins-hours).

1.	Rise in hydrostatic pressure
2.	Tissue and chondrocyte deformation
3.	Fluid flow resulting in streaming potentials
4.	Fluid loss
5.	Increased GAG concentration
6.	Changes to ionic composition of interstitial fluid; (a) *increased* cation (Na^+,K^+,Ca^{2+},H^+) concentrations (b) *increased* osmolality (c) *decreased* anion (Cl^-,HCO_3^-,SO_4^{2-}) concentrations

Hydrostatic Pressure

The changes in hydrostatic pressure during the loading of intervertebral disc and articular cartilage are surprisingly high. The intervertebral disc is exposed to a significant pressure when supine (about 0.2MPa; 1 atmosphere absolute = 1ATA = 0.1MPa), which can rise to 1-2MPa when lifting moderate weights (Nachemson and Elfstrom, 1970). Changes in pressure in articular cartilage are more impressive, and have been measured *in vivo* using a pressure-sensitive pseudo-femoral head (Hodge et al., 1986). This device implanted in a 73yr old woman measured increases in pressure to 18MPa within about 0.2 sec during standing from a 45 cm chair. Normal level walking caused pressure to cycle between near atmospheric and 3.5MPa. It seems reasonable to propose that these pressure changes are at the lower end of the scale and that in young athletic individuals the pressure levels are likely to be considerably higher. The physiological significance of these changes in pressure should be considered in the light of a wide range of other studies which show that the application of even relatively low levels of pressure can profoundly alter a range of cellular functions, some of which are summarised in Table 2. It is conceivable that chondrocytes are also sensitive to changes in hydrostatic pressure which will be changing routinely and which could represent a useful stimulus for detecting load.

The likelihood that pressure could be an important physiological signal for cartilages has been recognized for many years, and has accordingly attracted some interest. Early studies on epiphyseal embryo cartilage and chondrocytes (Bourret and Rodan, 1976; Veldhüijzen et al., 1979; Norton et al., 1977) using compressed air as the method for the application of pressure, showed marked effects on chondrocytes. Proteoglycan synthesis rates and rates of calcification were elevated, whereas levels of cAMP were decreased at pressures as low as 0.01MPa (Bourret and Rodan, 1976; Klein-Nulend et al., 1987; Van Kampen et al., 1985; Veldhuijzen et al., 1979). However because gas pressures were used it is not clear whether the cells were responding to hydrostatic pressure *per se* or to other factors for example changes to partial pressures of gas, pH, temperature etc. Lippiello et al., (1985) using hydrostatic pressure found that the response depended on the levels of pressure applied. Thus 0.5-2.4 MPa inhibited radiotracer ($^{35}SO_4$) incorporation in bovine and human articular cartilage whereas at 2.5 MPa there was a significant stimulation. In contrast, Kimura et al., (1985) using bovine cartilage found no effects of pressure up to 2.7MPa.

Table 2. Some effects of hydrostatic pressure on a variety of cellular processes. Refs; 1, Hall et al., 1982; 2, Goldinger et al., 1980a,b; 3, Conti et al., 1982a,b; 4, refs. in Hall and Ellory, 1987; 5, Hall et al., 1993; 6, Zimmerman et al., 1987; 7, Otter et al., 1987; 8, Heinemann et al., 1987a,b.

Process	System studied	Effect[(Ref)]
1. active cation transport	ouabain-sensitive Na, K fluxes	inhibited (50% at 20MPa)[1,2]
2. cation channels	Na, K, Ca currents	inhibited (20% at 20MPa)[3]
	Ca-activated K channel	stimulated (200% at 15MPa)[4]
3. amino acid transport	alanine, glycine, lysine uptake	inhibited (30% at 15MPa)[4,5]
4. ATPase activity	erythrocyte Na/K ATPase	stimulated (40% at 15MPa)[2]
5. cell morphology /cyto-skeletal elements	F-actin, tubulin, microfilaments	cell rounding, disruption of cytoskeleton (30min at 40MPa)[6]
6.macromolecular synthesis	DNA, RNA, protein synthesis	inhibited (20-40% at 30MPa)[7]
7. exocytosis	chromaffin cells	inhibited (50% at 10MPa)[8]

More recently, studies have been performed which ensures that the pressure applied is purely hydrostatic and the range of pressures cover and extend beyond the physiological range. Figure 1 shows that the pressure effect is biphasic and that pressures over the physiological range (0.1-20MPa) can stimulate matrix synthesis rates whereas higher pressures

(20-50MPa) cause inhibition. It can be seen that it is not only the levels of pressure that are important, but also the length of time for which pressure was applied (Hall et al, 1991). Other studies have investigated the frequency of loading and also reported a biphasic response such that there is significant depression of matrix synthesis at low frequencies, whereas synthesis is stimulated at high frequencies as shown in Figure 2 (Parkkinen et al, 1993a). The magnitude of the stimulation is of the same order as that seen with one 20sec pulse of pressure.

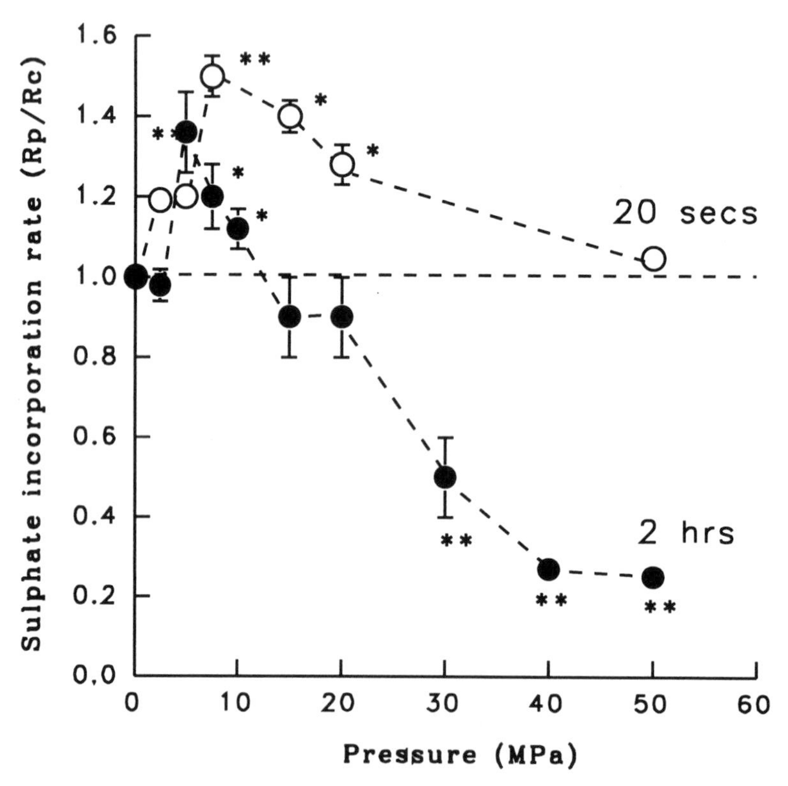

Figure 1. The effects of hydrostatic pressure on sulphate incorporation into cartilage explants (adapted from Hall et al, 1991). Pressure was applied for 20 secs (open symbols - rates measured for 2 hrs), or 2 hrs (closed symbols - rates measured over pressurization period); data expressed as rates at pressure/atmospheric rates).

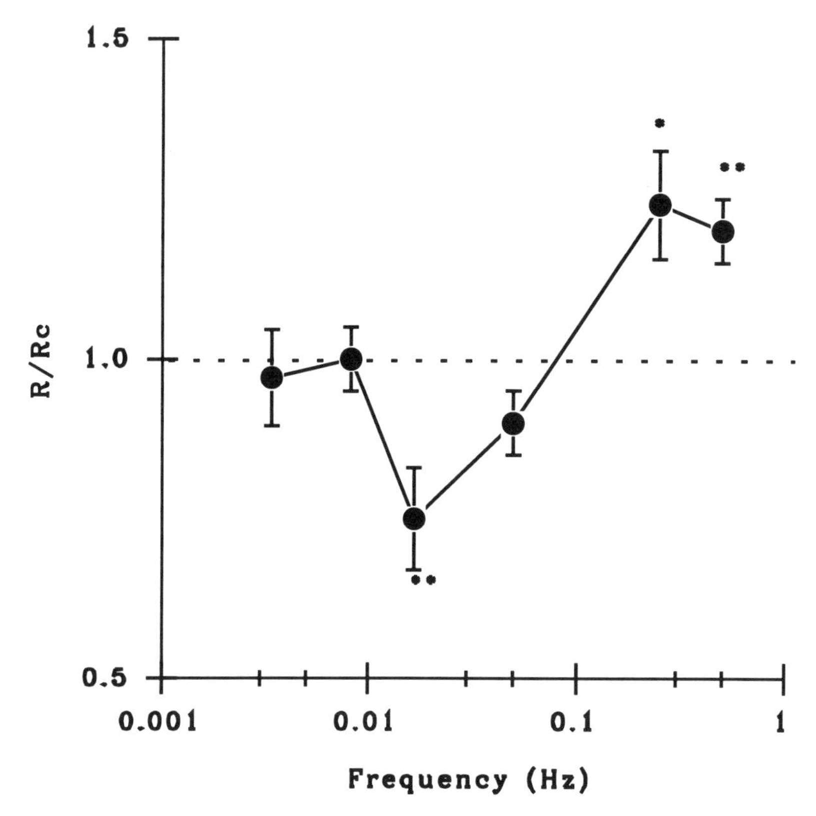

Figure 2. The effects of frequency on the response of cartilage plugs to 5MPa hydrostatic pressure for 20 hrs (adapted from Parkkinen et al, 1993a). Data are expressed as sulphate incorporation rates at pressure relative to those at atmospheric pressure.

From studies on pressure effects it appears that while long exposures to high pressure always inhibit synthesis, tests done under physiological applications of pressure do not produce such consistent results. There appear to be several factors which influence the pressure response including frequency, pressure level and possibly composition of medium used during the experiment (Lippiello et al, 1985; Parkkinen et al, 1993a; Hall and Urban, 1989, 1990, 1991). Biological variability also influences the sensitivity of articular cartilage to pressure. For example Table 3 shows that even in apparently identical experiments performed on the same day with cartilage obtained from 6 different feet, pressure had no effect on some feet but stimulated synthesis rates significantly in others. There are thus undoubtably other variables such as previous loading history, which are important in determining the response to physiological pressures, but are very difficult to control. Taken together it seems likely that pressure is a signal which can influence synthesis rates, but its influence can be over-ridden by other factors.

Table 3. The variation in the response of synthesis rates of bovine articular cartilage explants to pressure. Cartilage slices from the bovine metacarpo-phalangeal joint were obtained from the legs of 6 different animals, and synthesis rates measured at normal and at elevated pressure (2hrs at 5MPa)(data from Hall et al, 1991).

Leg number;	Rate at pressure/control rate
1	2.3**
2	1.1
3	1.3*
4	1.0
5	1.4*
6	1.4*
Pooled data	1.4 ± 0.2**

In view of these problems it might be expected that studies on the pressure response of cultured isolated chondrocytes might be fruitful since control of the experimental conditions is reproducible. There has been little work on articular chondrocytes, however the response of isolated chondrocytes to pressure appears to be very different from that of chondrocytes *in situ*. Experiments suggest that pressures which stimulate synthesis rates in cartilage inhibit rates in isolated chondrocytes (Figure 3). The causes of the difference in pressure-sensitivity over the physiological range are unknown, however high pressures (>20 MPa) consistently depress synthesis in chondrocytes, isolated or *in situ*. Parkkinen et al, (1993b) have suggested that alterations to cytoskeletal elements and the Golgi are involved in the pressure response. Their studies used pressure levels within the range where disruption of cytoskeletal elements are expected (Table 2). There is no evidence that the inhibition of synthesis at lower pressures (10 MPa) in isolated chondrocytes is due to alterations to the cytoskeleton.

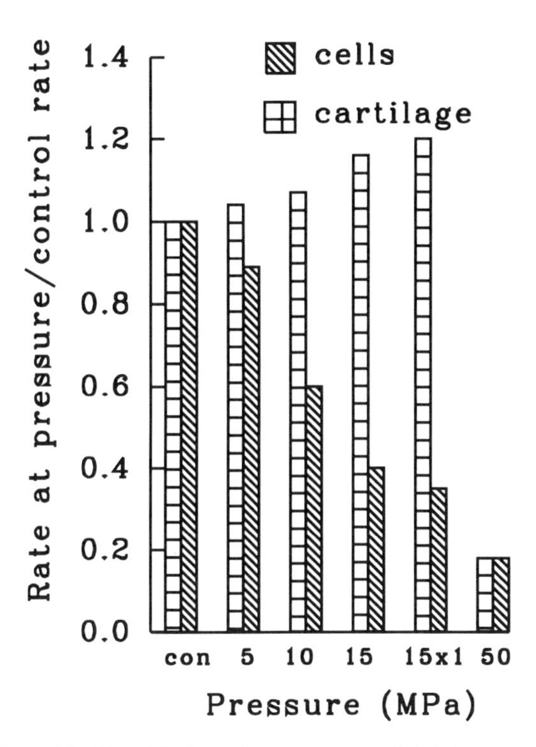

Figure 3. The differential effect of hydrostatic pressure on sulphate incorporation in cartilage and isolated chondrocytes. Pressure was applied for 5 minutes (5,10,15 MPa), 20 secs (15 MPa x 1 cycle) or 2hrs (50 MPa) and synthesis measured over the 2hrs following initial pressurization. Rates reported relative to rates at atmospheric pressure (adapted from Hall and Urban, 1989).

There have been suggestions, although very few supporting experiments, that perturbations to the functioning of the chondrocyte plasma membrane (e.g. membrane transport pathways, second messenger cascades) might underly the pressure-sensitivity of chondrocytes (Stockwell, 1991). Previous studies support the notion that the membrane is also a putative pressure-sensitive site since a range of membrane-mediated functions are known to be influenced by pressure (Table 2). Wright et al, (1992) studied some electrophysiological properties of sheep articular chondrocytes and human skin fibroblasts grown on plastic culture dishes. They found that although continuous pressure, applied through a gas phase, caused depolarization of both cell types, low frequency pressure applications (<0.08 Hz) depolarized chondrocytes, but hyperpolarized fibroblasts (Figure 4). From pharmacological evidence, they proposed a role for Ca-activated K channels and the chondrocyte cytoskeleton in mediating the pressure response. However it should be noted that because of their culture conditions and the use of flexible plastic Petri dishes, there may have been cell deformation during pressure application and therefore membrane stretch in addition to pressure might have been a signal. Deformation has been shown to activate a similar channel in other cell types (Christensen, 1987). Nevertheless it is of interest that chondrocytes appear to behave very differently to the same stimulus as a closely related cell type.

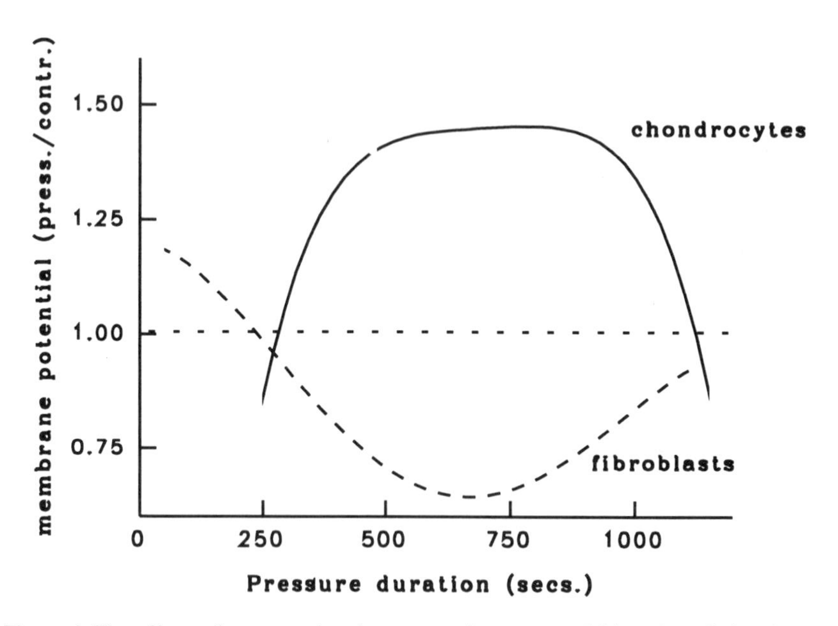

Figure 4. The effects of pressure duration on membrane potential in cultured chondrocytes and fibroblasts (adapted from Wright et al, 1992).

To summarise this section, it is clear that pressure can alter chondrocyte metabolism, however the response depends critically on the level and duration of the applied pressure. Pressure applications above those expected physiologically appear to inhibit synthesis rates in a reasonably consistent manner. Nevertheless, over the physiological range, the response is more complex and unpredictable, and depends not only on the subtleties of the way pressure is applied, but also on other factors which are not understood. In the following section we consider changes to the ionic/osmotic environment, which as well as influencing cell behaviour, could have an influence on the pressure response.

Osmotic Pressure

The extracellular ionic environment of chondrocytes is very different from that of most other cells and alters during loading. Chondrocytes are embedded in a high concentration of proteoglycans; since these impart a net negative charge to the matrix they influence the extracellular ionic environment, with the concentration of ions very sensitive to proteoglycan concentration because that directly affects the fixed negative charge density (Table 4).

Table 4 shows that in the matrix, chondrocytes are surrounded by an extracellular fluid whose cation concentration is considerably higher than in synovial fluid, and whose anion concentration is considerably lower; the ionic concentrations are well predicted by Donnan equilibrium conditions (Maroudas and Evans, 1974). The total concentration of ions in the extracellular fluid of cartilage is considerably higher than in standard medium, serum or synovial fluid, and the osmolality of the interstitial fluid in cartilage is correspondingly greater. It is this osmotic difference which maintains cartilage hydration under high external loads (Maroudas, 1979; Grodzinsky, 1983; Lai et al., 1991).

Chondrocytes are now isolated from the matrix for many experimental purposes and placed in standard tissue culture medium such as Dulbecco's modified Eagle's medium (DMEM). One consequence of this process, is that the extracellular osmolality and ionic environment are altered significantly (Table 4). Since cells respond to changes in extracellular osmolality by losing or gaining water to maintain osmotic equilibrium across the cell membrane, such changes lead to alterations in cell volume (Hoffman and Simonsen, 1989). Figure 5 shows how the volume of chondrocytes changes with extracellular osmolality, the average cell size being considerably greater in 280 mOsm medium (DMEM) than in 450 mOsm medium. Since 450 mOsm is within the extracellular range found *in vivo* (Table 4), chondrocytes will swell considerably during the isolation process.

Table 4. Extracellular environment of articular chondrocytes from the surface and deep zones of the human femoral head compared to the composition of tissue culture medium or synovial fluid (data from Maroudas and Evans, 1974).

	Human Hip: Surface zone	Human Hip: Deep zone	DMEM, synovial fluid
Fixed charge density (meq/l H_2O)	100-180	150-240	0
[Na^+] mM	200-240	220-300	130-150
[K^+] mM	6-8	8-12	5
[Ca^{++}] mM	4-8	6-20	2
[Cl^-] mM	80-100	60-90	150
pH	7-7.2	7.2-6.9	7.4
osmolality (mOsm)	350-400	350-470	280-300

Examination of chondrocyte volumes *in situ* in the matrix by confocal microscopy has shown that in the matrix too, chondrocyte volumes shrink or swell in proportion to the increase or decrease in extracellular osmolality (Deshayes et al, 1993). Many cells, including chondrocytes are able to regulate their volume to some extent after extracellular osmotic shock by expelling or taking up extracellular solutes (Hoffman and Simonsen, 1989; Deshayes et al, 1993). However this process, while returning the cell towards its initial volume also alters intracellular composition and thus may affect cellular metabolism.

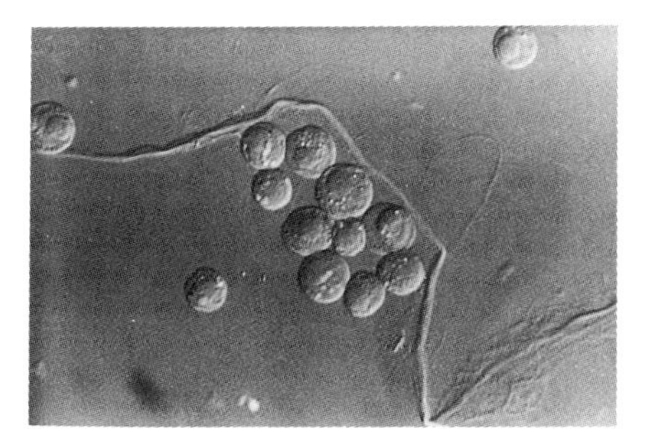

150 mOsm

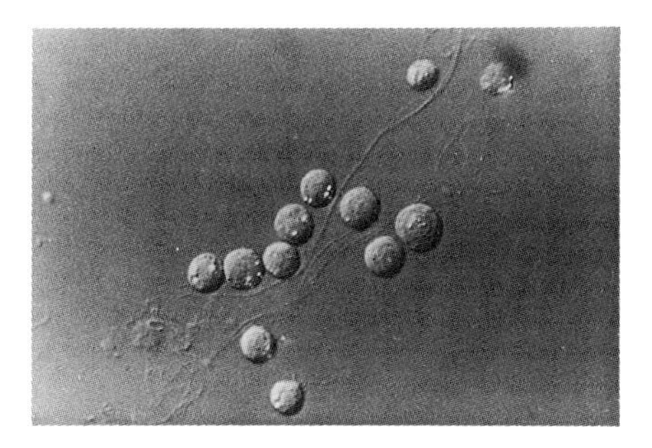

280 mOsm (DMEM)

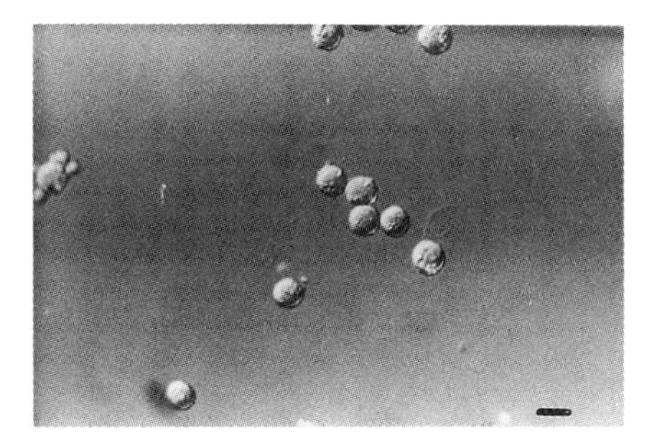

450 mOsm

Figure 5. The effect of changes in medium osmolality on chondrocyte volume. Chondrocytes were isolated from bovine articular cartilage by enzyme digestion in DMEM using standard procedures (Urban et al, 1993) and resuspended in medium whose osmolality was changed to 150, 280 or 450 mOsm by water or sucrose addition as required; the scale bar represents 10 microns. The cells were photographed 5-10 mins after osmotic shock.

When cells are isolated from the matrix and their synthesis rates measured, the rates are only around one tenth the rate/cell in the matrix (Urban et al, 1993). Of course, many aspects of the chondrocyte's environment change when it is isolated. For instance direct cell-matrix interactions are lost, cell shape changes, the concentration of cytokines and growth factors is altered; all of these changes could affect the chondrocyte's metabolism and thus lead to fall in synthesis rates. One factor which seems to be a significant regulator of matrix synthesis rates is the change in extracellular osmolality, which as shown above alters with cell isolation and affects chondrocyte volumes and thus the intracellular composition of chondrocytes.

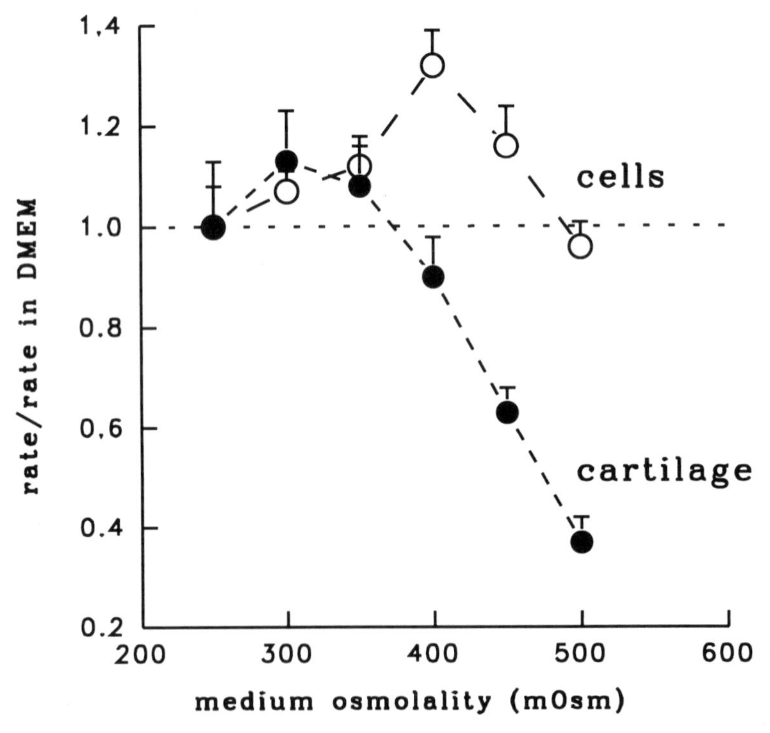

Figure 6. The effect of extracellular osmolality (NaCl addition) on ^{35}S-sulphate incorporation rates in bovine articular cartilage slices and isolated bovine articular chondrocytes. The rates were normalised to the rate in DMEM (adapted from Urban et al, 1993).

Figure 6 shows that if the extracellular osmolality of chondrocytes is altered by adding NaCl to the medium, proteoglycan and protein synthesis rates are affected. For isolated cells, synthesis rates increased as the osmolality increased, peaked between 350-450 mOsm and then decreased as the osmolality was further increased. Normalised rates in cartilage followed a similar pattern, but the relative peak in synthesis rates was around 300 mOsm and was far lower than in isolated cells.

The extracellular osmolality for the isolated cells is of course the same as that of the medium but this is not true in cartilage (Table 4). Here the osmolality of the interstitial fluid bathing the chondrocytes is raised as a result of the Donnan equilibria. Thus when cartilage and cells are incubated in the same medium, the extracellular osmolality for the chondrocytes in the cartilage slices is always greater than that of the isolated cells. If the curves of Figure 6 are plotted against extracellular osmolality rather than medium osmolality, they superimpose (Urban et al, 1993) suggesting direct cell matrix interactions have little effect on the response of chondrocytes to extracellular osmolality. For both cartilage and isolated cells, it thus appears that a change in osmolality affects synthesis and that the peak in synthesis rates occurs when extracellular osmolality (as opposed to medium osmolality) is at *in vivo* values.

When cartilage is loaded, fluid is expressed. The amount of fluid lost depends on the magnitude and duration of the load and on the proteoglycan concentration of the cartilage (Maroudas, 1979). Loading will influence fixed charge density since fluid expression under load increases proteoglycan concentration in proportion to the amount of extrafibrillar fluid expressed. Thus under load, cation concentrations and extracellular osmolality will tend to increase, and anion concentrations decrease. Many studies have found that if proteoglycan synthesis rates are measured in cartilage under static loads, synthesis rates fall in proportion to applied stress. Gray et al, (1989) has summarised the results of several studies. Jones et al, (1982) suggested that this effect was related to loss of fluid.

This suggestion is supported by several studies. Schneiderman et al, (1986) for instance, placed cartilage plugs under a mechanical load or alternatively equilibrated plugs in a solution of polyethylene glycol of high extracellular osmolality. In both cases fluid was lost in proportion to the applied mechanical or osmotic stress. In the first case, the plugs deformed anisotropically and hydrostatic pressure increased. In the second case, there was no increase in hydrostatic pressure, the stress was isotropic and the shape of the cartilage plugs and cells at equilibrium was very different from those which had been loaded (Broom and Myers, 1980; Guilak and Bachrach, 1993). If synthesis rates were plotted against fluid loss however, there was no apparent difference between the two curves (Figure 7).

Fluid loss will of course increase extracellular osmolality, and at equivalent degrees of fluid loss, the change in extracellular osmolality will be similar. This result is thus in agreement with that of Figure 6, indicating again that a change in extracellular osmolality will lead to a proportional change in synthesis rates and that this effect is not strongly influenced by the chondrocyte shape or cell-matrix interactions.

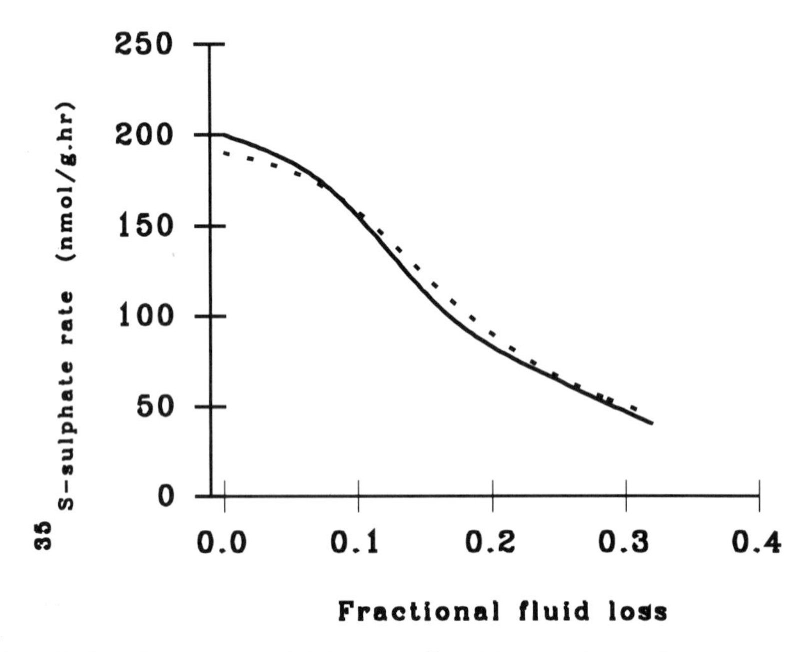

Figure 7. The effect of fractional fluid loss on ^{35}S-sulphate uptake rates in human femoral head cartilage. Cartilage plugs were either loaded mechanically or equilibrated in polyethylene glycol solutions (adapted from Schneiderman et al, 1986).

A change in fluid content will affect the overall ionic composition of the cartilage as well as increasing extracellular osmolality. Cation concentrations will rise; [Ca^{++}] concentrations are particularly sensitive to fixed charge density (Maroudas and Evans, 1974) and can increase to 20mM free Ca^{++} in some regions of cartilage. A change in extracellular proteoglycan concentration will also affect pH as [H^+] also varies with fixed charge density (Maroudas, 1979; Gray et al, 1988). Anion concentrations will however fall. Since intracellular pH, Ca^{++}, and K^+ are known to be important regulators of cellular behaviour, one might expect these ions to influence chondrocytes and consequently there have been several studies on the effects of these ions on cartilage metabolism.

Intracellular K^+ controls many aspects of protein synthesis as enzyme activities are sensitive to intracellular potassium levels (Kernan, 1980). An indication of this sensitivity is given by the study of Horowitz and Lau (1988); they measured intracellular potassium concentrations ($[K^+]_i$) and protein synthesis rates in the same oocyte and found that rates increase 4-5 fold for a 20 to 30% change in $[K^+]_i$. It is thus not very surprising that extracellular potassium ($[K^+]_o$) alters chondrocyte behaviour since one would predict that an increase in $[K^+]_o$ would lead to a rise in $[K^+]_i$ via specialised membrane transport systems (Stein, 1986); $[K^+]_o$ was found alter matrix synthesis and deposition by chondrocytes in long-term culture (Daniel et al, 1974).

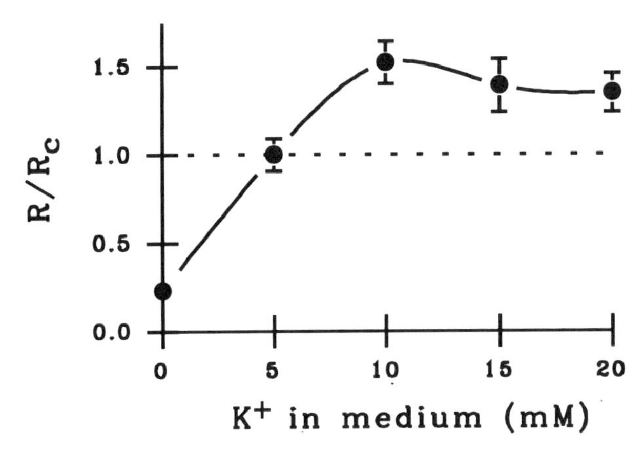

Figure 8. The effect of extracellular potassium concentration on ^{35}S-sulphate incorporation in isolated chondrocytes (adapted from Urban et al, 1993).

The immediate effect of a change in $[K^+]_o$ on synthesis in isolated chondrocytes is shown in Figure 8. Medium concentrations of potassium were altered using sucrose as an osmotic replacement (Urban et al, 1993). Synthesis rates were lowest in potassium-free media. Synthesis rates in 10mM K^+ (physiological levels) were 50% higher than in 5mM K^+ (i.e. DMEM). Increases in K^+ beyond 10mM had no further effect suggesting that loading does not influence synthesis via alteration to K^+ levels.

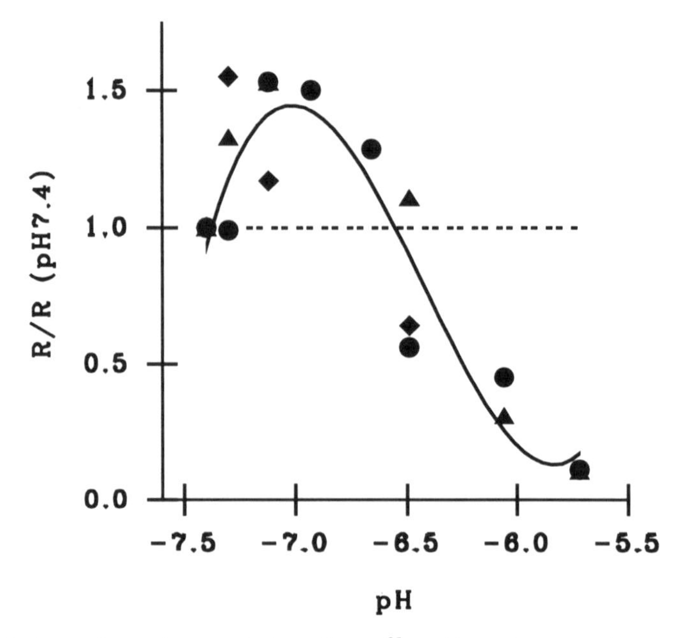

Figure 9. The effect of extracellular pH on ^{35}S-sulphate incorporation rates in the intervertebral disc (adapted from Ohshima and Urban, 1992). The disc slices were cultured in DMEM to which polyethylene glycol was added to maintain osmotic equilibrium according to the method of Bayliss et al (1986), and pH was altered by addition of HCl or lactate. Bovine disc (HCl addition-triangles); bovine disc (lactate addition-circles); human disc (HCl addition-diamonds).

The effect of other ions has also been investigated. Calcium ions are reported to affect phenotypic expression in cultured chondrocytes (Deshmukh et al, 1976). Patch-clamp studies and fluorescence measurements of intracellular free Ca^{++} have shown that the most marked changes occur at very low extracellular Ca^{++} levels and that little further effect is seen in the physiological range (Grandolfo et al, 1992; Ponte and Hall, 1993). This suggests that chondrocytes can regulate internal calcium efficiently, which is probably necessary in view of the high and changing calcium levels they experience. Extracellular Ca^{++} over the physiological range (2-20 mM) appears to have no immediate affect on synthesis rates in cartilage or chondrocytes (Schneiderman et al, 1986; Urban et al, 1993).

In contrast to calcium, extracellular pH has a potent effect on synthesis rates in cartilages. A change in pH in epiphyseal cartilage has been found to decrease proteoglycan synthesis in a dose-dependent manner (Gray et al, 1988). Figure 9 shows the effect of extracellular pH on synthesis rates in the intervertebral disc nucleus. The *in vivo* pH in the disc is around pH 6.9-7.1 (Eyre et al, 1989) and synthesis rates at that pH are 40-50% greater than at pH 7.4, again suggesting that chondrocyte synthesis rates are depressed by any change from their *in vivo* ionic and osmotic environment. Results were similar for lactate and protons, suggesting that pH was the effector. Extracellular pH may also be involved in the response of cartilage to load. Gray et al, (1988) found that in epiphyseal cartilage the decrease in synthesis rates with compression correlated well with a decrease in pH induced by altering medium pH. Thus a change in pH induced by compression could be important in regulating the response of chondrocytes to static loads.

In summary, any change in extracellular osmolality leads to a proportional depression of matrix synthesis. Loads which result in fluid loss and increase osmolality of cartilage, consistently depress synthesis in a similar fashion. It thus seems likely that the effects of static loading are mediated at least in part through their effects on extracellular ionic composition and osmolality. This suggestion is supported because there seems to be little difference in the response of cartilage and chondrocytes to such changes, suggesting that other extracellular factors such as cell shape or direct cell-matrix interactions are not significantly involved in this response.

Conclusions

Two factors in the extracellular environment of chondrocytes which can change markedly during loading, *viz* hydrostatic and osmotic pressure, both have significant effects on matrix synthesis in cartilage explants and on isolated chondrocytes. In general loading which leads to fluid expression leads invariably to a fall in synthesis rates, and this effect appears be modulated to a large extent by changes in ionic composition and hence in extracellular osmolality. High and prolonged pressure also invariably inhibit synthesis. While physiological pressures can stimulate synthesis by around 50%, factors which are not yet understood are involved in the response. It is also clear that chondrocytes respond to a range of signals arising from the matrix changes which occur during loading, and that their interaction determines the overall cellular response.

Acknowledgements

We would like to thank the Arthritis and Rheumatism Council (J.P.G.U.) and the Wellcome Trust (A.C.H.) for support.

References

Bayliss, M.T., Urban, J.P.G., Johnson, B. and Holm, S. In vitro method for measuring synthesis rates in the intervertebral disc. J. Orthop. Res. 4:10-17, 1986.

Bourret, L.A. and Rodan, G.A. The role of calcium in the inhibition of cAMP accumulation in epiphyseal cartilage cells exposed to cartilage pressure. J. Cell. Physiol. 88:353-362, 1988.

Broom, N.D. and Myers, D.B. A study of the structural response of wet hyaline cartilage to various loading stresses. Conn. Tiss. Res. 7:227-237, 1980.

Christensen, O. Mediation of cell volume regulation by Ca^{2+} influx through stretch-activated channels. Nature 330, 66-68, 1987.

Copray, J.C.V.M., Jansen, H.W.B. and Duterloo, H.S. The role of biomechanical factors in mandibular condylar cartilage growth and remodelling in vitro. In: *Developmental Aspects of Temporomandibular Joint Disorders*, edited by Carlson, D.S., McNamara, J.A. and Ribbens, K.A. Ann Arbor: University of Michigan Press, 1985, pp. 235-269.

Conti, F., Fioravanti, R., Segal, J.R. and Stuhmer, W. Pressure dependence of the sodium currents of squid giant axon. J. Membr. Biol. 69:23-34, 1982a.

Conti, F., Fioravanti, R., Segal, J.R. and Stuhmer, W. Pressure dependence of the potassium currents of squid giant axon. J. Membr. Biol. 69:35-40, 1982b.

Daniel, J.C., Kosher, R.A., Hamos, J.E. and Lash, J.W. Influence of external potassium on the synthesis and deposition of matrix components by chondrocytes in vitro. J. Cell. Biol. 63:843-854, 1987.

Deshayes, C.M.P., Hall, A.C. and Urban, J.P.G. Effects of extracellular osmolality on porcine articular chondrocyte volume. J. Physiol. 467:214P, 1993.

Deshmukh, K., Kline, W.G. and Sawyer, B.D. Role of calcium in the phenotypic expression of rabbit articular chondrocytes in culture. FEBS Letters 67:48-51, 1976.

Eyre, D.R., Caterson, B., Benya, P., et al. The intervertebral disc. In: *New Perspectives on Low Back Pain*, edited by Gordon, S. and Frymoyer, J. Philadelphia: Am. Inst. Orthop. Surg., 1991, pp. 147-208.

Goldinger, J.M., Kang, B.S., Choo, Y.E., Paganelli, C.V. and Hong, S.K. Effect of hydrostatic pressure on ion transport and metabolism in human erythrocytes. J. Appl. Physiol. 49:224-231, 1980a.

Goldinger, J.M., Kang, B.S., Morin, R.A., Paganelli, C.V. and Hong, S.K. Effect of hydrostatic pressure on active transport, metabolism and the Donnan equilibrium in human erythrocytes. In: *Proc. VIIth Symp. on Underwater Physiol.* Eds., A.J. Bachrach and M.M. Matzen. Undersea Med. Soc. Inc., Bethesda. pp. 589-599, 1980b.

Grandolfo, M., D'Andrea, P., Martina, M., Ruzzier, F. and Vittur, F. Calcium-activated potassium channels in chondrocytes. Biochem. Biophys. Res. Comms. 182:1429-1434, 1992.

Gray, M.L., Pizzanelli, A.M., Grodzinsky, A.J. and Lee, R.C. Mechanical and physiochemical determinants of the chondrocyte biosynthetic response. J. Orthop. Res. 6:777-792, 1988.

Gray, M.L., Pizzanelli, A.M., Lee, R.C., Grodzinsky, A.J. and Swann, D.A. Kinetics of the chondrocyte biosynthetic response to compressive loading and release. Biochim. Biophys. Acta. 991:415-425, 1989.

Grodzinsky, A.J. Electromechanical and physico-chemical properties of connective tissue. CRC Crit. Rev. Biomed. Eng. 9:133-199, 1983.

Guilak, F. and Bachrach, N.M. Compression-induced changes in chondrocyte shape and volume determined in situ using confocal microscopy. Trans. Am. Orthop. Res. Soc. 18:619, 1993.

Hall, A.C. and Ellory, J.C. Hydrostatic pressure effects on transport in liposomes and red cells. In: *Current Perspectives in High Pressure Biology*, Eds. H.W. Jannasch, R.E. Marquis, A.M. Zimmerman, Academic Press, London, 1987, pp. 191-206.

Hall, A.C., Ellory, J.C. and Klein, R.A. Pressure and temperature effects on human red cell cation transport. J. Membr. Biol. 68:47-56, 1982.

Hall, A.C., Pickles, D.M. and Macdonald, A.G. Effects of high pressure on biological systems: aspects of eukaryotic cells. In: *Advances in Comp. and Environ. Physiol.*, Vol. 17, Ed. A.G. Macdonald, Springer Verlag, Heidelberg, 1993, pp. 29-85.

Hall, A.C. and Urban, J.P.G. Responses of articular chondrocytes and cartilage to high hydrostatic pressure. Trans. Am. Orthop. Res. Soc. 14:49, 1989.

Hall, A.C. and Urban, J.P.G. The role of chondrocyte membrane transport processes in mediating the effect of fluid loss and hydrostatic pressure on matrix synthesis. Trans. Am. Orthop. Res. Soc. 15:340, 1990.

Hall, A.C. and Urban, J.P.G. Fast and slow responses of articular chondrocytes to hydrostatic pressure. Trans. Am. Orthop. Res. Soc. 16:240, 1991.

Hall, A.C., Urban, J.P.G. and Gehl, K.A. The effects of hydrostatic pressure on matrix synthesis in articular cartilage. J. Orthop. Res. 9:1-10, 1991.

Heinemann, S.H., Conti, F., Stuhmer, W. and Neher, E. Effects of hydrostatic pressure on membrane processes: sodium channels, calcium channels and exocytosis. J. Gen. Physiol. 90:765-778, 1987a.

Heinemann, S.H., Stuhmer, W. and Conti, F. Single acetylcholine receptor channel currents recorded at high hydrostatic pressures. Proc. Natl. Acad. Sci. 84:3229-3233, 1987b.

Hodge, W.A., Fijan, R.S., Carlson, K.L., Burgess, R.G., Harris, W.H. and Mann, R.W. Contact pressures in the human hip joint measured in vivo. Proc. Natl. Acad. Sci. USA 83:2879-2883, 1986.

Hoffman, E.K. and Simonsen, L.O. Membrane mechanisms in volume and pH regulation in vertebrate cells. Physiol. Rev. 69:315-382, 1989.

Horowitz, S. and Lau, Y-T. A function that relates protein synthetic rates to potassium activity in vivo. J. Cell. Physiol. 135:425-434, 1988.

Jones, I.L., Klamfeldt, A. and Sandstrom, T. The effect of continuous mechanical pressure upon the turnover of articular cartilage proteoglycans in vitro. Clin. Orthop. 165:283-289, 1982.

Kernan, R.P. Cell Potassium. John Wiley and Sons, New York, pp. 138-150, 1980.

Kimura, J.H., Schipplein, O.D., Kuettner, K.E. and Andriacchi, T.P. Effects of hydrostatic loading on extracellular matrix formation. Trans. Am. Orthop. Res. Soc. 10:365, 1985.

Klein-Nulend, J., Veldhuizen, J.P., van der Stadt, R.J., Van Kampen, G.P.J., Kuijer, R. and Burger, E.H. Influence of intermittent compressive force on proteoglycan content in calcifying growth plate cartilage in vitro. J. Biol. Chem. 262:15490-15495, 1987.

Lai, W.M., Hou, J.S. and Mow, V.C. A triphasic theory for the swelling and deformation behaviours of articular cartilage. J. Biomech. Eng. 113:245-258, 1991.

Lippiello, L., Kaye, C., Neumata, T. and Mankin, H.J. In vitro metabolic response of articular cartilage segments to low levels of hydrostatic pressure. Conn. Tiss. Res. 13:99-107, 1985.

Maroudas, A. Physico-chemical properties of articular cartilage. In: *Adult Articular Cartilage,* edited by Freeman, M., London: Pitman Medical, 1979, pp. 215-290.

Maroudas, A. and Evans, H. A study of ionic equilibria in cartilage. Conn. Tiss. Res. 1:69-79, 1974.

Mow, V.C., Holmes, M.H. and Lai, W.M. Fluid transport and mechanical properties of articular cartilage. J. Biomech. 17:377-394, 1984.

Nachemson, A. and Elfstrom, G. Intravital dynamic pressure measurements in lumbar discs. A study of common movements, manoeuvres and exercises. Scand. J. Rehabil. Med. 2 (suppl 1):1-40, 1970.

Norton, L.A., Rodan, G.A. and Bourret, L.A. Epiphyseal cartilage cAMP changes produced by electrical and mechanical perturbations. Clin. Orthop. 124:59-68, 1977.

Ohshima, H. and Urban, J.P.G. The effects of lactate and pH on sulphate and protein metabolism in the intervertebral disc. Spine 17:1079-1082, 1992.

Otter, T., Bourns, B., Franklin, S., Reider, C. and Salmon, E.C. Hydrostatic pressure effects on cytoskeletal organization and ciliary motion: a calcium hypothesis. In: *Current Perspectives in High Pressure Biology*, Eds. H.W. Jannasch, R.E. Marquis, A.M. Zimmerman, Academic Press, London, 1987, pp. 75-93.

Parkkinen, J.J., Ikonen, J., Lammi, M.J., Laakonen, J. and Helminen, H.J. Effects of cyclic hydrostatic pressure on proteoglycan synthesis in cultured chondrocytes and articular cartilage explants. Arch. Biochem. Biophys. 300:458-465, 1993a.

Parkkinen, J.J., Lammi, H.J., Helminen, H.J. and Tammi, M. Local stimulation of proteoglycan synthesis in articular cartilage explants by dynamic compression in vitro. J. Orthop. Res. 10:610-620, 1992.

Parkkinen, J.J., Lammi, M.J., Pelttari, A., Helminen, H.J., Tammi, M. and Virtanen, I. Altered Golgi apparatus in hydrostatically loaded articular cartilage chondrocytes. Ann. Rheum. Dis. 52:192-198, 1993b.

Ponte, M.R. and Hall, A.C. Effect of extracellular Ca and Na on $[Ca]_i$ of porcine articular chondrocytes. J. Physiol. (King's College Meeting, P43) In Press, 1993.

Sah, R.L., Grodzinsky, A.J., Plaas, A.H.K. and Sandy, J.D. Effects of static and dynamic compression on matrix metabolism in cartilage explants. In: *Articular Cartilage and Osteoarthritis*, edited by Kuettner, K., Shleyerbach, R., Peyron, J. and Hascall, V. New York: Raven Press, 1992, pp. 373-392.

Sah, R.L., Kim, Y.L., Doong, J-Y.H., Grodzinsky, A.J., Plaas, A.H.K. and Sandy, J.D. Biosynthetic response of cartilage explants to dynamic compression. J. Orthop. Res. 7:619-639, 1989.

Schneiderman, R., Keret, D. and Maroudas, A. Effects of mechanical and osmotic pressure on the rate of glycosaminoglycan synthesis in the human adult femoral head. J. Orthop. Res. 4:393-408, 1986.

Stein, W.D. Transport and diffusion across cell membranes. Academic Press, London, 1986.

Stockwell, R.A. Cartilage failure in osteoarthritis: relevance of normal structure and function. A review. Clin. Anat. 4:161-191, 1991.

Tammi, M., Paukkonen, K., Kiviranta, I., Jurvelin, J., Saamenen, A-M. and Helminen, H.J. Joint loading induced alterations in articular cartilage. In: *Joint Loading. Biology and Health of Articular Cartilage*, edited by Helminen, H.J., Kiviranta, I., Saamanen, A-M., Tammi, M., Paukkonen, K. and Jurvelin, J. Bristol: Wright, 1987, pp. 64-88.

Urban, J.P.G., Hall, A.C. and Gehl, K.A. Regulation of matrix synthesis rates by the ionic and osmotic environment of articular chondrocytes. J. Cell. Physiol. 154:262-270, 1993.

Van Kampen, G.P.J., Veldhuizen, J.P., Kuijer, R., van der Stadt, R.J. and Schipper, C.A. Cartilage response to mechanical forces studied in high density cultures. Arthr. Rheum. 28:419-424, 1985.

Veldhuijzen, J., Bourret, L. and Rodan, G. In vitro studies of the effect of intermittent compressive forces on cartilage cell proliferation. J. Cell. Physiol. 98:299-306, 1979.

Weightman, B. and Kempson, G. Load carriage. In: *Adult Articular Cartilage*, edited by Freeman, M.A.R. London: Pitman Medical, 1979, pp. 293-341.

Wright, M.O., Stockwell, R.A. and Nuki, G. Response of plasma membrane to applied hydrostatic pressure in chondrocytes and fibroblasts. Conn. Tiss. Res. 28:49-70, 1992.

Zimmerman, A.M., Tahir, S. and Zimmerman, S. Macromolecular synthesis under hydrostatic pressure. In: *Current Perspectives in High Pressure Biology*, Eds. H.W. Jannasch, R.E. Marquis and A.M. Zimmerman, London, Academic Press, 1987, pp. 49-63.

23
Proteoglycan Synthesis and Cytoskeleton in Hydrostatically Loaded Chondrocytes

J.J. Parkkinen, M.J. Lammi, M.I. Tammi,
H.J. Helminen

Introduction

The cartilage extracellular matrix, maintained by the articular chondrocytes, allows the tissue to withstand large compressive stresses. The proteoglycan (PG) matrix, together with the collagen fibril network, largely determines the biomechanical properties of the articular cartilage. The importance of joint motion and weight-bearing for the maintenance of normal, healthy articular cartilage has been shown in numerous studies. *In vivo* models have been useful in broadening our knowledge of cartilage tissue responses to the absence of joint movement, reduced or increased weight-bearing, and increased physical activity by running (Tammi et al. 1987; Helminen et al. 1992).

Extracellular matrix and loading are important modulators of the PG metabolism of chondrocytes. However, the mechanisms mediating loading effects on chondrocyte metabolism are not well understood although they are crucial to understand the regulation of the extracellular matrix composition. Using *in vivo* models, relevant alterations of articular cartilage can be reliably quantified, but the cellular mechanisms controlling those changes are difficult to examine. Experiments carried out using *in vivo* models, therefore, cannot provide the information about the influence of one regulating component among the several separate factors operating in the cartilage during normal loading. In addition, investigation of immediate cellular signals induced by joint loading is virtually impossible in animal experiments.

The *in vitro* models, both cartilage explant and chondrocyte cultures, offer a more direct way to study the cellular influences of articular cartilage loading. These models work under a defined environment and they can be utilized to reveal the effects of loading on the metabolism of chondrocytes. As hydrostatic pressure is one of the potential factors (besides mechanical loading and stretching) that controls the metabolism of the articular cartilage, we have examined cartilage explants and cultured chondrocytes after *in vitro* loading with continuous and cyclic hydrostatic pressures. The objective of our study was to investigate how loading of the chondrocytes affects the biosynthesis of PGs. The effects of hydrostatic pressure on PG structures were studied and the steady-state levels of PG mRNA transcripts were quantified. Also the structural changes in the Golgi apparatus and cytoskeleton due to pressure were assessed.

Regulation of Extracellular Matrix Components

Regulation of the genes encoding cartilage macromolecules is a complex process. As regards to developmental regulation, chondrocytes contain type I collagen mRNA only at the very early stages of chondrogenesis (Sandberg and Vuorio 1987). *In situ* hybridization for type II collagen transcripts shows that they are detectable in chick limb bud a few stages earlier than those for aggrecan and link protein transcripts, the latter two appearing at the stage when chondrogenesis begins morphologically (Goetinck et al. 1990). During human skeletal development, an identical pattern for aggrecan and link protein was noticed by *in situ* hybridization (Mundlos et al. 1991). With the small dermatan sulfate (DS)-PGs, remarkable differences were noticed at specific regions of developing human bones. High levels of biglycan mRNA were noticed in superficial chondrocytes, whereas decorin was strongly expressed in the deep zone of the cartilage (Bianco et al. 1990). Immunostaining for decorin and biglycan resulted in a similar distribution (Bianco et al. 1990).

In normal articular cartilage, the matrix components undergo continuous and balanced turnover. Fibrillar collagen is very resistant to proteolytic cleavage and has the slowest turnover rate. Due to the slow turnover, the mRNA levels for type II collagen transcripts in adults are often undetectable or very low. On the other hand, PGs have faster turnover rates and they respond more quickly to the chemical and environmental alterations of the tissue. In cartilage explant cultures, serum was required in the medium to maintain the normal PG concentration of the tissue (Hascall et al. 1983). Later, several individual hormones and growth factors present in serum have been shown to increase PG synthesis in explant cultures, e.g., transforming growth factor β (Morales 1992) and insulin-like growth factor 1 (Luyten et al. 1988) are capable of maintaining PG levels in explant cultures. In addition, bone-derived growth and differentiation factors can maintain PG synthesis in bovine articular cartilage explant cultures (Luyten et al. 1992).

In a diseased cartilage such as in osteoarthrosis, a net loss of large aggregating PGs (aggrecans) and small PGs occurs (Mow et al. 1992). In early osteoarthrosis, aggrecan molecules have a reduced molecular size, while in a more advanced stage they appear larger than those in a healthy tissue. This suggests increased degradation of PGs in the early disease and a later repair process replacing the matrix PGs by a population of newly-synthesized, larger molecules (Mow et al. 1992). Increased PG (Sandy et al. 1984) and collagen synthesis (Poole et al. 1993) is noticed in the beginning of experimental osteoarthrosis. Monoclonal antibodies recognizing neoepitopes in chondroitin sulfate (CS) produced during repair may be efficient biomarkers for the detection of early osteoarthrosis (Caterson et al. 1990; Poole et al. 1993).

In explant cultures, hyaluronic acid (HA) and aggrecan have the same turnover time (Morales and Hascall 1988; Ng et al. 1992). Addition of proteinase inhibitors into the culture decreased the loss of aggrecan and HA, suggesting that catabolism of HA is cell-mediated and depends on metabolically active chondrocytes (Ng et al. 1992). As the catabolism of aggrecan and HA occur at a similar rate, their degradation may be coupled to each other and involve internalization of HA (Hua et al. 1993), as previously suggested (Morales and Hascall 1988).

Mechanical Loading and Stretching Affect Cartilage and Chondrocytes *in Vitro*

In vitro models offer a direct way to study the cellular influences of loading on articular cartilage (van Kampen and van de Stadt 1987). An important advantage of *in vitro* conditions for biosynthetic experiments is that they form a closed system. The flux of isotope precursors and the passage of metabolic products can be measured precisely in different compartments. The conditions are manipulated easily and the variability between individual animals can be reduced.

Under *in vitro* conditions cartilage explants or chondrocyte cultures have been be exposed to direct mechanical load (Norton et al. 1977; Jones et al. 1982; Palmoski and Brandt 1984; Copray et al. 1985; Klämfeldt 1985; Schneiderman et al. 1986; Gray et al. 1988; Sah et al. 1989; Sah et al. 1990; Guilak et al. 1991; Larsson et al. 1991; Sah et al. 1991; Korver et al. 1992; Parkkinen et al. 1992) or a stretching force (Lee et al. 1982; de Witt et al. 1984; Uchida et al. 1988). Most of the pressures used have been low when one considers the loads estimated to act *in vivo*.

The principal mode of loading of articular cartilage is compression applied perpendicularly to the surface. With *in vitro* models, cartilage explants can be exposed to loads which mimic the *in vivo* loading conditions. Continuous, mechanical compression caused a remarkable decrease in PG synthesis as measured by [^{35}S]sulfate incorporation into the cartilage explants (Jones et al. 1982; Palmoski and Brandt 1984; Schneiderman et al. 1986; Gray et al. 1988; Sah et al. 1989; Sah et al. 1990; Guilak et al. 1991; Larsson et al. 1991; Sah et al. 1991). A decrease was also produced by osmotic stresses in explants dialyzed against polyethylene glycol (Bayliss et al. 1986; Schneiderman et al. 1986), reproducing this aspect of the continuously compressed state of cartilage. The time needed for recovery of the synthesis rate after release of this type of static load depended on the length of the loading period (Schneiderman et al. 1986). Prolonged compression impeded transport of macromolecules by diffusion, while a sustained increase in macromolecule release from the explants was noticed upon load removal (Sah et al. 1991). Since continuous compression would result in an increased PG concentration in the matrix and thereby a lower intra-tissue pH (Gray et al. 1988), it seems likely that pH may influence PG aggregate assembly by an effect on the HA-binding affinity of the PG monomer (Sah et al. 1990).

The cyclic loading increases the transport of large solutes (e.g., albumin, growth factors and cytokines) through cartilage tissue but has no significant influence on the movement of small solutes (Tomlinson and Maroudas 1980; O'Hara et al. 1990). The dynamic compression of cartilage explants has yielded somewhat variable results, with respect to its influence on the synthesis of matrix macromolecules. However, recent findings suggest that a particular frequency and amplitude of compression is required to elicit an anabolic response in the cartilage (Sah et al. 1989; Larsson et al. 1991; Korver et al. 1992; Parkkinen et al. 1992). High frequency dynamic compression (0.01 – 1 Hz) had a stimulating effect on the incorporation of both [^{3}H]proline and [^{35}S]sulfate by approx. 20 – 40% (Sah et al. 1989). Low frequency ($\leq$ 0.001 Hz) loading induced negligible effects at low amplitudes (forces), but stimulated [^{3}H]proline and [^{35}S]sulfate incorporation at medium amplitudes and gave highly variable effects at high amplitudes (Sah et al. 1989). *In vitro* loading of anatomically intact articular cartilage on calf sesamoid bones at 0.33 Hz for 7 days enhanced [^{35}S]sulfate incorporation by 45% (Korver et al. 1992). Increased amplitude of the load caused a shift toward synthesis of larger aggrecans, with longer glycosaminoglycan (GAG) chains. The synthesis of small PGs by cartilage explants was reduced during culture, but *in vitro* loading restored the production to the control level (Korver et al. 1992). Mechanical compressive forces possibly regulate synthesis of large PGs (Koob and Vogel 1987; Koob et al. 1992).

The effects of tensional forces have been investigated especially with cells and tissues which are tensioned *in vivo*. These studies have utilized myocytes (Komuro et al. 1991; Sadoshima et al. 1992), cardiac fibroblasts (Carver et al. 1991) and fibrous joints (Meikle et al. 1984), but less chondrocytes.

In stretching experiments, chondrocytes have been cultured on a membrane (Lee et al. 1982; de Witt et al. 1984) or coverslips (Uchida et al. 1988) which were then exposed to tensional force. The tension leads to direct deformation of the chondrocytes, similar to the distortion of the collagen network which occurs with a direct compressive load. Cyclic (1 Hz) 10% stretching of the supporting membrane for 8 h increased GAG synthesis 2-3 fold (Lee et al. 1982). Interestingly, agitation of the membranes produced a 3-fold increase in GAG synthesis over the control membranes (Lee et al. 1982). A strain of 5.5% at a frequency of 0.2 Hz increased [^{35}S]sulfate incorporation 1.4-fold (de Witt et al. 1984). The tension of coverslips also leads to a stimulation of radiosulfate incorporation with concomitant changes in cAMP and PGE_2 levels (Uchida et al. 1988).

Proteoglycan Synthesis in Different Zones of Cartilage Explants During Cyclic Mechanical Compression

In our previous studies, a mechanical loading apparatus with a non-porous loading head was used to cyclically compress full-depth cartilage discs which had a diameter larger than the loading head (Parkkinen et al. 1992). The non-porosity of the loading head and the large size of the explant were factors which appeared to retard fluid flow in the tissue and therefore increase hydrostatic pressure during dynamic loading of the cartilage. The [^{35}S]sulfate incorporation pattern in the directly loaded and the non-loaded cartilage of the explants was different.

In the directly loaded area, the highest stimulation in [^{35}S]sulfate incorporation during a short-term experiment was noticed with a 0.25 Hz cycle. The stimulation took place through the whole depth of the cartilage. Raising the pressure from 0.5 MPa to 1 MPa, or increasing the frequency to 0.5 Hz, reduced the stimulation particularly in the superficial parts of the articular cartilage. In the indirectly loaded area, the stimulation occurred with a faster (0.5 Hz) cycle, and only in the superficial zone (Parkkinen et al. 1992).

As the strain of the cartilage was minimal with the short loading pulse (50 ms) used, the stimulation of [^{35}S]sulfate incorporation was

Table 1. Effect of cyclic hydrostatic pressure on [^{35}S]sulfate incorporation in cultured cartilage explants or chondrocytes#. The results are expressed as a pressurized/control ratio.

Original publication	Pressure (MPa)	Loading Time (h)	Cycle	Pressurized / Control
van Kampen et al. # 1985	0.013	24	0.3 Hz	1.4
Klein-Nulend et al.‡ 1987	0.013	120	0.3 Hz	3.4
Hall et al. 1991	5.0		20s on / 2h off	1.2
	7.5			1.2
	10.0			1.5
	15.0			1.4
	20.0			1.3
	50.0			1.0
Lafeber et al. 1992	0.013	4	1 s on / 2 s off	1.1
	0.013	100		1.0
	0.013	196		0.9
Takano-Yamamoto # et al. 1991	0.0005		5 min on/ 27 h off	1. 0
	0.0025			1. 1
	0.005			1. 4
	0.01			1. 6
	0.02			1. 6
	0.0005		1 min on/ 27h off	1.0
	0.0025			0.9
	0.005			0.9
	0.01			0.9
	0.02			0.9

‡ Fetal mouse long bone rudiments were used in this experiment

probably due to enhanced hydrostatic pressure. Since all these alterations were observed within 1.5 h from the beginning of loading, it appeared probable that the speed of posttranslational processing of the core protein, including GAG assembly and sulfation in the endoplasmic reticulum and the Golgi apparatus, was involved, rather than influence on transcription.

Influences of Hydrostatic Pressure: Earlier Observations

Pressure is raised at every dynamic loading event in articular cartilage. The loads from the opposing joint surfaces create hydrostatic pressure which is immediately spread in the underlying cartilages around the contact sites. The hydrostatic pressure combined with mechanical loading induces a flow of fluid through the matrix, creating streaming potentials and currents. Eventually, a new equilibrium is achieved when the osmotic pressure of the matrix has gradually increased to match the external pressure. This state involves a local dehydration and deformation of the matrix and often causes a shape change of chondrocytes. By that time hydrostatic pressure has decreased to the level of the tissue swelling pressure generated by PGs (0.1-0.2 MPa in the uncompressed state)(Grushko et al. 1989). Utilizing a pressure chamber, the effects of hydrostatic pressure can be investigated independently of the other factors involved in joint loading. Increased hydrostatic pressure influences various cell types like bone-derived cells (Imamura et al. 1990; Ozawa et al. 1990) and osteoblastoma cells (Quinn and Rodan 1981). Hydrostatic pressure

Table 2. Effect of continuous hydrostatic pressure loading on [^{35}S]sulfate incorporation in cultured cartilage explants. The results are expressed as a pressurized-control ratio.

Original publication	Pressure (MPa)	Loading Time(h)	Pressurized / Control
Kimura et al. 1985	0.35-2.7	24	1.0
Lippiello et al. 1985	0.5	24	0.8
	1.0	24	0.8
	1.5	24	0.5
	2.0	24	0.5
	2.5	24	1.1
Hall et al. 1991	5.0	2	1.4
	7.5	2	1.2
	10.0	2	1.1
	15.0	2	0.9
	20.0	2	0.9
	30.0	2	0.5
	40.0	2	0.2
	50.0	2	0.2

has clear influences on cytoskeleton, integrins and cell organelles (Goldinger et al. 1980; Heremans 1982; Varga et al. 1986; Zimmerman et al. 1987; Heinemann et al. 1989; Kavecansky et al. 1992; Acevedo et al. 1993; Haskin and Cameron 1993) and presumably other proteins (Kim et al. 1993) and membrane systems, too.

Relatively few experiments have been done by applying hydrostatic pressure on cartilage explants or chondrocytes (Bourret and Rodan 1976; Veldhuijzen et al. 1979; Kimura et al. 1985; Lippiello et al. 1985; van Kampen et al. 1985; Klein-Nulend et al. 1987; Hall et al. 1991; Takano-Yamamoto et al. 1991; Lafeber et al. 1992; Wright et al. 1992). Continuous hydrostatic pressure ranging from 0.35 MPa to 2.75 MPa had no effect on [^{35}S]sulfate incorporation in explants of bovine articular cartilage (Kimura et al. 1985). In another study, the response of cartilage explants subjected to continuous hydrostatic pressure depended on the level of pressure and the presence of serum in the culture medium (Lippiello et al. 1985). Pressures between 75 and 300 psi (0.5 - 2.1 MPa) were inhibitory (50% of control level), but a pressure of 375 psi (2.6 MPa) stimulated [^{35}S]sulfate incorporation by 10 - 15%. If dialysed serum was present, the stimulation increased up to 50% (Lippiello et al. 1985). Intermittent hydrostatic pressure as low as 13 kPa increased both [^{35}S]sulfate incorporation and the aggregating capacity of PGs in high density chondrocyte cultures (van Kampen et al. 1985). The response of cartilage explants to hydrostatic pressure depended on the magnitude of the pressure and the time of exposure, the synthesis of matrix being most active between 5 and 15 MPa pressures (Hall et al. 1991). The response of chondrocytes to hydrostatic pressure seems to depend on the source of the cartilage (Takano-Yamamoto et al. 1991). A summary of the data available from previous literature on the effects of cyclic and continuous hydrostatic pressures on cartilage PG synthesis is presented in Tables 1 and 2.

Proteoglycan Synthesis of Chondrocytes Exposed to Hydrostatic Pressure: New Findings

To further investigate the role of hydrostatic pressure in the articular cartilage PG synthesis, we developed a novel apparatus for this purpose. The loading apparatus is composed of a water-filled loading cylinder, a reference cylinder of the same size and a computer-controlled hydraulic system to produce hydrostatic pressure, cyclic or continuous, in the loading cylinder (Fig. 1). The cartilage explants or monolayer cell cultures were exposed to pressure in petri dishes filled with medium and sealed by a special membrane after excluding air from the dish. The petri dishes were put into chambers

Figure 1. The apparatus designed for in vitro hydrostatic pressurization. The pressure chamber (PC) and reference chamber (RC) are maintained inside a temperature-controlled room at 37°C, while the control unit (an IBM-compatible computer) and the hydraulic machinery are situated in an adjacent room with normal temperature.

filled by pre-warmed (37°C) distilled water with a computer-controlled pump. The pressure was monitored by a sensor in the chamber and kept at the predetermined level at an accuracy of 0.1 MPa (Parkkinen et al. 1993a).

Cartilage explants and chondrocyte monolayers were subjected to short-term (1.5 h) or long-term (20 h) hydrostatic pressure in loading chamber. To confirm the correct phenotype of the chondrocytes cultured in monolayers, the PGs secreted were characterized by SDS-agarose gel electrophoresis and using specific glycosaminoglycan (GAG) degrading enzymes to identify the [^{35}S]sulfate labeled macromolecules. Two major bands were noticed in autoradiography of the gels. The macromolecules in the slow-mobility band (band I, Fig. 2A) were cleaved almost totally by chondroitinase AC (Fig. 2A, lane 5). The fast-mobility band (band II, Fig. 2A) was specifically eliminated by chondroitinase B (Fig. 2A, lane 3). In addition, Northern analyses of total RNA isolated from control and pressurized cultures expressed abundantly mRNA transcript of aggrecan, decorin and biglycan, while no detectable expression of versican was detected

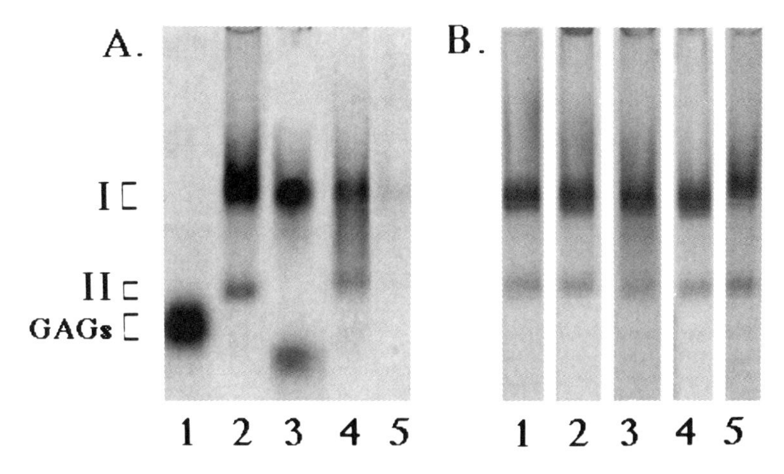

Figure 2. Characterization of PG subpopulations. PGs secreted into the control chondrocyte culture medium were electrophoresed on SDS agarose gels intact (lane 2) and after specific enzymic digestions: proteinase K, lane 1; chondroitinase B, lane 3; keratanase, lane 4; and chondroitinase AC, lane 5 (A). PGs secreted into the medium of control cultures (lane 1) and those pressurized either cyclically at 5 MPa (0.5 Hz, lane 2; 0.25 Hz, lane 3; 0.0167 Hz, lane 4) and continuously at 30 MPa (lane 5) (B).

(Fig. 3A). Total RNA from smooth muscle cells was used as a positive control for versican expression (Fig. 3A, lane 2). Therefore, it was concluded that the cells used in these experiments were chondrocytes synthesizing aggrecans (band I) and small interstitial DS-PGs (band II), typical of articular cartilage.

The response to the hydrostatic pressure depended on the frequency of the pressure, the length of pressurization and whether chondrocyte monolayer cultures or cartilage tissue explants were pressurized (Fig. 4). In short-term loading experiments (1.5 h), the [^{35}S] sulfate incorporation was stimulated in explants when 0.5 Hz cycle was applied with 5 MPa pressure. In chondrocyte cultures, however, [^{35}S]sulfate incorporation was inhibited with the same protocol, and also with 0.25 Hz and 0.05 Hz loading cycles. The response of the chondrocytes in the long-term experiments (20 h) was totally reversed; a stimulation in [^{35}S]sulfate incorporation was found with the same cycles and pressures that were inhibitory in short-term experiments. When the pressurization frequency was further decreased under 0.0167 Hz, there were no alterations in [^{35}S]sulfate incorporation.

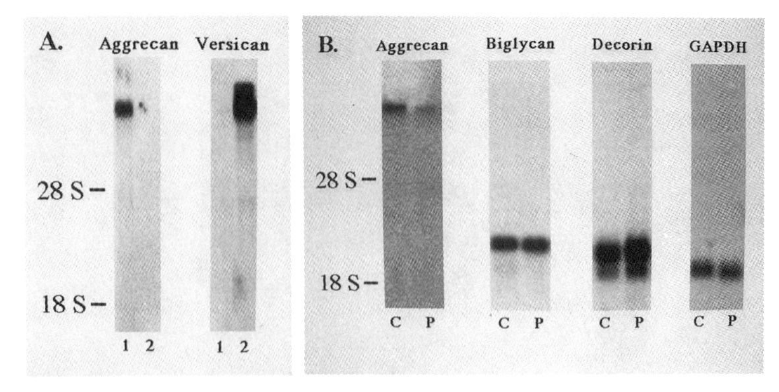

Figure 3. Northern blot analysis of PG mRNAs. Total RNAs from chondrocytes (lane 1) and human smooth muscle cells (lane 2) were electrophoresed on agarose, transferred onto nylon membrane, and hybridized with aggrecan and versican cDNA probes (A). Total RNAs from control cultures (C) and from chondrocytes exposed to 30 MPa continuous hydrostatic pressure (P) were transferred onto a nylon membrane, and hybridized with aggrecan, biglycan, decorin and GAPDH cDNA probes (B). The cDNA probes were generous gifts from Drs. V. Glumoff and E. Vuorio (aggrecan), Dr. L. Fisher (biglycan) and Dr. T. Krusius (decorin and versican).

The structure of PGs secreted by chondrocyte cultures after exposure to continuous (5 MPa and 30 MPa) and cyclic hydrostatic pressures (5 MPa) was investigated. Continuous 30 MPa hydrostatic pressure caused a retardation in the mobility of aggrecans in SDS-agarose gel electrophoresis (Fig. 2B, lane 5). The change was caused by the high magnitude of the pressure, and not by the static type of the loading, as 5 MPa continuous pressure did not have a similar effect on the electrophoretic mobility of aggrecan. A change into a higher molecular weight aggrecan was in line with the finding that in Sephacryl S-1000 gel chromatography the peak of the monomeric large PGs shifted into a smaller K_{av} value. A slight increase in the average chain length of GAGs was observed in the chromatographic analysis on Sepharyl S-300. No similar observations were found after cyclic pressurizations.

Continuous high (30 MPa) hydrostatic pressure remarkably inhibited sulfation, glycosylation and protein synthesis as measured by incorporations of [^{35}S]sulfate, [^{3}H]glucosamine and [^{14}C]leucine, respectively. Northern analyses were performed to determine the steady-state mRNA levels of aggrecan, decorin and biglycan as related to glyceraldehyde-3-phosphate dehydrogenase (GAPDH) levels. Aggrecan mRNA transcripts were slightly decreased, while decorin and biglycan levels were elevated. In the cyclic pressurizations, 0.5

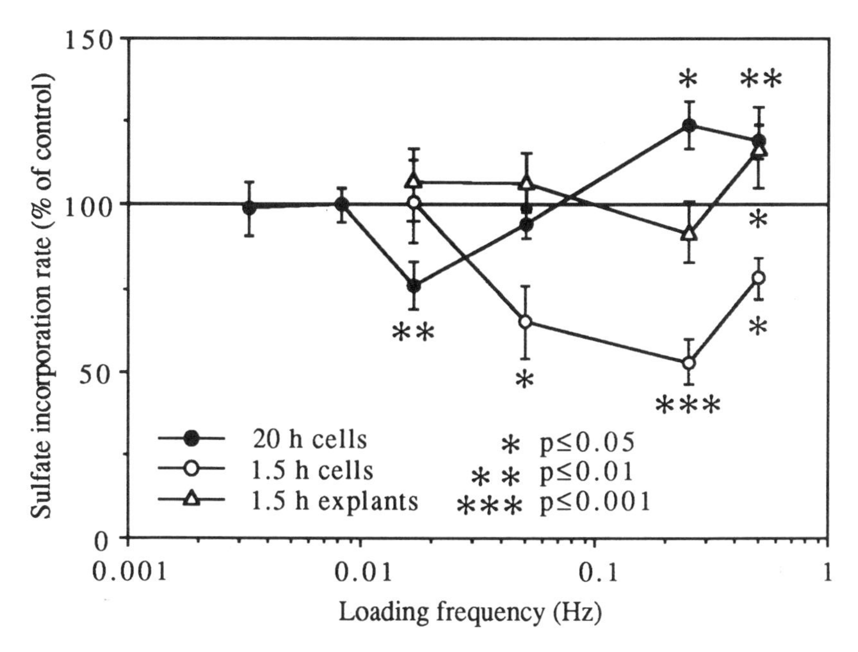

Figure 4. [^{35}S]sulfate incorporation vs. frequency of hydrostatic pressure on cartilage explants and chondrocyte cell cultures during cyclic hydrostatic pressurization in short-term (tissue and cell cultures) and long-term experiments (cell cultures).

Hz caused a small increase in [^{35}S]sulfate incorporations, whereas with 0.017 Hz pressurization it was slightly inhibited.

Changes in the Golgi Apparatus and Cytoskeleton of Chondrocytes under High Hydrostatic Pressure

After 30 MPa continuous hydrostatic pressure for 2 h, the morphology of the chondrocytes was altered to a more retracted form. The stress fibers disappeared in the centre of the cells, and only small punctually distributed material was stained (Fig. 5B). If the 30 MPa pressure was applied cyclically, the number of the stress fibers was reduced and the fibers remaining were thinner than in control cells. Decreasing the continuous pressure level to 15 MPa conveyed to a reduced number of thinner stress fibers. With lower levels of cyclic or continuous pressures, there were no changes in the organization of stress fibers, only some individual cells had thinner stress fibers.

After a 30 MPa continuous pressure, the stacks of Golgi apparatus were reversibly collapsed into a packed structure (Fig. 5D). The normal stacked appearance of Golgi apparatus was no more visible in the electron microscopic preparation of the pressurized chondrocytes. Similar changes, but to a minor extent, were seen after 15 MPa continuous pressure. After a 60 min recovery following pressurization, the organization of the Golgi apparatus was fully re-established. Lower pressures had no effects on the morphology of the Golgi apparatus.

The staining pattern of microtubules was also altered due to 30 MPa continuous hydrostatic pressure (Fig. 5F). The tubules were more kinky and did not extend so clearly and radially to the periphery of the cells, as in controls.

Nocodazole treatment, which causes disassembly of chondrocyte microtubules, led to fragmentation and dispersion of the Golgi apparatus throughout the cytoplasm and pressurization had no effect on its vesicular appearance, showing that the packing of Golgi apparatus required intact microtubules. The general morphology of the chondrocytes remained unchanged during nocodazole treatment.

Discussion

Studies addressing the mechanisms of signal transduction from physical factors of the environment to a relevant adjustment of cell metabolism are few, and the results obtained have been very complex (Stockwell 1987; Avecado et al. 1993; Komuro et al. 1991; Sadoshima et al. 1992). One of the problems in these studies has been to isolate from the often multiple physical factors a single candidate for closer metabolic examination.

In cartilage, for example, joint loading creates cell and matrix deformation, fluid flow, and electrical events induced by the flow of ions (Frank and Grodzinsky 1987; O'Connor et al. 1988), in addition to the initial increase of hydrostatic pressure. However, in the case of hydrostatic pressure it was possible to build a system that almost completely excludes the other physical phenomena involved in normal cartilage loading. Since water is virtually incompressible up to 50 MPa (Macdonald 1975), there is little if any flow of water around chondrocytes in a hydrostatic pressure chamber. Neither is there any cell shape change like in mechanically compressed tissues.

Hydrostatic pressure spreads in the intra- and extracellular fluids equally. Therefore, if a conceptual receptor for hydrostatic pressure is considered, it can exist not only in the pericellular or membrane position, but also anywhere intracellularly. Hydrostatic pressure has its most profound influence in the contacts of two phases, like those of lipid and water at membranes (Jannasch et al. 1987). Thus, membrane structures such as ion channels could serve as a sensor of ambient

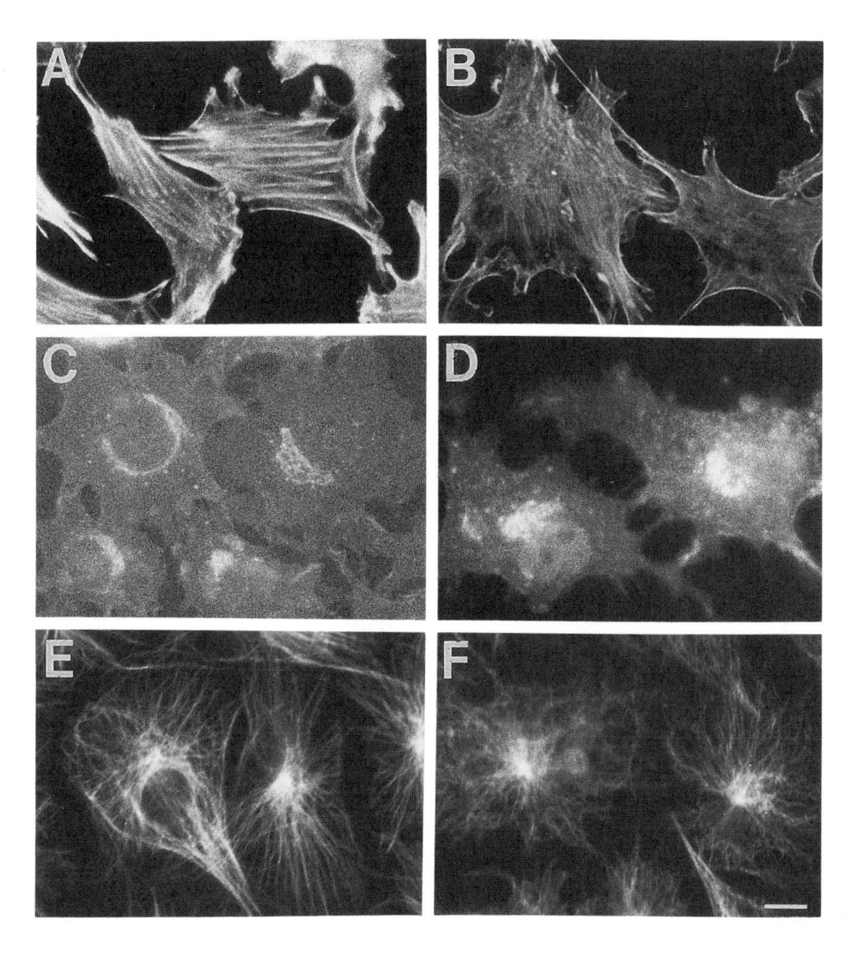

Figure 5. Effects of hydrostatic pressure on chondrocyte cytoskeleton and Golgi apparatus. The cells were exposed to 30 MPA continuous pressure for 2h before fixation and staining with TRITC-phalloidin (A,B), TRITC-WGA (C,D), antibody against MTs (E,F) and to visualize Golgi apparatus. Control cells (A,C,E); pressurized cells (B,D,F). The bar represents 10 μm.

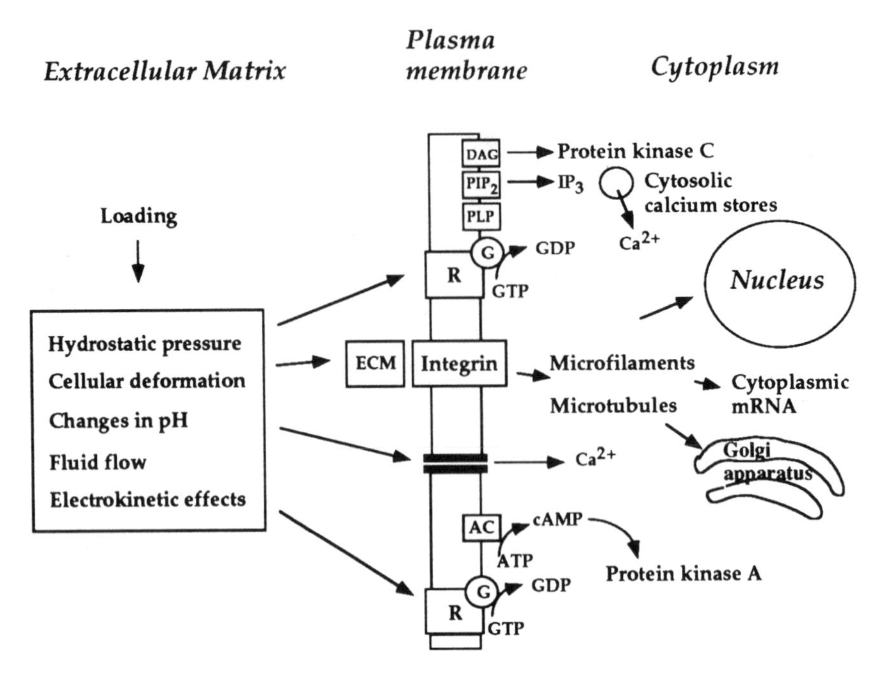

Figure 6. Some of the signal transduction pathways that may convert the mechanical and hydrostatic forces into a biochemical signal in the cell. Abbreviation used: DAG, diacylglycerol; PIP_2, phosphoinositolphosphate; PLP, phospholipase C; IP_3, inositol triphosphate; G, G protein; GDP, guanosine 5'-diphosphate; GTP, guanosine 5'-triphosphate; R, receptor; ECM, extracellular matrix; AC, adenylate cyclase; cAMP, cyclic adenosine 3'5'-monophosphate; ATP, adenosine triphosphate (modified from a poster presented by F. Guilak et al. 39th Annual Meeting of the Orthopaedic Research Society, San Francisco, 1993).

hydrostatic pressure (Urban et al. 1993). Another possible target is the conformation of proteins which can be influenced by hydrostatic pressure (Heremans 1982). Thus, little is known on the primary mechanism of action of hydrostatic pressure, and likely sites include the membranes of the cell and perhaps specific proteins sensitive to hydrostatic pressure. Furthermore, it is probable that there are several "receptors" activated at different levels of hydrostatic pressure. Some of the possible pathways for cellular signal transduction are presented in Fig. 6.

Influences of hydrostatic pressure on the metabolism of chondrocyte cultures were studied using an apparatus specially designed for this purpose. The experiments suggested that hydrostatic pressurization of articular cartilage and chondrocytes may influence the posttranslational processing of PGs, involving, e.g., the GAG chain assembly and sulfation, as well as the intracellular transport of PGs and secretion into the extracellular space. Whether the simultaneous changes in the morphology of Golgi apparatus and cytoskeleton contribute to the altered PG synthesis remains to be elucidated. Regulation by hydrostatic pressure also involved mRNA levels of the PG core proteins.

Continuous high (30 MPa) hydrostatic pressure clearly inhibited sulfation, glycosylation and protein synthesis. In addition, aggrecans secreted during the high pressure were larger, as indicated by retarded migration in SDS-agarose gel electrophoresis and earlier elution on Sephacryl S-1000 gel chromatography. A slight increase observed in the relative length of GAG chains could explain this finding. However, a higher substitution of the serine residues with GAG chains seems possible, too. Only 55% of the serine residues in rat chondrosarcoma are substituted with carbohydrate (Lohmander et al. 1989).

High hydrostatic pressure had profound effects on the structural organization of microfilaments and microtubules in epithelial cells (Bourns et al. 1988) and in chondrocytes (Parkkinen et al. 1993b; Parkkinen et al. 1993c). Along with the depolymerization of microtubules, high pressure caused a packed pattern of the Golgi apparatus close to the cell nucleus (Parkkinen et al. 1993c). Microtubules probably facilitate and direct the intracellular targeting and secretion of macromolecules (Kelly 1990), and help to maintain the structural integrity of the Golgi apparatus (Thyberg and Moskalewski 1985). Thus it seems possible that high hydrostatic pressure (causing depolymerization of microtubules) may slow down the intracellular traffic and thus the rate of the biosynthetic machinery. If the precursor spends relatively more time in a compartment responsible for chain initiation, perhaps more and longer carbohydrate chains per core protein are built, resulting in larger PG molecules. A microtubule disrupting agent, nocodazole, decreased [^{35}S]sulfate incorporation of chondrocyte cultures by 30-40%, but the incorporation was further decreased if nocodazole treated cultures were exposed to 30 MPa hydrostatic pressure (unpublished data). This supports the idea that microtubules facilitate the synthesis or secretion of PGs, while high hydrostatic pressure influences other aspects in the synthesis, perhaps also transcriptional or translational.

The steady-state mRNA levels of aggrecan, decorin and biglycan after continuous pressurization were quantified, too. A decrease (25%) was noticed in aggrecan mRNA, compatible with the reduced [^{35}S]sulfate incorporation rate. However, decorin and biglycan mRNA transcripts were increased (250-500%), demonstrating the specific character of the regulation of the different PG mRNAs. In accordance with an earlier report on cartilage explants (Curtis et al. 1992), mRNA levels did not consistently predict the rate of the synthesis of the corresponding PGs. At least certain mRNAs have been shown to associate with cytoskeletal elements (Bagchi et al. 1987), and disruption of cytoskeleton affected the stability of mRNAs (Symington et al. 1991). Amino acid incorporations into PG core proteins were not specifically measured, so it remains to be determined whether there were significant alterations in the rate of core protein biosynthesis.

Short-term pressurization (5 MPa, 0.5 Hz for 90 min) stimulated PG synthesis in explants but inhibited it in chondrocyte monolayers (Parkkinen et al. 1993a). This clearly demonstrates that the normal extracellular matrix modulates the response of the chondrocytes to hydrostatic pressure. Whether this involves specific bonds to, or molecular messengers from the matrix, is not known.

While the 90 min pressurization of chondrocyte monolayers inhibited their PG synthesis, a longer period of loading (20 h) in similar conditions brought about a stimulation (Parkkinen et al. 1993a), a finding stressing the importance to carefully examine the time dependence of these phenomena.

Hydrostatic pressurization thus offers an interesting experimental model to study the regulation of chondrocyte matrix production by a naturally occurring environmental factor.

Conclusions

Hydrostatic pressure, both continuous and intermittent, causes significant alterations in the metabolism and structure of the chondrocytes. The response depends on the frequency of the pressure, the length of pressurization and whether chondrocyte monolayer cultures or cartilage tissue explants were pressurized. In short-term (1.5 h) loading experiments, the [^{35}S]sulfate incorporation, giving an idea of the activity of PG synthesis, was stimulated in explants when 0.5 Hz cycle was applied with 5 MPa pressure. In chondrocyte cultures, however, [^{35}S]sulfate incorporation was inhibited with the same protocol. In long-term experiments (20 h), the response of the chondrocytes was totally reversed, the chondrocytes showed stimulated [^{35}S]sulfate incorporation with the same cycles and pressures that were inhibitory in short-term experiments.

Continuous 30 MPa hydrostatic pressure increased the size of aggrecans, at least partly due to a slight increase in the average chain length of GAGs. No similar observations were noted after cyclic pressurizations. Continuous high (30 MPa) hydrostatic pressure remarkably inhibited sulfation, glycosylation, and protein synthesis as measured by incorporation radioactive precursors. The aggrecan mRNA transcripts were slightly decreased, while decorin and biglycan levels were elevated. In the cyclic pressurizations, 0.5 Hz caused a small increase in [^{35}S]sulfate incorporations, wheras with 0.017 Hz pressurization it was slightly inhibited.

After 30 MPa continuous hydrostatic pressure for 2 h, the morphology of the chondrocytes was altered to a more retracted form. The stress fibers disappeared in the centre of the cells, and only small, evenly distributed points showed staining. The staining pattern of microtubules was also altered. The pressurization caused the Golgi apparatus to collapse into a packed structure. If the pressure was applied cyclically, the number of stress fibers was reduced and they were thinner than in control cells. The lower pressures or cyclic pressurization had no effect on Golgi apparatus structure.

At present, we do not know by which mechanism the change of hydrostatic pressure is sensed by the chondrocytes and how the pressure afffects the transcription, translation, and glycosylation of the PGs.

Acknowledgements

The skillful and efficient technical assistance of Mrs. Elma Sorsa and Ms. Eija Antikainen is gratefully acknowledged. This work was financially supported by grants from The University of Kuopio, The North Savo Fund of the Finnish Cultural Foundation, The Finnish Cultural Foundation, The Research and Science Foundation of Farmos, The Paulo Foundation, The Academy of Finland, The Sigrid Juselius Foundation, and The Finnish Research Council for Physical Education and Sports, Ministry of Education.

References

Acevedo, A. D.; Bowser, S. S.; Gerritsen, M. E.; Bizios, R. Morphological and proliferative responses of endothelial cells to hydrostatic pressure: role of fibroblast growth factor. J. Cell. Physiol. 157: 603-614; 1993.

Bagchi, T.; Larson, D. E.; Sells, B. H. Cytoskeletal association of muscle-specific mRNAs in differentiating L6 rat myoblasts. Exp. Cell Res. 168: 160-172; 1987.

Bayliss, M. T.; Urban, J. P. G.; Johnstone, B.; Holm, S. In vitro method for measuring synthesis rates in the intervertebral discs. J. Orthop. Res. 4: 10-17; 1986.

Bianco, P.; Fisher, L. W.; Young, M. F.; Termine, J. D.; Robey, P. G. Expression and localization of the two small proteoglycans biglycan and decorin in developing human skeletal and non-skeletal tissues. J. Histochem. Cytochem. 38: 1549-1563; 1990.

Bourns, B.; Franklin, S.; Cassimeris, L.; Salmon, E. D. High hydrostatic pressure effects in vivo: changes in cell morphology, microtubule assembly, and actin organization. Cell Motil. Cytoskeleton 10: 380-390; 1988.

Bourret, L. A.; Rodan, G. A. The role of calcium in the inhibition of cAMP accumulation in epiphyseal cartilage cells exposed to physiological pressure. J. Cell. Physiol. 88: 353-362; 1976.

Carver, W.; Nagpal, M. L.; Nacthigal, M.; Borg, T. K.; Terracio, L. Collagen expression in mechanically stimulated cardiac fibroblasts. Circ. Res. 69: 116-122; 1991.

Caterson, B.; Griffin, J.; Mahmoodian, F.; Sorrell, J. M. Monoclonal antibodies against chondroitin sulphate isomers: their use as probes for investigating proteoglycan metabolism. Biochem. Soc. Trans. 18: 820-823;1990.

Copray, J. C. V. M.; Jansen, H. W. B.; Duterloo, H. S. An in-vitro system for studying the effect of variable compressive forces on the mandibular condylar cartilage of the rat. Arch. Oral. Biol. 30: 305-311; 1985.

Curtis, A. J.; Devenish, R. J.; Handley, C. J. Modulation of aggrecan and link-protein synthesis in articular cartilage. Biochem. J. 288: 721-726; 1992.

de Witt, M. T.; Handley, C. J.; Oakes, B. W.; Lowther, D. A. *In vitro* response of chondrocytes to mechanical loading. The effect of short term mechanical tension. Connect. Tissue Res. 12: 97-109; 1984.

Frank, E. H.; Grodzinsky, A. J. Cartilage electromechanics I. Electrokinetic transduction and effects of electrolyte pH and ionic strength. J. Biomech. 20: 615-627; 1987.

Goetinck, P. F.; Kiss, I.; Deák, F.; Stirpe, N. S. Macromolecular organization of the extracellular matrix of cartilage. Ann. N.Y. Acad. Sci. 599: 29-38; 1990.

Goldinger, J. M.; Kang, B. S.; Choo, Y. E.; Paganelli, C. V.; Hong, S. K. Effect of hydrostatic pressure on ion transport and metabolism in human erythrocytes. J. Appl. Physiol. 49: 224-231; 1980.

Gray, M. L.; Pizzanelli, A. M.; Grodzinsky, A. J.; Lee, R. C. Mechanical and physicochemical determinants of the chondrocyte biosynthetic response. J. Orthop. Res. 6: 777-792; 1988.

Grushko, G.; Schneiderman, R.; Maroudas, A. Some biochemical and biophysical parameters for the study of the pathogenesis of osteoarthritis: a comparison between the processes of ageing and degeneration in human hip cartilage. Connect. Tissue Res. 19: 149-176; 1989.

Guilak, F.; Meyer, B. C.; Ratcliffe, A.; Mow, V. C. The effect of static loading on proteoglycan biosynthesis and turnover in articular cartilage explants. Trans. Orthop. Res. Soc. 16:50; 1991.

Hall, A. C.; Urban, J. P. G.; Gehl, K. A. The effects of hydrostatic pressure on matrix synthesis in articular cartilage. J. Orthop. Res. 9: 1-10; 1991.

Hascall, V. C.; Handley, C. J.; McQuillan, D. J.; Hascall, G. K.; Robinson, H. C.; Lowther, D. A. The effect of serum of biosynthesis of proteoglycans by bovine articular cartilage in culture. Arch. Biochem. Biophys. 224: 206-223; 1983.

Haskin, C.; Cameron, I. Physiological levels of hydrostatic pressure alter morphology and organization of cytoskeletal and adhesion proteins in MG-63 osteosarcoma cells. Biochem. Cell Biol. 71: 27-35; 1993.

Heinemann, S. H.; Conti, F.; Stuhmer, W.; Neher, E. Effects of hydrostatic pressure on membrane processes. J. Gen. Physiol. 90: 765-778; 1989.

Helminen, H. J.; Kiviranta, I.; Säämänen, A.-M.; Jurvelin, J. S.; Arokoski, J.; Oettmeier, R.; Abendroth, K.; Roth, A. J.; Tammi, M. I. Effect of motion and load on articular cartilage in animal models. In: Kuettner, K.; Schleyerbach, R.; Peyron, J. G.; Hascall, V. C., eds. Articular cartilage and osteoarthritis. New York: Raven Press; 1992:p. 501-510.

Heremans, K. High pressure effects on proteins and other biomolecules. Ann. Rev. Biophys. Bioengin. 11: 1-21; 1982.

Hua, Q.; Knudson, C. B.; Knudson, W. Internalization of hyaluronan by chondrocytes occurs via receptor-mediated endocytosis. J. Cell Sci. 106: 365-375; 1993.

Imamura, K.; Ozawa, H.; Hiraide, T.; Takahashi, N.; Shibasaki, Y.; Fukuhara, T.; Suda, T. Continuously applied compressive pressure induces bone resorption by a mechanism involving prostaglandin E_2 synthesis. J. Cell. Physiol. 144: 222-228; 1990.

Jannasch, H. W.; Marquis, R. E.; Zimmerman, A. M. Current perspectives in high pressure biology. London: Academic Press; 1987.

Jones, I. L.; Klämfeldt, A.; Sandström, T. The effect of continuous mechanical pressure upon the turnover of articular cartilage proteoglycans *in vitro*. Clin. Orthop. 165: 283-289; 1982.

Kavecansky, J.; Dannenberg, A. J.; Zakim, D. Effects of high pressure on the catalytic and regulatory properties of UDP-glucuronosyltransferase in intact microsomes. Biochemistry 31: 162-168; 1992.

Kelly, R. B. Microtubules, membrane traffic, and cell organization. Cell 61: 5-7; 1990.

Kim, K.; Homma, Y.; Ikeuchi, Y.; Suzuki, A. Effect of high hydrostatic pressure on the conversion of -connectin to ß-connectin. J. Biochem. 114: 463-467; 1993.

Kimura, J. H.; Schipplein, O. D.; Kuettner, K. E.; Andriacchi, T. P. Effects of hydrostatic loading on extracellular matrix formation. Trans. Orthop. Res. Soc. 10: 365; 1985.

Klein-Nulend, J.; Veldhuijzen, J. P.; van de Stadt, R. J.; van Kampen, G. P. J.; Kuijer, R.; Burger, E. H. Influence of intermittent compressive force on proteoglycan content in calcifying growth plate cartilage in vitro. J. Biol. Chem. 262: 15490-15495; 1987.

Klämfeldt, A. Continuous mechanical pressure and joint tissue. Effect of synovial membrane products and indomethacin *in vitro*. Scand. J. Rheumatol. 14: 431-437; 1985.

Komuro, I.; Katoh, Y.; Kaida, T.; Shibazaki, Y.; Kurabayashi, M.; Hoh, E.; Takaku, F.; Yazaki, Y. Mechanical loading stimulates cell hypertrophy and specific gene expression in cultured rat cardiac myocytes. Possible role of protein kinase C activation. J. Biol. Chem. 266: 1265-1268; 1991.

Koob, T. J.; Clark, P. E.; Hernandez, D. J.; Thurmond, F. A.; Vogel, K. G. Compression loading in vitro regulates proteoglycan synthesis by tendon fibrocartilage. Arch. Biochem. Biophys. 298: 303-312; 1992.

Koob, T. J.; Vogel, K. G. Site-related variations in glycosaminoglycan content and swelling properties of bovine flexor tendon. J. Orthop. Res. 5: 414-424; 1987.

Korver, T. H. V.; van de Stadt, R. J.; Kiljan, E.; van Kampen, G. P. J.; van der Korst, J. K. Effects of loading on the synthesis of proteoglycans in different layers of anatomically intact articular cartilage *in vitro*. J. Rheumatol. 19: 905-912; 1992.

Lafeber, F.; Veldhuijzen, J. P.; Vanroy, J. L. A. M.; Huber-Bruning, O.; Biljsma, J. W. J. Intermittent hydrostatic compressive force stimulates exclusively the proteoglycan synthesis of osteoarthritic human cartilage. Br. J. Rheumatol. 31: 437-442; 1992.

Larsson, T.; Aspden, R. M.; Heinegård, D. Effects of mechanical load on cartilage matrix biosynthesis *in vitro*. Matrix 11: 388-394; 1991.

Lee, R. C.; Rich, J. B.; Kelley, K. M.; Weiman, D. S.; Mathews, M. B. A comparison of *in vitro* cellular responses to mechanical and electrical stimulation. Am. Surg. 48: 567-574; 1982.

Lippiello, L.; Kaye, C.; Neumata, T.; Mankin, H. J. *In vitro* metabolic response of articular cartilage segments to low levels of hydrostatic pressure. Connect. Tissue Res. 13: 99-107; 1985.

Lohmander, L. S.; Shinomura, T.; Hascall, V. C.; Kimura, J. H. Xylosyl transfer to the core protein precursor of the rat chondrosarcoma proteoglycan. J. Biol. Chem. 264: 18775-18780; 1989.

Luyten, F. P.; Hascall, V. C.; Nissley, S. P.; Morales, T. I.; Reddi, A. H. Insulin-like growth factors maintain steady-state metabolism of proteoglycans in bovine articular cartilage explants. Arch. Biochem. Biophys. 267: 416-425; 1988.

Luyten, F. P.; Yu, Y. M.; Yanagishita, M.; Vukicevic, S.; Hammonds, R. G.; Reddi, A. H. Natural bovine osteogenin and recombinant human bone morphogenetic protein-2B are equipotent in the maintenance of proteoglycans in bovine articular cartilage explant cultures. J. Biol. Chem. 267: 3691-3695; 1992.

Macdonald, A. G. Physiological aspects of deep sea biology. Monographs of the Physiological Society no. 31. London: Cambridge University Press; 1975.

Meikle, M. C.; Heath, J. K.; Reynolds, J. J. The use of in vitro models for investigating the response of fibrous joints to tensile mechanical stress. Am. J. Orthod. 85: 141-153; 1984.

Morales, T. I. Polypeptide regulators of matrix homeostasis in articular cartilage. In: Kuettner, K. E.; Schleyerbach, R.; Peyron, J. G.; Hascall, V. C., eds. Articular Cartilage and Osteoarthritis. New York: Raven Press; 1992:p. 265-279.

Morales, T. I.; Hascall, V. C. Correlated metabolism of proteoglycans and hyaluronic acid in bovine cartilage organ cultures. J. Biol. Chem. 268: 3632-3638; 1988.

Mow, V. C.; Ratcliffe, A.; Poole, A. R. Cartilage and diarthrodial joints as paradigms for hierarchical materials and structures. Biomaterials 13: 67-97; 1992.

Mundlos, S.; Meyer, R.; Yamada, Y.; Zabel, B. Distribution of cartilage proteoglycan (aggrecan) core protein and link protein gene expression during human skeletal development. Matrix 11: 339-346; 1991.

Ng, C. K.; Handley, C. J.; Preston, B. N.; Robinson, H. C. The extracellular processing and catabolism of hyaluronan in cultured adult articular cartilage explants. Arch. Biochem. Biophys. 298: 70-79; 1992.

Norton, L. A.; Rodan, G. A.; Bourret, L. A. Epiphyseal cartilage cAMP changes produced by electrical and mechanical perturbations. Clin. Orthop. 124: 59-68; 1977.

O'Connor, P.; Orford, C. R.; Gardner, D. L. Differential response to compressive loads of zones of canine hyaline articular cartilage: micromechanical, light and electron microscopic studies. Ann. Rheum. Dis. 47: 414-420; 1988.

O'Hara, B. P.; Urban, J. P. G.; Maroudas, A. Influence of cyclic loading on the nutrition of articular cartilage. Ann. Rheum. Dis. 49: 536-539; 1990.

Ozawa, H.; Imamura, K.; Abe, E.; Takahashi, N.; Hiraide, T.; Shibasaki, Y.; Fukuhara, T.; Suda, T. Effect of a continuously applied compressive pressure on mouse osteoblast-like cells (MC3T3-E1) in vitro. J. Cell. Physiol. 142: 177-185; 1990.

Palmoski, M. J.; Brandt, K. D. Effects of static and cyclic compressive loading on articular cartilage plugs in vitro. Arthritis Rheum. 27: 675-681; 1984.

Parkkinen, J. J.; Ikonen, J.; Lammi, M. J.; Laakkonen, J.; Tammi, M.; Helminen, H. J. Effects of cyclic hydrostatic pressure on proteoglycan synthesis in cultured chondrocytes and articular cartilage explants. Arch. Biochem. Biophys. 300: 458-465; 1993a.

Parkkinen, J. J.; Lammi, M. J.; Helminen, H. J.; Tammi, M. Local stimulation of proteoglycan synthesis in articular cartilage explants by dynamic compression in vitro. J. Orthop. Res. 10: 610-620; 1992.

Parkkinen, J. J.; Lammi, M. J.; Pelttari, A.; Tammi, M.; Helminen, H. J. Hydrostatic pressure modulates the organization of the cytoskeleton and Golgi apparatus in cultured chondrocytes. Trans. Orthop. Res. Soc. 18: 617; 1993b.

Parkkinen, J. J.; Lammi, M. J.; Pelttari, A.; Tammi, M.; Helminen, H. J.; Virtanen, I. Altered Golgi apparatus in hydrostatically loaded articular cartilage chondrocytes. Ann. Rheum. Dis. 52: 192-198; 1993c.

Poole, A. R.; Rizkalla, G.; Ionescu, M.; Reiner, A.; Brooks, E.; Rorabeck, C.; Bourne, R.; Bogoch, E. Osteoarthritis in the human knee: a dynamic process of cartilage matrix degradation, synthesis and reorganization. In: van den Berg, W. B.; van der Kraan, P. M.; van Lent, P. L. E. M., eds. Joint destruction in arthritis and Osteoarthritis. Basel, Switzerland: Birkhäuser Verlag; 1993:p. 3-13.

Quinn, R. S.; Rodan, G. A. Enhancement of ornithine decarboxylase and Na^+, K^+ ATPase in osteoblastoma cells by intermittent compression. Biochem. Biophys. Res. Commun. 100: 1696-1702; 1981.

Sadoshima, J.-I.; Takahashi, T.; Jahn, L.; Izumo, S. Roles of mechanosensitive ion channels, cytoskeleton, and contractile activity in stretch-induced immediate-early gene expression and hypertrophy of cardiac myocytes. Proc. Natl. Acad. Sci. USA 89: 9905-9909; 1992.

Sah, R. L.; Grodzinsky, A. J.; Plaas, A. H. K.; Sandy, J. D. Effects of tissue compression on the hyaluronate-binding properties of newly synthesized proteoglycans in cartilage explants. Biochem. J. 267: 803-808; 1990.

Sah, R. L.; Kim, Y.-J.; Doong, J.-Y. H.; Grodzinsky, A. J.; Plaas, A. H. K.; Sandy, J. D. Biosynthetic response of cartilage explants to dynamic compression. J. Orthop. Res. 7: 619-636; 1989.

Sah, R. L.-Y.; Doong, J.-Y. H.; Grodzinsky, A. J.; Plaas, A. H. K.; Sandy, J. D. Effects of compression on the loss of newly synthesized proteoglycans and proteins from cartilage explants. Arch. Biochem. Biophys. 286: 20-29; 1991.

Sandberg, M.; Vuorio, E. Localization of types I, II and III collagen mRNAs in developing human skeletal tissues by in situ hybridization. J. Cell Biol. 104: 1077-1084; 1987.

Sandy, J. D.; Adams, M. E.; Billingham, M. E. J.; Plaas, A.; Muir, H. In vivo and in vitro stimulation of chondrocyte biosynthetic activity in early experimental osteoarthritis. Arthritis Rheum. 27: 388-397; 1984.

Schneiderman, R.; Keret, D.; Maroudas, A. Effects of mechanical and osmotic pressure on the rate of glycosaminoglycan synthesis in the human adult femoral head cartilage: an in vitro study. J. Orthop. Res. 4: 393-408; 1986.

Stockwell, R. A. Structure and function of chondrocyte under mechanical stress. In: Helminen, H. J.; Kiviranta, I.; Säämänen, A.-M.; Tammi, M.; Paukkonen, K.; Jurvelin, J., eds. Joint Loading — Biology and Health of Articular Structures. Bristol: John Wright & Sons, Ltd.; 1987:p. 126-148.

Symington, A. L.; Zimmerman, S.; Stein, J.; Stein, G.; Zimmerman, A. M. Hydrostatic pressure influences histone mRNA. J. Cell Sci. 98: 123-129; 1991.

Takano-Yamamoto, T.; Soma, S.; Nakagawa, K.; Kobayashi, Y.; Kawakami, M.; Sakuda, M. Comparison of the effects of hydrostatic compressive force on glycosaminoglycan synthesis and proliferation in rabbit chondrocytes from mandibular condylar cartilage, nasal septum, and spheno-occipital synchondrosis in vitro. Am. J. Orthod. Dentofac. Orthop. 99: 448-455; 1991.

Tammi, M.; Paukkonen, K.; Kiviranta, I.; Jurvelin, J.; Säämänen, A.-M.; Helminen, H. J. Joint loading-induced alterations in articular cartilage. In: Helminen, H. J.; Kiviranta, I.; Tammi, M.; Säämänen, A.-M.; Paukkonen, K.; Jurvelin, J., eds. Joint Loading — Biology and Health of Articular Structures. Bristol: John Wright & Sons, Ltd.; 1987:p. 64-88.

Thyberg, J.; Moskalewski, S. Microtubules and the organization of the Golgi complex. Exp. Cell Res. 159: 1-16; 1985.

Tomlinson, N.; Maroudas, A. The effect of cyclic and continuous compression on the penetration of large molecules into articular cartilage. J. Bone Joint Surg. [Br] 62: 251; 1980.

Uchida, A.; Yamashita, K.; Hashimoto, K.; Shimomura, Y. The effect of mechanical stress on cultured growth cartilage cells. Connect. Tissue Res. 17: 305-311; 1988.

Urban, J.P.G; Hall, A.C.; Gehl, K.A. Regulation of matrix synthesis rates by the ionic environment of articular chondrocytes. J. Cell. Physiol. 154: 262-270; 1993.

van Kampen, G. P. J.; van de Stadt, R. J. Cartilage and chondrocyte responses to mechanical loading in vitro. In: Helminen, H. J.; Kiviranta, I.; Säämänen, A.-M.; Tammi, M.; Paukkonen, K.; Jurvelin, J., eds. Joint Loading — Biology and Health of Articular Structures. Bristol: John Wright & Sons, Ltd.; 1987:p. 112-125.

van Kampen, G. P. J.; Veldhuijzen, J. P.; Kuijer, R.; van de Stadt, R. J.; Schipper, C. A. Cartilage response to mechanical force in high-density chondrocyte cultures. Arthritis Rheum. 28: 419-424; 1985.

Varga, S.; Mullner, N.; Pikula, S.; Papp, S.; Varga, K.; Martonosi, A. Pressure effects on sarcoplasmic reticulum. J. Biol. Chem. 261: 13943-13956; 1986.

Veldhuijzen, J. P.; Bourret, L. A.; Rodan, G. A. In vitro studies of the effect of intermittent compressive forces on cartilage cell proliferation. J. Cell. Physiol. 98: 299-306; 1979.

Wright, M. O.; Stockwell, R. A.; Nuki, G. Response of plasma membrane to applied hydrostatic pressure in chondrocytes and fibroblasts. Connect. Tissue Res. 28: 49-70; 1992.

Zimmerman, A. M.; Tahir, S.; Zimmerman, S. Macromolecular synthesis under hydrostatic pressure. In: Jannasch, H. W.; Marquis, R. E.; Zimmerman, A. M., eds. Current perspectives in high pressure biology. London: Academic Press; 1987:p. 49-63.

24

Altered Chondrocyte Gene Expression in Articular Cartilage Matrix Components and Susceptibility to Cartilage Destruction

T. Kimura, K. Nakata, N. Tsumaki, K. Ono

Introduction

Articular cartilage destruction results from a number of degenerative and inflammatory joint diseases. Among these, osteoarthritis (OA) accounts for a majority of symptomatic arthritis in middle aged persons. OA is a multifactorial disease which is characterized by the degeneration of articular cartilage with widely varying progression and severity. The cartilage shows fibrillation, erosion, and cracking indicating the disruption of cartilage matrix integrity. Biochemically, these histological changes are associated with increased water content, loss of proteoglycan, and minimal change in collagen content. The chondrocytes, partly responding to the physical and biochemical changes of the matrix, show increased metabolic activity (Mankin and Brandt 1992) and occasionally undergo cell proliferation to form clusters. Following cartilage degeneration, changes in bony structure, such as osteophyte formation, subchondral bone sclerosis, and sometimes cyst formation become evident.

Various pathomechanisms have been implicated in the development of OA and the importance of mechanical and metabolic factors as well as of the aging process has been documented (for review see Peyron and Altman 1992). However, it is also evident that OA is not the result of simple "wear and tear" of cartilage during the aging process. There are many elderly people who do not show any evidence of degenerative joint diseases. It is also known that members of certain families are predisposed to degenerative joint disease. In fact, familial aggregation and dominant inheritance are evident in certain forms of primary generalized OA (Kellgren et al. 1963). Genetic factors, therefore, should play an important role in the onset of OA and it is reasonable to assume that the causative mutations reside in genes expressed as components of the cartilage matrix.

This chapter outlines mutations in cartilage collagens as predisposing factors to cartilage degeneration or destruction. First, the mutation in one of the major cartilage components, type II collagen, is reviewed, followed by a discussion of type IX collagen mutation as a cause of OA and as a risk factor for cartilage vulnerability.

Mutations in the Genes for Matrix Components: A True Cartilage Matrix Failure

Collagenous Components of Articular Cartilage

Articular cartilage consists of a highly hydrated extracellular matrix and cellular components; the latter make up less than 5% of the tissue volume. The matrix consists mostly of a mixture of collagen fibrils and proteoglycan aggregates, which provides the cartilage with its unique physiological properties. The collagen fibrils provide the cartilage with shape and high tensile strength, while the proteoglycan aggregates create a large osmotic pressure and help give cartilage its resilience by resisting compressive loads (Kempson et al. 1973; Roth and Mow 1980; Mow et al. 1984, 1992). The collagen fibrils are mainly composed of type II collagen with small amounts of types IX and XI collagen (Figure 1). Type II collagen is the main building block of the fibrils; the diameter of the fibrils themselves increases from less than 20 nm in fetal tissue to a range of 50 to 100 nm in adult human articular cartilage (Lane and Weiss 1975). Type IX collagen molecules are associated with the surface of type II-containing collagen fibrils and type XI collagen seems to form the central core of the fibrils (Eyre et al. 1987; van der Rest and Mayne 1988; Mendler et al. 1989). These collagens, by forming a fibrillar network and interacting with other matrix components, play a critical role in the maintenance of biomechanical properties of the cartilage (Mow et al. 1992). Since the collagen fibrils act as a tension-resistant network against the swelling pressure of charged proteoglycans (Maroudas, 1976), a molecular defect that causes a failure of this fibrillar network would result in the loss of the unique physiological properties of cartilage. Softening of the cartilage matrix, for example, may occur, leading to cartilage degeneration under repeated stress and as a result of aging process (Mow et al. 1992).

Mutation in Type II Collagen Gene and Cartilage Degeneration

Since type II collagen is the major collagenous component of articular cartilage, it is reasonable to hypothesize that mutations in the gene for type II collagen

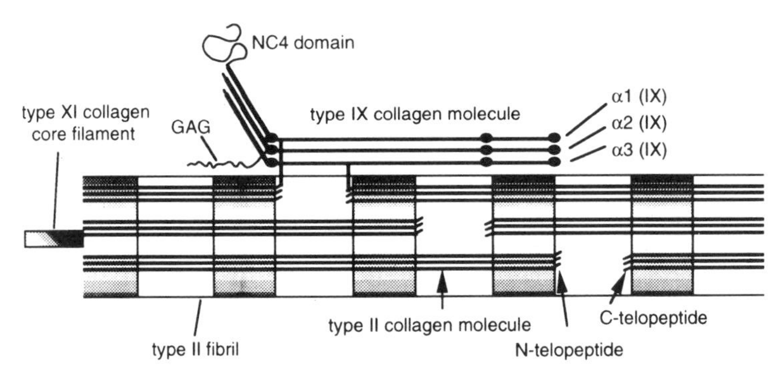

Figure 1. Diagram showing collagen arrangement in cartilage fibrils. Type II collagen is the major component of the fibrils. Type IX collagen molecule covalently cross-links to collagen type II and decorates the surface of the fibril. The N-terminal globular domain (NC4 domain) is cationic and potentially interacts with other matrix components. Type XI collagen resides in the core of the fibril.

(COL2A1) cause predisposition to OA. In fact, alleles of the COL2A1 gene were shown to be linked to familial OA (Palotie et al. 1989; Knowlton et al. 1990) and a mutation in the COL2A1 gene was found to cause primary generalized OA. Ala-Kokko et al. (1990) demonstrated a single base mutation that converted the codon for arginine at position 519 in the triple helical region to a codon for cysteine in a family with primary generalized OA with mild chondrodysplasia. Recently, two more families were reported to have the same Arg-Cys ^{519} mutation in COL2A1 (Holderbaum et al. 1993). Affected members of such families usually begin to develop joint pain and stiffness in the second and third decades and thereafter develop typical radiographic evidence of OA in multiple joints. In addition, a recent report by Vikkula et al. (1993) demonstrates that there are early-onset OA patients who show no evidence of chondrodysplasia but of tight linkage to the COL2A1 gene. These observations imply that OA is a consequence of collagen molecular abnormality and partly confirm the view that the main function of type II collagen is to provide the cartilage with long-term internal resistance.

It should be noted here that a series of mutations in COL2A1 gene has been shown to result in various forms of chondrodysplasias. Mutations introducing a premature termination codon into the COL2A1 gene were identified as a cause of Stickler syndrome, which is characterized by OA combined with ophthalmopathy (Ahmad et al. 1991). Mutations in the COL2A1 gene were also identified in chondrodysplasias with more severe phenotypes, such as spondyloepiphyseal dysplasia congenita, achondrogenesis, and hypochondrogenesis (Lee et al. 1989; Vissing et al. 1989; Bogaert et al. 1990). Thus, the data currently available about mutations suggest that type II collagen gene may harbor mutations that cause a wide spectrum of both rare and common diseases (Kuivaniemi et al. 1991).

Type IX Collagen Mutation and OA: Transgenic Mouse Model

Based on our present knowledge of the structure and function of type IX collagen, it is very likely that type IX collagen molecules play a role in the interaction between collagen fibrils and other matrix components in cartilage tissue. Type IX collagen molecules are composed of α1(IX), α2(IX), and α3(IX) chains that form a long and a short triple-helical arm connected by a flexible non-triple-helical hinge (Ninomiya and Olsen 1984; van der Rest et al. 1985; Irwin and Mayne 1986). The long arm is associated with type II-containing collagen fibrils in cartilage (Vaughan et al. 1988; Eyre et al. 1992), and the short arm projects into the perifibrillar matrix (see Figure 1). In cartilage, a large globular domain (NC4 domain) is located at the amino terminus of the short arm, and this domain could potentially participate in interactions with other matrix components such as proteoglycan (Vasios et al. 1988). Thus, type IX collagen molecules on the surface of fibrils have been implicated in determining the surface properties of type II-containing collagen fibrils and their interactions with other components of the matrix. As discussed previously, these interactions are likely to provide the molecular basis for the normal supramolecular assembly of the matrix components and the unique physiological properties of cartilage. Therefore, mutations affecting type IX collagen should lead to a failure in matrix architecture and result in eventual cartilage degeneration, such as OA.

To demonstrate the involvement of type IX collagen gene mutation in OA, we have generated transgenic mice bearing type IX collagen mutations and analyzed their phenotypic abnormalities. As reported previously (Nakata et al. 1993), the initial gene construct was designed with an in-frame deletion in the central part of α1(IX) cDNA and tissue specific expression in cartilage was obtained by using the type II collagen promoter-enhancer. The construct insert was released from the vector and microinjected into fertilized mouse eggs. The eggs were then transferred to pseudopregnant mice and several transgenic founder mice bearing the mutant α1(IX) gene were obtained. Because the presence of the repetitive Gly-Xaa-Yaa amino acid sequence is essential for the formation and stability of the triple helix of collagen molecules (Prockop 1990, Kuivaniemi et al. 1991), the synthesis of a shortened α1(IX) chain can prevent folding into a stable type IX collagen triple helix. In fact, our mRNA and protein analysis of the mice indicated that the transgene was preferentially expressed in cartilage and the translated product was detected as a shortened α1(IX) chain that was disulfide-bonded with endogenous α chains to form abnormal type IX collagen molecules (Figure 2). What, then, is the consequence of abnormality in type IX collagen? As expected, we found pathological changes similar to OA in the transgenic mice. The articular cartilage of the knee joint looked normal until 6 weeks of age (Figure 3a). The earliest lesion was loss of proteoglycan from the matrix

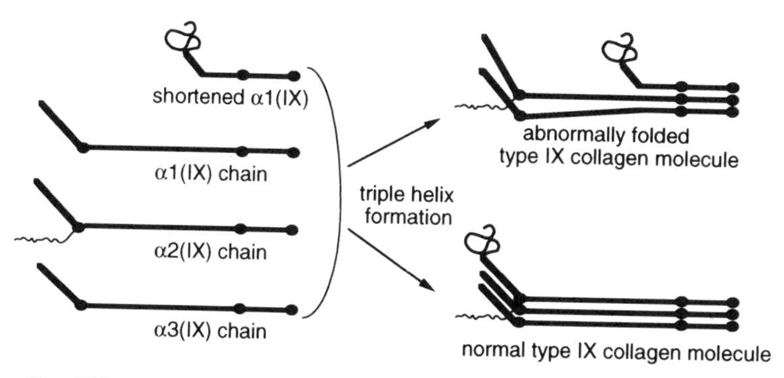

Figure 2. Effects of mutation that changes the structure of type IX collagen. Shortened α1(IX) chain associates with other α chains by competing with normal endogenous α1(IX) chain. The presence of shortened α1(IX) chain disturbs proper zipper-like folding of the triple helix. Such abnormally folded type IX collagen molecule may become unstable and/or form abnormal conformation on the surface of type II-containing collagen fibrils.

and a slight roughening of the articular surface, which was first observed at 5-6 months of age (Figure 3b, c). Polarized light micrographs showed thinning and discontinuity of the articular cartilage surface layer at this stage (Figure 3e). From then on, cartilage erosion and loss of proteoglycan progressed and this process was occasionally accompanied by chondrocyte degeneration and fissure formation (Figure 3f-h). The superficial fibrillation was observed in limited areas (Figure 3i). The histological score for OA changes, obtained with a modification of Mankin's grading (Kim et al. 1991), indicated onset of OA changes after the mice reached maturity, and that the changes progressed with aging (Figure 4). Regenerative or reparative changes were not remarkable at first. However, X-ray and histological analysis of 12- to 18-month-old transgenic mice showed development of marginal osteochondrophytes, subchondral bone sclerosis, and chondrocyte clusters (Figure 3j and Figure 5). Such cellular responses following matrix degeneration in the mice have not been widely investigated yet. Nevertheless, it can be safely assumed that the altered extracellular environment caused by the presence of abnormal type IX collagen and following matrix degeneration seemed to have enhanced the proliferation and biosynthetic activity of the cells.

The two heterozygous transgenic mouse lines presented here did not show any signs of skeletal dysplasia, and the growth plates appeared normal (Figure 3i). Why, then, was the joint cartilage predominantly affected? As shown in Figure 6, the expression of the shortened α1(IX) chains affected the assembly of collagen II-containing fibrils in the articular cartilage because individual fibrils in the transgenic mice were thinner and more disorganized than those in normal mice (Nakata et al. 1993). Since type IX collagen is normally associated with

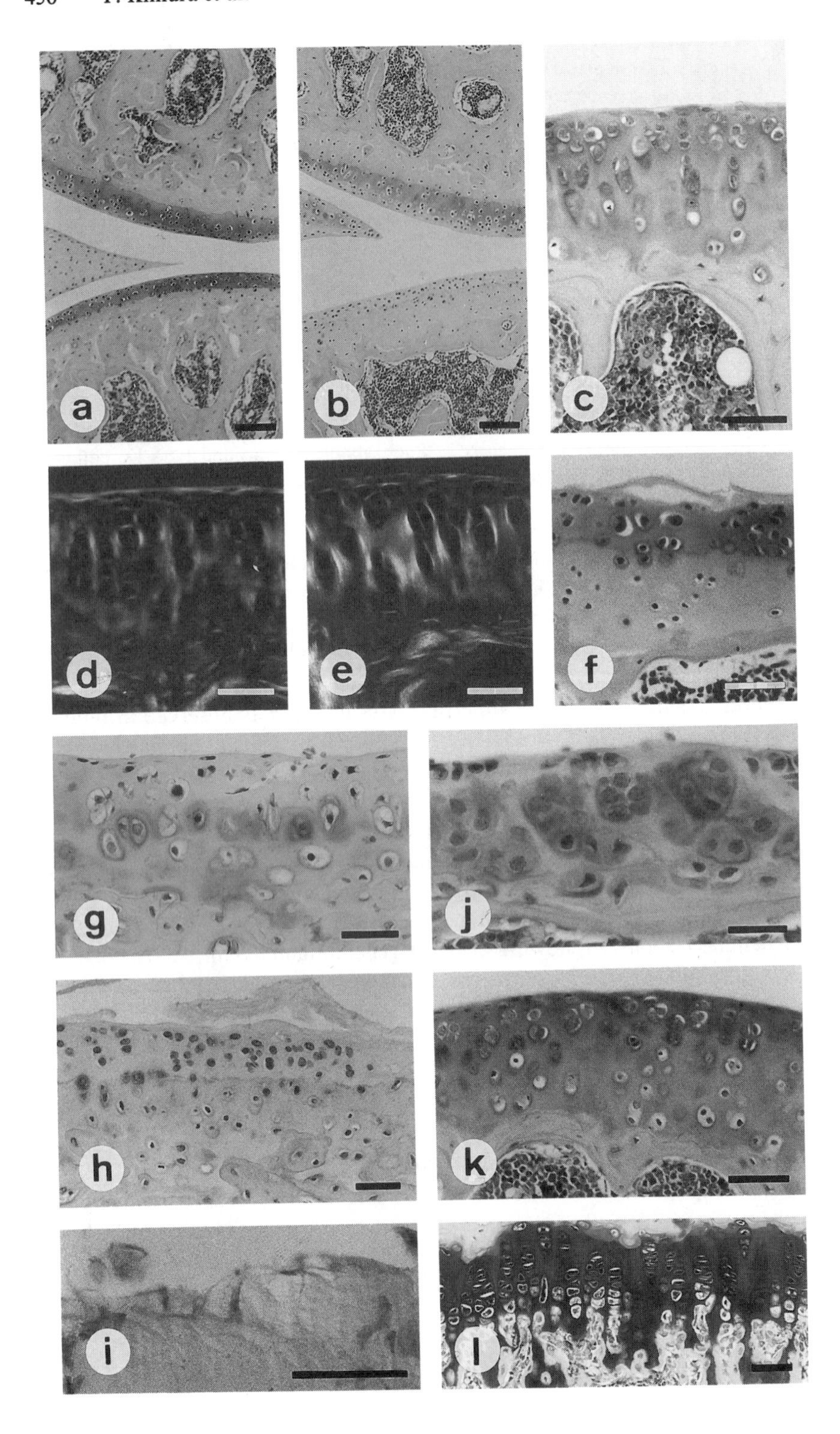
a
b
c
d
e
f
g
j
h
k
i
l

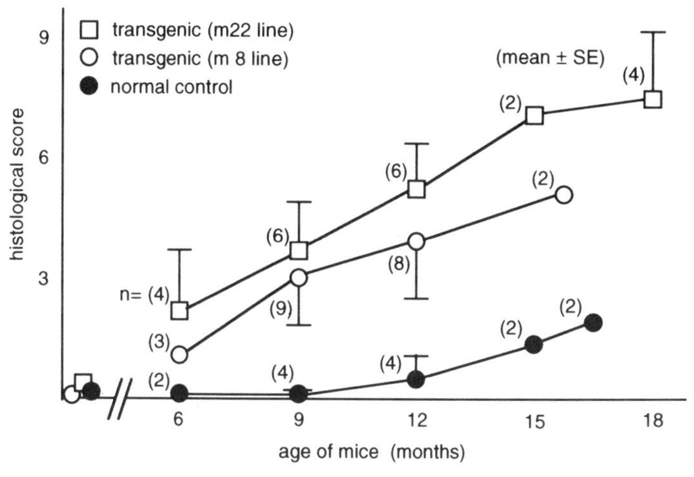

Figure 4. Histological score of OA changes. The total score ranges from 0 to 9, with 0 indicating normal cartilage and 9 indicating severe degeneration.

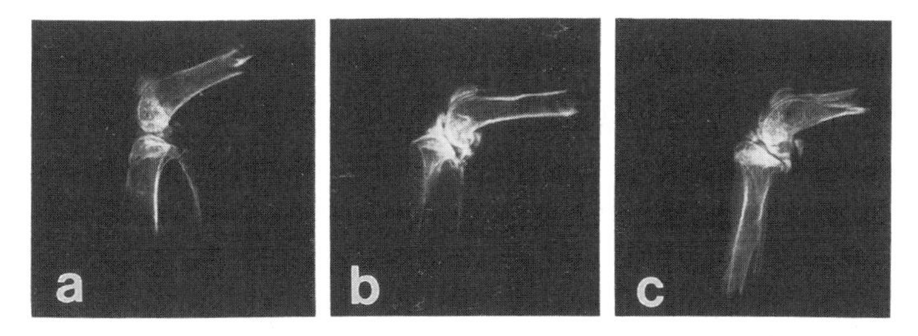

Figure 5. X-ray pictures from mouse knee joints. *(a)* Normal joint from 16-month-old control mouse. Ossification centers within the menisci, sesamoid bones in the popliteal region, and slight quadriceps tendon calcification are present. *(b)(c)* Severe osteoarthritis in 16- and 18-month-old transgenic mouse, respectively. Large marginal osteophytes and sclerotic subchondral bone are present.

Figure 3. Photomicrographs from the knee joints. *(a)(b)* Articular cartilage of femur and tibia from 6-week- and 6-month-old transgenic mouse, respectively. *(c)* The tibial articular cartilage from 6-month-old transgenic mouse. *(d)(e)* Polarized light micrograph of the articular cartilage from 6-week- and 6-month-old transgenic mouse, respectively. *(f)* Nine-month-old transgenic mouse. *(g)-(i)* Twelve-month-old transgenic mice. *(j)* Chondrocyte proliferation, or chondrone formation in 12-month-old transgenic mouse. *(k)* Twelve-month-old control mouse. *(l)* Growth plate cartilage from 8-month-old transgenic mouse. (safranin O staining except for *d, e*; Bar in *a, b* =100 μm, *c-l* =50 μm).

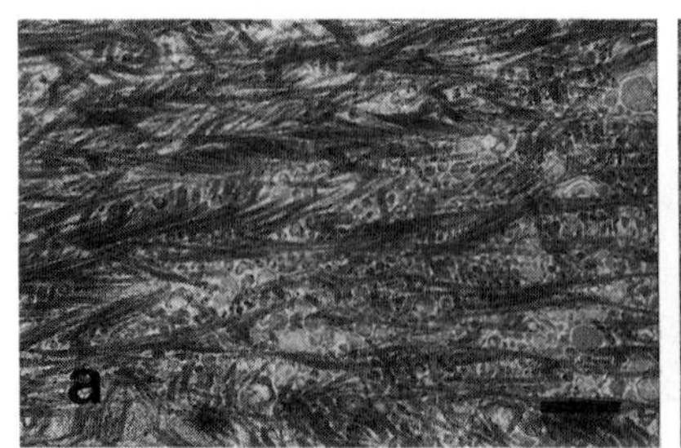

Figure 6. Electron micrographs of the articular cartilage from 6-week-old mice. *(a)* Normal control mouse. *(b)* Transgenic mouse. In the transgenic mouse, the network of type II-containing collagen fibril is more disorganized and the fibrils are thinner than those in normal control. (Bar = 350 nm)

the surface of collagen II-containing fibrils, it is quite possible that the synthesis of abnormally assembled type IX collagen altered the surface properties of the collagen fibrils in the cartilage and/or deteriorated the collagen network structure necessary to oppose the swelling pressure of the proteoglycans (Maroudas, 1976). We believe, therefore, that genetic abnormality in type IX collagen and consequent alteration in cartilage matrix architecture are important predisposing factors to OA.

It should be pointed out, however, that the OA change observed in the transgenic mice is not simply the result of "mutant genes." As described above, the OA change became obvious only after the mice had reached maturity and the changes progressed with aging. In addition, the OA change was more obvious in the weight-bearing joints. For these reasons, it is likely that the aging process as well as repetitive and long-term mechanical stress applied to the articular cartilage played an important role during progression of the OA change. Therefore, the "biology" of OA seems to consist of a combination of genes and environment.

Susceptibility to Cartilage Destruction Resulting from Inflammatory Insults

Severe destruction of articular cartilage is one of the major consequences of chronic arthritis, such as rheumatoid arthritis. The severity of cartilage matrix breakdown is largely dependent on the severity and chronicity of the inflammation. The biological processes that contribute to this severe destruction involve interactions among many cytokines and enzymatic breakdown of cartilage matrix components (Sandy et al. 1981; Campbell et al. 1988; MacNaul et al. 1990; Hasty et al. 1990; Woessner 1991). The susceptibility to cartilage destruction is also determined by various other factors,

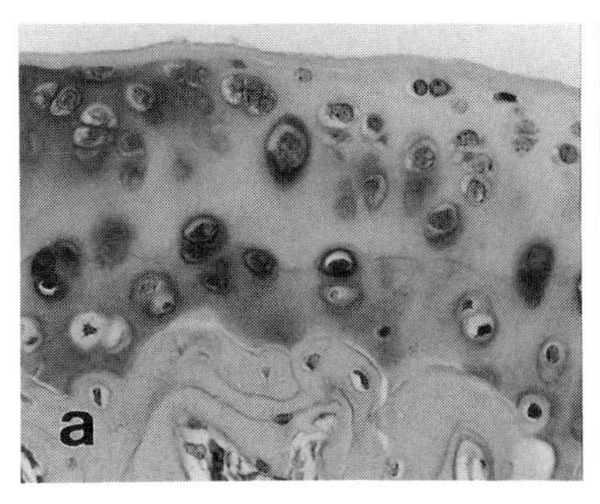

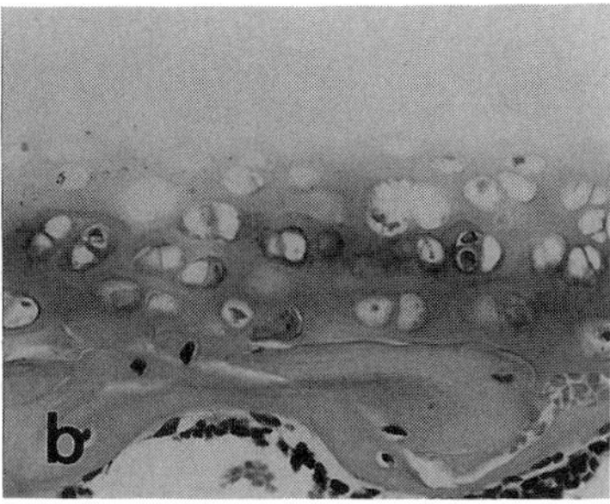

Figure 7. Safranin O-stained sections of knee joint cartilage from 10-week-old mice injected 3 times with interleukin-1. *(a)* Normal mouse. *(b)* Transgenic mouse. (x 117)

such as individual immunological background as well as sex and age (Lawrence 1970; Nunez et al. 1984; van Beuningen et al. 1989). In addition, it is possible that genetic abnormality or heterogeneity in the structure of cartilage matrix components may influence the susceptibility to arthritic damage.

When the transgenic mice with type IX collagen mutation and the normal control mice were exposed to antigen-induced arthritis, both of them developed synovial inflammation in the antigen-injected knee joint. Although the degree of inflammation was similar in both groups, resultant articular cartilage destruction was more severe in the transgenic mice (Tsumaki et al. 1993). Similar findings were observed following repeated intra-articular injection of interleukin-1 (IL-1). Such administration of IL-1 usually increases expression of matrix-destroying enzymes, suppresses biosynthetic activity of the chondrocytes and causes the loss of proteoglycans. In the normal mice the reduced proteoglycan content partly recovered after interruption of IL-1 injection (Figure 7a). On the other hand, the articular cartilage of the transgenic mice continued to show progressive destruction. The cartilage showed surface erosion and proteoglycan loss accompanied by severe degeneration and loss of chondrocytes (Figure 7b). These results suggest that synthesis of abnormal type IX collagen and subsequent alterations in the cartilage matrix architecture are responsible for the higher vulnerability to arthritic insults.

We speculate, therefore, that not only a major genetic abnormality but probably also a minor genetic heterogeneity in the cartilage matrix components are important risk factors responsible for the susceptibility to articular cartilage destruction during arthritis.

Summary

Articular cartilage destruction under various pathological conditions is the result of multiple causes. The relative importance of each of these causes depends on the type of disease. Increasing evidence suggests, at least in some familial forms of OA, that mutation in the gene for type II collagen is an important predisposing factor that causes accelerated degeneration of cartilage in response to environmental and extrinsic factors such as mechanical loading. In addition, molecular defects in other components of cartilage matrix are now waiting to be elucidated. It is still unclear to what extent the mutations of cartilage matrix components act as predisposing factors to cartilage destruction in a wide range of joint diseases. However, a combination of genetic analysis of human cases and development of genetically controlled animal models should make it possible to define the exact molecular mechanisms of cartilage destruction.

References

Ahmad, N.N.; Ala-Kokko, L.; Knowlton, R.G.; Jimenez, D.A.; Weaver, E.J.; Maguire, J.I.; Tasman, W.; Prockop, D.J. Stop codon in the procollagen II gene (COL2A1) in a family with the Sticker syndrome (arthro-ophthalmopahty). Proc. Natl. Acad. Sci. USA 88:6624-6627;1991.

Ala-Kokko, L.; Baldwin, C.T.; Moskowitz, R.W.; Prockop, D.J. Single base mutation in the type II procollagen gene (COL2A1) as a cause of primaty osteoarthritis with a mild chondrodysplasia. Proc. Natl. Acad. Sci. USA 87: 6565-6568;1990.

Bogaert, R.; Tiller, G.E.; Weis, M.A.; Gruber, H.E.; Rimoin, D.L.; Cohn, D.H.; Eyre, D.R. An amino acid substitution (Gly853 to Glu) in the collagen $\alpha 1$(II) chain produces hypochondrogenesis. J. Biol. Chem. 267:22522-22526;1992.

Campbell, I.K.; Piccoli, D.S.; Butler, D.M.; Singleton, D.K.; Hamilton, J.A. Recombinant human interleukin-1 stimulates human articular cartilage to undergo resorption and human chondroctyes to produce both tissue and urokinase-type plasminogen acitivator. Biochem. Biophys. Acta 967:183-194;1988.

Eyre, D.R.; Apone, S., Wu, J.-J.; Ericksson, L.H.; Walsh, K.A. Collagen type IX: evidence for covalent linkages to type II collagen in cartilage. FEBS Lett. 220: 337-341;1987.

Eyre, D.R.; Wu, J.-J.; Woods, P. Cartilage-specific collagens: structural studies. In: Kuettner, K.E.; Schleyerbach, R.; Peyron, J.G.; Hascall, V.C., eds. Articular cartilage and osteoarthritis. New York: Raven Press; 1992: p.119-131.

Hasty, K.A.; Reife, R.A.; Kang, A.H.; Stuart, J.M. The role of stromelysin in the cartilage destruction that accompanies inflammatory arthritis. Arthritis Rheum. 33:388-397;1990.

Holderbaum, D.; Malemud, C.J.; Moskowitz, R.W.; Haqqi, T.M. Human cartilage from late stage familial osteoarthritis transcribes type II collagen mRNA encoding a cysteine in position 519. Biochem. Biophys. Res. Commun. 192: 1169-1174;1993.

Irwin, M.H.; Mayne, R. Use of monoclonal antibodies to locate the chondroitin sulfate chain(s) in type IX collagen. J. Biol. Chem. 261:16281-16283;1986.

Kellgren, J.H.; Lawrence, J.S.; Bier, F. Genetic factors in generalized osteo-arthrosis. Ann. Rheum. Dis. 22:237-255;1963.

Kempson, G.E.; Muir, H.; Pollard, C.; Tuke, M. The tensile properties of the cartilage of human femoral condyles related to the content of collagen and glycosaminoglycans. Biochem. Biophys. Acta 297:456-472;1973.

Kim, H.K.W.; Moran, M.E.; Salter, R.B. The potential for regeneration of articular cartilage in defects created by chondral shaving and subchondral abrasion. J. Bone Joint Surg. 73A:1301-1315;1991.

Knowlton, R.G.; Katzenstein, P.L.; Moskowitz, R.W.; Weaver, E.J.; Malemud, C.J.; Pathria, M.N.; Jimenez, S.A.; Prockop, D.J. Genetic linkage of a polymorphism in the type II procollagen gene (COL2A1) to a primary osteoarthritis associated with mild chondrodysplasia. N. Engl. J. Med. 322:526-530;1990.

Kuivaniemi, H.; Tromp, G.; Prockop, D.J. Mutations in collagen genes: causes of rare and some common diseases in humans. FASEB J. 5:2052-2060;1991.

Lane, J.M.; and Weiss, C. Review of articular cartilage collagen research. Arthritis Rheum. 18:553-562;1975.

Lawrence, J.S. Rheumatoid arthritis: nature or nurture? Ann. Rheum. Dis. 29:357-359;1970.

Lee, B.; Vissing, H.; Ramirez, F.; Rogers, D.; Rimoin, D. Identification of the molecular defect in a family with spondyloepiphyseal dysplasia. Science 244: 978-980;1989.

MacNaul, K.L.; Chartrain, N.; Lark, M.; Tocci, M.J.; Hutchinson, N.I. Discoordinate expression of stromelysin, collagenase, and TIMP-1 in rheumatoid synovial fibroblasts. J. Biol. Chem. 265:17238-17245;1990.

Mankin, H.J.; Brandt, K.D. Biochemistry and metabolism of articular cartilage in osteoarthritis. In: Moskowitz, R.W.; Howell, D.S.; Goldberg, V.M.; Mankin, H.J., eds. Osteoarthritis- diagnosis and medical/surgical management. Philadelphia: WB Saunders Co.; 1992:p.109-154.

Maroudas, A. Balance between swelling pressure and collagen tension in normal and degenerate cartilage. Nature 260:808-809;1976.

Mendler, M.; Eich-Bender, S.G.; Vaughan, L.; Winterhalter, K.H.; Bruckner, P. Cartilage contains mixed fibrils of collagen types II, IX and XI. J. Cell Biol. 108:191-197;1989.

Mow, V.C.; Mak, A.F.; Lai, W.M.; Rosenberg, L.C.; Tang, L.-H. Viscoelastic properties of proteoglycan subunits and aggregates in varying solution concentrations. J. Biomech. 17:325-338;1984.

Mow, V.C.; Ratcliffe, A.; Poole, A.R. Cartilage and diarthrodial joints as paradigms for hierarchical materials and structures. Biometerials 13:67-97;1992.

Nakata, K.; Ono, K.; Miyazaki, J.; Olsen, B.R.; Muragaki, Y.; Adachi, E.; Yamamura, K.; Kimura, T. Osteoarthritis associated with mild chondrodysplasia in transgenic mice expressing $\alpha 1$(IX) collagen chains with a central deletion. Proc. Natl. Acad. Sci. USA 90:2870-2874;1993.

Ninomiya, Y.; Olsen, B.R. Synthesis and characterization of cDNA encoding a cartilage-specific short collagen. Proc. Natl. Acad. Sci. USA 81:3014-3018; 1984.

Nunez, G.; Moore, S.E.; Ball, G.V.; Hurd, E.R.; Stastny, P. Studies of HLA antigens in ten multiple case rheumatoid arthritis families. J. Rheumatol. 11:129-135; 1984.

Palotie, A.; Vaisanen, P.; Ott, J.; Ryhanen, L.; Elima, K.; Vikkula, M.; Cheah, K.; Vuorio, E.; Peltonen, L. Predisposition to familial osteoarthrosis linked to type II collagen gene. Lancet i: 924-927; 1989.

Peyron, J.G.; Altman, R.D. The epidemiology of osteoarthritis. In: Moskowitz, R.W.; Howell, D.S.; Goldberg, V.M.; Mankin, H.J., eds. Osteoarthritis-diagnosis and medical/surgical management. Philadelphia: WB Saunders Co.; 1992:p.15-37.

Prockop, D.J. Mutations that alter the primary structure of type I collagen. The perils of a system for generating large structures by the principle of nucleated growth. J. Biol. Chem. 265:15349-15352;1990.

Roth, V.; Mow, V.C. The intrinsic tensile behavior of the matrix of bovine articular cartilage and its variation with age. J. Bone Joint Surg. 62A:1102-1117;1980.

Sandy, J.D.; Sriratana, A.; Brown, H.L.G.; Lowther, D.A. Evidence for polymorphonuclear-derived proteinases in arthritic cartilage. Biochem. J. 193:193-202;1981.

Tsumaki, N.; Nakata, K.; Ochi, T.; Ono, K.; Kimura, T. Transgenic mice expressing truncated α1(IX) collagen chains show higher susceptibility to articular cartilage damage in the course of antigen-induced arthritis. Trans. 39the Annu. Meet. Orthopaed. Res. Soc.18 (sect.1): 121; 1993.

van Beuningen, H.M.; van den Berg, W.B.; Schalkwijk, J.; Arntz, O.J.; van de Putte, L.B.A. Age- and sex-related differences in antigen-induced arthritis in C57Bl/10 mice. Arthritis Rheum., 32: 789-794;1989.

van der Rest, M.; Mayne, R.; Ninomiya, Y.; Seidah, N.G.; Chretien, M.; Olsen, B.R. The structure of type IX collagen. J. Biol. Chem. 260:220-225;1985.

van der Rest, M.; Mayne, R. Type IX collagen proteoglycan from cartilage is covalently cross-linked to type II collagen. J. Biol. Chem. 263:1615-1618; 1988.

Vasios, G.; Nishimura, I.; Konomi, H.; van der Rest, M.; Ninomiya, Y.; Olsen, B.R. Cartilage type IX collagen-proteoglycan contains a large amino-terminal globular domain encoded by multiple exons. J. Biol. Chem. 263:2324-2329; 1988.

Vaughan, L.; Mendler, M.; Huber, S.; Bruckner, P.; Winterhalter, K.H.; Irwin, M.; Mayne, R. D-periodic distribution of collagen type IX along cartilage fibrils. J. Cell Biol. 106:991-997;1988.

Vikkula, M.; Palotie, A.; Ritvaniemi, P.; Ott, J.; Ala-Kokko, L.; Sievers, U.; Aho, K.; Peltonen, L. Early-onset osteoarthritis linked to the type II procollagen gene. Arthritis Rheum. 36:401-409;1993.

Vissing, H.; D'Alessio, M.; Lee; G.B.; Ramirez, F.; Godfrey, M.; Hollister, D.W. Glycine to serine substitution in the triple helical domain of proα1(II) collagen results in a lethal perinatal form of short-limbed dwarfism. J. Biol. Chem. 264: 18265-18267;1989.

Woessner J.F. Matrix metalloproteinases and their inhibitors in connective tissue remodeling. FASEB J. 5:2145-2154;1991.

Part VII

Mechanics of Cell Motility and Morphogenesis

25
Mechanics of Cell Locomotion

R.Skalak, B.A.Skierczynski, S.Usami, S.Chien

Introduction

Cell motility is essential in development and growth, defense against diseases, wound healing, tissue inflammation, and in other physiological and pathological processes. One form of cell locomotion is chemotaxis, in which the cell exhibits polarized pseudopod extension and migration when exposed to a gradient of a chemoattractant (Condeelis, 1992, Condeelis et al., 1992, Skierczynski et al., 1993b) .

In inflammatory responses, polymorphonuclear leukocytes (PMNs) can roll along the vessel wall in postcapillary venules (Abbassi et al., 1993, Buttrum et al., 1993, Dore et al., 1993, Kansas et al., 1993, Muller et al., 1993), and adhere to the microvascular endothelium in areas of tissue damage (Suchard, 1993) In the early stages of acute inflammation, the rolling cells frequently stop, change shape, and migrate through the vessel wall into surrounding tissue. This is generally a response to a chemoattractant. Studies on many cells types have shown that a chemoattractant elicits a series of biochemical and biomechanical events .

The precise sequence and regulation of these biochemical and morphological changes in the cell are not completely understood, but locomotion in response to chemoattractant requires at least five events;

(1) ligand-receptor interaction at the membrane surface, (2) intracellular signal transduction, (3) modification of the cytoskeleton, (4) force generation, and (5) cyclic establishment and release of adhesive contacts (Skierczynski et al., 1993c).

In recent years, several models have been proposed to explain the motility of various types of cells. These models include the "solation-expansion" hypothesis (Condeelis, 1992, Condeelis et al., 1990, Oster, 1988, Oster, 1984, Oster and Odell, 1984a, Oster and Odell, 1984b, Oster and Perelson, 1985, Oster and Perelson, 1987), myosin-actin contraction (Dembo, 1986, Dembo, 1989, Dembo and Harlow, 1986, Stossel, 1993), cyclic motion (DiMilla et al., 1991, Lauffenburger, 1989) and "Brownian ratchet" (Peskin et al., 1993). These models, however, have not captured the kinematics often observed in cell spreading and locomoting (Evans, 1993).

The focus of this paper is to present a model which can describe the main features of cell locomotion and have qualitative agreement with experiments on PMNs actively moving through a pipette in response to a chemical stimulus (fMLP) (Usami et al., 1992)

The rolling of PMNs on venular endothelim *in vivo* is mediated by L-selectins. When PMNs encounter a chemoattractant gradient due to an inflammatory stimulus, mediated by the integrins, especially CD11b/CD18 (Mac-1), which are mainly stored intracellularly and rapidly translocated to the cell surface upon activation (von_Andrian and Arfors, 1993). The integrins can bind a ligand only after a conformational change of the dimeric molecule to a transiently active stage has been induced by cell activation (Godin et al., 1993).

Expression of the integrins on the cell membrane is regulated by a complex interplay of different pathways. Distinct stimuli appear to affect different pathways. Microtubules and microfilaments may also be involved in modulation of the integrins expression (Volz, 1993). The integrins also behave as a signal transducer. Actin polymerization is elicited in PMNs upon adherence (Southwick et al., 1989). Direct engagement of β_2 integrins in non-adherent PMNs has also been shown to induce actin polymerization (Lofgren et al., 1993, Petty and Todd III, 1993).

The Locomotion Model

The locomoting PMN in the micropipette during active motion exhibits three distinct regions, Fig.1. The leading zone (Zone A) is clear and devoid of granules and other organelles. This region of the cell is a F-actin network filled with cytosol containing sequestered G-actin. At the front of zone A, the polymerization and formation of the actin gel takes place.

Zone B, which is located immediately behind Zone A, appears to be quite fluid with granules having a rapid, chaotic motion. The depolymerization of the F-actin gel takes place between Zones A and B. Zone C is the rear portion of the cell containing the cell nucleus and organelles. There is a slight degree of relative motion of the constituents within Zone C as the cell moves forward.

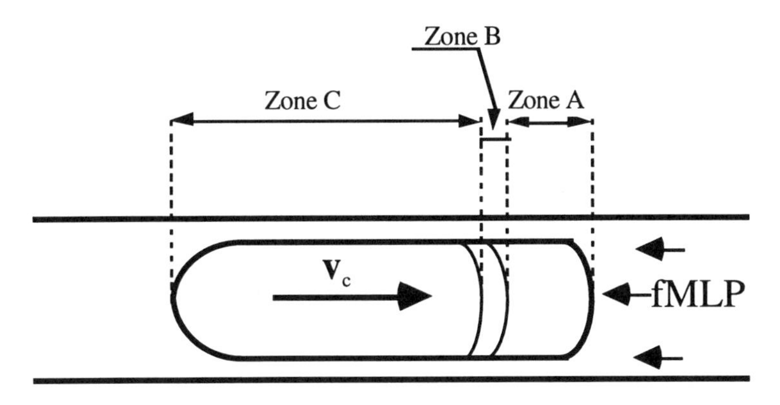

Fig.1 Sketch of the locomoting cell

The main assumptions of the model presented here are:

- Monomeric G-actin is polymerized to form F-actin filaments at the leading edge of the pseudopod. Depolymerization of the network takes place at the rear of the pseudopod, Zone B.
- Actin filaments are crosslinked to form a stiff network which remains stationary relative to the substratum during locomotion. Forces opposing the forward motion of the cell are transmitted through the actin gel and the adhesion bonds to the substratum.
- The lipid bilayer of the cell membrane is passively pulled past the adhesion bonds and anchoring proteins.
- Two mechanisms of force generation are postulated: One due to actin polymerization and the second with its origin in actin-myosin contraction.
- The chemotactic and adhesion receptors are recycled. They are expressed on the front surface of the cell membrane at the leading edge and move backwards relative to the cell. At the rear portion of the cell (Zone C) they are internalized and passively transported through the cell and the actin gel of the pseudopod.

Actin Gelation

Actin is a ubiquitous, highly conserved polymeric protein. Actin filaments provide structural support and are the key components in determining the motile and viscoelastic properties of the cell. The monomeric form of the actin (G-actin) possesses a nucleotide-binding site and one high-affinity divalent-cation-binding site. Under physiological salt concentration, G-actin spontaneously polymerizes into long helical filaments of F-actin (Dufort and Lumsden, 1993a, Dufort and Lumsden, 1993b). A number of proteins such as profilin, gelsolin, thymosin β4, α-actinin, and filamin can alter the fraction of polymerized actin and/or state of gelation in various cells (Abbassi et al., 1993, Cassimeris et al., 1992, Cooper, 1991, Goldschmidt-Clermont et al., 1992, Pollard and Goldman, 1993, Safer, 1992, Stossel, 1989, Stossel, 1990, Stossel, 1993, Wachsstock et al., 1993, Wang et al., 1993a, Wang et al., 1993b).

The polymerization and depolymerization of actin filaments inside nonmuscle cells are highly regulated, both spatially and in time, allowing drastic rearrangement of the cytoskeleton within minutes in responses to external stimuli. Most cells maintain a pool of the unpolymerized actin in which the protein thymosin β4 (Tβ4) seems to play a crucial role. Tβ4 does not bind to F-actin, shows no actin cutting, capping or nucleation activity (Cassimeris et al., 1992). The profilin protein discovered by Carlsson and coworkers in 1977 is assumed to be the major protein responsible for the regulation of F-actin assembly (Pantaloni and Carlier, 1993). Profilin uses the free energy of ATP hydrolysis associated with actin polymerization to promote the rapid assembly of F-actin filaments in the presence of Tβ4 and activated capping proteins.

Polymerization is known to be able to exert forces, as in the sickling of red blood cells due to hemoglobin polymerization, or during shape change in actin-containing vesicles (Cortese et al., 1989). Bacterium *Listeria monocytogenes* utilizes polymerization of the actin to locomote through the host cell cytoplasm (Theriot et al., 1992)

The polymerization of actin is assumed to be one of the sources for force generation in pseudopod extension. The actin filaments are extended by adding G-actin to the barbed ends, pushing the cell membrane outward. The energy release is due to a change in the chemical potential of actin from monomeric to polymeric states and is utilized to perform work against compressive external forces and drag forces generated in the cell membrane during advancing of the pseudopod. The detailed mechanism which is responsible for converting free chemical energy to the mechanical work still remains unknown. One possibility is the rectified Brownian motion proposed by Peskin et al. (1993), but in this model it is not immediately clear how the energy release in the polymerization process could be utilized to generate propulsion forces.

We have postulated that the free chemical energy released during polymerization of G-actin into F-actin filaments may be partially converted into mechanical work through the conformational change of the actin at the time of incorporation into filaments. This proposition is based on the observation of the energy release upon switching of a protein between two physiologically defined states (Kremen, 1992, Steven, 1993). Conformational changes within the F-actin structure after polymerization involving different states of ATP in actin monomers (Schutt et al., 1989) and conformational changes of actin monomers after reacting with proflin have been demonstrated *in vitro* (Theriot and Mitchison, 1993).

Osmotic pressure as a driving mechanism for pseudopod protrusion has been suggested by several researchers based on experimental observations on locomoting PMNs. Although bleb formation seems to be driven by the osmotic stress and actin polymerization follows the expansion of the membrane (Keller and Niggli, 1993). it seems unlikely that osmotic forces are involved in pseudopod formation except at the very early stage when the process is similar to the bleb formation.

We assume that when no external forces are acting on a filament during polymerization, the released energy is dissipated into heat and the rate of elongation of the filament can be described by the following equation (Frieden, 1985),

$$\frac{dL}{dt} = l(k_+ c_a - k_-) \tag{1}$$

where c_a is the concentration of G-actin in the vicinity of the leading edge of the extended pseudopod, k_+, k_- are association and dissociation rate constants, respectively, l is the length by which the F-actin filament will extend after the addition of one G-actin.

Dependence of the association and dissociation constants on external forces can be expressed as follows (Hill and Kirschner, 1982):

$$k_+ = k_{+0}\, exp(-lF/2kT) \tag{2}$$

$$k_- = k_{-0}\, exp(lF/2kT) \tag{3}$$

where k is the Boltzman constant, T is the absolute temperature, F represents the opposing force, and k_{+0}, k_{-0} are association and dissociation constants in the absence of forces, respectively.

The opposing force at the front of a locomoting cell may arise from various sources. One is the drag on the lipid membrane as it is pulled past the anchoring transmembrane adhesion proteins. This force can be described by the expression:

$$F_d = \frac{ds}{dt}[s(t) - w(t)] \frac{8\pi\mu_s n_2}{r_c n\left(\log\frac{\mu_s}{\mu' a} - 0.0772\right)} \tag{4}$$

where n_2 is a surface concentration of protein generating the drag, r_c is the radius of the cell in the micropipette, s(t)-w(t) is the length of the pseudopod, ds/dt is the velocity of the leading edge of the pseudopod (Fig.1), a is the radius of the model of the protein, n is the surface concentration of the F-actin ends at the leading edge, μ_s is a surface viscosity of the lipid bilayer and μ' is the viscosity of water.

A second source of opposing forces is the stress exerted by external tissue. In the experiment described here ,the external pressure was applied via the fluid at the front of the cell:

$$F_p = \frac{\Delta p}{C_{sF}} \tag{5}$$

where Δp is the difference between the external pressure at the front of the cell and the internal pressure at the front inside the pseudopod, and C_{sF} is the surface concentration of the F-actin ends.

The Actin-Myosin Motor

The model presented here assumes that there is an active contraction in the depolymerization zone at the rear end of the pseudopod. In unstimulated spherical PMNs, myosin II shows a diffuse cytoplasmic distribution and is responsible for generating the small tension in cortical layer of the cytoskeleton underlying the cell membrane. After activation, the PMNs extend pseudopods which are primarily an actin gel constituting a porous medium. In the time course the actin gel in the pseudopod is stabilized by the actin binding proteins that

provide the rigidity needed for force transmission to the adhesion bonds and thence to the substratum. The actin gel is concomitantly infiltrated with myosin II (Coates et al., 1992, Keller and Niggli, 1993).
It is assumed here that myosin exerts contraction forces in the rear part of the pseudopod which are balanced by the rigidity of the actin network. In the same region, calcium-dependent severing proteins cut up the actin gel. Cutting and depolymerization processes lead to decreased resistance of the gel against contraction forces generated by the myosin. Depending on the asymmetry in the number of adhesion bonds between the pseudopod and the main cell body, and asymmetry of the rheological properties between these two regions of the cell, a net forward movement of the cell body or retraction of the pseudopod will result. In any case it is assumed that there is a mechanical continuity between the F-actin network in the pseudopod and the cortical layer in the main cell body. The contraction due to myosin can then pull the rear portion of the cell (Zone C) forward. Depolymerization of the actin gel and its contraction occur in the same location and at the same time.

In spite of numerous studies, the working principles of the molecular myosin II motor remain elusive. This molecular motor can operate under the influence of thermal noise, with a high efficiency of the chemo-mechanical energy conversion. The actin-myosin motor is driven by the chemical energy of the ATP hydrolysis. Hydrolysis of one single ATP molecule can release about 20kT of energy (Yanagida et al., 1993).

The depolymerization process is a complicated series of molecular events that are only partially understood. When the concentration of calcium is sufficiently high, cutting proteins will be activated and fragment actin chains. The activation of cutting protein is a first order reaction. Taking into account that the cutting process is much faster that the overall time of the cell locomotion, the concentration of the active cutting protein can by expressed;

$$c_{wa} = \frac{c_{wT} c_{Caw}}{c_{Caw} + \dfrac{k_{-w}}{k_{+w}}} \tag{6}$$

where c_{Caw} is the calcium concentration above the threshold value, k_{-w}, k_{+w} are dissociation and association constants for this reaction, respectively, and c_{wT} is the total concentration of the cutting protein.

The cutting process for one F-actin filament is described by the probability of binding of the cutting protein to a site on the F-actin strand and is proportional to the number of the structural units in the chain. The overall velocity of the depolymerization can be described by the equation;

$$\frac{dw}{dt} = l\left(nKk_{-dep} - c_{Gf}k_{+dep}\right) \tag{7}$$

where n is the number of the F-actin filaments before the cutting process occurs, k_{-dep}, k_{+dep} are dissociation and association constants for depolymerization process, respectively, and K is the number of cuts per F-actin filament which, in quasi steady-state, is expressed by:

$$K = \frac{c_{wa}\left(r_0 - 1\right)}{c_{wa} + \dfrac{k_{-cu}}{k_{+cu}}} \tag{8}$$

where r_0 is the average number of unbonded sites on one F-actin filament, and k_{+cu}, k_{-cu} are association and dissociation constants for the cutting reaction.

The model assumes that the pseudopod behaves as porous, elastic solid and the cytosol which fills the pores behaves as a Newtonian fluid. Such a system can be described as a first approximation, on the basis of a multi-phase mixture theory (Lai et al., 1991, Mow et al., 1992), that includes the active stress generated by the actin-myosin motor.

The equations of motion for the fluid phase can be expressed by:

$$\frac{\partial\left(2\mu_s E_{ij}\right)}{\partial x_j} + \phi\kappa\left[V_i^s - V_i^f\right] - \frac{\partial p}{\partial x_i} - \frac{\partial \pi}{\partial x_i} = 0 \tag{9}$$

where μ_s is the viscosity of the solution, κ is the permeability of the porous medium , and ϕ is fraction of the volume occupied by the fluid. The pressure p in equation (9) is the hydrodynamic pressure. Equation (9) also includes the local osmotic pressure because the actin gel is considered to be a distributed osmometer. The osmolarity of the solution is changed during polymerization and depolymerization. The interphase drag is assumed to be proportional to the network density and relative network-solution velocity. E_{ij} is the strain rate for the fluid phase.

The network motion is described by the equation:

$$\frac{\partial \sigma_{ij}^e}{\partial x_j} + \phi\kappa\left[V_i^s - V_i^n\right] + \frac{\partial \sigma_{ij}^c}{\partial x_j} = 0 \tag{10}$$

which includes the elastic stress of the actin network, the active stress generated by myosin, and the interphase drag. Inertial effects are neglected in Eqs. (9) and (10).

Studies of *in vitro* rheology of actin networks have shown that the elastic stress can be a very complicated function of the deformation and the rate of deformation and that it is also strongly dependent on the biochemical processes. As a first approximation, we assume that the elastic stress can be expressed as:

$$\sigma_{ij}^{e} = e^{-(n_o - n_f)}\left[\lambda_0 e_{kk}\delta_{ij} + 2\mu_0 e_{ij}\right] \tag{11}$$

where λ_0, μ_0 are Lame' coefficients of the solid phase before activation of the cutting protein, n_0 and n_f are the numbers of the actin filaments per unit volume before and after activation, respectively, e_{ij} is the strain tensor in the network. It is assumed that the active stress can be described as,

$$\sigma_{ij}^{c} = \delta_{ij}\xi c_{ma} f_m \tag{12}$$

where f_m is the force generated by one myosin filament, ξ is the coefficient describing a fraction of the myosin heads interacting with the actin network. The is the concentration of the activated myosin (c_{ma}) can be written, in the quasi steady-state in the form:

$$c_{ma} = \frac{c_{mT}\chi(c_{CaT})\left[c_{Ca} - c_{CaT}\right]}{\left[c_{Ca} - c_{CaT}\right] + \dfrac{k_{-m}}{k_{+m}}} \tag{13}$$

where c_{mT} is the total concentration of the myosin, C_{Ca} is the concentration of calcium, k_{-m}, k_{+m} are the dissociation and association constants for activation of the myosin, respectively and C_{CaT} is the threshold concentration of the calcium required to activate the myosin.

The mass conservation for the fluid phase is given by:

$$\frac{\partial}{\partial x_j}\left[\phi\left(V_j^f - V_j^s\right) + V_j^f\right] = 0 \tag{14}$$

and for the solid phase by:

$$\frac{\partial\phi}{\partial t} = -\frac{\partial(\phi V_j^s)}{\partial x_j} - \frac{V_{Gact}}{\phi V_c}\left[nKk_{-dep} - c_{Gf}k_{+dep}\right] \tag{15}$$

where V_C is the volume of the reaction chamber, and V_{Gact} is the volume of one G-actin monomer. The last term on the right side in Eq.(15) is the rate at which F-actin filaments are subtracted from the gel. The active stress and elastic coefficients are calcium dependent, and they may generate spatial gradients due to the spatial distribution of calcium throughout the pseudopod.

Treadmilling of G-actin.

The treadmilling of the G-actin is assumed to take place within the cell and the cell membrane is treated as impermeable to G-actin. G-actin is transported from the main cell body to the leading edge as the G-actin-Tβ4 complex by diffusion and convection with a velocity equal to the velocity of the fluid within the F-actin network, Thus:

$$\frac{\partial c_G}{\partial t} = v_f\frac{\partial c_G}{\partial x} + D_G\frac{\partial^2 c_G}{\partial x^2} \qquad w(t) < x < s(t) \tag{16}$$

where D_G is the diffusion coefficient for the G-actin-Tβ4 complex, c_G is the concentration of the complex, s(t) and w(t) are the coordinates of the leading edge of the cell and the depolymerization front, respectively, and v_f is the velocity of the fluid inside the pseudopod which is described by the equation:

$$v_f = \frac{ds}{dt} = \frac{C_{sF}}{C_{sT}}l\left(k_{+0}e^{\frac{-lF_d - A_1\frac{ds}{dt}(s(t)-w(t))}{2kT}}c_G(x=s,t) - k_{-0}e^{\frac{lF_d + A_1\frac{ds}{dt}(s(t)-w(t))}{2kT}}\right) \tag{17}$$

At the leading edge of the cell, the conservation of G-actin leads to:

$$\frac{\partial c_G}{\partial x} = -\frac{C_{sF}}{\phi D_G}\left(k_{+0} e^{\frac{-lF_d - A_1 \frac{ds}{dt}(s(t)-w(t))}{2kT}} c_G - k_{-0} e^{\frac{lF_d + A_1 \frac{ds}{dt}(s(t)-w(t))}{2kT}} \right) \qquad x = s\,(t) \qquad (18)$$

As mentioned before, at the rear part of the pseudopod the depolymerization of the actin network takes place and serves as a source of the G-actin monomers. We assume, as a first approximation, that the released G-actin monomers are sequestered in a very short time by the Tβ4 molecules. Thus

$$c_G + \frac{C_{sT}}{l}\chi = c_{Gb}$$
$$\chi = \begin{cases} 0 & if \quad \dfrac{dw}{dt} = 0 \\ 1 & if \quad \dfrac{dw}{dt} \neq 0 \end{cases} \qquad x = w(t) \qquad (19)$$

and

$$c_{Gb}\frac{ds}{dt} - D_G \frac{\partial c_{Gb}}{\partial x} - \frac{dw}{dt}\left(c_{Gb} - \phi c_G\right) + \frac{C_{sT}}{l}\frac{dw}{dt} = \phi\left(c_G \frac{ds}{dt} - D_G \frac{\partial c_G}{\partial x}\right) \qquad x = w(t) \qquad (20)$$

where c_{Gb} is the concentration of the G-actin complex in the main cell body.
The initial condition are:

$$\text{At} \quad t=0 \qquad c_G = c_{Gb} = c_0 \qquad \text{and} \qquad w = s = 0 \qquad (21)$$

Interaction of the Cell with the Substratum

Adhesion to the substratum is a prerequisite for the locomotion of the cells and is a finely regulated process, controlled by the membrane receptors involved in cell-cell and cell-matrix recognition by several different pathways including, (a) modulation of the number of receptors expressed on the cell surface, (b) modification or removal of the molecules by activation processes or enzymes, (c) specific activation of the receptors to affect their ligand binding affinity, (d) changes in the organization of the cytoskeleton and of adhesion receptors at the cell surface, and (e) changes in concentration of ligands.

A long-standing issue in cell adhesion is how receptor-mediated cell/cell or cell/substratum adhesion strength is related to receptor/ligand binding affinity. There are several models proposed for a variety of cell behaviors (Bell et al., 1984, Dembo et al., 1988, DiMilla et al., 1991). Association of the ligand and cell cytoskeleton is a dynamic process in which individual receptors are continuously forming and breaking their association with cytoskeleton proteins (Miettinen et al., 1992, Ward and Hammer, 1993). In the model of the adhesion process proposed (Skierczynski et al., 1993a), the adhesion receptors can be found in four states: (1) receptors bound to the ligand but mobilized in respect to the cell cytoskeleton, (2) receptors bound to the cytoskeleton but not bound to the ligand, (3)receptors bound to both the ligand and the cytoskeletom "the true adhesion bonds", which are capable of transmitting forces, and (4) free receptors which are free to diffuse in the cell membrane.

The binding of the ligand to the receptor and the cytoskeleton can induce conformational changes of the receptor, as well as the binding affinity which is expressed as a change of the energy barrier which must be overcome during transition to different states of the adhesion receptor. The energy barrier can also be altered through the application of external forces in the course of separating molecules. In the model of the adhesion process involving the four states of receptors, we have proposed that the energy barriers for the formation of ligand-receptor complex, receptor-cytoskeleton complex, true adhesion bonds - ligand-receptor-cytoskeleton complex, as well as that for the dissociation of the receptor-cytoskeleton complex, are not dependent on the change of the potential energy delivered to the cell-substratum system through stretching by tension.

Cell locomotion over an adhesive surface is a dynamic phenomenon involving the formation and breakage of the attachment bonds with the substratum. The model of cell locomotion presented here assumes the existence of an asymmetry in the overall strength of the adhesion between the pseudopod and the main cell body. The locomoting cell can fulfill the asymmetry requirement, in several different ways e.g., by:

- asymmetry in spatial distribution of the adhesion receptors due to their preferential trafficking.
- asymmetry in the overall strength of adhesion due to variations in receptor-cytoskeleton interaction. The pseudopod has a well developed actin gel which has different rheological properties compared to the main cell body.
- asymmetry in dynamic properties of the cytoskeleton between the main cell body and the pseudopod.

Results

Simultaneous solution of the full set of the governing equations of the present model is under current investigation. Here we present some solutions of the locomotion equations, i.e., Eqs. (1-8), (16-21), together with correlated control mechanisms discussed in Skierczynski et. al., (1993).

The truncated set of equations was solved numerically using Gear's method. The results of this computation are presented in Fig.2. Figure 2A shows the history of an applied external pressure at the front of an actively locomoting cell in the micropipette in response to a chemoattractant (fMLP). The calculated time-dependent extension of the pseudopod and advancing of the depolymerization front are presented in Fig. 2B. Figure 2C shows computed velocities of the leading edge and the depolymerization front. The computations reported in Fig. 2 are based on a pipette with a diameter 2.5 μm and the cell as an initial sphere with a diameter 8 μm. Other parameters are: $k_{+0} = 1.93\ 10^{-14}$ cm $^3s^{-1}$, $k_{-0} = 1.4\ s^{-1}$,$l = 2.7\ 10^{-7}$ cm, $\mu_s = 10^{-3}$ dyn s cm^{-1}, $n = 4\ 10^{10}\ cm^{-2}$, $\mu' = 10^2$ dyn s cm^{-2}, $c_0 = 10^{17}\ cm^{-3}$, $D_G = 10^{-8}\ cm^2\ s^{-1}$, $a = 3\ 10^{-7}$ cm. The computed results in Fig. 2 correspond well qualitatively and quantitativly to the experimental results reported previously (Usami et. al. 1992). Without any opposing fluid pressure, the experimental mean velocity was approximately 0.3 μm/sec. As in the experiments, an application of a positive pressure to the front of the cell reduces the cell velocity and a negative pressure (suction) increases the cell velocity. During such manipulations, the computations show that the length of the pseudopod (Fig.2B) is variable, as observed also in the experiments (Skalak et al., 1993).

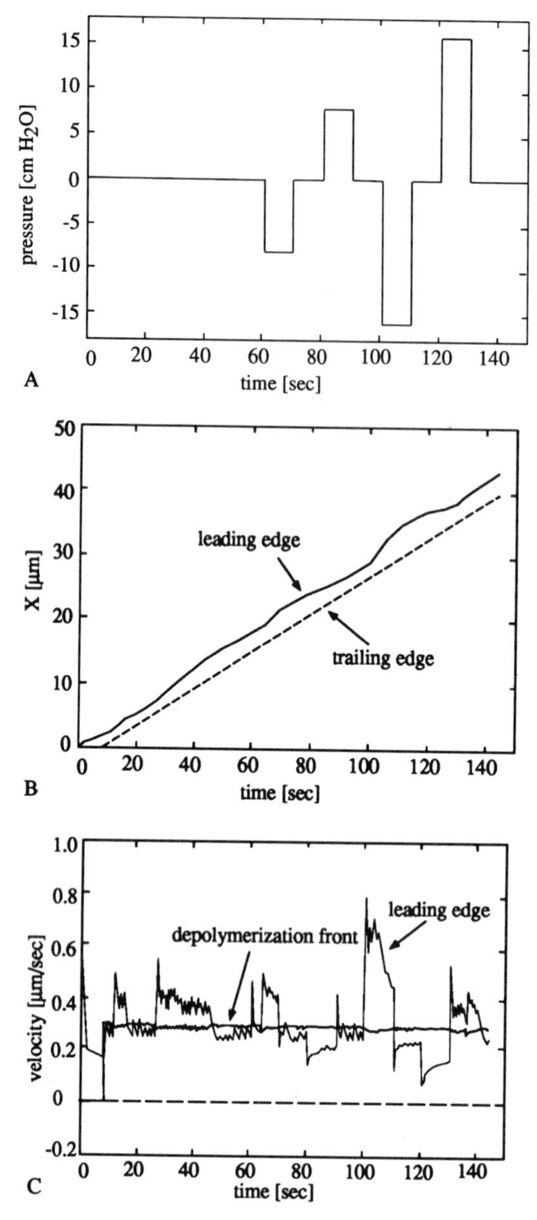

Fig.2 Computational results. A. The history of applied pressure, B. Time-dependent extension of the pseudopod and advancing of the depolymerization front, C. Velocities of the leading edge of the pseudopod and the depolymerization front.

Discussion

The model of the cell locomotion presented here emphasizes the mechanics and mechanisms underlying the active cell motion. The general mechanisms involved in the locomotion may be applicable to other types of mammalian cells. Some of the proposed mechanisms require further experimental and theoretical verifications and may require some corrections, and they may be quite different for plant cells and single animal cells such as the amoeba (Dembo, 1989).

The present model is based on observations on the PMN active moving in a micropipette experiments in response to chemical stimuli, as well as on the experimental and theoretical results available in the literature.

The model of PMN locomotion on a plane substrate may be interpreted as depicted in Fig. 3. It is proposed that in this type of cell locomotion the same two mechanisms of force generation are involved. One is the conversion of the chemical energy released during polymerization of the actin into mechanical work at the leading edge of the pseudopod. The other is the generation of force by the activation of the myosin II in the rear part of the pseudopod. These two processes can be independently controlled, resulting in the possibility that the length of the pseudopod can vary. The adhesion and depolymerization processes can be controlled to some extent independently from each other and from the processes described above. During locomotion, the pseudopod remains stationary with respect to the substratum because of the connection of the adhesion bonds which are connected by transmembrane proteins. The forces generated at the front of the cell need only to overcome the drag of the membrane pulled over the stationary gel. In locomotion through a solid tissue, the pseudopod must also overcome external forces required to deform the tissue through which the cell is moving. Contraction generated in the depolymerization zone pulls the main body forward and must overcome resistance generated by adhesion bonds on the rear portion of the cell.

For a cell in free suspension, the mechanisms described here do not result in cell locomotion because no force based on adhesion is possible in the fluid. Leukocytes in free suspension can protrude pseudopods by polymerization of actin at the membrane. However, if a cell is spherical, the polymerized region may grow into the cell itself. In either case, polymerization takes place at the membrane and there is a retrograde flow of F-actin filaments relative to the leading edge of the pseudopod.

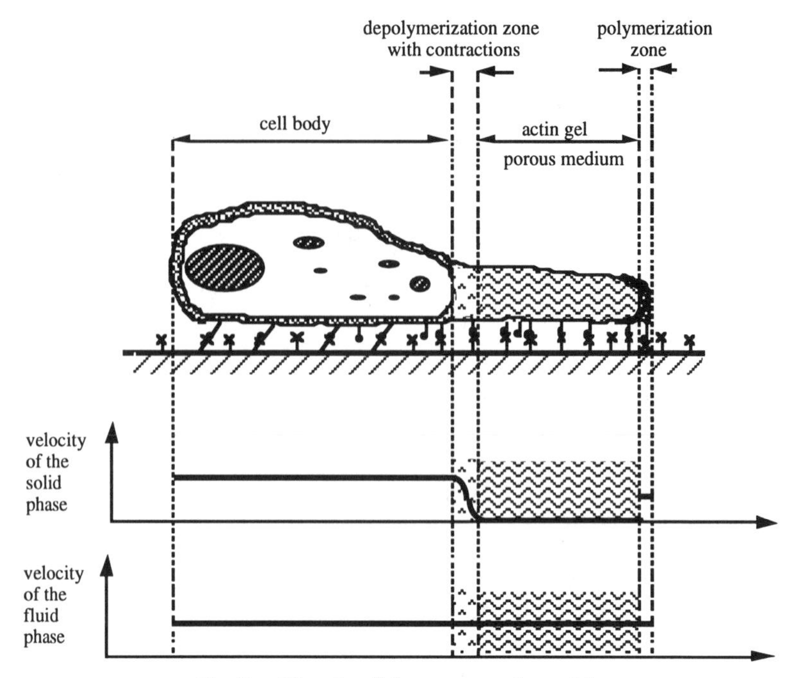

Fig.3 Sketch of the proposed model.

When the coordination between depolymerization and contraction is perturbed, necking can result at the base of the pseudopod. There are some experimental results (Usami et al., 1993) in which the pseudopod is detached from the main cell body, and the detached pseudopod fragment is still able to locomote. This may be interpreted to indicated that polymerization plus depolymerization alone are sufficient for locomotion and that contraction is not required. Adhesion bonds may be broken due to the depolymerization.

Another effect which is explainable in terms of the present model is that if the adhesion of the cell body is stronger then that of the pseudopod, the end result will be a retraction of a pseudopod, rather then locomotion. This effect has been previously suggested also by the model of Lauffenburger (1989).

In the present work the control mechanisms of the cell locomotion are not fully discussed. The complete set of control pathways are very complicated phenomena and must be included in order to have a comprehensive model for the cell motility. These processes include the interaction of the specialized receptors with external stimuli, signal generation at the leading edge of the cell, transduction of the signals to the main cell body, generation of signals inside the main cell body, and positive and negative feedback loops which affect contraction and adhesion, as well as morphological and rheological properties of the cell. The model presented here is intended primarily to bring out the mechanical mechanisms.

References

Abbassi O.; Kishimoto T.K.; McIntire L.V.; Smith C.W. Neutrophil Adhesion to Endothelial Cells. Blood Cells. 19: 245-260; 1993.

Bell G.I.; Dembo M.; Bongrand P. Cell Adhesion. Biophysical Journal. 45: 1051-1064; 1984.

Buttrum S.M.; Hatton R.; Nash G.B. Selectin-Mediated Rolling of Neutrophils on Immobilized Platelets. Blood. 82(4): 1165-1174; 1993.

Cassimeris L.; Safer D.; Nachmias V.T.; Zigmond S.H. Thymosin β_4 Sequesters the Majority of G-actin in Resting Human Polymorphonuclear Leukocytes. The Journal of Cell Biology. 119: 1261-1270; 1992.

Coates T.D.; Watts R.G.; Hartman R.; Howard T.H. Relationship of F-actin Distribution to Development of Polar Shape in Human Polymorphonuclear Neutrophils. The Journal of Cell Biology. 117: 765-774; 1992.

Condeelis J. Are All Pseudopods Created Equal? Cell Motility and the Cytoskeleton. 22: 1-6; 1992.

Condeelis J.; Bresnick A.; Demma M.; Dharmawardhane S.; Eddy R.; Hall A.L.; Sauterer R.; Warren V. Mechanisms of Amoeboid Chemotaxis: An Evaluation of the Cortical Expansion Model. Developmental Genetics. 11: 33-340; 1990.

Condeelis J.; Jones J.; Segall J.E. Chemotaxis of Metastatic Tumor Cells: Clues to mechanisms from the Dictyostelium paradigm. Cancer and Metastasis Reviews. 11: 55-68; 1992.

Cooper J.A. The Role of Actin Polymerization in Cell Motility. Ann. Rev. Physiol. 53: 585-605; 1991.

Cortese J.D.; Schwab_III B.; Frieden C.; Elson E.L. Actin Polymerization Induces a Shape Change in Actin-containing Vesicles. Proc. Natl. Acad. Sci. USA. 86: 5773-5777; 1989.

Dembo M. The Mechanics of Motility in Dissociated Cytoplasm. Biophysical Journal. 50: 1165-1183; 1986.

Dembo M. Mechanics and Control of the Cytoskeleton in *Amoeba proteus*. Biophysical Journal. 55: 1053-1080; 1989.

Dembo M.; Harlow F. Cell Motion, Contractile Networks, and the Physics of Interpenetrating Reactive Flow. Biophysical Journal. 50: 109-121; 1986.

Dembo M.; Torney D.C.; Saxman K.; Hammer D. The Reaction-limited Kinetics of Membrane-to-surface Adhesion and Detachment. Proc. R. Soc. Lond. B 234: 55-83; 1988.

DiMilla P.A.; Barbee K.; Lauffenburger D.A. Mathematical Model for the Effects of Adhesion and Mechanics on Cell Migration Speed. Biophysical Journal. 60: 15-37; 1991.

Dore M.; Korthius R.J.; Granger D.N.; Entman M.L.; Smith C.W. P-selectin Mediates Spontaneous Leukocyte Rolling In Vivo. Blood. 82(4): 1308-1316; 1993.

Dufort P.A.; Lumsden C.J. Cellular Automaton Model of the Actin Cytoskeleton. Cell Motility and the Cytoskeleton. 25: 87-104; 1993a.

Dufort P.A.; Lumsden C.J. High Microfilament Concentration Results in Barbed-end ADP Caps. Biophys. J. 65: 1757-1766; 1993b.

Evans E. New Physical Concepts for Cell Amoeboid Motion. Biophysical Journal. 64: 1-16; 1993.

Frieden C. Actin and Tubulin Polymerization: The Use of kinetic Methods to Determine Mechanisms. Ann. Rev. Biophys. Biophys. Chem. 14: 189-210; 1985.

Godin C.; Caprani A.; Dufaux J.; Flaud P. Interactions Between Neutrophils and Endothelial Cells. J. Cell Sci. 1067: 441-452; 1993.

Goldschmidt-Clermont P.J.; Furman M.I.; Wachsstock D.; Safer D.; Nachmias V.T.; Pollard T.D. The Control of Actin Nucleotide Exchange by Thymosin β_4 and Profilin. A Potential Regulatory Mechanisms for Actin Polymerization in Cells. Molecular Biology of the Cell. 3: 1015-1024; 1992.

Hill T.L.; Kirschner M.W. Subunit Treadmilling of Microtubules or Actin in the Presence of Cellular Barriers: Possible Conversion of Chemical Free Energy Into Mechanical Work. Proc. Natl. Acad. Sci. USA. 79: 490-494; 1982.

Kansas G.s.; Ley K.; Munro J.M.; Tedder T.F. Regulation of Leukocyte Rolling and Adhesion to High Endothelial Venules Through the Cytoplasmic Domain of L-Selectin. J. Exp. Med. 177: 833-838; 1993.

Keller H.U.; Niggli V. Colchicine-Induced Stimulation of PMN Motility Related to Cytoskeletal Changes ib Actin, α-Actinin and Myosin. Cell Motility and the Cytoskeleton. 25: 10-18; 1993.

Kremen A. Biological Molecular Energy Machines as Measuring Devices. Journal of Theoretical Biology. 154: 405-413; 1992.

Lai W.M.; Hou J.S.; Mow V.C. A Triphasic Theory for the Swelling and Deformation Behaviors of Articular Cartilage. Journal of Biomechanical Engineering. 113: 245-258; 1991.

Lauffenburger D. A Simple Model for the Effects of Receptor-mediated Cell-substratum Adhesion on Cell Migration. Chemical Engineering Science. 44(9): 1903-1914; 1989.

Lofgren R.; Ng-Sikorski J.; Sjolander A.; Andersson T. β_2 Integrin Engagement Triggers Actin Polymerization and Phosphatidylinositol Triphosphate Formation in Non-adherent Human Neutrophils. The Journal of the Cell Biology. 21(6 Part1): 1597-1605; 1993.

Miettinen H.M.; Matter K.; Hunziker W.; Rose J.K.; Mellman I. Fc Receptor Endocytosis Is Controlled by a Cytoplasmic Domian Determinant That Actively Prevents Coated Pit Localization. The Journal of Cell Biology. 116(4): 875-888; 1992.

Mow V.C.; Ratcliffe A.; Poole A.R. Cartilage and Diarthordial Joints as Paradigme for Hiererchical Materials and Structures. Biomaterials. 13(2): 67-98; 1992.

Muller W.A.; Weigl S.A.; Deng X.; Phillips D.M. PECAM-1 IS Required for Transedothelial Migration of Leukocytes. J. Exp. Med. 178: 449-460; 1993.

Oster G. Biophysics of the Leading Lamella. Cell Motility and the Cytoskeleton. 10: 164-171; 1988.

Oster G.F. On the Crawling of Cells. J. Embryol. exp. Morph. 83 Supplement: 329-364; 1984.

Oster G.F.; Odell G.M. Mechanics of Cytogel I: Oscilation in Physarum. Cell Motility. 4: 469-503; 1984a.

Oster G.F.; Odell G.M. The Mechanochemistry of Cytogels. Physica. 12D: 333-350; 1984b.

Oster G.F.; Perelson A.S. Cell Spreading and Motility: A Model Lamellipod. Journal of Mathematical Biology. 21: 383-388; 1985.
Oster G.F.; Perelson A.S. The Physics of Cell Motility. Journal of Cell Science. Suppl. 8: 35-54; 1987.
Pantaloni D.; Carlier M.-F. How Profilin Promotes Actin Filament Assembly in the Presence of Thymosin β4. Cell. 75: 1007-1014; 1993.
Peskin C.S.; Odell G.M.; Oster G.F. Cellular Motion and Thermal Fluctuations: The Brownian Ratchet. Biophysical Journal. 65: 316-324; 1993.
Petty H.R.; Todd_III R.F. Receptor-receptor Interactions of Complement Receptor Type 3 in Neutrophil Membranes. 54: 492-494; 1993.
Pollard T.D.; Goldman R.D. Cytoplasm and Cell Motility. Current Opinion in Cell Biology. 5: 1-2; 1993.
Safer D. The Interaction of Actin with Thymosin β_4. Journal of Muscle Research and Cell Motility. 13: 269-271; 1992.
Schutt C.E.; Linderberg U.; Myslik J.; Strauss N. Molecular Packing in Profilin: Actin Crystals and its Implications. Journal of Molecular Biolology. 209: 735-746; 1989.
Skalak R.; Skierczynski B.A.; Wung S.-L.; Chien S.; Usami S. Mechanical Models of Pseudopod Formation. Blood Cells. 19: 389-399; 1993.
Skierczynski B.A.; Skalak R.; Chien S. Energetics of Adhesion of Cells. 15th Annual International Conference of the IEEE Engineering in Medicine and Biology Society, San Diego, CA. Part3: 1120-1121; 1993a.
Skierczynski B.A.; Usami S.; Chien S.; Skalak R. Active Motion of Polymorphonuclear Leukocytes in Responce to Chemoattractant in a Micropipette. J. Biomech. Eng. 115: 503-508; 1993b.
Skierczynski B.A.; Usami S.; Skalak R. A Model of Leukocyte Migration Through Solide Tissue. NATO Advances Study Institute. in press: 1993c.
Southwick F.S.; Dabiri G.A.; Paschetto M.; Zigmond S.H. Polymorphonuclear Leukocyte Adherence Induces Actin Polymerization by a Transduction Pathway wich Differs from that Used by Chemoattractants. The Journal of Cell Biology. 109: 1561-1569; 1989.
Steven A.C. Conformational Change- An Alternative Energy Source?: Exothermic Phase Transition in Phase Capsid Maturation. Biophysical Journal. 65: 5-6; 1993.
Stossel T.P. From Signal to Pseudopod. The Journal of Biological Chemistry. 264(No.31): 18261-18264; 1989.
Stossel T.P. How Cells Crawl. American Scientist. 78: 408-423; 1990.
Stossel T.P. On the Crawling of Animal Cells. Science. 260: 1086-1094; 1993.
Suchard S.J. Interaction of Human Neurophils and HL-60 Cells with the Extracellular Matrix. Blood Cells. 19: 197-223; 1993.
Theriot J.A.; Mitchison T.J. The Three Faces of Profilin. Cell. 75: 835-838; 1993.
Theriot J.A.; Mitchison T.J.; Tilney L.G.; Portnoy D.A. The Rate of Actin-based Motility of Intracellular *Listeria monocytogenes* Equals the Rate of Actin Polymerization. Nature. 357: 257-260; 1992.
Usami S.; Wung S.-L.; Skalak R.; Chien S. Human Cytoplast Locomotion in Chemotaxis. The FASEB Journal. Part II: A906; 1993.
Usami S.; Wung S.-L.; Skierczynski B.A.; Skalak R.; Chien S. Locomotion Forces Generated by a Polymorphonuclear Leukocyte. Biophysical Journal. 63: 1663-1666; 1992.
Volz A. Regulation of CD18 Expression in Human Neutrophils as Related to Shape Changes. Biophys. J. 106: 493-502; 1993.
von_Andrian U.H.; Arfors K.-E. Neutrophil-Endothelial Cell Interactions *In Vivo* : A Chain of Events Characterized by Distinct Molecular Mechanisms. Inflammatory Disease Therapy. AAS 41: 153-164; 1993.

Wachsstock D.H.; Schwarz W.H.; Pollard T.D. Afinity of α-Actinin for Actin Determines the Structure and Mechanical Properties of Actin Filament Gels. Biophysical Journal. 65: 205-214; 1993.

Wang J.-S.; Pavlotsky N.; Tauber A.I.; Zaner K.S. Assembly Dynamics of Actin in Adherent Human Neutrophils. Cell. Motil. Cytoskel. 26: 340-348; 1993a.

Wang N.; Butler J.P.; Ingber D.E. Mechanotransduction Across the Cell Surface and Through the Cytoskeleton. Science. 260: 1124-1127; 1993b.

Ward M.D.; Hammer D.A. A Theoretical Analysis for the Effect of Focal Contact Formation on Cell-substrate Attachment Strength. Biophysical Journal. 64: 936-959; 1993.

Yanagida T.; Harada Y.; Ishijima A. Nano-manipulation of Actomyosin Molecular motors *in vitro*: A New Working Principle. Trends in Biomedical Sciences. 18: 319-324; 1993.

26

The Correlation Ratchet: A Novel Mechanism For Generating Directed Motion By ATP Hydrolysis

C. S. Peskin, G. B. Ermentrout, G. F. Oster

Introduction

Progressive enzymes are macromolecules which hydrolyze ATP while moving unidirectionally along a linear macromolecular "track". Examples include motor molecules such as myosin, kinesin and dynein, RNA and DNA polymerases, and chaperonins. We propose a specific mechanical model for transduction of phosphate bond energy during ATP hydrolysis into directed motion. This model falls within the class we call "Brownian ratchets" since the model cannot function without the Brownian motion and since it derives its directionality from the rectification of such motion. (Peskin et al., 1994; Peskin et al., 1993; Simon et al., 1992). The model depends on two features of the motor system. First, there is a molecular asymmetry in the binding between the motor and its track which determines the direction the motor will move (Magnasco, 1993). Second, the nucleotide hydrolysis cycle pumps energy into the system by providing an alternating sequence of strong and weak binding states. The novelty of this mechanism is how these effects of ATP hydrolysis conspire to bias the Brownian motion of the motor. Removing either feature (alternation or asymmetry), or removing the thermal motion of the enzyme eliminates the transduction of the chemical cycle into directed motion. We call this mechanism a "correlation ratchet", since its operation depends upon a correlation between the spatial position of the molecular motor and its binding state (strong/weak). When this correlation is lost the motor ceases to function (Feynman et al., 1963). We first present the model equations, and then supply an intuitive explanation for how the ratchet works.

Biasing Diffusion With ATP Hydrolysis

Consider the situation illustrated in Figure 1. As the motor diffuses along the polymer it encounters periodically spaced binding sites, described by the potential $V(x) = V(x + L)$, which exerts a force $-dV(x)/dx$ on the motor. The crucial feature of these binding sites is their asymmetry; for kinesin, there is some evidence that this asymmetry is partly a property of the motor head, rather than the track; however, since both microfilaments and microtubles are intrinsically asymmetrical, it probably involves structural asymmetries in both proteins (Stewart et al., 1992). Since the binding between the motor and the track is electrostatic, the size of the binding site is limited by the Debye length (< 1 nm), and so is much less than the distance between sites (5-10 nm)(Cordova et al., 1990; Vale et al., 1989). Therefore, for illustrative purposes we will use the piecewise linear potential shown in Figure 1. As the motor diffuses, it continuously hydrolyzes ATP, which alternates the binding affinity between two values: weak ($V_0 \approx 1\ k_BT$) and strong ($V_1 \approx$ 5-10 k_BT. k_B is Boltzmann's constant, T the absolute temperature; $k_BT \approx 4.1$ pN-nm at room temperature).

Denote by x(t) the position of the enzyme on its track; the stochastic equation governing the motion is:

$$\underbrace{\frac{dx(t)}{dt}}_{\text{velocity}} = \frac{D}{k_BT}\left[-\underbrace{A(t)}_{\text{amplitude}} \cdot \underbrace{V'(x)}_{\substack{\text{Force due to}\\ \text{binding potential}}} - \underbrace{\lambda}_{\text{load}}\right] + \underbrace{\sqrt{2D} \cdot w(t)}_{\text{white noise}} \tag{1}$$

where D is the diffusion coefficient of the enzyme, λ is a constant load force resisting the motion, and w(t) is uncorrelated white noise (Doering, 1990). The hydrolysis cycle controls the magnitudes, A(t), of the binding force. We model this as a 2-state Markov chain whose states are $z = 0$ (weak binding) and $z=1$ (strong binding); there is no difficulty in including more states,(Leibler; Huse, 1993) but this is not necessary to illustrate the principle. The rate constants for the transitions are $k_{10} = r \cdot f$, $k_{01} = r \cdot (1-f)$, where f is the fraction of time in the weakly bound state and $r \sim 1/\text{mean cycle time}$.

Figure 2 shows a numerical simulation of a typical trajectory of equation (1). The motor moves stochastically to the right, towards the *steep* end of the potential wells—i.e. opposite to the direction one would expect if it were a macroscopic ratchet. Its mechanical effectiveness can be characterized by the load-velocity curve shown in Figure 2 (see below).

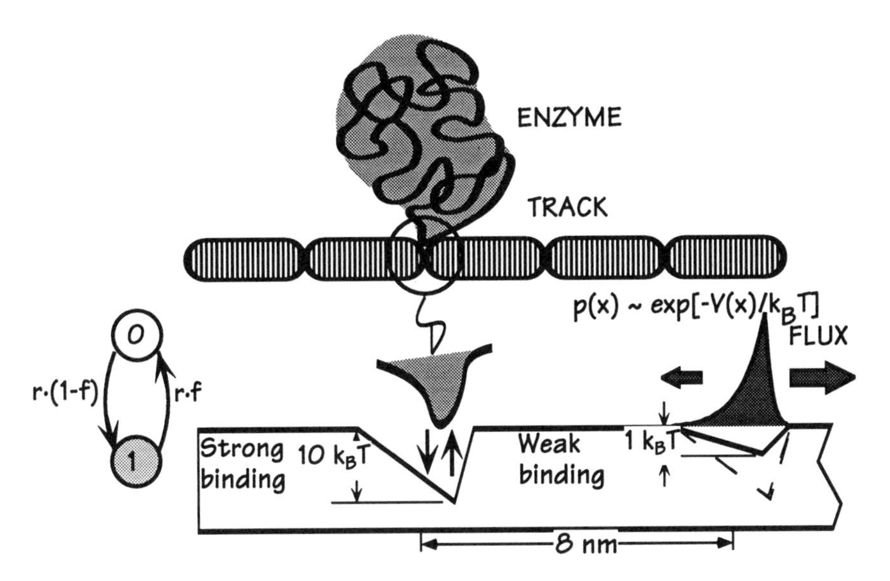

Figure 1. A protein diffusing along a linear polymer encounters electrostatic binding sites spaced 8 nm apart. In order to prevent diffusing away from the track the enzyme is either sleeve-shaped (RNA polymerase, GroEL), or works in tandem with another enzyme (myosin, kinesin). The ATP cycle alternates the binding affinity between strong and weak binding states with rates r·f and r·(1-f), where f is the fraction of time in the weak binding state and r is the overall cycle rate. The depth of the strong binding state is about 10 k_BT, while the weak binding state is about 1 k_BT. The binding potential wells are spaced L = 8nm apart and have asymmetry $\alpha = (L_1-L_2)/(L_1+L_2)$, where L_1 and L_2 are the lengths of the left and right sides of the well, respectively. While in the strong binding state the protein equilibrates in the deep well so that its position is given by a Boltzmann distribution, $p(x) \sim \exp[-V(x)/k_BT]$, where p(x) is the probability distribution of the enzyme's location at the moment of switching into the weak binding state. The asymmetry of this distribution is the source of the diffusion bias towards the steep side of the potential ratchet.

The explanation for the propulsive force lies in the statistics of the process. Consider the state of the enzyme when it is strongly bound. If it remains in the well long enough to thermally equilibrate ($< 10^{-9}$ s), its probability distribution relaxes to the Boltzmann form. That is, the probability, p(x), of being at position x is $p(x) \sim \exp[-A_1V_1(x)/k_BT]$, which is asymmetric because V(x) is. When the well is suddenly lifted to its weakly bound level, this probability cloud begins to diffuse away. On a population basis, there is more motion towards the steep side of the potential, since the probability gradient is steepest there. For the well shown in Figure 1, motion is to the right. If the switching rate of the well is approximately the mean time to diffuse between wells ($1/r \sim L^2/2D$), then the next well to the right will have a good opportunity to capture those enzymes that have moved in its direction. Because of the exponential Boltzmann factor the bias is quite sensitive to the asymmetry in the well, and considerable motion can be generated with only modest asymmetries.

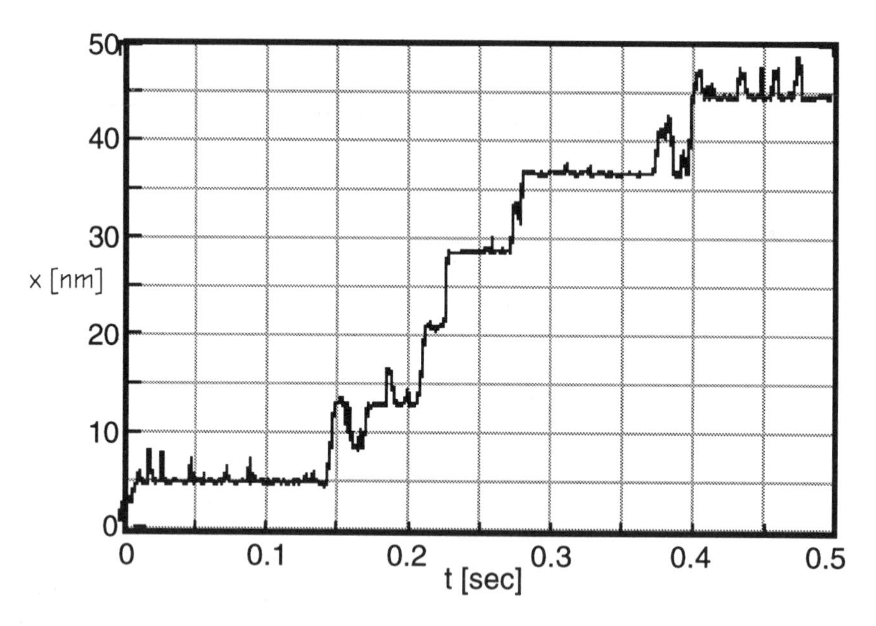

Figure 2. A typical sample path for equation (1). The time scale for a diffusion step is much shorter than the ATP cycle time; therefore, in order to compute a sample path the viscosity of the medium must be artificially increased so that $D = 10^3$ nm²/s. The other parameters employed in the simulation are: $V_0 = 0.5\ k_BT$, $V_1 = 10\ k_BT$, $L = 8$ nm, $L_1 = 0.8$ nm, $L_2 = 0.2$ nm, $r = 100$/s, $f = 0.1$, load = 0 .

We analyze the statistical behavior of equation (1) by converting it to a pair of diffusion (Fokker-Planck) equations (Doering, 1990):

$$\frac{\partial p_0}{\partial t} = -\frac{\partial}{\partial x}\left[-D\frac{\partial p_0}{\partial x} - \frac{D(\lambda + A_0 V'(x))}{k_B T} p_0\right] - r(1-f)p_0 + r f p_1 \quad (2)$$

$$\frac{\partial p_1}{\partial t} = -\frac{\partial}{\partial x}\left[-D\frac{\partial p_1}{\partial x} - \frac{D(\lambda + A_1 V'(x))}{k_B T} p_1\right] + r(1-f)p_0 - r f p_1 \quad (3)$$

where $p_0(x,t)$ and $p_1(x,t)$ are the probability density functions for the states 0 and 1, respectively. The steady state solution to equations (2) and (3) yields the load-velocity curve shown in Figure 3. Although Figure 3 is computed for a fixed value of r, we remark that the force and velocity developed by the motor both vanish when the hydrolysis rate, r, is zero or in the limit $r \rightarrow \infty$.

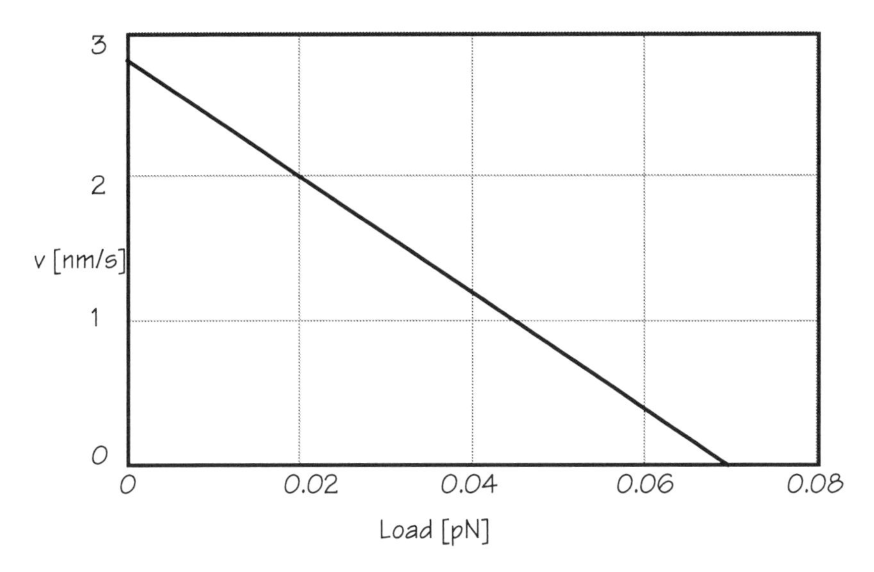

Figure 3. A load-velocity curve computed from equations (2) and (3) for $D = 10^4$ nm^2/s, $f = 0.1$, $r = 10^3$, $L_1 = 0.8$ nm, $L_2 = 0.2$ nm. Velocity decreases linearly, with a stall load much less than 1 pN.

Dimensional analysis of equations (2) and (3) shows that the mechanical behavior of the motor is governed by four dimensionless parameters: $\beta = rL^2/D$ measures the diffusion time between binding sites relative to the ATP cycle time, and three parameters describing the depths and asymmetry of the binding site: $\mathcal{V}_0 \equiv V_0/k_BT$, $\mathcal{V}_1 \equiv V_1/k_BT$, and $\alpha \equiv (L_1 - L_2)/(L_1+L_2)$, where L_1 and L_2 are the lengths of the shallow and steep sections of the potential, respectively.

In the experimental situation described by Svoboda, et al. (1993), $\beta \sim 100[1/s]\cdot 8^2[nm]/7.2\times10^5\ [nm^2/s] < 10^{-2}$. In the Appendix we show that, in the limit of small b, one can derive the following approximate expression for the unloaded velocity as a function of the reaction rate:

$$v = r \cdot f(1 - f) \cdot L \cdot F_0 \tag{4}$$

where F_0 is a dimensionless constant that depends only on the form of the two potential wells.

Discussion

Several authors have suggested that Brownian motion plays a crucial role in the operation of molecular motors (Cordova et al., 1991; Huxley, 1957; Meister et al., 1989). Magnasco was the first to point out the importance of asymmetric potentials in biasing thermal motion (Magnasco, 1993) , and Doering et al. solved a model similar to equation (1) but with additive noise (Doering et al., 1994).

We have shown here that one of the characteristic properties of ATP hydrolysis, its ability to alter the binding affinity of a macromolecule at a different site—so called, noncompetitive inhibition—can bias its diffusive motion providing the binding affinity at the track-binding site is not symmetrical. This diffusive bias generates a very weak force, incapable of driving much of a load. However, it can be turned into a powerful directional motor by coupling with a second ratchet mechanism (Peskin et al., 1994; Peskin et al., 1993; Simon et al., 1992). For example, a pair of kinesin ATPases can drive a vesicle against a considerable load (Svoboda et al., 1993), but there is no firm evidence that a single head can do so. However, if the two heads alternate binding to the microtubule lattice then the bound head can hold fast the load while the other head diffuses to its next binding site. A model incorporating two coupled heads, symmetrical in all respects save the potential well asymmetry, can drive a load of several piconewtons. We will report on such a model in a separate publication. There are other cases in which the motion of a progressive enzyme may work by a Brownian ratchet mechanism of the type described in [2]: An RNA polymerase, by adding ribonucleotide subunits to the posterior 3'-hydroxyl end of the RNA chain, could ratchet the progressive diffusion of the enzyme (Schafer et al., 1991). The progressive action of the chaperonin ATPase GroEL in facilitating polypeptide folding may work by a similar mechanism (Ellis, 1993).

Acknowledgments

GO and CSP were supported by NSF Grant FD92-20719; BE was supported by NSF Grant DMS9002028. The authors profited greatly from conversations with S. Block, D. Brillinger, C. Doering, S. Evans, W. Horsthemke, M. Magnasco, and K. Svoboda. We thank T. Ryan, K. Svoboda and P. Janmey for critical reading of the manuscript.

Note: Aftter completion of this manuscript our attention was called to two papers which propose models whose mathematical aspects are similar to ours, but with quite different biological assumptions: (Ajdari; Prost, 1992; Astumian; Bier, 1993).

References

Ajdari, A.; J. Prost: Mouvement induit par un potentiel périodique de basse symétrie: diélectrophorése pulsée. C. R. Acad. Sci. Paris. **315**: 1635-1639; 1992

Astumian, D.; M. Bier: Fluctuation driven ratchets—Molecular motors. Phys. Rev. Lett. (in press): 1993

Cordova, N.; B. Ermentrout; G. Oster: The mechanics of motor molecules I. The thermal ratchet model. Proc. Natl. Acad. Sci. (USA). **89**: 339-343; 1991

Cordova, N.; G. Oster; R. Vale. Dynein-microtubule interactions. In: L. Peliti, eds. *Biologically Inspired Physics*. New York: Plenum.; 1990: p. 207-215.

Doering, C. Modeling complex systems: Stochastic processes, stochastic differential equations, and Fokker-Planck equations. In: L. Nadel; D. Stein, eds. 1990 Lectures in Complex Systems. Redwood City, CA: Addison-Wesley; 1990: p. 3-51.

Doering, C.; W. Horsthemke; J. Riordan: Nonequilibrium fluctuation-induced transport. Phys. Rev. Lett. In press: 1994

Ellis, R.J.: Chaperonin duet. Nature. **366**: 213-214; 1993

Feynman, R.; R. Leighton; M. Sands.The Feynman Lectures on Physics. Reading, MA: Addison-Wesley; 1963

Huxley, A.F.: Muscle structure and theories of contraction. Prog. Biophys. biophys. Chem. **7**: 255-318; 1957

Leibler, S.; D. Huse: Porters versus rowers: a unified stochastic model of motor proteins. J. Cell Biol. **121**: 1357-1368; 1993

Magnasco, M.O.: Forced thermal ratchets. Phys. Rev. Lett. **71**: 1477-1481; 1993

Meister, M.; S.R. Caplan; H.C. Berg: Dynamics of a Tightly Coupled Mechanism for Flagellar Rotation: Bacterial Motility, Chemiosmotic Coupling, Protonmotive Force. Biophys. J. **55**: 905-914; 1989

Peskin, C.; V. Lombillo; G. Oster. A depolymerization ratchet for intracellular transport. In: M. Millonas, eds. Fluctuations and Order: The New Synthesis. New York: Springer-Verlag; 1994.

Peskin, C.; G. Odell; G. Oster: Cellular motions and thermal fluctuations: The Brownian ratchet. Biophys. J. **65**: 316-324; 1993

Schafer, D.A.; J. Gelles; M.P. Sheetz; R. Landick: Transcription by single molecules of RNA polymerase observed by light microscopy. Nature. **352**: 444-448; 1991

Simon, S.; C. Peskin; G. Oster: What drives the translocation of proteins? Proc. Natl. Acad. Sci. USA. **89**: 3770-3774; 1992

Stewart, R.J.; J.P. Thaler; L.S.B. Goldstein: Direction of microtubule movement is an intrinsic property of the motor domains of kinesin heavy chain and Drosophila ncd protein. Proc. Natl. Acad. Sci. **90**: 5209-5213; 1992

Svoboda, K.; C.F. Schmidt; B.J. Schnapp; S.M. Block: Direct observation of kinesin stepping by optical trapping interferometry. Nature. **365**: 721-727; 1993

Uyeda, T.; S. Kron; J. Spudich: Myosin step size: estimation from slow sliding movement of actin over low densities of heavy meromyosin. J. Mol. Biol. **214**: 699-710; 1990

Uyeda, T.; H. Warrick; S. Kron; J. Spudich: Quantal velocities at low myosin densities in an *in vitro* motility assay. Nature. **352**: 307-311; 1991

Vale, R.; D. Soll; I. Gibbons: One-dimensional diffusion of microtubules bound to flagellar dynein. Cell. **59**: 915-925; 1989

Appendix

In this Appendix we derive the following asymptotic formula for the velocity of a 1-legged ratchet in the limiting case where $\beta \equiv rL^2/D \to 0$, i.e. the reaction rate is much slower than diffusion:

$$v = R \cdot L \cdot F_0[V] \tag{A.1}$$

where R is the overall rate of hydrolysis, $R = r \cdot f \cdot (1-f)$, and $F_0[V]$ depends on the form of the two potential wells, $V_0(x)$ and $V_1(x)$.

The situation is this: since the reaction rate is much slower than the time for the probability distribution of particle locations, P(x,t) to reach equilibrium, then we know that the distributions in the strong and weakly bound states are Boltzmann. Thus we need only compute the net average transfer of probability between potential wells per hydrolysis cycle.

We begin with the conservation equation for an ensemble of points, c(x,t):

$$\frac{\partial c}{\partial t} = -\frac{\partial J}{\partial x} \tag{A.2}$$

where

$$J = -D\left(\frac{\partial c}{\partial x} + \frac{1}{k_B T}\frac{dV}{dX}c\right), \tag{A.3}$$

and the boundary condition on c(x, t) is periodic:

$$c(x + L, t) = c(x, t). \tag{A.5}$$

If the initial distribution, c(x,0), is given we wish to compute the total fllux past a given point x during the relaxation of $c(x, 0) \to c(x, \infty)$:

$$F(x) = \int_0^\infty J(x,t)dt \tag{A.6}$$

Integrating the conservation equation over all times yields:

$$\frac{\partial F}{\partial x} = c(x,0) - c(x,\infty) \tag{A.7}$$

This determines F(x) up to an additive constant, which can be computed as follows. Multiplying (A.3) by $\exp(V(x)/k_BT)$ and integrating first over x and then over t we obtain

$$\int_0^L F(x) \cdot exp\left(\frac{V(x)}{k_BT}\right) dx = 0 \tag{A.8}$$

which determines the additive constant.

The system switches back and forth between V_0 and V_1, and the corresponding equilibrium distributions in each state are obtained by setting J = 0 in (A.3):

$$c_i(x) = \frac{\exp(-V_i(x)/k_BT)}{\int_0^L \exp(-V_i(x')/k_BT)dx'}, \; i = 1,2 \tag{A.9}$$

Let $F_2(x)$ be the value of F(x) for the switch from $V_1 \rightarrow V_2$, and $F_1(x)$ be the value of F(x) for the switch form $V_2 \rightarrow V_1$. Then

$$c_2(x) - c_1(x) + \frac{\partial F_2}{\partial x} = 0 \tag{A.10a}$$

$$c_1(x) - c_2(x) + \frac{\partial F_1}{\partial x} = 0 \tag{A.10b}$$

Thus $F_0 = F_1(x) + F_2(x) =$ constant. From (A10a,b) we have

$$F_1(x) = F_1(0) + \Phi(x) \tag{A.11a}$$

$$F_2(x) = F_2(0) - \Phi(x) \tag{A.11b}$$

where $\Phi(x) = \int_0^x [c_2(x') - c_1(x')]dx'$. But from (A.8) we also have the identities

$$\begin{aligned} &\int_0^L F_1(x) b_1(x) dx = 0 \\ &\int_0^L F_2(x) b_2(x) dx = 0 \end{aligned} \tag{A.12}$$

where $b_i = \exp(V_i(x)/k_BT)/\int_0^L \exp(V_i(x')/k_BT)dx'$. Note that b_i has the same definition as c_i except that the sign of V has been reversed. From this we conclude that

$$F_1(0) = -\int_0^L \Phi(x) b_1(x) dx$$
$$F_2(0) = \int_0^L \Phi(x) b_2(x) dx \qquad (A.13)$$

Thus

$$F_0 = F_2(0) + F_1(0) = \int_0^L \Phi(x)\big(b_2(x) - b_1(x)\big)dx$$
$$= \int_0^L \big(b_2(x) - b_1(x)\big)\int_0^x \big(c_2(x') - c_1(x')\big)dx'\,dx \qquad (A.14)$$

Now the average hydrolysis cycle time for the 2-state model is $1/\tau_1 + 1/\tau_2 = 1/(r \cdot f) + 1/(r \cdot (1 - f))$; therefore, the mean cycle rate is the reciprocal of this: $R = r \cdot f \cdot (1-f)$. Since $F_0 \cdot L$ is the mean distance moved per cycle, so multiplying by the cycle rate, R, we obtain equation (4):

$$v = r \cdot f(1 - f) \cdot L \cdot F_0 \qquad (A.15)$$

Note that in this limit, the velocity is maximum at $f = 1/2$, and increases linearly with reaction rate, in accordance with the numerical solution to the full boundary value problem shown in Figure A1. Figure A.1 gives the impression that the velocity increases indefinitely with r, but this is not the case. As $r \to \infty$, the velocity has to approach zero because the motor then sees only a single average potential, but this is outside the domain of validity of the asymptotic method used here, and it is also far beyond the physiological range of hydrolysis rates. Note also that in the slow reaction rate limit the velocity is independent of the diffusion coefficient, D. This is because all the motion occurs on a time scale much shorter than the hydrolysis rate, and so the system is essentially switching instantly between equilibrium states.

An equation similar to (1) has been used to estimate the mean step size of single, unloaded, motors from *in vitro* assays wherein the velocity of a microtubule or vesicle is measured, and the stepsize inferred by dividing by the hydrolysis rate: $L \approx v/R$ (Uyeda et al., 1990; Uyeda et al., 1991). Aside from the experimental error of using hydrolysis rates obtained in separate assays, we see that there is an additional factor, F_0, that should also be included in the relationship between v, L, and R (c.f. equation A.15).

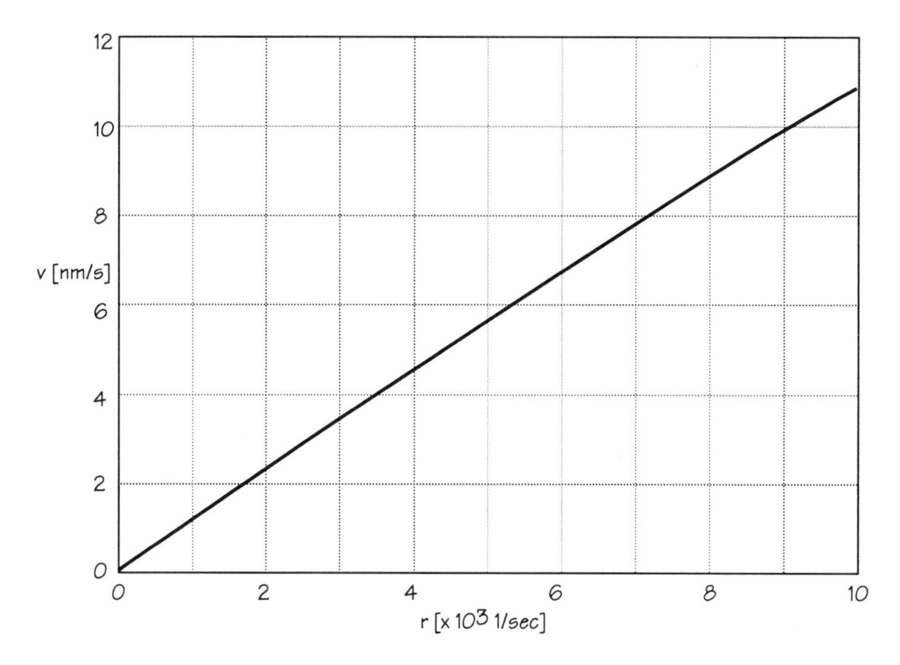

Figure A1. Velocity as a function of the hydrolysis rate, r, when $\beta = rL^2/D << 1$. The curve was computed from equations (2, 3) solved numerically by a shooting method.

27
Receptor-Mediated Adhesive Interactions at the Cytoskeleton/Substratum Interface During Cell Migration

P. A. DiMilla

Introduction

The migration of mammalian cells plays a critical role in a diverse array of physiological and pathological phenomena, including the proper development and repair of organs, inflammation, angiogenesis, and metastasis (Trinkaus, 1984). Understanding the molecular mechanisms responsible for movement can provide insight into how the clinician can control these phenomena and requires identifying quantitative relationships between molecular-level properties and cellular- and tissue-level function. Both mathematical models of motility -- in terms of fundamental chemical and physical processes -- and experimental measurements are instrumental for achieving these goals.

The phenomenon of cell crawling over adhesive substrata is complex and depends on coordinated interactions among numerous underlying biophysical and biochemical processes. In particular, three general characteristics have been identified for persistent motility in this situation: (1) cytoskeletal elements generate intracellular mechanical stresses (Stossel, 1993); (2) traction created by the dynamic formation and breakage of attachments with the underlying substratum (Regen and Horwitz, 1992) transforms these stresses into a displacement force on the cell body; and, (3) morphological polarization channels this force unidirectionally for net translocation. Cytoskeletal contraction is achieved by the reversible polymerization of actin monomers into filaments.

Although many of the components involved in the transmission of force from the contractile apparatus to the underlying substratum have not been identified, integrins -- heterodimeric adhesion receptors on the cell surface which bind adhesion ligands (e.g., extracellular matrix (ECM) proteins like fibronectin (Fn)) on opposing substrata (Hemler, 1991; Hynes, 1992) -- likely are part of a chain of transient molecular linkages (Wang et al., 1993): integrins bind both ECM proteins and cytoskeletal elements with low affinity ($K_D \approx 10^{-6}$ M for integrin/fibronectin (Akiyama and Yamada, 1985) and integrin/talin interactions (Horwitz et al., 1986)). Further, both short time-scale binding between individual receptors and ligands and longer time-scale events, such as the clustering of receptors into focal contacts (Burridge et al., 1988), may be relevant to movement occurring over periods of many hours. The relative contributions of these different time-scale adhesive interactions to cell locomotion, however, are poorly understood.

The overall goal of our studies has been to clarify the relationship between molecular-level components and mechanisms and the functional characteristics of cell adhesion and migration. In particular, we have focused on the dependence of motility on the strength of the transient cell-substratum attachments. Three regimes of adhesive, morphological, and motile behavior can be envisioned for a cell interacting with a surface (Fig. 1). This model suggests that if the strength of the adhesive interaction between the cell and substratum balances traction forces generated by intracellular contraction, net translocation is possible. In this paper we examine this hypothesis by combining experimental and modeling approaches -- designed to provide both predictive and interpretive information -- to determine quantitatively the role of linkages between adhesion receptors and ligands on the rate of cell movement.

Experimental Relationships Between Density of Adhesion Ligand, Adhesion, and Migration

In our experimental studies we focused on measuring the motile and adhesive properties of human smooth muscle cells (HSMCs). Alterations in the motile properties of these cells have been implicated in the development of atherosclerotic plaques (Ross, 1986) and in deleterious vascular remodeling after injury (Clyman et al., 1990). These cells also could be maintained in serum-free MCDB-104 medium during our experiments (DiMilla et al., 1993); use of a serum-free medium is important because serum contains ill-defined and variable types and concentrations of ECM proteins which can adsorb and alter the composition of a surface. HSMCs express on their surface the integrins VLA-2 and VLA-5 (DiMilla et al., in preparation), which bind the adhesion ligands collagen type IV (CnIV) and Fn, respectively (Hemler, 1991). These results suggested

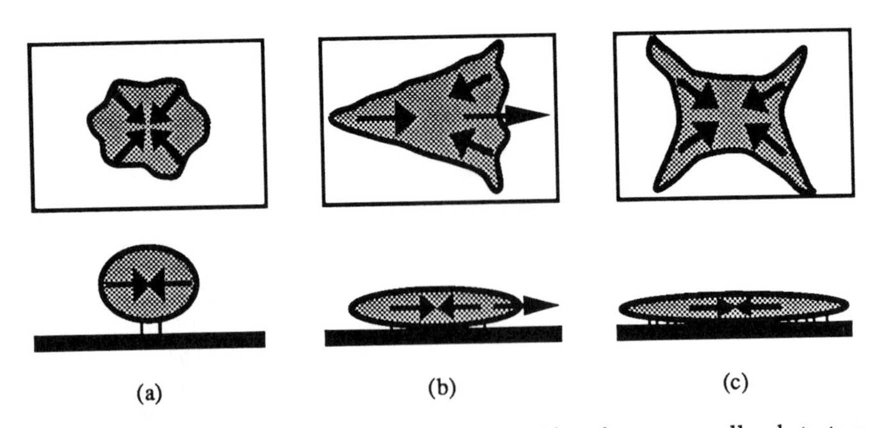

Fig. 1. Schematic of hypothetical relationships between cell-substratum adhesiveness, intracellular contractile force, cell morphology, and migration. (a) Adhesive force less than contractile force; neither spreading nor net movement. (b) Adhesive force balances contractile force; both spreading and net movement. (c) Adhesive force greater than contractile force; spreading but no net movement.

that the adhesion and migration of HSMCs might depend on the density of CnIV and Fn molecules on an underlying substratum.

Preparation and Characterization of Adsorbed Adhesion Ligands

Rigorous examination of relationships between the properties of a substratum, cell adhesion, and cell motility requires both the preparation of substrata with different surface densities of adhesion ligand and the determination of the density of active ligand that is accessible for binding with cell-surface adhesion receptors. The simplest and most common method for preparing substrata for experiments with cells is to nonspecifically adsorb ECM protein on polystyrene petri dishes. We have demonstrated that radiolabeling with ^{125}I and counting of gamma radiation provides a simple method for directly measuring total densities of adsorbed protein on these substrata (DiMilla et al., 1992c). Bacteriological petri dishes (35-mm diameter, Falcon #1008) containing removable 8 mm by 15 mm sections were incubated with 1.5 mL solutions of labeled and unlabeled CnIV or Fn in PBS for 24 hrs at 4°C. After incubation with 1.5 mL of 1 wt% heat-denatured bovine serum albumin (BSA) for 45-50 min at room temperature to block nonspecific interactions between cells and substrata (Basson et al., 1990) and washing with PBS, the radioactivity of removable sections was measured and adsorbed surface densities of adhesion ligand calculated with the specific activity of the label, the area of each section, and the ratio of labeled to unlabeled protein. Previous studies have verified that labeling does not affect adsorption (Grinnell and Feld, 1981).

We found that the surface density of adsorbed ECM protein depended not only on the soluble bulk concentration of protein but also on the type of protein: roughly one order of magnitude more CnIV than Fn adsorbed at the same soluble concentration (Fig. 2). Further, adsorption of CnIV from solutions with concentrations greater than ca. 30 μg/mL resulted in surface densities greater than the maximum coverage predicted for close-packed monolayers, but each density for Fn was less than or within the range of densities for a monolayer of this protein (DiMilla et al., 1992c).

Assay for Migration of Individual Cells

We assessed the motile properties of HSMCs by tracking the paths of individual cells using time-lapse videomicroscopy and image analysis and analyzing these paths with a model for a persistent random walk (DiMilla et al., 1992a; DiMilla et al., 1992b; DiMilla et al., 1993). This approach allowed variations in the speed of movement to be distinguished unambiguously from variations in the rate of turning. Cells at a density of $1\text{-}2{\bullet}10^3$ cells/cm^2 in supplemented MCDB-104 medium were allowed to spread overnight at 37°C on petri dishes coated with adhesion ligand and BSA and observed at 37°C and a magnification of 10x under phase-contrast optics using a Nicholson ASI 400 Air Stream Incubator, a Zeiss Axiovert 10 inverted microscope, and a Hamamatsu C2400 videocamera. The behavior of cells in 30 fields -- scanned sequentially once every 15 min using a Mertzhauser IM-EK32 motorized stage -- was recorded continuously at 1/120 of real time with a JVC BR-9000U VCR. For representative fields the paths of the centroids of isolated and spread cells were tracked at intervals of Δt = 15 min for up to 36 hrs using an Imaging Technology Series 151 image processor. Fig. 3a depicts the paths for three cells, based on plots of {x, y} position as a function of elapsed tracking time, $n\Delta t$. It is obvious that cell B did not move compared to cells A and C and that cell C moved further than cell A.

In uniform extracellular environments isolated HSMCs move as persistent random walkers: the mean-squared displacement ($\langle d^2 \rangle$), of each motile cell is a function of time (t), cell speed (S), and persistence time (P) (DiMilla et al., 1992b):

$$\langle d^2 \rangle = 2S^2P[t - P(1 - e^{-\frac{t}{P}})]. \tag{1}$$

The parameters S and P represent measures of the "rate" of movement and the time between "significant" changes in the direction of movement, respectively. For a cell tracked a total of $t_{max} = N\Delta t$ min, we calculated the value of $\langle d^2 \rangle$ as a function of time interval $t = n\Delta t$ as:

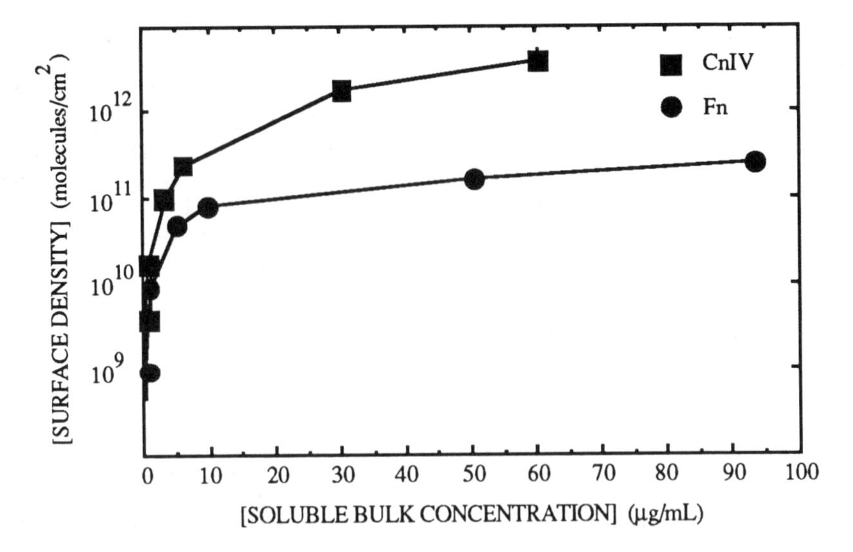

Fig. 2. Relationship between soluble bulk concentration and adsorbed surface density of adhesion ligands CnIV and Fn on bacteriological polystyrene surfaces as measured by radiolabeling with ^{125}I (based on data presented in DiMilla et al., 1992c).

$$< d^2(n\Delta t) > = \frac{1}{(N-n+1)} \sum_{i=0}^{N-n} \left[\left[x((n+i)\Delta t) - x(i\Delta t) \right]^2 + \left[y((n+i)\Delta t) - y(i\Delta t) \right]^2 \right]. \quad (2)$$

Values of S and P for each motile cell were determined by fitting Eqn. 1 to experimental data from Eqn. 2. This approach distinguished motile characteristics more properly than do other common methods, such as calculating the average distance migrated in an arbitrary interval of time (Dunn, 1983). For example, Fig. 3b demonstrates that there was excellent correspondence between experimental and fitted curves for cells A and C over a large range of times, from $t < P$ to $t \gg P$ (given values of P on the order of a few hrs). Further, although cell C moved more than three times as fast as cell A, as predicted intuitively from the corresponding paths, the statistically-significant difference in persistence time is not obvious from examination of the paths.

Values of S and P for cells with $t_{max} < 6$ hrs or $P > t_{max}/3$ were excluded from analysis. Immotile cells, such as cell B, were distinguished by negligible displacements relative to their body length over long times of observation and small persistence times relative to Δt. Because we have measured speeds as low as 1 µm/hr with our

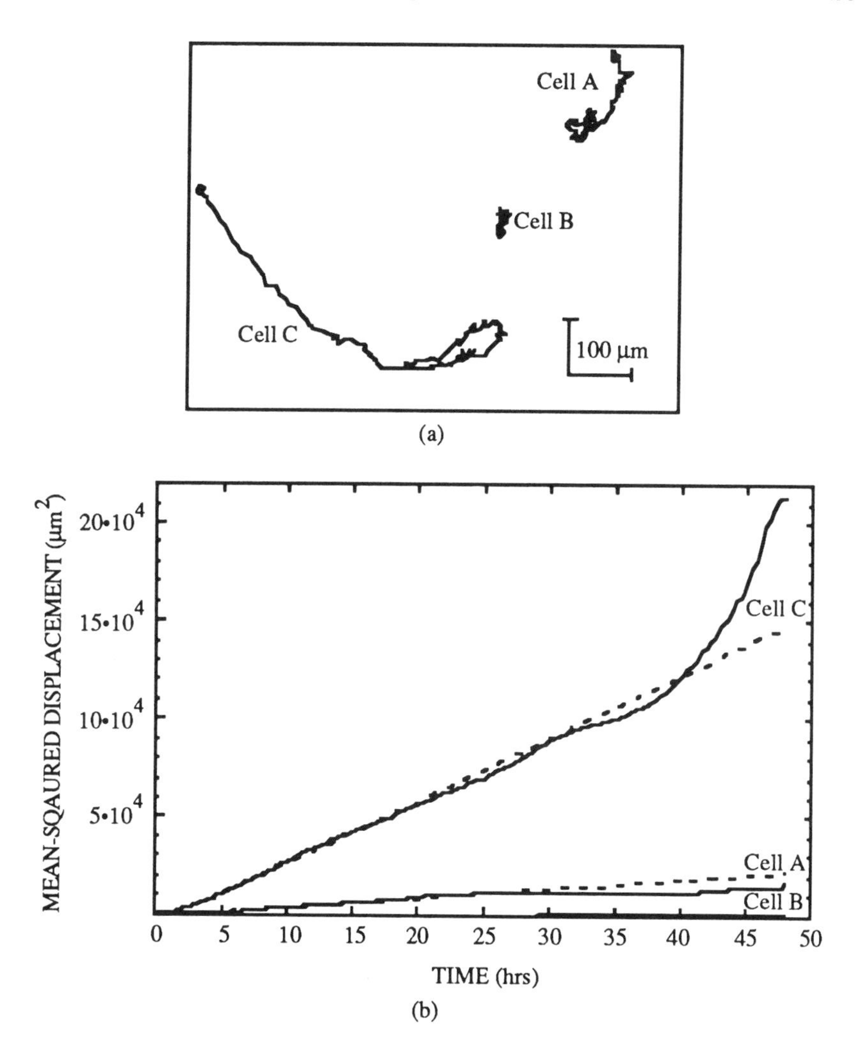

Fig. 3. Distinguishing motility of HSMCs from an assay for the migration of individual cells (based on data presented in DiMilla et al., 1992b). (a) Paths for three typical cells over 48 hrs. Diamonds represent the starting position for each cell. (b) Relationship between mean-squared displacement and time for cells in (a). Solid curves depict experimental mean-squared displacements calculated with Eqn. 2. Fits of model for a persistent random walk (Eqn. 1) with best-fit values for cell speed, S, and persistence time, P (cell A: 8.4 μm/hr and 3.5 hrs; cell C: 27.3 μm/hr and 2.2 hrs) are represented by dashed curves. Cell B was classified as immotile (S = 0 and P undefined).

analysis, the possibility that "immotile" cells actually moved at a speed below some measurable threshold is unlikely.

Cell speed was greatest at intermediate densities of CnIV and Fn (Fig. 4). However, the density of Fn for maximum speed ($1.0 \cdot 10^{11}$

molecules/cm^2) was approximately an order of magnitude less than corresponding density for CnIV ($7.0 \cdot 10^{11}$ to $2.6 \cdot 10^{12}$ molecules/cm^2), and the maximum speed on Fn was half that on CnIV. A similar trend was observed for the morphology of HSMCs on CnIV and Fn -- cells were poorly spread and unpolarized at low densities of adhesion ligand, became bipolar at intermediate densities, and were found to be spread laterally at high densities -- but the density of CnIV had to be increased over two orders of magnitude (from $3.0 \cdot 10^{10}$ to $4.0 \cdot 10^{12}$ molecules/cm^2) to produce the same range of changes in morphology achieved by a one order of magnitude increase in the density of Fn (DiMilla et al., 1993). Mean values of persistence time also were maximum at intermediate densities of CnIV; the variation of persistence time with the density of Fn was less significant.

Assay for the Strength of Cell/Substratum Adhesion

We measured the strength of initial attachment of HSMCs to CnIV and Fn using a radial-flow detachment chamber (DiMilla et al., 1993; Stone, 1993). In this assay 10^5 cells/cm^2 were incubated for 30 min at room temperature on bacteriological petri dishes (60 mm-diameter, Falcon #1007) coated sequentially with adhesion ligand and BSA; a volume of 4.4 mL was used for protein adsorption to expose these dishes to the same number of protein molecules per area at each bulk concentration as the dishes with 35-mm diameters. Incubating cells with surfaces for 30 min before initiation of detachment also allowed stable attachment while limiting cell spreading and reorganization of the cytoskeleton and cell surface. Adherent cells were exposed for 10-20 min to an axisymmetric flow in which the hydrodynamic shear stress on the cells, S, decreased with the radial distance, r, from a central inlet point:

$$S = \frac{3\mu Q}{\pi r h^2}, \tag{3}$$

where μ is the fluid viscosity (controlled by adding dextran), Q the volumetric flow rate, and h the height of the gap through which fluid flowed. For small values of r cells detached because the fluid shear forces exceeded the cell/substratum adhesive force; cells remained attached for large values of r where the shear forces were smaller. The radial position at which 50% of the cells detached at pseudo-steady state was defined by convention as the critical radius, r_c (Cozens-Roberts et al., 1990; Stone, 1993) The critical shear stress for detachment, S_c, was defined as the result of Eqn. 3 with $r = r_c$. Critical radii were measured from patterns of detachment of stained cells.

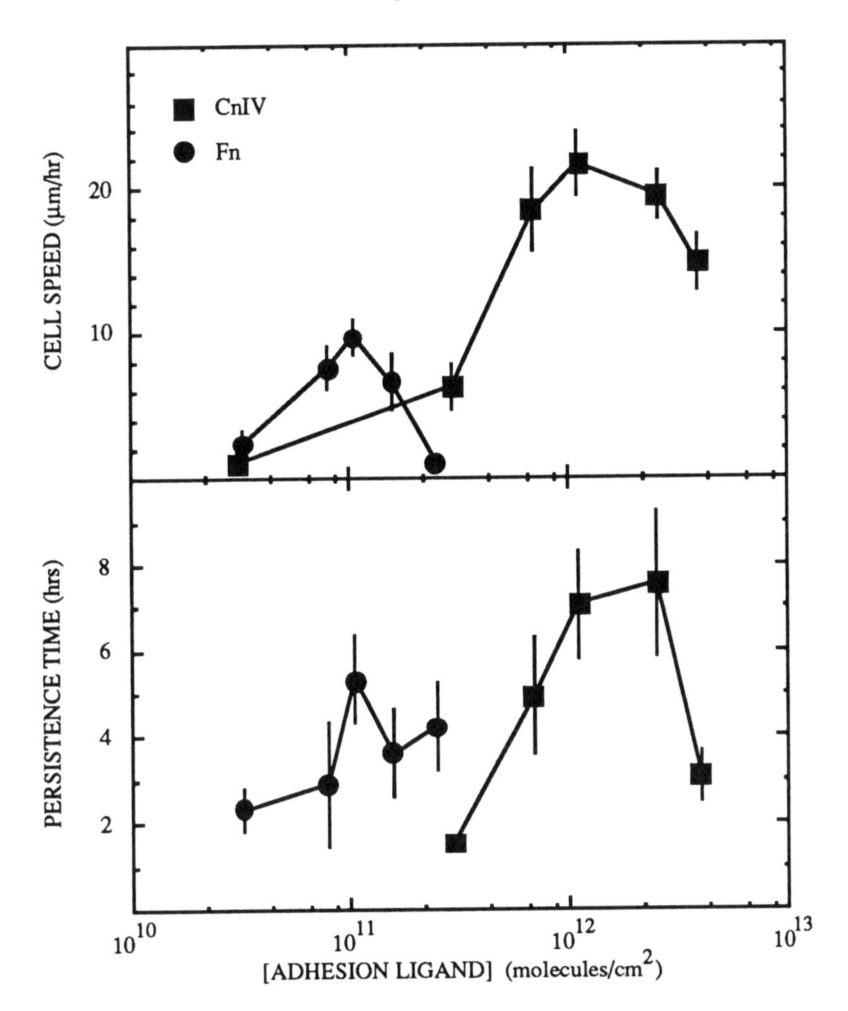

Fig. 4. Effect of surface density of CnIV and Fn on the motility of HSMCs as measured in an assay for the migration of individual cells (DiMilla et al., 1993). Results represent means and standard errors for 28-76 cells at each density of adhesion ligand.

To calculate the total force for detachment from S_c required a representation of the shape of an adherent cell. Because even for limited times of attachment -- as in our experiments -- some deformation in the shape of a cell occurs (e.g., a spherical cell upon contact decreases its hydrodynamic profile and increases its area of cell/substratum contact), we modeled adherent HSMCs as hemispheres with radii R_{cell} as a first approximation. Adhesion bonds experienced forces tangential ($F_{||}$) and normal ($F_{\perp}$) to the substratum due to the fluid shear force (F_S) and torque (T_S). Following the analysis proposed by Hammer and Lauffenburger (1987) for a spherical cell and assuming that all bonds in the area of cell/substratum contact were equally stressed, these forces were determined from balances on the forces on the hemisphere parallel to the substratum:

$$F_{\parallel} + F_S = 0, \tag{4a}$$

and the torques on the hemisphere:

$$T_S + \frac{4R_{cell}}{3\pi} F_{\perp} = 0. \tag{4b}$$

where the fluid shear force and torque were estimated by Price (1985):

$$F_S = 4.3\pi R_{cell}^2 S \tag{5a}$$

$$T_S = 2.44\pi R_{cell}^3 S. \tag{5b}$$

The critical force for detachment, F_{Tc}, was the total force on the adhesion bonds for a cell at the critical radius:

$$F_{Tc} = \sqrt{F_{\parallel c}^2 + F_{\perp c}^2} = 52 A_{proj} S_c, \tag{6}$$

where $A_{proj} = \pi R_{cell}^2 = 460\ \mu m^2$ is the projected area of cell/substratum contact for HSMCs (Stone, 1993). F_{Tc} was our measure of the strength of initial HSMC/substratum attachment and is proportional to the critical shear stress.

The critical force for detachment increased linearly with the density of adsorbed CnIV and Fn (Fig. 5). Linearity was independent of the model assumed for the shape of an adherent cell (although the absolute magnitude of the detachment force calculated varied slightly with the shape of the cell). The difference in strength per molecule, directly reflected in the larger range of forces required for detachment on a smaller range of densities of Fn compared to CnIV, suggests that HSMCs adhered approximately seventy times more tightly per Fn molecule than per CnIV molecule.

Although the absolute density of adhesion ligand by itself does not completely correlate motility, a plot of cell speed versus the strength of attachment for different densities of CnIV and Fn (Fig. 6) demonstrates that not only was motility greatest at intermediate strengths of detachment but also that the optimal critical forces for detachment for maximum speed for CnIV ($4 \cdot 10^4$ μdynes) and Fn

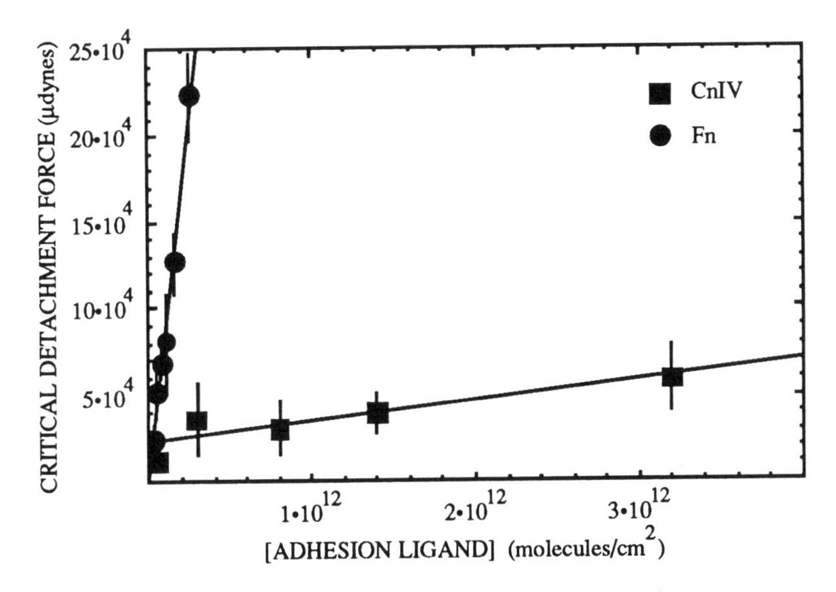

Fig. 5. Effect of surface density of CnIV and Fn on the strength of initial cell/substratum attachment of HSMCs (based on data presented in DiMilla et al., 1993). The critical force for cell detachment (after 30 min incubation at room temperature), F_{Tc}, was calculated using Eqns. 3 and 6. Lines represent linear least-squares fits of data with slopes of $7.9 \pm 2.2 \times 10^{-9}$ µdynes-cm²/CnIV molecule and $5.7 \pm 0.32 \cdot 10^{-7}$ µdynes-cm²/Fn molecule.

($9 \cdot 10^4$ µdynes) differed by only a factor of ca. 2. Further, the same minimum strength of attachment (corresponding to a force of detachment of $\approx 2.5 \cdot 10^4$ µdynes) was required for migration on both of these adhesion ligands. These results suggest that variations in the strength of short time-scale adhesive events -- mediated by individual receptor/ligand bonds -- are relevant to movement. This plot also offers a possible explanation for why the decreasing portion of the data for speed as a function of density of ligand appears incomplete for CnIV: sufficiently great strengths of attachment were not achieved using the limited range of densities of CnIV available.

Models for the Effects of Receptor-Mediated Interactions on Adhesion and Migration

In our modeling studies we focused on predicting relationships between receptor/ligand-mediated adhesive interactions, the strength of cell-substratum adhesion, and the speed of cell migration (Lauffenburger, 1989; DiMilla et al., 1991; DiMilla and Lauffenburger, 1991; DiMilla, 1994). These studies had two goals: 1) to rationalize our cellular-level experimental observations in terms

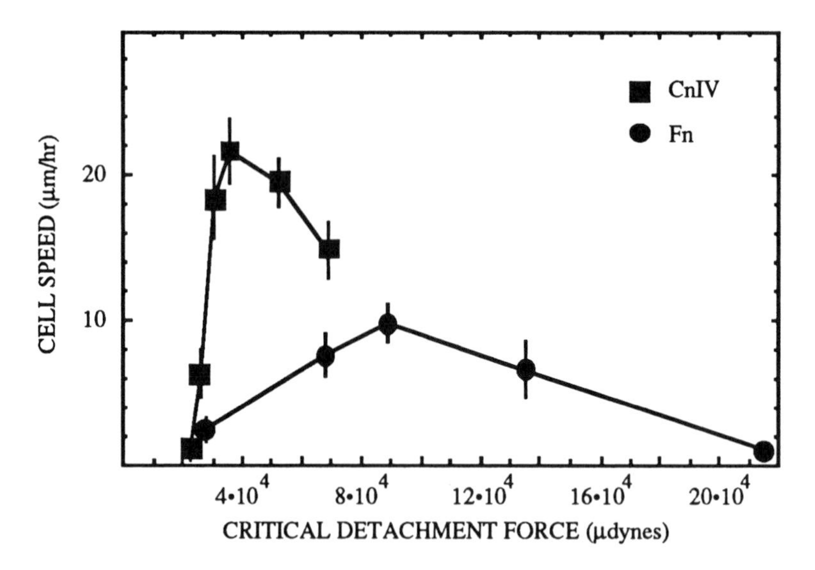

Fig. 6. Relationship between cell speed and strength of initial cell-substratum attachment for HSMCs on CnIV and Fn. Data from Fig. 4 are plotted against the critical force for cell detachment, F_{Tc}, (determined from linear least-squares fits of data in Fig. 5), at each surface density examined in assays of migration.

of molecular-level properties, such as number of receptors and the affinity of binding between receptors and ligands, and 2) to provide novel and non-intuitive predictions useful for developing new hypotheses concerning the actual molecular mechanisms of migration and adhesion.

Cell Migration as a Chronological Cycle

We -- and many others (e.g., Trinkaus, 1984, Harris, 1990) -- have observed in time-lapse studies that the movement of cells over a time scale of minutes appears to follow a saltatory cycle of extension at the front of the cell (the lamellipod) and retraction at the rear of the cell (the uropod) (DiMilla et al., 1993). The body of the cell also appeared to move most dramatically during uropodal retraction. These descriptions suggest modeling unidirectional cell movement as a continuously-repeating two-stage cycle of lamellipodal extension with characteristic time t_l and cytoskeletal contraction with characteristic time t_c. The distribution and dynamics of adhesion bonds, which are responsible for traction forces at the cell/substratum interface (Wang et al., 1993), varies during this cycle. During both stages of the cycle adhesion receptors reversibly bind with immobilized adhesion ligands, but only during lamellipodal extension is an equilibrium distribution of bonds achieved (the kinetics of binding are much faster than the dynamics of extension). In contrast, cytoskeletal contraction results in an enhanced breakage of bonds as bonds must withstand an applied traction force. If an asymmetry exists in the distribution or strength of adhesion bonds between the lamellipod and uropod, contraction results in the generation of a net traction force on the substratum and net translocation of the cell's center of mass.

Two types of approaches are applicable for mathematically modeling cell-surface and cytoskeletal events during extension and contraction (DiMilla et al., 1991). The dynamics of adhesion receptors, ligands, and bonds is a classic problem in chemical dynamics featuring reaction, diffusion, and convection. This type of approach is less feasible for representing cellular mechanics during cytoskeletal contraction because many of the molecular components, mechanics, and mechanisms responsible for these events currently are ill-defined. As a simple alternative, we model the deformation of the cell and exertion of traction forces during contraction by describing the cytoskeleton and individual adhesion bonds as viscoelastic solids. Because the traction generated at the cell/substratum interface also depends on the distribution of adhesion bonds, the two problems are coupled. The net translocation determined from cell displacement during the contraction phase can be averaged over the entire cycle to obtain an observable speed. Of course, individual cells typically move such that it may be difficult to continuously identify discrete stages of extension and contraction, but this two-stage cycle does provide a mathematical basis for breaking up the complex process of migration into simpler events which can be analyzed.

Trafficking of Adhesion Receptors During Migration

In our model for the distribution of adhesion bonds on the ventral surface of a migrating cell (DiMilla et al., 1991), we restrict ourselves to situations in which ligand is immobilized on the substratum at a uniform density, n_s; Dickinson and Tranquillo (1993) have discussed the implications of relaxing this assumption. We model the surface of the cell as two flat sheets of length L and width W "sewn" together at finite "edges." Only free receptors, n_r^d, exist on the dorsal surface, but on the ventral surface both free receptors, n_r^v, and receptors bound to ligand immobilized on the underlying substratum, n_b, are present. The total number of adhesion receptors on the surface is R_T:

$$\int_0^L \int_{-\frac{W}{2}}^{\frac{W}{2}} \left(n_r^d + n_r^v + n_b \right) dy\, dx = R_T \tag{7}$$

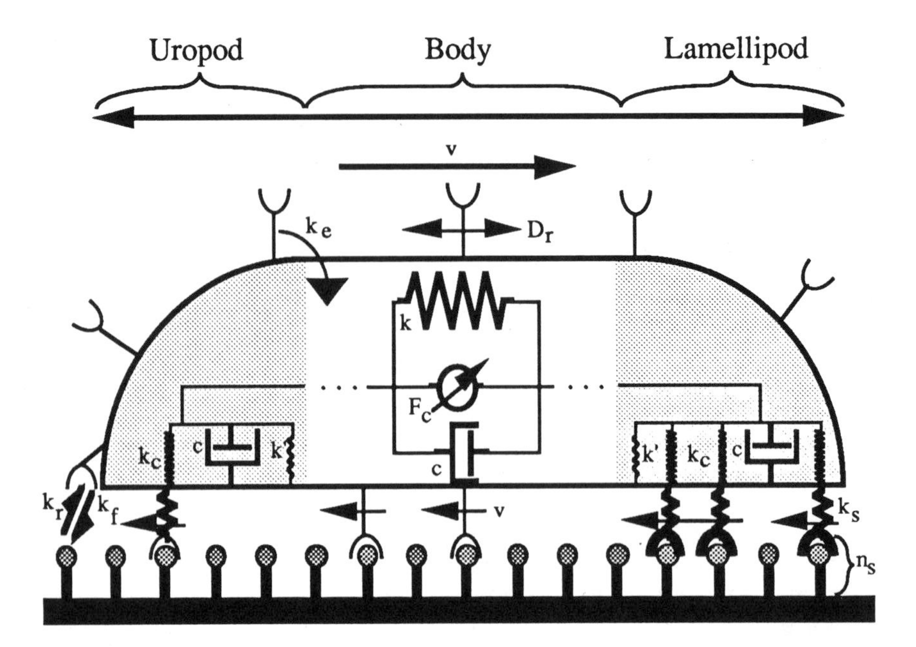

Fig. 7. Schematic of trafficking events for adhesion receptors and a viscoelastic-solid model for intracellular and bond mechanics during migration over adhesive substrata. Free receptors reversibly bind immobilized adhesion ligand with rate of association k_f and rate of dissociation k_{ro}, diffuse with diffusivity D_r, and are internalized at rate k_e; bound receptors drift backwards with respect to the cell's forward motion, v. The cell body is divided into four compartments, which each contain Hookean springs (with spring constant k) and viscous dashpots (with dashpot constant c) in parallel with contractile elements (with force constant F_c); the lamellipodal and uropodal compartments are modeled with sets of springs and a dashpot in parallel. Asymmetry in traction at the cytoskeleton/substratum interface is generated by a spatial variation in affinity between adhesion receptor and ligand.

A variety of trafficking events affect the distribution of adhesion molecules (Fig. 7), including reversible association between free receptors on the ventral surface and immobilized ligands, diffusion of free receptors over and between the ventral and dorsal surfaces, endocytosis and exocytosis of free receptors, and the "drift" of adhesion bonds backwards relative to the speed of overall cell translocation. To model the effects of focal contacts we also can allow adhesion bonds to dynamically organize into clusters (DiMilla and Lauffenburger, 1991). Assuming that ligand is present in excess, mass balances for free receptor on the dorsal and ventral surfaces, respectively, are written as:

$$\frac{\partial n_r^d}{\partial t} = D_r \nabla^2 n_r^d - k_e n_r^d + \frac{k_e}{2LW} \int_0^L \int_{-\frac{W}{2}}^{\frac{W}{2}} \left(n_r^d + n_r^v\right) dy dx \tag{8a}$$

$$\frac{\partial n_r^v}{\partial t} = D_r \nabla^2 n_r^v - k_f n_s n_r^v + k_{r0} n_b - k_e n_r^v + \frac{k_e}{2LW} \int_0^L \int_{-\frac{W}{2}}^{\frac{W}{2}} \left(n_r^d + n_r^v\right) dy dx. \tag{8b}$$

The balance on adhesion bonds is:

$$\frac{\partial n_b}{\partial t} = k_f n_s n_r^v - k_{r0} n_b + v \frac{\partial n_b}{\partial x}. \tag{8c}$$

Relevant boundary conditions include conservation of total receptors (Eqn. 7), continuity for the densities of free receptor, and continuous total fluxes of receptor between dorsal and ventral surfaces.

Cellular Mechanics During Migration and the Role of Asymmetry

In our model for the mechanics of a migrating cell (DiMilla et al., 1991), we divide the cell into six compartments of equal length, limit deformation to the axis of motion, and assume that the mechanical properties of each compartment are spatially and temporally homogeneous. Four middle compartments, representing the cell body, are identical and individually represented by a Hookean spring, linear dashpot, and contractile element in parallel (Fig. 7). The compartment representing the lamellipod contains a dashpot in parallel with both R_{bl} springs (where R_{bl} is the number of adhesion bonds in the lamellipod; these springs model the connections between the cytoskeleton and substratum), and a spring representing the intrinsic stiffness of the lamellipod. The uropodal compartment is similar except it contains R_{bu} adhesion bonds

The boundaries at either end of the cell are fixed, but the motion of the free nodes separating each compartment, u_i (i = 1-5), is described by a series of force balances on the nodes:

$$c\frac{du_1}{dt} + k'u_1 + k_cR_{bu}u_1 = c\frac{d(u_2-u_1)}{dt} + k(u_2-u_1) + F_c \tag{9a}$$

$$c\frac{d(u_i-u_{i-1})}{dt} + k(u_i-u_{i-1}) = c\frac{d(u_{i+1}-u_i)}{dt} + k(u_{i+1}-u_i) \tag{9b}$$

$$c\frac{d(u_5-u_4)}{dt} + k(u_5-u_4) + F_c = -c\frac{d(u_5)}{dt} - k'u_5 - k_cR_{bl}u_5. \tag{9c}$$

We also define a frame of reference for each node and constrain them to remain within the cell ($u_i(t = 0) = 0$ and $iL/6 \leq u_i(t) \leq (6 - i)L/6$ for $i = 1$-5). The cell's observable speed is the average speed of the nodes over a full cycle of extension and contraction.

Assuming that adhesion bonds within the lamellipodal and uropodal compartments are distributed uniformly, are at pseudo-steady state (Lauffenburger, 1989), and are stressed homogeneously during contraction, we can write R_{bl} and R_{bu}, respectively, as:

$$R_{bl} = \left[1 + \frac{K_{Dl}}{n_s}\right]^{-1} \int_0^{\frac{L}{6}} \int_{\frac{-W}{2}}^{\frac{W}{2}} \left(n_r^{\,v} + n_b\right) dy\,dx \tag{10a}$$

$$R_{bu} = \left[1 + \frac{K_{Du}}{n_s}\right]^{-1} \int_{\frac{5L}{6}}^{L} \int_{\frac{-W}{2}}^{\frac{W}{2}} \left(n_r^{\,v} + n_b\right) dy\,dx, \tag{10b}$$

where K_{Dl} and K_{Du} are the equilibrium receptor/ligand dissociation constants for the lamellipod and uropod, respectively, equal to the ratios of the rate of dissociation to rate of association.

Although several molecular-level mechanisms for generating a spatial asymmetry in traction at the cell/substratum interface have been suggested (see DiMilla (1994) for a review), current experimental results (e.g., Marks et al., 1991; Schmidt et al., 1993) favor a mechanism based on a spatial asymmetry in adhesion receptor/ligand affinity between the lamellipod and uropod. At the molecular level, affinity can be decreased by phosphorylation of adhesion receptors (Tapley et al., 1989) or increased by the exchange of Ca^{2+} for Mn^{2+} (Gailit and Ruoslahti, 1988). To model a spatial variation in the strength of adhesion bonds, we define ψ as the ratio of the equilibrium receptor/ligand dissociation constant in the absence of applied stress for the lamellipod (K_{Dl}) to the corresponding constant for the uropod (K_{Du}):

$$\psi = \frac{K_{Dl}\{F_c = 0\}}{K_{Du}\{F_c = 0\}}. \tag{11}$$

ψ is a measure of the asymmetry in traction between lamellipod and uropod in this scheme. An analogous approach can be applied if asymmetry in traction results from a spatial asymmetry in the organization of adhesion bonds between lamellipod and uropod, generated by preferential clustering of bonds with the cytoskeleton. The significance of this mechanism is the subject of ongoing work in our lab.

A number of researchers (Bell, 1978; Dembo et al., 1988) have suggested that the equilibrium dissociation constant between adhesion receptors and ligands increases exponentially with the energy applied by contractile forces, scaled by the thermal energy. With this model and assuming that asymmetry in traction is present only during contraction and that adhesion bonds behave as springs which are stiffer than their cytoskeletal connections, the total numbers of bonds in each compartment are:

$$R_{bl} = \left[1 + \frac{e^{\frac{k_c^2 u_5^2}{k_s k_b T}}}{\kappa}\right]^{-1} \int_0^{\frac{L}{6}} \int_{-\frac{W}{2}}^{\frac{W}{2}} \left(n_r^v + n_b\right) dy\,dx \tag{12a}$$

$$R_{bu} = \left[1 + \frac{e^{\frac{k_c^2 u_1^2}{k_s k_b T}}}{\psi\kappa}\right]^{-1} \int_{\frac{5L}{6}}^{L} \int_{-\frac{W}{2}}^{\frac{W}{2}} \left(n_r^v + n_b\right) dy\,dx, \tag{12b}$$

where $\kappa = \frac{k_f n_s}{k_{ro}}$ is the dimensionless affinity between receptor and ligand in the absence of applied stress.

Predictions of Model for Speed of Migration

Predictions for the effects of both interactions between adhesion receptors and ligands and cellular mechanics on the migration speed of individual cells are generated by numerically solving Eqns. (7)-(9) and (12) after dedimensionalization (DiMilla et al., 1991). Because κ represents the product of the affinity between adhesion receptors and ligand (the inverse of the equilibrium dissociation constant between these species) and the ligand density, it is equally appropriate to interpret this parameter as the dimensionless density of adhesion ligand.

One of the fundamental predictions of the model is that dimensionless cell speed should exhibit a biphasic dependence on κ in the presence of asymmetry in the cell/substratum interaction (DiMilla et al., 1991). No net movement occurs for low values of κ because contraction dissociates the few adhesion bonds at the lamellipod and uropod. In contrast, the cell experiences a net forward translocation for moderate values of κ because enough adhesion bonds exist at the pseudopodia to resist contraction but the lamellipod has more bonds to withstand contractile forces than the uropod. Net translocation ceases for high values of κ because now both pseudopodia possess enough adhesion bonds to fully resist contractile forces and limit deformation. This prediction is consistent with our experimental data.

Our modeling predicts that at least three types of properties of the receptor/ligand interaction -- affinity, total number of receptors, and strength of the spatial asymmetry in affinity -- affect the range of densities of adhesion ligand over which significant migration can occur (DiMilla et al., 1991; DiMilla, 1994). The inverse relationship between density of ligand and affinity is supported by experimental data from Duband et al. (1991) demonstrating that the motility of neural crest cells increases with increasing surface concentration of low-affinity ligands but decreases with increasing surface concentration of high-affinity ligands over the same range of concentrations. The densities of adhesion ligand allowing movement also depends on the number of adhesion receptors on the cell surface (Fig. 8): increasing the number of receptors shifts the range of densities supporting movement to larger values. These predictions also demonstrate that maximum speed is proportional to the product of κ and R_T. Thus, the existence of different optimal densities of CnIV and Fn for greatest speed for HSMCs (Fig. 4) may have resulted from either non-identical numbers of receptors for each ligand or distinct affinities for each receptor/ligand pair. We can not distinguish between these possibilities, however, based solely on our data for migration. Finally, the range of densities of ligand permitting migration is proportional to ψ^{-1}, the strength of polarization in adhesion-receptor/ligand affinity between lamellipod and uropod.

Note that this framework assumes that detachment only occurs by dissociation of adhesion bonds. Events during migration are,

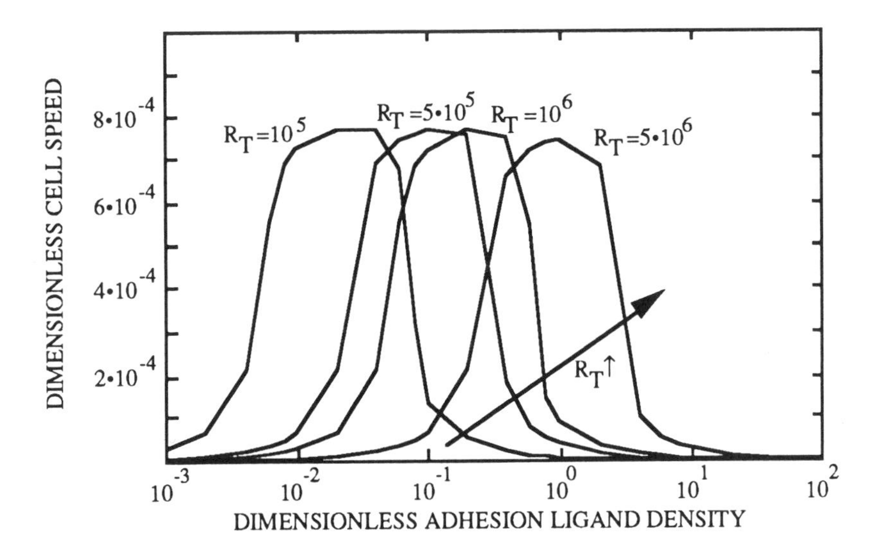

Fig. 8. Model predictions for the effect of number of adhesion receptors, R_T, on the biphasic relationship between dimensionless cell speed and dimensionless adhesion ligand density for a relatively low level of spatial asymmetry in traction ($\psi = 0.1$).

unfortunately, more complex: Regen and Horwitz (1992) recently have shown that locomotion can involve distraction of adhesion receptors from the cell membrane as well as dissociation of adhesion-receptor/ECM linkages. We also occasionally observed (especially on relatively highly-adhesive substrata) that the retraction of the uropod during contraction resulted in the deposition of cellular fragments (DiMilla et al., 1993). The basic trends predicted by the model, are not affected by the presence of this alternative mechanism. This phenomena does, however, deserve further study.

Modeling the Strength of Receptor/Ligand-Mediated Adhesion

Our experimental data for the critical forces for detachment of HSMCs on CnIV and Fn provide indirect and functional estimates of the strength of interactions between these ligands and their corresponding adhesion receptors. Mathematical models, such as the one developed by Cozens-Roberts et al. (1990), also are useful for interpreting this data in terms of molecular-level properties. Assuming that adhesion ligand is present in excess and neglecting the diffusion of receptors into the contact area, for a hemispherical cell at pseudo-steady state:

$$n_b = \frac{R_T}{3A_{proj}\left[1 + \frac{K_D}{n_s}\right]}. \tag{13}$$

Assuming that the equilibrium dissociation constant increases exponentially with applied force, as proposed by Bell (1978):

$$K_D = K_{Do}\{F_T = 0\}\exp\left[\frac{\gamma F_T}{A_{proj} n_b k_b T}\right], \tag{14}$$

and that all bonds in the area of contact are equally stressed, the critical force for detachment is predicted to be proportional to the dimensionless density of adhesion ligand and the number of adhesion receptors (for relatively small κ and relatively large R_T):

$$F_{Tc} = \left[\frac{k_b T}{3e\gamma}\right]\kappa R_T + \lambda, \tag{15}$$

where γ is the characteristic bond length and λ a measure of nonspecific forces. In dimensionless form Eqn. 15 becomes:

$$\alpha_{Tc} = \kappa R_T + \Lambda, \tag{16}$$

where α_{Tc} is the dimensionless critical force for detachment -- the dimensionless strength of adhesion -- and Λ is the dimensionless nonspecific force. Eqn. 15 expressed in terms of S_c has explained satisfactorily several sets of experimental results to date (Cozens-Robert et al., 1990; Stone, 1993).

The linear relationship between critical force for detachment and dimensionless density of adhesion ligand predicted by Eqn. 15 is consistent with our experimental observations (Fig. 4). Further, the model suggests that the steeper slope observed on Fn compared to CnIV may be due to a greater number of receptors for Fn than for CnIV and/or a stronger receptor/ligand affinity for Fn than for CnIV. These trends are identical to those predicted for cell speed, and it is impossible to estimate values for K_D and R_T from even the combined data because both the adhesive and motile responses exhibit the same functionality with respect to numbers of receptors and affinity.

Relationships Between Adhesive and Traction Forces

The models for migration and adhesion demonstrate that $R_T\kappa$ is a single functional parameter affecting both dimensionless cell speed (Fig. 8) and dimensionless strength of attachment (Eqn. 16). Thus, combining these models allows us to compare predictions for dimensionless speed as a function of dimensionless force of detachment at different levels of asymmetry in traction with dedimensionalized experimental data without having to assume values for the properties of receptors and ligands (Fig. 9). The actual values of the maximum dedimensionalized speed and its corresponding dedimensionalized strength of adhesion depend on assumed values for the rate of receptor/ligand dissociation, the length of the cell, and the characteristic length of the bond (the values chosen are characteristic for HSMCs (DiMilla et al., 1991)); the experimental range of dimensionless strengths of adhesion over which movement was observed, however, is independent of the procedure for dedimensionalization. The correspondence between experimental data and model predictions suggests that a relatively low level of asymmetry ($\psi \approx 0.5$) is necessary for the movement of HSMCs, consistent with the low levels of asymmetry in integrin-cytoskeletal attachments for fibroblasts observed directly by Schmidt et al. (1993).

The correspondence between the predictions and experimental data is further revealed by similarities between the traction forces predicted by the model (DiMilla et al., 1991) and estimated critical forces of detachment permitting cell movement. Both James and Taylor (1969) and Harris et al. (1980) have measured traction forces on the order of 10^4 µdynes/cell; these values also compare favorably with our strengths of attachment and offer additional support that transient, short-time scale interactions between adhesion receptors and ligands can transduce the traction necessary for movement.

Perspectives

Our studies demonstrate that details of the underlying molecular-level mechanisms responsible for the crawling of cells over adhesive surfaces can be revealed by a combination of experimental and modeling approaches focusing on measuring and predicting quantitative properties -- such as the critical fluid shear forces necessary for detachment and the speed of movement of individual cells -- that reflect the intrinsic relationship between adhesive and migratory processes. Fig. 10 summarizes our findings. In particular, our results suggest that variations in the strength of adhesion mediated by transient, individual adhesion bonds -- a single functional parameter incorporating the effects of surface density, affinity, and receptor number -- are more critical to movement than the density of the adhesion ligand or affinity between adhesion receptors and

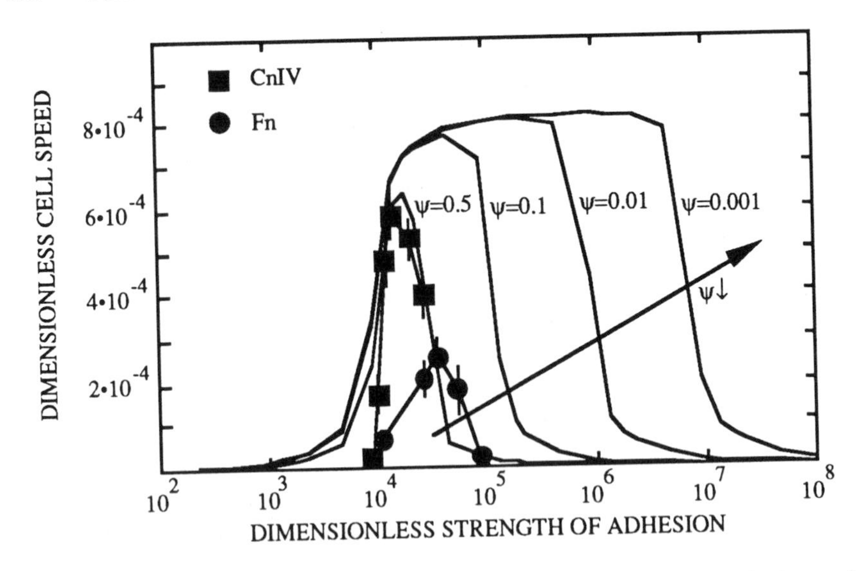

Fig. 9. Model predictions for the biphasic relationship between dimensionless cell speed and dimensionless critical detachment force for spatial asymmetry in traction generated by a spatial variation in the strength of adhesion bonds (ψ = ratio of bond strength in uropod to bond strength in lamellipod). Dedimensionalized data for HSMCs on CnIV and Fn from Fig. 6 is superimposed.

ligands *per se*. Our combination of approaches also provides a method to indirectly evaluate the level of asymmetry in traction at the cell/substratum interface: low levels of asymmetry in which the lamellipod is approximately twice as adhesive for the substratum as the uropod are consistent with both our cellular-level and more direct (but experimentally more-intensive) subcellular-level observations (Schmidt et al., 1993).

Although it should be clear that short time-scale adhesive events are important processes governing motility, it is important to recognize the roles of the generation of intracellular force, mechanotransduction at the cytoskeleton/receptor linkage, and the organization of adhesion bonds at the cell/substratum interface in descriptions of movement. For example, both the maximum speed allowed and the minimum strength of adhesion permitting movement appear to depend primarily on cellular mechanics, including the ability of adhesion receptors to interact with the cytoskeleton for transmission of intracellular motile forces and the level of these forces themselves. Thus, the greater value of maximal speed observed for HSMCs on CnIV compared to on Fn may reflect an underlying difference in cytoskeletal interactions or signaling by the corresponding integrins. Ongoing efforts in our group are focusing on the role of the dynamics of adhesion receptor and bond organization and the strength of the cytoskeleton/receptor connection in the regulation of cell migration and adhesion.

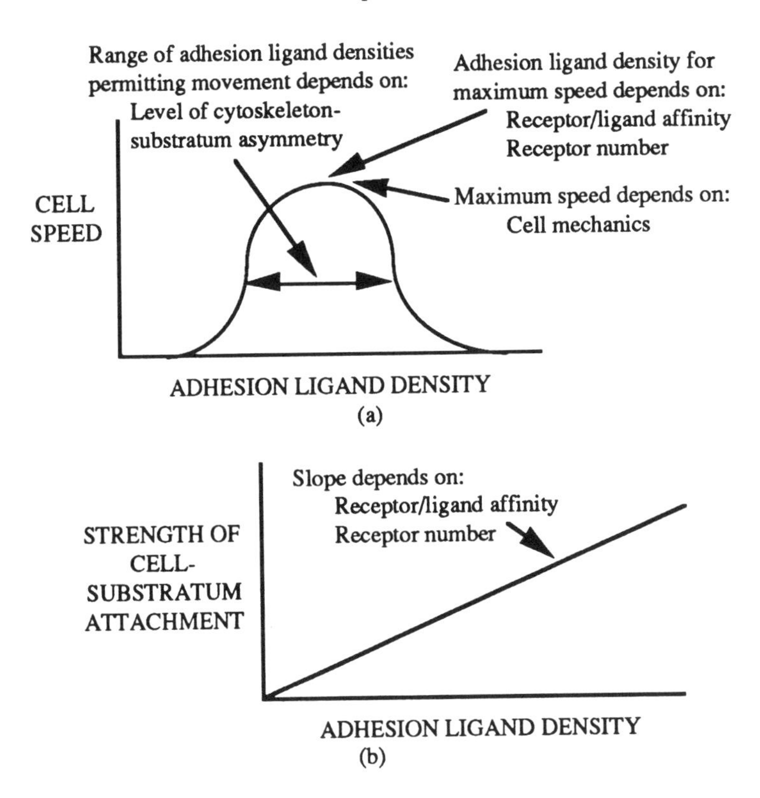

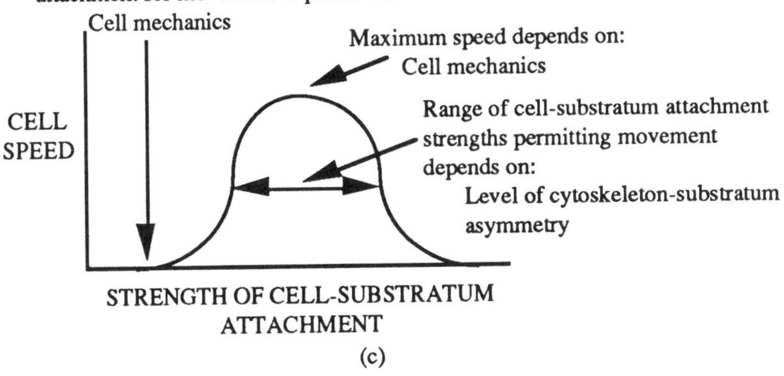

Fig. 10. Schematic of relationships between surface density of adhesion ligand, strength of cell-substratum attachment, and cell speed, based on experimental observations and model predictions. (a) An optimal surface density of adhesion ligand exists for maximum cell speed (compare with Figs. 4 and 8). (b) Strength of cell-substratum attachment increases linearly with surface density of adhesion ligand (compare with Fig. 5 and Eqn. 15). (c) An optimal strength of cell-substratum attachment exists for maximum cell speed (compare with Figs. 6 and 9).

Acknowledgments

The experiments and modeling described here were initiated as part of my doctoral thesis with Doug Lauffenburger and John Quinn at the University of Pennsylvania. Several other collaborators also were instrumental in this work: Steve Albelda for support for cell culture, identification of integrins on HSMCs, and iodination of adhesion ligands, Julie Stone for measurements of adhesive properties, and Ken Barbee for mathematical modeling of migration. This work has been supported by a grant from the Whitaker Foundation, NIH grant GM-41476, and NSF grant BIR-8920118 to the Center for Light Microscope Imaging and Biotechnology.

References

Akiyama, S. K.; Yamada, K. M. The interaction of plasma fibronectin with fibroblastic cells in suspension. J. Biol. Chem. 260:4492-4500; 1985.

Bell, G. I. Models for the specific adhesion of cells to cells. Science 200:618-627; 1978.

Burridge, K.; Faith, K.; Kelly, T.; Nuckolls, G.; Turner, C. Focal adhesions: transmembrane junctions between the extracellular matrix and the cytoskeleton. Ann. Rev. Cell Biol. 4:487-525; 1988.

Clyman, R. I.; Turner, D. C.; Kramer, R. H. An α_1/β_1-like integrin receptor on rat aortic smooth muscle cells mediates adhesion to laminin and collagen types I and IV. Arteriosclerosis 10:402-409; 1990.

Cozens-Roberts, C.; Lauffenburger, D. A.; Quinn, J. A. Receptor-mediated adhesion phenomena. Model studies with the radial-flow detachment assay. Biophys. J. 58:107-125; 1990.

Dickinson, R. B.; Tranquillo, R. T. A stochastic model for cell random motility and haptotaxis based on adhesion receptor fluctuations. J. Math. Biol. 31:563-600; 1993.

DiMilla, P. A.; Barbee, K.; Lauffenburger, D. A. Mathematical model for the effects of adhesion and mechanics on cell migration speed. Biophys. J. 60:15-37; 1991.

DiMilla, P. A.; Lauffenburger, D. A. Models for integrin clustering dynamics, anti-integrin antibody binding, and cell migration speed. J. Cell Biol. 115:114a; 1991.

DiMilla, P. A.; Stone, J. A.; Albelda, S. M.; Lauffenburger, D. A.; Quinn, J. A. Measurement of cell adhesion and migration on protein-coated surfaces. In:Cima, L. G.; Ron, E. S., eds. Tissue-Inducing Biomaterials. Pittsburgh, PA: Mater. Res. Soc. Proc. Vol. 252; 1992a:p. 205-212.

DiMilla, P. A.; Albelda, S. M.; Lauffenburger, D. A.; Quinn, J. A. Measurement of individual cell migration parameters for human tissue cells. AIChE J. 37:1092-1104; 1992b.

DiMilla, P. A.; Albelda, S. M.; Quinn, J. A. Adsorption and elution of extracellular matrix proteins on non-tissue culture polystyrene petri dishes. J. Colloid. Interface Sci. 153:212-225; 1992c.

DiMilla, P. A.; Stone, J. A.; Quinn, J. A.; Albelda, S. M.; Lauffenburger, D. A. Maximal migration of human smooth muscle cells on type IV collagen and fibronectin occurs at an intermediate initial attachment strength. J. Cell Biol. 122:729-737; 1993.

DiMilla, P. A. Adhesion and traction forces in migration: Insights from mathematical models and experiments. In:Akkas, N., ed. NATO Advanced Study Institute on Biomechanics of Active Movement and Division of Cells. Berlin: Springer-Verlag; 1994: in press.

Dembo, M.; Torney, D. C.; Saxman, K.; Hammer, D. The reaction-limited kinetics of membrane-to-surface adhesion and detachment. Proc. Roy. Soc. Lond. B 234:55-83; 1988.

Duband, J.-L., Dufour, S.; Yamada, S. S.; Yamada, K. M.; J. P. Thiery, J. P. Neural crest cell locomotion induced by antibodies to β_1 integrins. A tool for studying the roles of substratum molecular avidity and density in migration. J. Cell Sci. 98:517-532; 1991.

Dunn, G. A. Characterizing a kinesis response: time averaged measures of cell speed and directional persistence. Agents and Actions [Suppl.] 12:14-33; 1983.

Gailit, J.; Ruoslahti, E. Regulation of the fibronectin receptor affinity by divalent cations. J. Biol. Chem. 263:12927-31292; 1988.

Grinnell, F.; Feld, M. K. Adsorption characteristics of plasma fibronectin in relationship to biological activity. J. Biomed. Mater. Res. 15:363-381; 1981.

Hammer, D. A.; Lauffenburger., D. A. A dynamic model for receptor-mediated cell adhesion to surfaces. Biophys. J. 52:475-487; 1987

Harris, A. K. Protrusive activity of the cell surface and the movements of tissue cells. In:Akkas, N., ed. NATO Advanced Study Institute on Biomechanics of Active Movement and Deformation of Cells. Berlin: Springer-Verlag; 1990:p. 249-294.

Harris, A. K.; Wild, P.; Stopak, D. Silicone rubber substrata: a new wrinkle in the study of cell locomotion. Science 208:177-179; 1980.

Hemler, M. E. VLA proteins in the integrin family: structures, functions, and their roles on leukocytes. Ann. Rev. Immunol. 8:365-400; 1990.

Hemler, M. E. Structures and functions of VLA proteins and related integrins. In:Mecham, R. P.; McDonald, J. A., eds. Receptors for Extracellular Matrix Proteins. San Diego, CA: Academic Press, Inc.; 1991: p. 255-299.

Horwitz, A. F.; Duggan, K.; Buck, C.; Beckerle, M. C.; Burridge, K. Interaction of plasma membrane fibronectin receptor with talin. A transmembrane linkage. Nature (London) 320:531-532; 1986.

Hynes, R. O. Integrins: versatility, modulation, and signaling in cell adhesion. Cell 69:11-25; 1992.

James, D. W.; Taylor, J. F. The stress developed by sheets of chick fibroblasts in vitro. Exp. Cell Res. 54:107-110; 1969.

Lauffenburger, D. A simple model for the effects of receptor-mediated cell-substratum adhesion on cell migration. Chem. Eng. Sci. 44:1903-1914; 1989.

Marks, P. W.; Hendley, B.; Maxfield, F. R. Attachment to fibronectin or vitronectin makes human neutrophil migration sensitive to alterations in cytosolic free calcium concentration. J. Cell Biol. 112:149-158; 1991.

Price, T. C. P. Slow linear flow past a hemispherical bump on a plane wall. Q. J. Mech. Appl. Math. 38:93-104; 1985.

Regen, C. M.; Horwitz, A. F. Dynamics of β_1 integrin-mediated adhesive contacts in motile fibroblasts. J. Cell Biol. 119:1347-1359; 1992.

Ross, R. The pathogenesis of atherosclerosis - an update. New Eng. J. Med. 314:488-500; 1986.

Schmidt, C. E.; Horwitz, A. F.; Lauffenburger, D. A.; Sheetz, M. P. Integrin-cytoskeletal interactions in migrating fibroblasts are dynamic, asymmetric, and regulated. J. Cell Biol. 123:977-991; 1993.

Stone, J. A. Quantitative analysis of receptor-mediated adhesion of mammalian cells. Ph. D. Thesis; University of Pennsylvania; 1993.

Stossel, T. P. On the crawling of animal cells. Science 260:1086-1094; 1993.

Tapley, P.; Horwitz, A.; Buck, C.; Duggan, K.; Rohrschneider, L. Integrins isolated from Rous sarcoma virus-transformed chicken embryo fibroblasts. Oncogene 4:325-333; 1989.
Trinkaus, J. P. Cells Into Organs. The Forces that Shape the Embryo. Englewood Cliffs, NJ: Prentice-Hall, Inc.: 1984.
Wang, N.; Butler, J. P.; Ingber, D. E. Mechanotransduction across the cell surface and through the cytoskeleton. Science 260:1124-1127; 1993.

28

Actin Polymerization and Gel Osmotic Swelling in Tumor Cell Pseudopod Formation

C. Dong, J. You, S. Aznavoorian, D. Savarese, L.A. Liotta

Introduction

Active tumor cell motility has long been appreciated to play a major role in invasion and metastasis. Tumor cells exhibit an amoeboid movement similar to that of polymorphonuclear (PMN) leukocytes and the lower eukaryote *Dictyostelium*, characterized by pseudopod protrusion at the leading edge of the cell (Oster and Perelson 1987; Guirguis et al 1987; Condeelis 1992; Condeelis et al 1988, 1990, 1992; Stossel 1989, 1990, 1993; Usami et al 1992; Liotta 1992). Most of the current knowledge of mechanisms of amoeboid chemotaxis is derived from studies of PMN leukocytes and *Dictyostelium* amoebae, which have revealed many similarities between the two different eukaryotic models.

Metastasizing tumor cells are likely to encounter both soluble and insoluble extracellular matrix (ECM) components as they traverse vascular basement membranes. Some chemotactic factors, including ECM molecules, have been shown to stimulate the intrinsic motility of tumor cells *in vitro* (Liotta 1986; Aznavoorian et al 1990). These factors are believed to influence both the extent and the direction of tumor cell movement *in vivo* to specific target organs. Extension of the pseudopod is one of the earliest morphological responses of these tumor cells to soluble chemoattractants. The pseudopod is a specialized region of the cell cortex, consisting primarily of a network of cross-linked actin filaments (Condeelis 1992; Stossel 1990). It is generally believed that chemoattractant stimulation causes remodeling of the cell's peripheral actin network at the location of the stimulus, which results in protrusion of pseudopod and subsequent cell locomotion (Cunningham 1992). However, the manner in which pseudopod extension is coupled to the intracellular signals generated by chemoattractant stimulation is not yet completely understood, and is currently under active investigation in these systems.

Actin polymerization plays an important role in cell motility (Condeelis et al 1988, 1990; Cooper 1991; Watts et al 1991) and actin microfilament dynamics of locomoting cells (Theriot and Mitchison 1991; Wang 1985, 1991; Stossel 1993). The current literature also shows that many cellular motions are associated with the behavior of actin and its interaction with myosin I and II (Adams and Pollard 1989; Janson et al 1992; Giuliano et al 1992; Titus et al 1993). In addition to actin polymerization, there are several additional possible mechanisms which have been implicated as contributors to cell motility. These include the development of forces from hydrostatic pressure, gel osmotic pressure, localized transmembrane osmotic pressure, interfacial tension or interfacial pressure. It has been suggested that gel and transmembrane osmotic pressures are probably the most likely to be involved in cell active motion (Oster and Perelson, 1987; Stossel, 1990; Cunningham 1992).

To understand the physical aspects of tumor cell motility, we have employed a micropipette system to study the kinetics and extent of pseudopod protrusion in individual human A2058 melanoma cells in response to focal stimulation with soluble type IV collagen, one of the ECM components, as a chemoattractant for these cells (Dong et al 1993). Pretreatment of A2058 tumor cells with pertussis toxin (PT), which inactivates the G-protein pathway in signal transduction, was previously shown in a microchemotaxis chamber assay to inhibit ~90% chemotaxis to type IV collagen (Aznavoorian et al 1990). Type IV collagen also induces a transient increase in the concentration of intracellular calcium (Ca^{2+}), which is not inhibited by PT, but is inhibited by the calcium chelator BAPTA [bis-(amino-phenoxy)ethane tetraacetic acid] (Savarese et al 1992). These studies concluded that type IV collagen induces an inositol 1,4,5-trisphosphate-independent release of intracellular Ca^{2+} stores which appears to play a necessary role in the chemotactic response of A2058 cells. The possible involvement of osmotically generated pressure in driving actin-rich extensions of the cell surface was also studied by using sucrose (Bray et al 1991), which showed cell responses to changes in osmolality of the surrounding medium. Use of PT, BAPTA and sucrose, therefore, would allow us to differentially inhibit G protein activation and Ca^{2+} release, to assess the effect on actin polymerization and depolymerization, as well as osmotic force generation during pseudopod formation.

To analyze the relative role of actin polymerization and actin gel osmosis, which both may modulate the process of pseudopod protrusion, a one-dimensional pseudopod model is proposed. The model is based on the previous studies in actin transport and polymerization using a moving boundary problem to represent the characteristic growth in pseudopod formation (Perelson and Coutsias 1986; Oster and Perelson 1987; Zhu and Skalak 1988; Evans and Dembo 1990). To explain the experimental observations on pseudopod protrusion due to the osmotic effect, the model further adds the possible contribution of local osmotic pressure to pseudopod formation, which is generated from the actin gel swelling. The model, therefore, aims to predict the sequence in which polymerization and osmotic forces come into play. Recent analyses of *Dictyostelium* migration have led to the cortical expansion model (Condeelis et al 1988, 1990), in which focused actin polymerization and cross-linking are the immediate, primary driving forces for pseudopod extension during chemotaxis. In contrast, pseudopod extension in PMNs is thought to be driven initially by localized Ca^{2+}-activated disassembly of the actin gel, generating osmotic forces that distend and protrude out against the membrane (Cunningham 1992; Stossel 1989, 1990). The two models may not be mutually exclusive; certain aspects of these models can be manipulated in our experimental system. The knowledge gained may reveal more similarities than differences in migration mechanisms among the different cell types.

Experimental Methods

Materials

Type IV collagen was purchased from Collaborative Research, Inc. (Bedford, MA). For assays, the collagen was diluted to 100 μg/ml in Dulbecco's modified eagle medium (DMEM) with 0.1% bovine serum albumin (BSA), and the solution was brought to a pH of 7.4. Pertussis toxin (PT) was from List Biological Laboratories, Inc. (Campbell, CA). BAPTA-AM was from Molecular Probes (Eugene, OR).

Cell Culture

The human melanoma cell line, A2058 was maintained in culture in DMEM supplemented with 10% fetal bovine serum (FBS). Medium and serum were from Biofluids, Inc. (Gaithersburg, MD).

Micropipette Assay

Experiments with micropipette manipulation were performed with single cells. Fig. 1 is a schematic diagram of the micropipette apparatus.

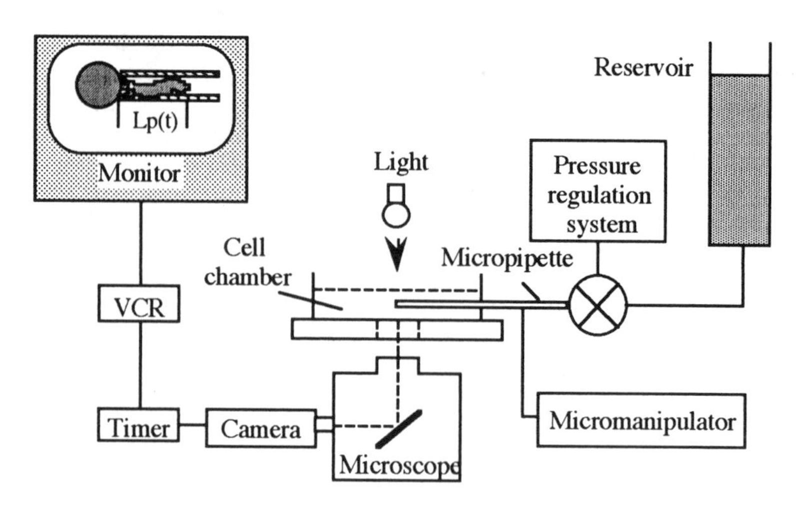

Fig. 1. Schematic diagram of the micropipette set-up.

Tumor cells (from A2058 human melanoma cell line) for micropipette assay were detached when subconfluent by brief exposure to 0.05% trypsin / 0.02% EDTA, allowed to regenerate for 1 hour in DMEM / 10% FBS, centrifuged at 800g for 5 min, and then resuspended in DMEM / 0.1% BSA (2 x 10^6/ml concentration).

Micropipettes with an internal diameter of about 5-7 μm at the tip, were prepared using a pipette puller (Narishige, Tokyo, Japan) and a microforge (Stoelting, Wood Dale, IL). The micropipette was made in such a way that the internal diameter of the micropipette was uniform for at least 20 μm from the tip. The whole micropipette was then filled with soluble type IV collagen solution (100 μg/ml) serving as the chemoattractant in the micropipette assay. The wide end of the micropipette was connected to a syringe reservoir through a three-way stop-cock. A serum-free DMEM was collected in the reservoir and filled the whole Teflon tubing. The pressure level at the micropipette tip (positioned adjacent to the cell surface) was adjusted and controlled by a pressure regulation system. Individual tumor cells were suspended in DMEM / 0.1% BSA in a small chamber mounted on the stage of an inverted microscope (Olympus, New Hyde Park, NY). The cells were viewed through the bottom of the chamber with the use of an oil lens objective (100x, NA 1.25) and a 10x eyepiece. The image of cell pseudopod formation taking place during the experiment was taken by a video camera (Dage MTI, Michigan City, IN), recorded using a Super-VHS video recorder (Panasonic, Secaucus, NJ) with a video timer (For.A, Boston, MA), and displayed on a video monitor (Panasonic, Secaucus, NJ). The magnification of the system was calibrated before each experiment with the use of a micron scale (50x2 μm, Graticules Ltd, Towbridge, Kent, England).

The detailed descriptions regarding the procedures involved in the micropipette assay have been recently published (Dong et al 1993).

Microchemotaxis Chamber Assay

The motility of tumor cells was measured by using a 48-well microchemotaxis chamber (Neuro Probe Inc., Cabin John, MD). The device consists of both top and bottom wells separated by a 8-μm pore diameter polycarbonate Nuclepore filter (Neuro Probe Inc.), which was precoated by soaking overnight in 0.01% gelatin, 0.02M acetic acid, then air-dried prior to use. A2058 human tumor cells were prepared for chamber assay similar to that for micropipette assay and placed in the top wells. Type IV collagen was diluted in DMEM / 0.1% BSA (100μg/ml) and placed in the bottom wells serving as the chemoattractant. Cells migrated through the filter pores to the opposite side over a standard period of time in response to soluble type IV collagen. The entire apparatus was incubated at 37°C. At the end of the assay, filters were removed, fixed, stained with Diff-Quik (Baxter Scientific, McGaw Park, IL), and mounted on glass slides. Cells that had migrated were quantitated by light microscopy under a high power field (HPF) (500x). This microchemotaxis chamber assay was performed in parallel with each micropipette assay using the same batches of cells, following the procedures described previously (Aznavoorian et al 1990).

Measurement of Intracellular Ca^{2+}

To observe the intracellular Ca^{2+} response in A2058 human melanoma cells associated with the type IV collagen stimulation, Fura-2 fluorescence was measured using video microscopy (Savarese et al 1992). Tumor cells prepared from the cell culture were plated on sterile 25-mm round glass coverslips in 35-mm tissue culture dishes (Costar, Cambridge, MA) and incubated at 37°C for 18-48 hr. Fura-2 acetoxymethyl ester (Fura-2 AM, Molecular Probes Inc., Eugene, OR) was reconstituted in anhydrous Me_2SO and added to the cell monolayers on the coverslips at a final concentration of 5 mM for 10 min at 37°C. The Fura-2 was diluted 2.3-fold, and the coverslips were then mounted on the stage of the microscope and perfused with DMEM / 0.1% BSA with phenol red. Fura-2-loaded cells were illuminated with a mercury arc lamp and imaged with an inverted microscope, as described in detail by Savarese et al (1992). Output fluorescence images were digitized and averaged.

Pretreatment of Cells with PT

In experiments with PT, A2058 tumor cells in suspension were preincubated with the toxin at 0.5 μg/ml for an additional 2 hr at room temperature before they were used for all experimental assays.

Pretreatment of Cells with BAPTA

In experiments using BAPTA-AM for chelation of intracellular Ca^{2+}, regenerated A2058 tumor cells from trypsin / EDTA detachment in cell culture were washed in Ca^{2+}-free DMEM / 0.1% BSA / 1mM EGTA, and then incubated in the same medium added with 75 μM BAPTA-AM for 30 min at 37°C in the dark.

Experimental Results

Pseudopod Protrusion Revealed by a Micropipette

Experiments were performed at room temperature on individual A2058 tumor cells ranging from 11.4-15.3 μm, using micropipettes with tip diameters ranging from 5-7 μm. A micropipette filled with soluble type IV collagen solution (100μg/ml) was positioned so that the tip was adjacent to a cell suspended in the medium described earlier. Measurements of pseudopod protrusive length (L_p) vs. time were made from the series of photographs of video microscopic images (Figs. 2-4) and used to generate growth curves $L_p(t)$.

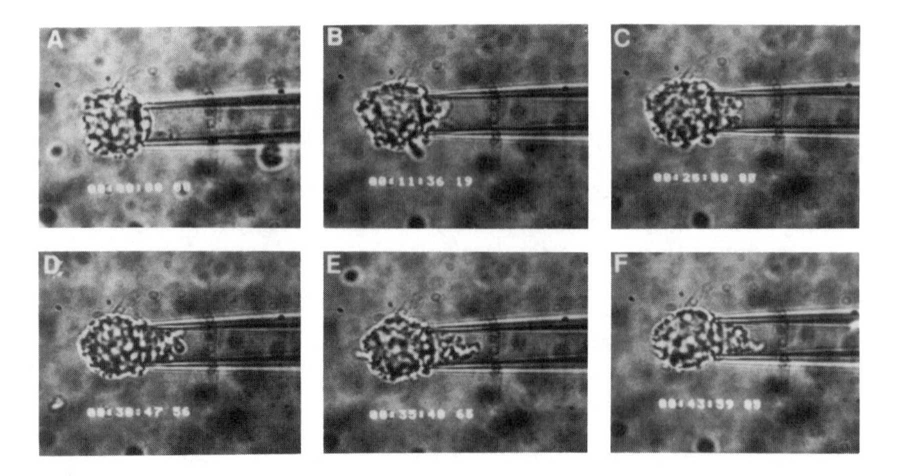

Fig. 2. Sequence of light video microscopic images showing progressive pseudopod protrusion of an untreated human A2058 melanoma cell at the indicated time intervals (A-F), in response to soluble type IV collagen. Pipette inner diameter is 7.0 μm. From Dong et al (1993), by permission.

For untreated cells, tumor cells started to generate a pseudopod within 10 min of contact between the cell surface and the micropipette tip containing the attractant (Fig. 2B). The pseudopod then entered the micropipette and proceeded to lengthen for about 25 min (Fig. 2B-E). Pseudopods from individual cells ranged from 7.5-10 μm at that time, and were characterized by an irregular shape which did not fill the lumen of the micropipette. At the last observed time point in Fig. 2F, the pseudopod of that particular cell had reached a length of 10.2 μm. The average velocity during the first 44 min of protrusion of this pseudopod is estimated to be 0.24 μm/min.

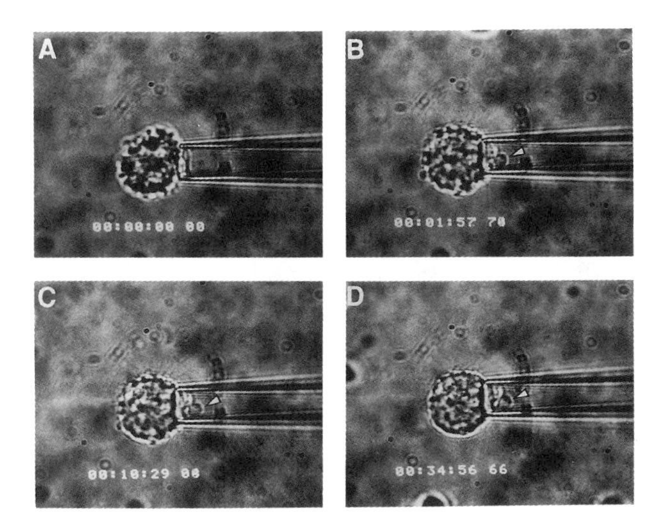

Fig. 3. Sequence of light video microscopic images showing bleb formation (by arrows) of a PT-treated human A2058 melanoma cell at the indicated time intervals (A-D), in response to soluble type IV collagen. Pipette inner diameter is 5.6 μm. From Dong et al (1993), by permission.

Pretreatment of cells with PT (0.5 μg/ml), which catalyzes the ADP-ribosylation of a class of G proteins and uncouples them from their receptors and attenuates receptor-generated signals, blocked formation of the irregular, extended pseudopod, while allowing a much smaller outpouching, or bleb to form (Fig. 3, A-D). Lengths of such blebs from individual PT-treated cells reached a plateau at ~20 min, and ranged from 2.2-4 μm at the endpoint. From the particular cell shown in Fig. 3, a rounded bleb formed from the cell surface (indicated by arrows in Fig. 3 (B-D)). This protrusion reached a length of ~3.4 μm within 35 min, which did not change significantly over the duration of the observation period (up to 1 hour). This bleb formation are usually associated with a phase of submaximal motility as it has been described in studies on PMNs (Cassimeris and Zigmond 1990; Watts et al 1991) and human melanoma cells (Cunningham et al 1992). The diminished pseudopod protrusion and alteration in pseudopod morphology observed in the single PT-treated cell and untreated cell presumably reflects the cell's inability to translocate. This agrees with the chemotaxis assay (Aznavoorian et al 1990) performed in parallel with the micropipette assay using the same batches of cells, in which the treatment with PT inhibited stimulated chemotaxis to type IV collagen by ~90%.

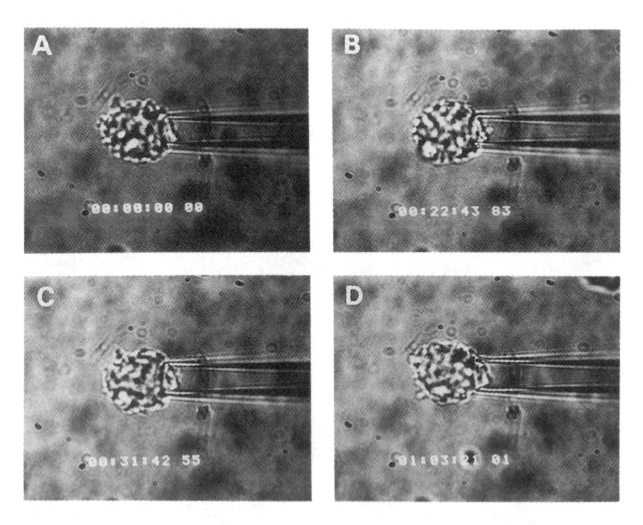

Fig. 4. Sequence of light video microscopic images showing lack of protrusion of a BAPTA-treated human A2058 melanoma cell at the indicated time intervals (A-D), in response to soluble type IV collagen. Pipette inner diameter is 5.0 μm. From Dong et al (1993), by permission.

Treatment of cells with BAPTA (75 μM), an intracellular Ca^{2+} chelator, blocked pseudopod protrusion. Neither bleb formation nor irregular extension was observed in any cell of this treatment group over a period of 1 hour (Fig. 4, A-D).

Chemotactic Migration of Tumor Cells

A2058 human melanoma cells migrated in a concentration-dependent manner to soluble type IV collagen. The response was maximal at the highest dose tested, 1000 μg/ml (Aznavoorian et al 1990). Increasing the gradient resulted in increased migration until a plateau was reached, from which we chose the dose of 100 μg/ml for all our studies.

Chemotaxis to ECM proteins may be stimulated through different pathways, including the immediate post-receptor signal transduction. Using optimally stimulating concentration of type IV collagen, we tested the effect of a wide range of PT concentrations on chemotactic responses, which showed that chemotaxis to type IV collagen was virtually inhibited ($\sim$ 90%) by the toxin ($>$ 0.1 μg/ml) (Aznavoorian et al 1990). The result indicates the involvement of a second messenger pathway regulated by a PT-sensitive G protein in the stimulation of chemotaxis by type IV collagen. Treatment of cells with BAPTA-AM (75 μM) abolished A2058 cell chemotaxis to type IV collagen completely in the migration assay (Dong et al 1993), which suggests that intracellular Ca^{2+} can play a potential role in tumor cell motility.

Intracellular Calcium Response

Intracellular Ca^{2+} response of A2058 cells to type IV collagen was studied in parallel by Savarese et al (1992) using Fura-2 and digital imaging fluorescence microscopy. Type IV collagen, at a dose 100 μg/ml, induced a significant rise in cytosolic free Ca^{2+} concentration within 100 sec. This response was not inhibited by PT. Removal of extracellular Ca^{2+} failed to inhibit the type IV collagen-stimulated rise in Ca^{2+} in the cells. However, depletion of intracellular Ca^{2+} inhibited both chemotaxis and the type IV collagen-induced increase in intracellular Ca^{2+}.

In comparison with the micropipette assay in PT-treated cells, the bleb formation in response to type IV collagen stimulation could be mediated by localized Ca^{2+}-activated actin depolymerization and osmotic flux. This bleb formation was not able to lead the whole cell migration easily as observed from the chemotactic migration assay. For BAPTA-treated cells, cell motility was completely abolished.

Two Phases of Pseudopod Protrusion

The micropipette system has revealed that pseudopod protrusion in tumor cells may have two distinct phases, membrane blebbing and extension, which are regulated by separate intracellular signals. Soluble type IV collagen activates a G protein-coupled receptor (unidentified) on A2058 melanoma cells to stimulate chemotaxis, and induces a transient rise in the concentration of intracellular Ca^{2+} by an as yet unknown mechanism. The use of PT to inhibit the G protein-linked effectors without inhibiting the rise in intracellular Ca^{2+} has revealed what may be a distinct early phase of pseudopod formation, separable from the G protein-mediated events. The smaller, smooth-edged bleb in PT-treated, type IV collagen-stimulated cells, may be formed as a result of localized Ca^{2+}-activated actin disassembly and subsequent fluid flow into the region of higher osmotic pressure. This is also supported by the observation that BAPTA-treated cells did not form even a small bleb in response to type IV collagen. Extension of the pseudopod beyond a bleb is likely to require G protein-mediated actin polymerization, or reassembly, at the leading front (Zigmond 1989; Stossel 1990). The Ca^{2+}-activated process leading to blebbing should occur in both untreated and PT-treated cells, however, blebs may only be observable as a distinct morphological phase when G protein-linked actin polymerization is inhibited.

Biophysical Analyses

Physical Models of Pseudopod Protrusion

Models have been developed to explain the driving force behind pseudopod protrusion, based on *in vitro* studies of actin biochemistry (Forscher 1989; Janmey et al 1985; Stossel 1989, 1990) and on studies with PMNs and *Dictyostelium* in which cell shape was correlated with biochemical changes in the cytoskeleton upon chemoattractant stimulation (Coates et al 1992; Condeelis et al 1988, 1990; Condeelis 1992; Watts et al 1991). Two of the prevailing models (Stossel 1989, 1990; Condeelis et al 1988, 1990, 1992) have proposed that there is a linkage of pseudopod extension to a local, receptor-mediated chemotactic stimulus and to forces driven by a combination of actin polymerization, cross-linking, and osmotic swelling at the site of a chemotactic stimulus. Actin polymerization has indeed been demonstrated in the leading edge of pseudopods of chemoattractant-stimulated amoebae (Condeelis et al 1988) and neutrophils (Cassimeris et al 1990; Coates et al 1992). However, the two models differ somewhat as to the sequence in which actin polymerization and osmotic forces (subsequent to depolymerization) occur. In the cortical expansion model, formulated through studies with synchronized *Dictyostelium* amoebae (Condeelis et al 1988, 1990), pseudopod protrusion is driven by focal actin polymerization and cross-linking which occurs immediately after chemotactic receptor stimulation. Transient severing of actin filaments followed by osmotic swelling of the actin gel are thought to occur at a subsequent step, to further expand the pseudopod. The model for PMNs (Stossel 1989, 1990) makes a clear distinction between separate "disassembly" and "assembly" phases of pseudopod protrusion which are linked to stages of phosphatidylinositol (PI)-cycle turnover. Receptor-generated signals are thought to initially cause localized actin disassembly, mediated by calcium-activated severing proteins (i.e., gelsolin). The transiently disrupted actin gel at the leading edge imbibes water and swells outward, pushing against the membrane. Thus, osmotic forces are hypothesized to be the primary driving forces of pseudopod protrusion. Subsequent actin reassembly and cross-linking them stabilize the newly formed protrusion. Actin reassembly is thought to be initiated by metabolites in the PI cycle (Stossel 1989, 1990), which is activated by a receptor-coupled G protein in neutrophils (Lassing and Lindberg 1988).

Analysis of the sequence and relative contributions of actin polymerization vs. gel osmotic swelling in pseudopod protrusion of tumor cells requires an experimental system in which these forces can be selectively manipulated. Our experimental observations of pseudopod formation in melanoma cells using a micropipette assay, in which a stimulus-induced intracellular Ca^{2+} burst is correlated with protrusion of a membrane bleb (observed in PT-treated cells), is consistent with a theory of osmotic pressure-mediated initial pseudopod growth. The irregular appearance of pseudopods from untreated cells seems to indicate that actin polymerization is not merely "filling in" a formed membrane bleb, but rather provides a more active protrusive force. It is possible that, in the continuous presence of type IV collagen, pseudopod extension is initiated by Ca^{2+}-activated actin depolymerization and localized blebbing, followed by a more prolonged phase of G protein-mediated actin polymerization at the distal tip of the pseudopod.

Previous studies in modeling actin transport and polymerization used a moving boundary problem to represent the characteristic growth in pseudopod formation (Perelson and Coutsias 1986; Oster and Perelson 1987; Zhu and Skalak 1988; Evans and Dembo 1990), in which a pseudopod was treated as a separated region from the main cell body. The main cell body contains mostly aqueous solution of water, physiological ions, and macromolecular proteins in addition to other elements (Alberts et al 1983). Likewise, our analysis treated the main cell body as a pool containing all unpolymerized actin monomers (G-actin) in a tight complex formation of profilactin to provide pseudopod region free monomers when the chemotactic signal is "switched on" (Fig. 5). The pseudopod region is considered as an actin network with solid and fluid phases. The solid phase represents the actin filament framework (or F-actin bundle). The fluid phase represents the solute transporting free G-actin by convection and diffusion in the spaces among actin filament bundles. During polymerization, it is assumed that the length of F-actin grows by adding more G-actin onto the filament tip at the moving front. The possibility of subsequent fluid flow into the region of higher osmolality due to localized Ca^{2+}-activated actin disassembly is considered. The pseudopod tip is treated as a moving front varying with time denoted as L_p. The geometry of the pseudopod is confined by the pipette size in the micropipette manipulation assay, which can be considered as simple as a cylinder (Fig. 5).

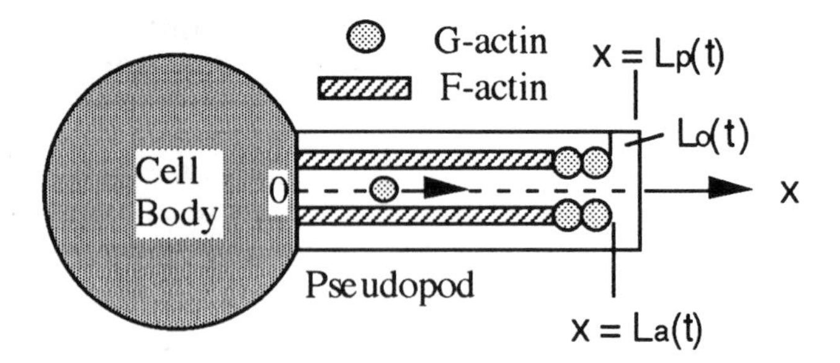

Fig. 5. A sketch of a one-dimensional model of pseudopod protrusion with a coordinate x. Pseudopod length $L_p=L_a+L_o$, in which L_a is the length for actin filament and L_o is the gap between the actin filament tip and the pseudopod tip.

Osmotic Pressure

One compelling idea, based on what we have learned about actin gels, has been proposed by Oster and Perelson (1987), which provided an argument linking the disassembly of the actin gel with the generation of osmotic force.

The effect of osmotic pressure difference generated across a semi-permeable membrane has been well-documented (Hobbie 1978). Oster and Perelson (1987) used this transmembrane osmotic effect to model the fluid flow and actin gel swelling at the pseudopod tip in the acrosomal process. The localized transmembrane osmosis was assumed to cause a water flux from the cell exterior to interior regions at the pseudopod tip. A recent report by Stossel (1990) has discussed the possibility that the local actin gel swelling at the pseudopod tip is induced by the water flux from the regions inside the cell body. The osmotic pressure due to the actin gel swelling depends on the molar concentration of solutes and/or counter-ions in the fluid phase of the gel. The theories proposed by either Oster and Perelson or Stossel have suggested the potential role of actin gel swelling in cell pseudopod protrusion.

Let J_v be the volume flow per unit area per second through a membrane. The simplest relationship between J_v and the driving-pressure difference Δp_d across the membrane is given by:

$$J_v = K_p\, \Delta p_d \qquad \text{at } x = L_p \tag{1}$$

where the proportionality constant K_p is called the hydraulic conductivity. The driving-pressure difference is defined by:

$$\Delta p_d = \Delta p - \Delta \pi \qquad \text{at } x = L_p \tag{2}$$

If the gel osmotic swelling is induced by a local fluid flow from the cell exterior at the pseudopod tip, then Δp is the total hydrostatic pressure difference across the membrane and $\Delta\pi$ is the total osmotic pressure difference across the membrane. In the approximation that van't Hoff's law holds $\Delta\pi = kT\Delta c$, where k is the Boltzmann's constant; T is the absolute temperature; and $\Delta c = c_L - c_e$, in which c_L is the G-actin concentration at the pseudopod tip inside cell membrane and c_e is a reference value of concentration for all osmotically active solutes at the pseudopod tip outside of the cell membrane at $x=L_p$. An equation governing the actin gel swelling, denoted by a length L_o, at the pseudopod tip becomes:

$$\frac{dL_o}{dt} = -K_p(\Delta p - kT\Delta c) \qquad \text{at } x = L_p \tag{3}$$

Note that Eq. (3) treats G-actin that was transported from the base of the pseudopod as the *only* contribution of osmotically active solute in generating the total osmotic pressure difference at the pseudopod tip; it neglects the contributions of other osmotically active solutes in the reaction. This approximation can be adjusted by multiplying c_L by a fraction coefficient α if it is necessary, which gives $\Delta c = \alpha c_L - c_e$. The fraction coefficient α will then depend on the concentration of other solutes which may generate osmotic pressure, such as profilin (a protein that prevents filament elongation by binding to individual actin molecules), profilactin (an actin-profilin complex), gelsolin (a protein that breaks apart the actin filaments into smaller filament fragments and individual actin molecules) and calcium ions (Stossel 1990).

In contrast, the fluid flow from the pseudopod base to the pseudopod tip through the entire pseudopod region due to the osmotic pressure gradient is governed by Darcy's law described in the following text.

Actin Transport and Polymerization

An article by Stossel (1990) described the actin in cytoplasmic structure and function in quite detail. Based on those reviews, the pseudopod is modeled as a solid-fluid network. The porosity ϕ for this network is denoted by the volume fraction of the fluid, which can be defined by:

$$\phi = \frac{V_p}{V_{total}} \tag{4}$$

where V_p is the volume of the fluid contained in a sample of bulk volume V_{total}. (Biot 1955).

Actin monomers have been modeled as dispersing solutes while being transported through diffusion and convection from the base to the tip of the pseudopod. The monomers are the supplies of the solid component for the growth actin filament upon polymerization (Perelson and Coutsias 1986; Zhu and Skalak 1988). Using this concept, the concentration of the solute, c, may be defined as the number of G-actin molecules per unit fluid volume which is a function of space and time. The governing equation for the actin monomers being transported in the network is then generalized to:

$$\frac{\partial c}{\partial t} + v\frac{\partial c}{\partial x} = D\frac{\partial^2 c}{\partial x^2} \tag{5}$$

Here D is the diffusion coefficient. v is the fluid convective velocity driven by both hydrostatic and osmotic pressure gradients inside the pseudopod actin network ($0<x<L_a$). In the one dimensional case, this convective velocity obeys the modified Darcy's law ($0<x<L_a$):

$$v = -\frac{K_d}{\phi\mu}\left(\frac{\partial p}{\partial x} - KT\frac{\partial c}{\partial x}\right) \tag{6}$$

where μ is the fluid viscosity and K_d is a permeability coefficient that accounts for the average resistance of the solid to the fluid in the actin network.

G-actin is assumed to undergo change from a monometric to a polymeric state at the barbed ends of the actin filaments ($x=L_a$) where polymerization takes place. The mass conservation of actin monomers at the F-actin tip is, therefore, related to polymerization kinetics at the moving front. The extension of the F-actin network is assumed to be a result of actin polymerization, which occurs at the filament tip. Let L_a be a total length of the filament (L_a as a function of time), then the velocity of the moving tip of the filament is $\mathrm{d}L_a(t)/\mathrm{d}t$. The rate of actin monomers supplied to the filament tip in the pseudopod through diffusion and convection is $\phi(vc\text{-}D\partial c/\partial x)$ per unit cross-section area. The amounts of monomers that deposit into (or withdraw from) the F-actin during polymerization (or depolymerization) are $(n/\lambda)\{\mathrm{d}L_a/\mathrm{dt}\}$, whereas those monomers that disperse into the new-born spaces due to the moving front are $\phi c(\mathrm{d}L_a/\mathrm{d}t)$, where n is the total number of actin filament ends per unit area and λ is the length that a filament extends when one monomer adds to it. In other words, n/λ represents the concentration of filament growing ends at the moving front. In addition, the amounts of monomers that disperse into the swelling region between the filament tip and pseudopod membrane front are $c(\mathrm{d}L_o/\mathrm{d}t)$. The conservation of actin monomers therefore leads to:

$$\phi \left(vc - D\frac{\partial c}{\partial x}\right) = \frac{n}{\lambda}\left(\frac{dL_a}{dt}\right) + \phi c\frac{dL_a}{dt} + c\frac{dL_o}{dt} \qquad \text{at } x = L_a \qquad (7)$$

Note that $(1/\lambda)\{dL_p/dt\}$ represents the number of actin monomers added to (or released from) an individual filament per unit time. Wegner (1976) and Pollard and Mooseker (1981) had developed a kinetic equation which described the problem for reversible polymerization in *in vitro* experiments. Applying their equation to the actin filament undergoing polymerization and depolymerization, an expression for the kinetics becomes:

$$\frac{1}{\lambda}\left(\frac{dL_a}{dt}\right) = k_1 c - k_2 \qquad \text{at } x = L_a \qquad (8)$$

where k_1 and k_2 are called association and dissociation rate constants, respectively, whose values depend upon which end of the actin filament monomers are reacting.

Summary

Motility in tumor cells is stimulated *in vitro* by a wide variety of agents which fall into three general categories, including growth factors, extracellular matrix (ECM) components, and motility factors, some of which are secreted by the tumor cells and thus act in an autocrine fashion (Stracke et al 1991; Gherardi 1991; Lester and McCarthy 1992; Nabi et al 1992). *In vivo*, each type of motility-stimulating factor may have a predominant role at different stages of the metastatic cascade; for example, motility factors may be important for the initial dissemination of tumor cells away from the primary tumor, ECM components during intravasation and extravasation of blood vessels and during migration through tissue stroma, and finally growth factors for "homing" of tumor cells to secondary sites where they grow and establish new colonies.

The amoeboid movement exhibited by tumor cells requires the coordinated action of interrelated steps, including pseudopod protrusion at the leading edge, new adhesion formation on its substrate, and release of "old" adhesions at the trailing edge. Since pseudopod protrusion is a prominent feature of this type of migration, understanding the biochemical and biophysical mechanisms regulating the assembly of this motility "organ" is very important and may lead to development of therapeutic agents designed to block this component of metastasis.

A micropipette system has been used to study the dynamics of pseudopod protrusion in individual human melanoma cells in response to type IV collagen, a component of basement membranes. Soluble type IV collagen stimulates chemotaxis of A2058 melanoma cells through a G protein-coupled receptor, and induces an early burst of intracellular calcium. Tumor cells generate a pseudopod that enters the micropipette and is characterized by an irregular shape which does not fill the lumen of the micropipette. We have not seen yet that tumor cells crawl entirely into the micropipette, as has been observed for leukocytes (Usami et al 1992), although these same tumor cells can deform enough to traverse a filter with pores as small as 3 μm in a microchemotaxis chamber assay. This is likely due to the apparent lack of adhesion of the pseudopod to the pipette wall, which would prevent the necessary traction for movement. It is well known that cells require an optimal adhesiveness for locomotion (DiMilla et al 1991). Leukocytes, in contrast, may require less adhesion than tissue cells, and can easily crawl into a micropipette.

Pretreatment of tumor cells with pertussis toxin (PT), which inhibits cell migration by ~90% but not the intracellular calcium burst, blocked formation of the irregular, extended pseudopod, while allowing a much smaller outpouching, or bleb to form in response to type IV collagen stimulation. Treatment of cells with BAPTA, an intracellular calcium chelator, blocked initial bleb formation and prevented extension. From these observations we hypothesize that tumor cell pseudopod protrusion induced by soluble type IV collagen takes place in distinct, separable phases: an initial convex, symmetrical bleb, caused by localized Ca^{2+}-activated actin depolymerization and osmotic flux, followed by an extension with an irregular shape, which require G protein-mediated actin polymerization.

To analyze the relative role of actin polymerization and gel osmotic swelling in tumor cell pseudopod formation, we are developing an experimental system and biophysical models which will allow us to selectively manipulate these factors (You and Dong 1993). A complete theory requires knowledge of the sequence in which polymerization and osmotic forces come into play, as well as the cellular motion associated with the actin-myosin interactions. Such studies are currently underway in our laboratories.

Acknowledgment

This research was supported in part by the U.S. National Science Foundation Research Initiation Award BCS-9308809 (C.D.).

References

Adams, R.J.; Pollard, T.D. Membrane-bound myosin-I provides new mechanisms in cell motility. *Cell Motil. Cytoskel.* 14:178-182; 1989.

Alberts, B.; Bray, D.; Lewis, J.; Raff, M.; Roberts, K.; Watson, J.D. *Molecular Biology of the Cell*. Garland Publishing, Inc., New York; 1983.

Aznavoorian, S.; Stracke, M.L.; Krutzsch, H.; Schiffmann, E.; Liotta, L.A. Signal transduction for chemotaxis and haptotaxis by matrix molecules in tumor cells. *J. Cell Biol.* 110:1427-1438; 1990.

Biot, M.A. Theory of elasticity and consolidation for a porous anisotropic solid. *J. Applied Physics* 26:182-185; 1955.

Bray, D.; Money, N.P.; Harold, F.M.; Bamburg, J.R. Responses of growth cones to changes in osmolality of the surrounding medium. *J. Cell Sci.* 98:507-515; 1991.

Cassimeris, L.; Zigmond, S.H. Chemoattractant stimulation of polymorphonuclear leucocyte locomotion. *Cell Biol.* 1:125-134; 1990.

Coates, T.D.; Watts, R.G.; Hartman, R.; Howard, T.H. Relationship of F-actin distribution to development of polar shape in human polymorphonuclear neutrophils. *J. Cell Biol.* 117:765-774; 1992.

Condeelis, J. Are all pseudopods created equal? *Cell Motil. Cytoskel.* 22:1-6; 1992.

Condeelis, J.; Hall, A.; Bresnick, A.; Warren, V.; Hock, R.; Bennett, H.; Ogihara, S. Actin polymerization and pseudopod extension during amoeboid chemotaxis. *Cell Motil. Cytoskel.* 10:77-90; 1988.

Condeelis, J.; Bresnick, A.; Demma, M.; Dharmawardhane, S.; Eddy, R.; Hall, A.L.; Sauterer, R.; Warren, V. Mechanisms of amoeboid chemotaxis: An evaluation of the cortical expansion model. *Developmental Genetics* 11:333-340; 1990.

Condeelis, J.; Jones, J.; Segall, J.E. Chemotaxis of metastatic tumor cells: Clues to mechanisms from the *Dictyostelium* paradigm. *Cancer Met. Rev.* 11:55-68; 1992.

Cooper, J. The role of actin polymerization in cell motility. *Annu. Rev. Physiol.* 53:585-605; 1991.

Cunningham, C.C. Actin structural proteins in cell motility. *Cancer Met. Rev.* 11:69-77; 1992.

DiMilla, P.A.; Barbee, K.; Lauffengurger, D.A. Mathematical model for the effects of adhesion and mechanics on cell migration speed. *Biophys. J.* 60:15-37; 1991.

Dong, C.; Aznavoorian, S.; Liotta, L.A. Two phases of pseudopod protrusion in tumor cells revealed by a micropipette. *Microvasc. Res.* 47:55-67, 1994.

Evans, E.A.; Dembo, M. Physical model for phagocyte motility: Local growth of a contractile network from a passive body. *Biomechanics of Active Movement and Deformation of Cells* (Ed. Akkas, N.), NATO ASI Series H42:185-214, Springer-Verlag, Berlin-New York; 1990.

Forscher, P. Calcium and polyphosphoinositide control of cytoskeletal dynamics. *Trends Neuro. Sci.* 12:468-474; 1989.

Gherardi, E. Growth factors and cell movement. *Eur. J. Cancer* 27:403-405; 1991.

Giuliano, K.A.; Kolega, J.; DeBiasio, R.L.; Taylor, D.L. Myosin II phosphorylation and the dynamics of stress fibers in serum-deprived and stimulated fibroblasts. *Mol. Biol. Cell* 3:1037-1048; 1992.

Guirguis, R.; Margulies, I.; Taraboletti, G.; Schiffmann, E.; Liotta, L. Cytokine-induced pseudopodial protrusion is coupled to tumour cell migration. *Nature* 329:261-263; 1987.

Hobbie, R.K. Transport through neutral membranes. In: *Intermediate Physics for Medicine and Biology*, John Wiley-Sons, New York; 1987.

Janmey, P.A.; Chaponnier, C.; Lind, S.E.; Zaner, K.S.; Stossel, T.P.; Yin, H.L. Interactions of gelsolin and gelsolin-actin complexes with actin. *Biochem.* 24:3714-3723; 1985.

Janson, L.W.; Sellers, J.R.; Taylor, D.L. Actin-binding proteins regulate the work performed by myosin II motors on single actin filaments. *Cell Motil. Cytoskel.* 22:274-280; 1992.

Lassing, I.; Lindberg, U. Evidence that the phosphatidylinositol cycle is linked to cell motility. *Exp. Cell Res.* 174:1-15; 1988.

Lester, B.R.; McCarthy, J.B. Tumor cell adhesion to the extracellular matrix and signal transduction mechanisms implicated in tumor cell motility, invasion and metastasis. *Cancer Met. Rev.* 11:31-44; 1992.

Liotta, L.A.; Mandler, R.; Murano, G.; Katz, D.A.; Gordon, R.K.; Chiang, P.K.; Schiffmann, E. Tumor cell autocrine motility factor. *Proc. Natl. Acad. Sci.* 83:3302-3306; 1986.

Liotta, L.A. Cancer cell invasion and metastasis. *Scientific American* 266:54-63; 1992.

Nabi, I.R.; Watanabe, H.; Raz, A. Autocrine motility factor and its receptor: Role in cell locomotion and metastasis. *Cancer Met. Rev.* 11:5-20; 1992.

Oster, G.F.; Perelson, A.S. The physics of cell motility. *J. Cell Sci.* 8:35-54; 1987.

Perelson, A.S.; Coutsias, E.A. A moving boundary model of acrosomal elongation. *J. Math. Biol.* 23:361-378; 1986.

Pollard, T.P.; Mooseker, M.S. Direct measurement of actin polymerization rate constants by electron microscopy of actin filaments nucleated by isolated microvillus cores. *J. Cell Biol.* 88:654-659; 1981.

Savarese, D.M.F.; Russell, J.T.; Fatatis, A.; Liotta, L.A. Type IV collagen stimulates an increase in intracellular calcium: Potential role in tumor cell motility. *J. Biol. Chem.* 267:21928-21935; 1992.

Stossel, T.P. From signal to pseudopod: How cells control cytoplasmic actin assembly. *J. Biol. Chem.* 264:18261-18264; 1989.

Stossel, T.P. How cells crawl. *American Scientist* 78:408-423; 1990.

Stossel, T.P. On the crawling of animal cells. *Science* 260:1086-1094; 1993.

Stracke, M.L. Aznavoorian, S.; Beckner M.; Liotta, L.; Schiffmann, E. Cell motility, a principle requirement for metastasis. In: *Cell*

Motility Factors (Goldberg, I.D. Ed.), Birkhauser Verlag Basel/Switzerland, pp.147-162; 1991.
Theriot, J.A.; Mitchison, T.J. Actin microfilament dynamics in locomoting cells. *Nature* 352:126-131; 1991.
Titus, M.A.; Wessels, D.; Spudich, J.A.; Soll D. The unconventional myosin encoded by the *myo*A gene plays a role in *Dictyostelium* motility. *Mol. Biol. Cell* 4:233-246; 1993.
Usami, S.; Wang, S.L.; Skierczynski, B.A.; Skalak, R.; Chien, S. Locomotion forces generated by a polymorphonuclear leukocyte. *Biophys. J.* 63:1663-1666; 1992.
Wang, Y.L. Exchange of actin subunits at the leading edge of living fibroblasts: Possibly role of treadmilling. *J. Cell Biol.* 101:597-602; 1985
Wang, Y.L. Dynamics of the cytoskeleton in live cells. *Current Opinion in Cell Biology* 3:27-32; 1991.
Watts, R.G.; Crispens, M.A.; Howard, T.H. A quantitative study of the role of F-actin in producing neutrophil shape. *Cell Motil. Cytoskel.* 19:159-168; 1991.
Wenger, A. Head to tail polymerization of actin. *J. Mol. Biol.* 108:139-150; 1976.
You, J.; Dong, C. Analysis of pseudopod formation during tumor cell migration. Submitted; 1994.
Zhu, C.; Skalak, R. A continuum model of protrusion of pseudopod in leukocytes. *Biophys. J.* 54:1115-1137; 1988.
Zigmond, S.H. Chemotactic response of neutrophils. *Am. J. Respir Cell Mol. Biol.* 1:451-453; 1989.

29

Biomechanical Model for Skeletal Muscle Microcirculation with Reference to Red and White Blood Cell Perfusion and Autoregulation

S. Lee, D. Sutton, M. Fenster, G. W. Schmid-Schönbein

Introduction

While an expanding body of experimental observations on blood flow in skeletal muscle is accumulating (Granger et al., 1984), few efforts have been made to integrate these data into a unifying picture of the circulation based on microanatomy and properties of microvessel and blood. An analysis is useful since it unifies experimental observations and serves to interpret microvascular hemodynamics in terms of the properties of components that make up the microcirculation. A continuum analysis allows not only to explore the validity of numerous hypothesis but it also is a testing ground for our actual understanding of organ perfusion.

Our objective is to study the microcirculation in resting skeletal muscle under different physiological conditions. Many factors influence blood flow in microvessels and therefore we have found it useful to proceed in increasing steps of complexity, starting from simpler cases even though they may be of limited physiological significance. Complexities are introduced one at a time and wherever possible, predictions are tested by experiments.

We start with a description of the microvascular network anatomy and passive blood vessel properties and then formulate the equations of motion. Several different cases will be treated, perfusion with different blood cell suspensions, the myogenic control, and perfusion under pulsatile arterial pressure. We will discuss the 'critical closing pressure' observed in skeleltal muscle circulation. Finally, we will study pressure and flow distributions in different hierarchies of the microcirculation by means of a network analysis.

Vascular Microanatomy

There exists a deeply rooted impression that arteries or veins form trees, with a single root at the heart and progressively increasing number of bifurcations. While microvascular network reconstructions in the pulmonary circulation support this picture (Yen et al., 1984), in muscle even early investigations have contradicted it and have indicated the existence of arteriolar meshworks in addition to vascular trees (Saunders et al., 1957).

Reconstruction of the microvascular anatomy in rat skeletal muscle shows an interconnected arteriolar mesh, designated as *arcade arterioles* (Engelson et al., 1985), that spans the entire muscle. The arcade arterioles are supplied with blood from several central arteries (Schmid-Schönbein et al., 1986), two of which form a centrally positioned arcade that constitutes the largest artery in the muscle, the *arcade bridge* (Schmid-Schönbein et al., 1987). The connections from the arcade arterioles into the capillary network is provided by side branches to the arcades, the *transverse* or *terminal arterioles* (Engelson et al., 1986; Koller et al., 1987) (Figure 1). The transverse arterioles form asymmetric bifurcating trees with a single root at the arcade arterioles and multiple capillary endings. They vary in size based on position in the muscle and on rat strain and may have as few as 10 to 20 or as many as several hundred capillary endings. Several topological branching schemes have been proposed for quantitative description of these networks (Chen, 1983; Chen, 1984; Koller et al., 1987; Koller and Johnson, 1986; Popel et al., 1988; Woldenberg, 1986).

Arcade and transverse arterioles are positioned inside the skeletal muscle parenchyma and sandwiched between several layers of skeletal muscle fibers. Diffusion distances from the muscle fibers to the arteriolar smooth muscle are short. The majority of arcade arterioles and virtually all transverse arterioles have one smooth muscle layer in the adventitia, larger arcade arterioles may have a media with two to three layers. The smooth muscle layer tapers towards the capillary endings and frequently terminates at the entry to capillaries in form of a single coil which forms a neck at the entry into the capillaries. In arcade and transverse arterioles the elastica intima is reduced to a set of longitudinally oriented collagen fibers positioned between the endothelium and the smooth muscle cells (Schmid-Schönbein et al., 1990). The network of arterioles is innervated with adrenergic nerve fibers at the interface between the media and the adventitia. A remarkable feature of the innervation pattern in rat skeletal muscle is the high density of adrenergic fibers at the root of the transverse arterioles, at levels which are higher than in the more proximal arcade arterioles or in the more distal ramifications of the transverse arteriolar tree (Saltzman, 1993). At the same location, the arterioles exhibit the highest level of smooth muscle tone (Schmid-Schönbein et al., 1987) and they control the perfusion of the capillary network. Many arcade arterioles are accompanied by initial lymphatics positioned in the adventitia (Skalak et al., 1984).

The capillaries in muscle are aligned predominantly with the direction of the muscle fibers and form *bundles* with a *modular* structure (Skalak and Schmid-Schönbein, 1986a). Each capillary bundle consists of a repeating sequence of modules of parallel capillaries with capillary cross connections. The capillary endings of transverse arterioles and collecting venules connect into modules on

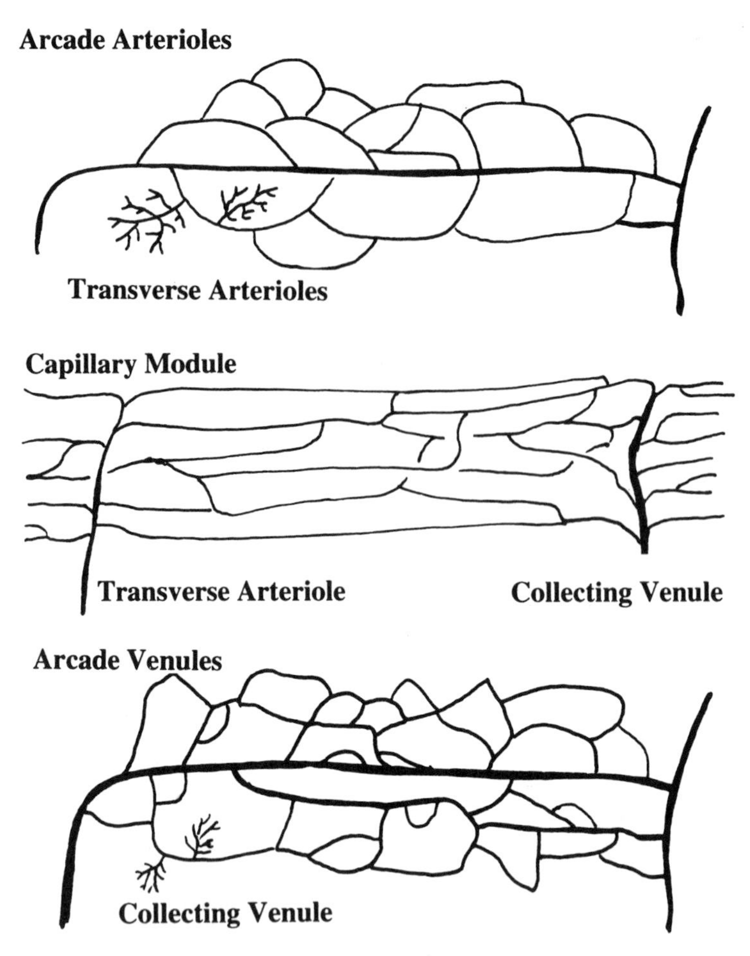

Figure 1. Schematic of arcade arterioles (top panel), capillary module (middle), and arcade venule (bottom) in skeletal muscle. Transverse arterioles and collecting venules (examples are shown in top and bottom panel, respectively) form regular side branches to the arcades reaching all regions of the tissue parenchyma.

either side, such that capillary bundles are supplied and drained by an alternating sequence of arterioles and venules (Figure 1). In spite of the fact that the capillaries inside a bundle are interconnected, each red cell that enters a transverse arteriole passes only the length of a single module along a capillary bundle and then leaves the capillary network via a *collecting venule*. A detailed quantitative network schema for skeletal muscle with measurements has been proposed by Skalak and Schmid-Schönbein (1986a). The capillaries consist of a single lining of endothelial cells with pericytes that share a basement membrane. The endothelial cells have the ability to undergo spontaneous shape changes by actin polymerization with projection of pseudopods into the lumen of the capillaries (Lee, 1990). Such cytoplasmic projections are stiff and have a pronounced influence on capillary perfusion.

The collecting venules form the counterpart to the transverse arterioles. Their branching pattern is qualitatively similar to that found in transverse arterioles. But individual venules (taken from branchpoint to branchpoint) tend to be shorter and have larger diameters (Engelson et al., 1985). Collecting venules drain into a network of *arcade venules* which is denser than its arteriolar counterpart. The density of the arcade venules is about 50% higher than the arcade arterioles. The combination of larger diameters with denser network pattern in collecting and arcade venules yields the characteristic low hemodynamic resistance on the venous side of the microcirculation (Zweifach et al., 1981). Larger arcade arterioles are paired with arcade venules while smaller arterioles and venules are unpaired (Schmid-Schönbein et al., 1987).

While the microvascular network in rat skeletal muscle has been reconstructed in detail, similar reconstructions in other species or in human muscle are incomplete at this time in spite of the fundamental importance of this information for understanding of organ perfusion in health and disease.

Flow Kinematics

The flow in the microcirculation at low Reynolds numbers is dominated by viscous forces and governed by the Stokes approximation of the equation of motion (Fung, 1984):

$$0 = \frac{\partial p}{\partial x_i} + \mu \frac{\partial^2 v_i}{\partial x_j^2} \, . \tag{1}$$

Since plasma is incompressible at physiological pressures

$$\frac{\partial v_i}{\partial x_i} = 0 \, . \tag{2}$$

v_i is the velocity, p is pressure, x_i are spatial coordinates and μ is the plasma viscosity. The Womersley parameter

$$\alpha = a\sqrt{\frac{\omega}{\nu}} \tag{3}$$

is significantly less than 1 at typical physiological heart rates with angular frequencies $\omega = 2\ \pi/T$, plasma kinematic viscosity $\nu = \mu/\rho$, and vessel radius a. The low Womersley parameter indicates that the velocity profile in the microcirculation is fully developed at any instant of time. In spite of this, as outlined below, significant time dependencies exist within the microvessel network due to distensibility of the blood vessels. Integration of the above equations over a circular vessel crossection of radius a with uniaxial velocity v along the axis z of the vessel gives Poiseuille's law

$$Q = -\frac{\pi a^4}{8 \mu} \frac{\partial p}{\partial z} \tag{4}$$

where $\partial p/\partial z$ is the pressure gradient along the length of the vessel, and

$$Q = \pi a^2 \bar{v} \tag{5}$$

is the volumetric flow rate. $\bar{v}$ is the mean velocity over the lumen crossection.

Since blood vessels are distensible, the radius a is a function of the stresses applied to the blood vessel wall. In skeletal muscle, these stresses are due to intravascular stresses and the stresses applied by the surrounding tissue parenchyma, including connective tissue and skeletal muscle fibers. In resting non-contracting muscle the dominating force responsible for deformation of the blood vessel is the intravascular pressure, p. In rat skeletal muscle the relationship between p and radius a is non-linear and time dependent, i. e. a(t), indicating viscoelastic wall properties. Direct measurements can be represented by a quasi-linear standard solid model of the form

$$p + \frac{\beta}{\alpha_1} \frac{\partial p}{\partial t} = \alpha_2 E + \beta \left(1 + \frac{\alpha_2}{\alpha_1}\right) \frac{\partial E}{\partial t} \tag{6}$$

where

$$E = \frac{1}{2} \left(\left(\frac{a}{a_o}\right)^2 - 1 \right) \tag{7}$$

is a non-linear measure of diameter, and a_o is a reference diameter. α_1, α_2, and β are empirical coefficients which have been determined for all hierarchies of microvessels in rat skeletal muscle (Skalak and Schmid-Schönbein, 1986b).

At steady state ($\partial(\ )/\partial t = 0$) the relation between pressure and vessel radius, according to equations 6 and 7, reduces to

$$p = \frac{\alpha_2}{2} \left(\left(\frac{a}{a_o}\right)^2 - 1 \right) \tag{8}$$

which after substitution into Poiseuille's equation (4) gives the basic pressure-flow relationship for a single microvessel in skeletal muscle

$$Q = \frac{\pi \alpha_2 a_o^4}{48 \mu l} \left(P_A^3 - P_V^3\right). \tag{9}$$

P_A and P_V are non-dimensional pressures at the upstream arterial end ($z = 0$) and at the downstream venous end of the vessel ($z = l$), respectively, normalized by the vessel elastic modulus

$$P = 1 + \frac{2p}{\alpha_2}. \tag{10}$$

If the flow is expressed in form of the upstream and downstream vessel radii a_A, a_V, respectively, then

$$Q = \frac{\pi \alpha_2}{48 \mu l a_o^3} \left(a_A^6 - a_V^6\right). \tag{11}$$

Equation (9) shows that flow of plasma in skeletal muscle microvessels without the effects of autoregulation depends on plasma viscosity μ and vessel distensibility α_2. The fact that the flow depends on the sixth power of upstream and downstream vessel radius is a consequence of the specific vessel distensibility (equation 8) encountered in skeletal muscle. In lung capillaries, which are more distensible than skeletal muscle capillaries, the flow is a function of the capillary sheet thickness h with distensibility α in the form of a 4th power relationship (Fung, 1984)

$$Q = \frac{1}{48 \mu l \alpha} \left(h_A^4 - h_V^4\right). \tag{12}$$

The consequence of the power pressure-flow relationships in skeletal muscle or in the lung is that the pressure-flow curve at low arterial pressures is nonlinear, while at higher pressures it becomes almost linear. Equation (9) was confirmed by perfusion of rat muscle with plasma of different viscosities by addition of high molecular weight dextran (Sutton and Schmid-Schönbein, 1991). Fixation of the microvessels with glutaraldehyde, which rigidifies the tissue and blood vessels, eliminates the curvature in the pressure-flow curve and converts it to a linear relationship, as predicted by Poiseuille's law (equation 4) (Sutton and Schmid-Schönbein, 1991).

Red Blood Cell Perfusion

In the presence of blood cells, additional factors start to play a role in microvascular perfusion, cell deformability, concentration, aggregation, and adhesion to the endothelium. In the case of red cells, the increase of hematocrit leads to a systematic right shift in the pressure-flow curve which is remarkably reproducible (Figure 2).

In the presence of red cells the fluid becomes non-Newtonian. A number of blood viscosity measurements in rigid glass tubes and in-vivo can be closely fitted over a range of velocities and hematocrits by a Casson's relationship

$$\sqrt{\mu^{app}} = k_1 + \frac{k_2}{\sqrt{\bar{v}/d}} \tag{13}$$

where k_1 and k_2 are empirical constants which depend on hematocrit (Schmid-Schönbein, 1988). The equation of motion (4) for distensible blood vessels with non-Newtonian blood of apparent viscosity μ^{app} assumes the form

$$\frac{dp}{dt} = - Q \frac{8}{\pi a^4} \left(k_1 + k_2 \left(\frac{2\pi a^3}{Q}\right)^{1/2}\right)^2 \tag{14}$$

with distensible radius a = a(p) given by

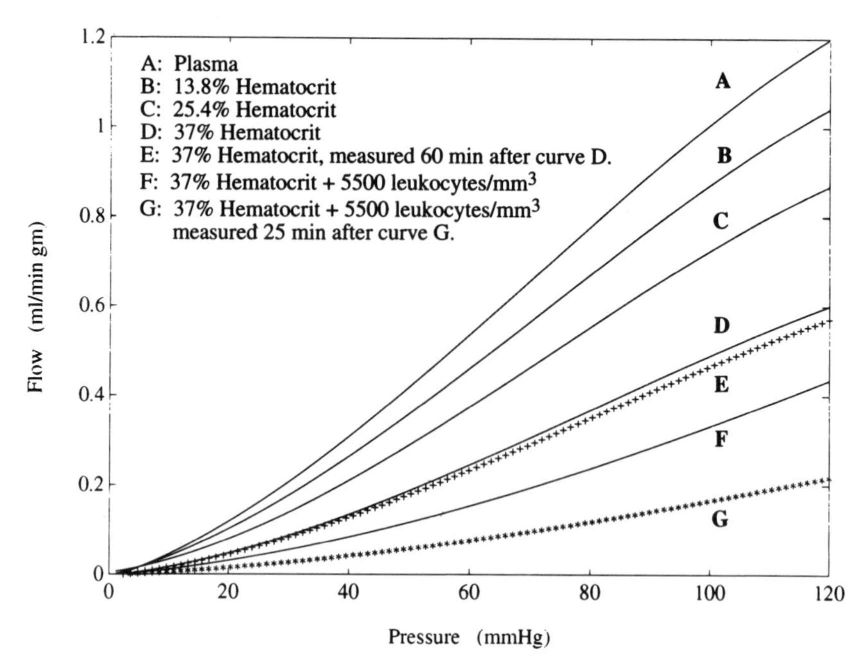

Figure 2. Experimental pressure-flow curves in rat gracilis muscle for pure red cell suspensions (top four curves) and mixed red and white cell suspensions (bottom two curves). The arterial feed pressure was maintained at 80 mmHg in the period between measurement of curves D and E, and curves F and G.

$$a = a_o \left(1 + \frac{2\,p}{\alpha_2}\right)^{1/2} \tag{15}$$

according to equation (8).

Current predictions by equation (14) with actual network anatomy and rheological models derived from measurements of apparent viscosity in straight glass tubes tends to underestimate the measured perfusion pressure-flow curves in spite of the fact that pressure-flow curves for plasma agree with the experiments (Fenster, 1992; Secomb et al., 1989). The discrepancy between predictions and measurements may in part be due to underestimation of the additional pressure drop for red cells (pressure drop in the presence of a red cell minus the pressure drop for plasma only) in capillaries and due to radial displacement of blood cells in larger vessels. During passage through microvascular bifurcations, blood cells are displaced into asymmetric positions (Nobis et al., 1984) where higher additional pressure drops are encountered than in corresponding symmetric positions. Furthermore, in the presence of red cell sedimentation the apparent viscosity may shift significantly due to phase separation (Reinke et al., 1987).

Leukocyte Perfusion

The discrepancy between current models of microvascular perfusion with blood cells and the actual experimental result in-vivo becomes more apparent in the presence of leukocytes. In spite of their relative small number and almost negligible volume fraction in whole blood, both bolus perfusions (Braide et al., 1984) or continuous perfusion of leukocytes (Sutton and Schmid-Schönbein, 1992) reveal a significant influence by leukocytes on whole organ resistance. In skeletal muscle with dilated arterioles, the additional pressure drop across the microvasculature due to the presence of the blood cells tends to be about 750 times higher for a single leukocyte than for a single red cell. Consequently, the relative few leukocytes in the buffy coat of a blood sample (<1%) may have almost the same effect on the pressure drop than all of the approximately 40% red cells. When white blood cells become activated and polymerize their cytoplasmic actin while projecting relative rigid pseudopods or adhering to the endothelium, the additional pressure drop for leukocytes will rise to even higher values. At the same time the pressure-flow curves become non-reproducible and exhibit a shift in time towards higher resistances, a process that is essentially irreversible (Sutton and Schmid-Schönbein, 1992). The relative high additional pressure drops encountered in experiments exceed current theoretical estimates (Warnke and Skalak, 1990) and may be associated with hydrodynamic interaction of white and red cells in narrow capillaries (Thompson et al., 1989). Further exploration of the rheological effect by the circulating leukocytes may have important influence on understanding of pathophysiological states where significant activation of circulating leukocytes may be encountered.

Under conditions of low flow, the relative large and stiff leukocytes may become entrapped in skeletal muscle capillaries (Bagge, 1980). During reperfusion this leads to actual obstruction of microvascular pathways, e.g. a capillary no-reflow phenomenon (Kurthius et al., 1988). Immunoprotection of leukocyte membrane integrins (CD18) with monoclonal antibodies serves to abolish the capillary no-reflow, suggesting that the membrane adhesion of circulating leukocytes to microvascular endothelium may have an important influence on perfusion and organ pathology (Jerome et al., 1993). Leukocytes may also become entrapped in capillaries and postcapillary venules after activation in the central circulation even in the presence of normal perfusion pressures (Schmid-Schönbein, 1987).

Myogenic Arteriolar Contraction

Among the local mechanisms determining the contraction of the arterioles - metabolic, flow and endothelium dependent smooth muscle contraction - the pressure dependent mechanism is the most fundamental (Kuo et al, 1992; Johnson, 1991). The pressure dependent contraction of the arterioles is an inherent feature of arteriolar smooth muscle (Falcone et al., 1991) which can be enhanced in the case of neurogenic mechanisms or attenuated in the case of endothelium derived dilators (Koller et al., 1991; Kuo et al., 1990). Numerous experimental observations have been reported (Bohlen and Harper, 1984; Davis and Sikes, 1990; Jackson and Duling, 1989; Johnson, 1980; Kuo et al., 1988). In addition to the exploration of the cellular and molecular contractile mechanisms and their control, there is a need to develop biomechanical models for analysis of blood flow in the presence of autoregulation.

Most mathematical models of the myogenic response provide qualitative descriptions for steady state situations (see Koch, 1964). The formulation of the vascular smooth muscle properties were largely based on Hill's model and its modifications (Johnson, 1980). It consists of a contractile element, and a series and parallel elastic element. The difficulty with Hill's model and its modifications is that the division of stress between the parallel and contractile elements (τ_P and τ_S, respectively) has no clearly identifiable structural counterpart and the division of extensions between the contractile and the series elements are arbitrary (Fung, 1993). Representation of the resultant stress in muscle as

$$\tau(t) = \tau_P(t) + \tau_s(t) \qquad (16)$$

presumes a time sequence in which passive and active stress occur simultaneously. There exists, however, no experiment on muscle tissue in which τ_P and τ_S are determined simultaneously, since introduction of a passive state requires cessation of normal muscle contraction, be it by interruption of the normal stimulation or by application of a muscle relaxant. In a strict sense, the time specification in equation (16) is not correct and therefore alternatives to Hill's model are required. Borgström et al. (1982) described a time dependent myogenic response in form of segmental resistance during an ambient pressure change. Braakman et al. (1989) and Ursino and Fabbri (1992) formulated lumped element electrical analog models. Iida (1989) proposed a model with pressure dependent vessel stiffness.

One of the fundamental issues in developing a basic concept in smooth muscle (or other muscle) mechanics is the choice of a *reference state* in order to define strain. Usually the zero-stress configuration serves as the reference state and it is presumed to be constant. But it has been well documented that blood vessels have different zero-stress states and muscle may have a range of unstretched lengths, depending on the contractile state. A shift in the reference state is a requirement that needs to be accounted for in any model of muscle mechanics. Apter and Graessley (1970) have considered the variation of initial length in a model of muscle mechanics.

In the following we will describe an alternative approach for active contractile muscle that accounts for both of these concerns about the definition of stress and strain (Lee, 1993). The model is based on two assumptions.

1) The arterioles maintain viscoelastic properties during active myogenic vasoconstriction qualitatively similar to those seen in passive non-contracting arterioles.

2) The active myogenic vasoconstriction or dilation is the result of a change in the *reference radius* of the arteriole which depends on the pressure in the arteriole.

Let a be the instantaneous lumen radius of a pressurized arteriole, and a_{Ref} the reference diameter. Local details of lumen shape changes, such as protrusion of endothelial cells (Schmid-Schönbein and Murakami, 1985) are neglected and we consider only the resultant stress in the arteriolar wall due to the transmural pressure. The average circumferential wall stretch ratio is defined as

$$\lambda(t) = \frac{a(t)}{a_{Ref}} . \tag{17}$$

The viscoelastic behavior of microvessels can be expressed in the form of a hereditary integral (Fung, 1993)

$$p(t) = p^{(e)}[\lambda(t)] + \int_0^t p^{(e)}[\lambda(t-\tau)] \frac{dG_d(\tau)}{d\tau} d\tau \tag{18}$$

where $G_d(t)$ is called the *reduced relaxation function* and $p^{(e)}(\lambda)$ the *elastic response*. The stretch ratio λ is a function of time in general.

When smooth muscle is distended by application of a vasodilator, $a_o = a_{Ref}$, the passive reference state at zero pressure. Passive viscoelasticity of microvessels in rat skeletal muscle microcirculation (equations 6 and 7) can be summarized in form of the reduced relaxation function for a quasi linear standard solid as

$$G_d(t) = \frac{1}{\alpha_2} [1 - (1 - \frac{\tau_\varepsilon}{\tau_\sigma}) e^{-t/\tau_\sigma}] H(t) \tag{19}$$

with the elastic response

$$p^{(e)}(\lambda) = \frac{\alpha_2}{2} [\lambda^2 - 1] . \tag{20}$$

$H(t)$ is the unit step function, and $\tau_\varepsilon = \beta/\alpha_1$ and $\tau_\sigma = \beta(\alpha_1+\alpha_2)/\alpha_1\alpha_2$ are relaxation times for the parallel Maxwell fluid and standard solid model.

If the elastic response is exponential, as suggested by Fung (1993), equation (20) may be replaced by

$$p^{(e)}(\lambda) = \gamma [e^{\delta(\lambda - 1)} - 1] \tag{21}$$

where γ and δ are elastic constants.

The discussion above refers to the passive response. When arterioles contract in the *active* state, the reference radius a_{Ref} will change due to smooth muscle shortening in circumferential direction and therefore in general $a_{Ref} \leq a_o$. According to assumption 1) above, the arteriolar wall remains viscoelastic during contraction and equations (17) to (21) are still valid under active smooth muscle contraction. It is possible that the viscoelastic coefficients α_1, α_2, and β vary with muscle shortening, but the general form (equation 18) remains the same. This assumption needs to be tested by experiments and if necessary may be relaxed to accomodate more general viscoelastic properties in the activated state. According to assumption 2), the definition of the stretch ratio, λ, needs to account for muscle shortening. During contraction, a_o in the passive state needs to be replaced by $a_{Ref}(t)$ so that the stretch ratio in equation (17) becomes

$$\lambda(t) = \frac{a(t)}{a_{Ref}(t)} \tag{22}$$

which after expansion by means of the reference state in the passive distended state, a_o, assumes the form

$$\lambda(t) = \lambda_p(t)\, \lambda_a(t) \tag{23}$$

with

$$\lambda_p(t) = \frac{a(t)}{a_o} \quad \text{and} \quad \lambda_a(t) = \frac{a_o}{a_{Ref}(t)}. \tag{24 a,b}$$

$\lambda_a(t)$ is the *active stretch ratio*. During isometric contractions, when the overall length of the arteriolar smooth muscle is constant, stress is generated by shortening of the reference state $a_{Ref}(t)$ so that $\lambda_a(t)$ increases while λ_p stays constant. The *apparent stretch ratio*, λ_p, can be regarded as a dimensionless measure of active vessel diameter.

In the case of myogenic contraction, the reference radius a_{Ref} depends on the transmural pressure. Specifically, as first approximation, λ_a is assumed to be a linear function of the transmural pressure according to:

$$\lambda_a = \frac{p + a_m}{b_m} \tag{25}$$

where a_m and b_m are two empirical contractile constants. Comparison with currently available experimental results suggests that this linear form may be sufficient (see below). As more experimental details become available, equation (25) may be replaced by a non-linear approximation. If other than myogenic control mechanisms are acting on the arteriolar smooth muscle, equation (25) needs to be expanded and other variables introduced, such as shear stress acting on the endothelium, stresses exerted on smooth muscle cells by collagen fibers in the adventitia, concentration of metabolic mediators, or molecular details of the smooth muscle contractile machinery. The smooth muscle contraction postulated by a change of reference length requires energy.

When arteriolar pressures varies with time, vascular smooth muscle does not instantly reach full response due to internal signal transmission. Instead, a delay between applied pressure and change of the reference diameter occurs. Then

$$\lambda_a(t) = \frac{1}{b_m}\left(\int_0^t G(t-\tau)\frac{dp}{d\tau}\, d\tau + a_m\right) \tag{26}$$

where G(t) is a sigmoidal delay function

$$G(t) = \frac{(t/t_D)^n}{1 + (t/t_D)^n} \tag{27}$$

in which t_D is a time constant measuring the length of the delay. The integer n is an exponent that serves to determine the average rate of signal transmission between pressure and change in reference diameter. The delay function G(t) represents a gradual shift of the reference diameter after a step pressures.

The assumption underlying the change of reference radius as a function of the transmural pressure under steady state (equation 25) can be tested in the following way. For a passive vessel $\lambda_a = 1$. Therefore from equation (24)

$$\lambda = \lambda_{dilated} = \frac{a_{passive}(p)}{a_o} \tag{28}$$

where $a_{passive}$ is the diameter in the passive dilated state. According to assumption 1) the mechanical properties of the blood vessel remain unchanged in first approximation after contraction, so that λ will maintain the same relation with respect to pressure in both passive elastic and active myogenic state. Consequently, when the arteriole is under active myogenic contraction with radius a_{active}, λ_a can be obtained from equation (23) in the form

$$\lambda_a = \frac{\lambda_{dilated}}{\lambda_p} = \frac{a_{passive}/a_o}{a_{active}/a_o} = \frac{a_{passive}}{a_{active}} . \tag{29}$$

Equation (29) states that λ_a may be obtained by measuring the passive and active myogenic pressure-diameter relationships from the same vessel. If the hypothesis in equation (25) is true, then the ratio of passive diameter and active diameter will be a linear function of pressure. Figure 3 shows a graphical display of this hypothesis from the data of Kuo et al. (1990). The linear relationship can be observed with many other experimental data in the literature (Lee, 1993).

Under steady state pressures, the active myogenic response for a quadratic elastic response (equation 20) is

$$\lambda_p = \frac{b_m \sqrt{(1 + 2\, p/\alpha_2)}}{p + a_m} . \tag{30}$$

When an exponential elastic response (equation 21) is applied, the steady state pressure-diameter relation assumes the form

$$\lambda_p = \frac{b_m\left[1 + \frac{1}{\delta}\ln\left(1 + \frac{p}{\gamma}\right)\right]}{p + a_m} . \tag{31}$$

Figure 4 shows experimental results from Kuo et al. (1990) on single isolated arterioles and the theoretical approximations. At low pressures the exponential elastic response fits the experiments better than the quadratic or a linearized

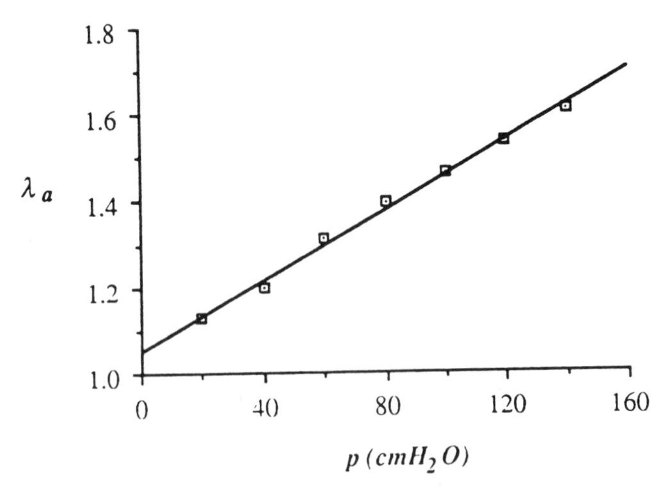

Figure 3. The stretch ratio λ_a, according to equation (29), as a function of the vascular pressure p. The linear regression is $\lambda_a = 1.05 + .004$ p.

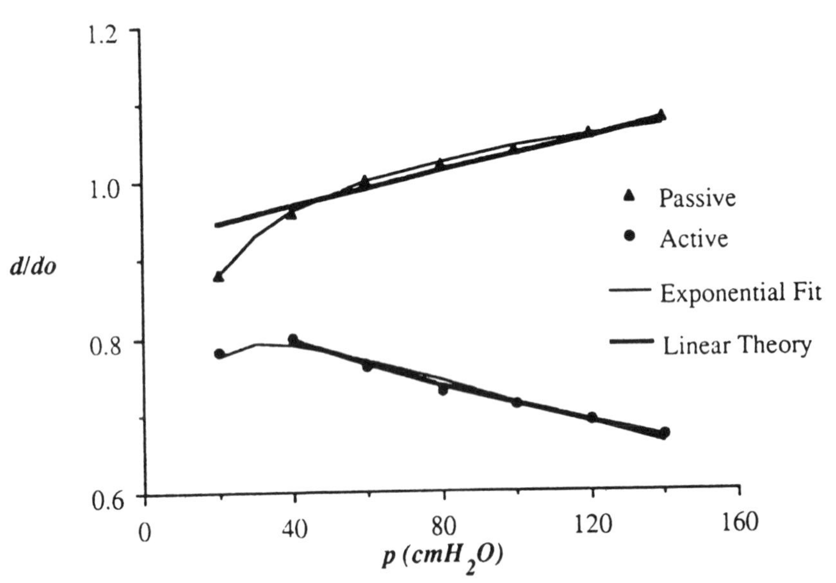

Figure 4. Diameter ratio $\lambda_p = d/d_o$ for passive distensible and active contractile isolated arteriole. Experimental results for coronary arterioles by Kuo et al. (1990). Linear theory according to equation (32) and exponential theory according to equation (31). $\alpha_2 = 857$ cmH_2O, A = - 0.868, B = 0.274 for linear elastic response and $\delta = 11.6$, $\gamma = 10.4$ cmH_2O, $a_m = 258$ cmH_2O, and $b_m = 251$ cmH_2O for exponential elastic response.

response (see below), at higher pressures all responses agree with measurements.

This observation lead us to reinvestigate the pressure-radius relationship at low pressures. In many in-vitro experiments, the pressure-radius curves show an increase with p in the low pressure range ("passive" behavior) and reach a maximum value ("threshold" for the myogenic response), in order to decrease with increased pressure (myogenic phase). This phenomenon can be simulated by equation (31) in which an exponential elastic response (21) was assumed. The quadratic form of the elastic response (equation 30) can closely approximate the experimental data at pressures above about 40 mmHg. At lower pressures the quadratic form overestimates the measured diameters. Closer inspection of the underlying cause indicates that measurements of vessel diameter in the presence of an immissible silicone polymer (in order to minimize fluid filtration during vessel distension) (Skalak and Schmid-Schönbein, 1986b) may interfere with shape changes of the endothelial cells at low distending pressures due to the presence of a small surface tension. Distension of microvessels at low transmural pressures with plasma shows distinct shape changes of the capillary lumen at the ultrastructural level (Lee, 1990) with cytoplasmic projections into the lumen. Such projections are less developed in the presence of a polymer. Resolution of this issue requires introduction of detailed mechanical models of endothelial cells in the future with three-dimensional stresses and stains.

During a transient pressure change the myogenic response becomes time dependent. Equation (25) has to be replaced by equation (26) with delay function G(t) (equation 27)). For a step from the initial pressure P_o, the pressure history can be expressed by means of the Heavyside unit step function H(t) as $p(t) = P_o + \Delta P\ H(t)$. For small pressure steps ΔP the elastic response $p^{(e)}$ (equation 20) can be replaced in first approximation by its linearized form

$$p^{(e)} = \alpha_2 (\lambda - 1) \ . \tag{32}$$

Differentiation of equations (18) and (19), and introduction of equation (32) togther with non-dimensional forms gives

$$P + C_1 \frac{dP}{dt'} = \lambda + \frac{d\lambda}{dt'}. \tag{33}$$

with

$$P = 1 + \frac{p}{\alpha_2},\quad t' = \frac{t}{\tau_\sigma},\ \text{and}\ C_1 = \frac{\tau_\varepsilon}{\tau_\sigma}. \tag{34 a,b,c}$$

From equation (26) and (27) we obtain

$$\lambda_a(t') = \frac{1}{B}\left(\int_0^{t'} G(t' - \tau)\frac{dP}{dt}\, d\tau + A \right) \tag{35}$$

with

$$A = \frac{a_m}{\alpha_2} - 1\ ,\ B = \frac{b_m}{\alpha_2} \tag{36a,b}$$

and the delay function

$$G(t') = \frac{(t'/t'_D)^n}{1 + (t'/t'_D)^n} \tag{37}$$

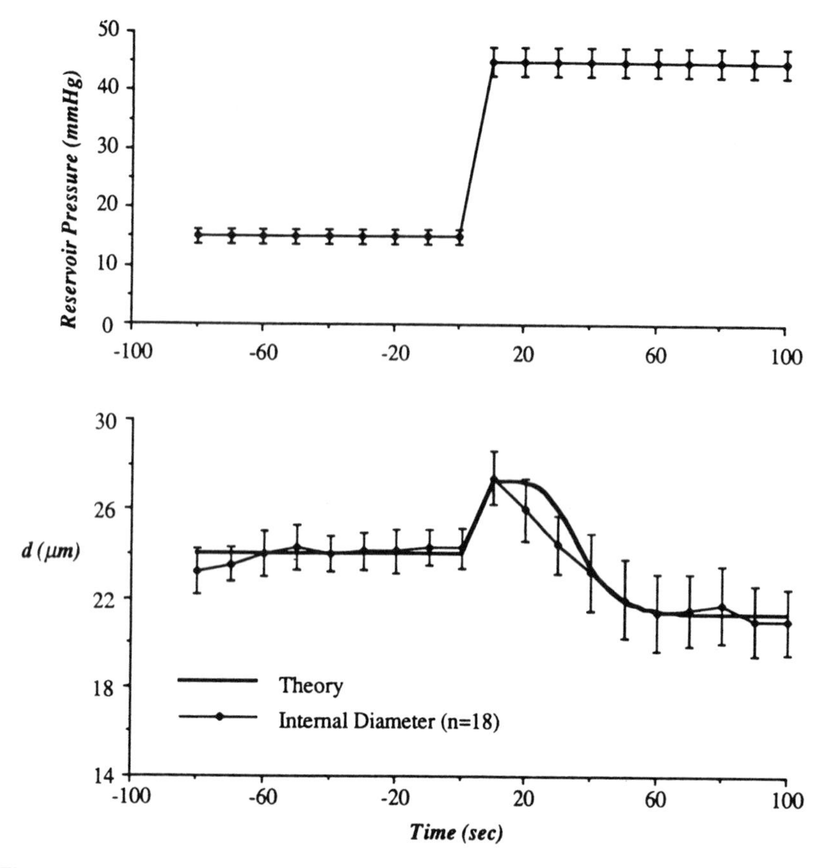

Figure 5. Experimental results (Johnson and Intaglietta, 1976) and theoretical approximation (equation 38) of a transient myogenic response in-vivo. $\alpha_1 = 35$ mmHg, $\alpha_2 = 173$ mmHg, $\beta = 2940$mmHg sec, A = - 0.498, B = 0.542, $\tau_D = 33$ sec, and n=6.

Integration of equation (33) and (35) for the step pressure history and substitution into the definition of active stretch ratio (equation 23) yields the normalized radius

$$\lambda_p(t') = B \frac{P_o + \Delta P\left[1 + (C_1 - 1)\, e^{-t'}\right]}{P_o + \Delta P\left[\dfrac{(t'/t'_D)^n}{1 + (t'/t'_D)^n}\right] + A}. \tag{38}$$

Figure 5 shows λ_p in equation (38) togther with in-vivo observations by Johnson and Intaglietta (1976). The multiphase response of an arteriole to a step pressure can be interpreted by the current model in terms of viscoelastic and contractile deformation. At the time of the pressure step, the arteriolar smooth muscle cells are passively stretched but cannot instantly reach a fully contractile

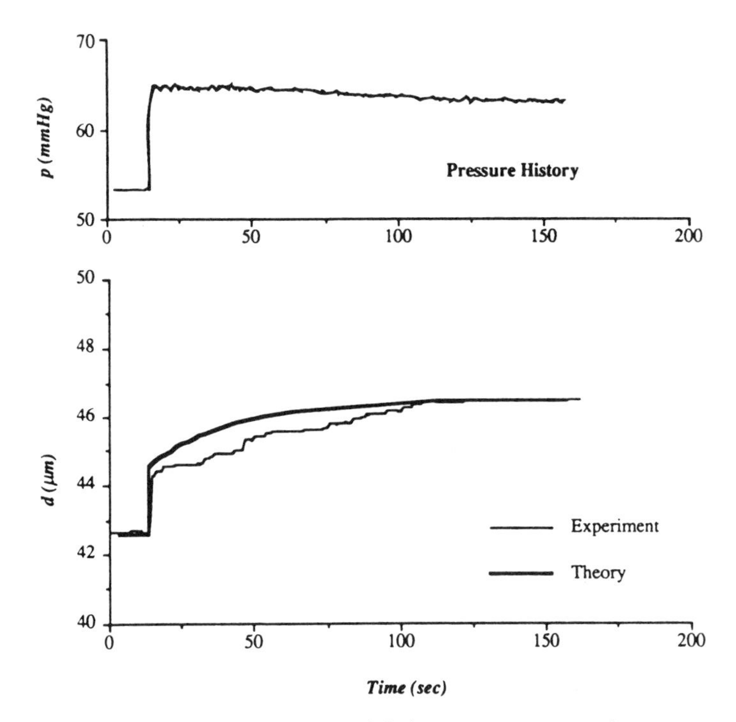

Figure 6a. The passive response of an arteriole in rat cremaster muscle to a step pressure history. Theory according to equation 6 with α_1 = 19 mmHg, α_2 = 173 mmHg, and β = 510 mmHg sec.

state due to delay in intracellular signal transmission. As shortening of reference length begins to dominate the vessel wall strain, vasoconstriction occurs. Viscoelastic creep continues at the same time, although it is overshadowed by the active contraction. Eventually the shift of the reference diameter is complete and muscle constriction stops. If the viscoelastic creep is not completed, the vessel will continue to show creep under the new reference state. The result is the phenomenon of secondary relaxation.

Figure 6 shows application of the current theory to experimental results in rat cremaster muscle with the *tissue in the box* technique which allows rapid changes in extravascular pressures on exteriorized muscle preparations (Lee, 1993). The results in Figure 6a,b,c serve to determine the viscoelastic coefficients α_1, α_2, and β, the myogenic constants A and B, and the delay constant t'_D, respectively. The results in Figure 6d represents an independent prediction of the experimental observations for the same arteriole with the same coefficients but a ramp pressure history.

The myogenic response to other pressure histories applied to the microcirculation can also be described by the current theory (Lee, 1993). In fact the model serves to predict arteriolar contractions for several types of pressure histories after determination of the empirical coefficients from steady state response and a step pressure. If, however, the shift in perfusion pressure is accompanied by a significant change in blood velocity, we detected that the myogenic model tends to overestimates the measured arteriolar radius. This

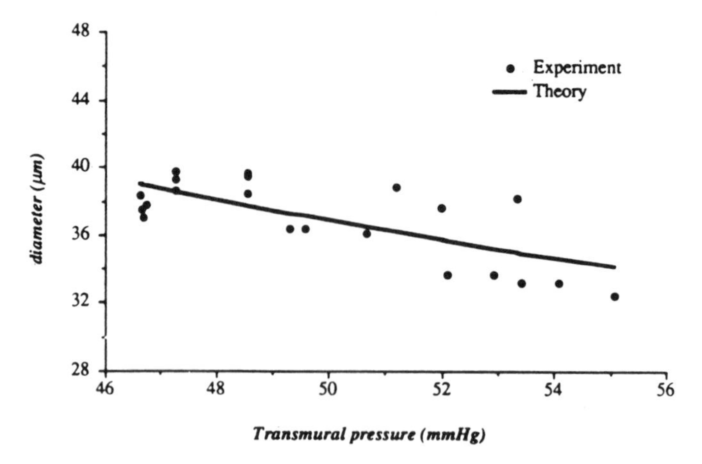

Figure 6b. Steady state myogenic response according to equation (38). A least square fit gives A =-1.01, B = 0.24 and α_1 , α_2, β remain the same as in Figure 6a. .

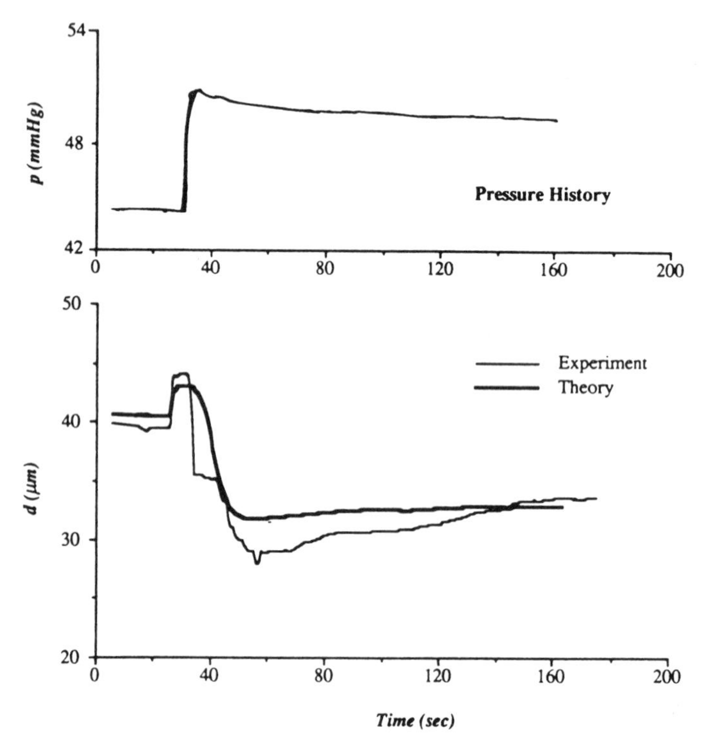

Figure 6c. Myogenic response to a step pressure (equation 32 -37). $t'_D = 0.52$ and n =6 in the delay function G(t').

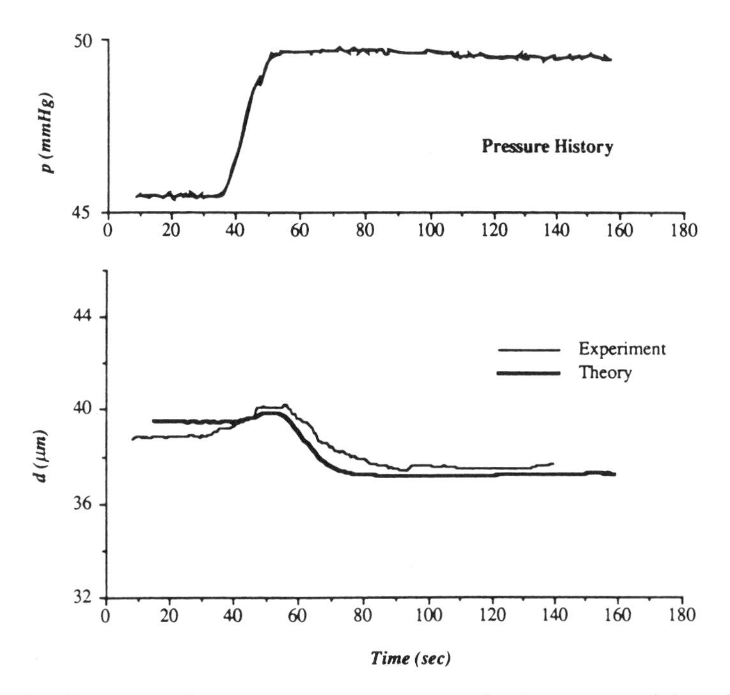

Figure 6d. Experimental response to a ramp pressure for the same arteriole and same theoretical coefficients as in Figure 6a - c.

observation supports the hypothesis, that flow dependent vasodilation may influence the arteriolar state of contraction by an endothelium and shear stress dependent mechanism (Koller and Kaley, 1990; Koller et al., 1991; Kuo et al., 1990; Kuo et al., 1992).

Zero-Flow Pressure

A long-standing issue in circulatory physiology concerns the pressure-flow relationship at low flow near zero pressures. Burton (1951) was one of the pioneers to specifically address a phenomenon in which arterial pressure may have a finite value although arterial flow is zero. He coined the term *critical closing pressure* or more generally *zero-flow pressure*. Several different mechanisms have been proposed that may be involved in the zero-flow pressure, such as a yield stress in blood, surface tension at the vascular interface, or arteriolar contraction (for summary see Sutton and Schmid-Schönbein, 1989).

In the analysis above for single vessels, there is only one mechanism that causes a finite arteriolar pressure at zero flow, i.e. by elevation of the venous pressure (see e.g. equation 9). But zero-flow pressure in-vivo is encountered without venous pressure elevation, so we need to search for other mechanisms.

In the in-situ perfused rat gracilis muscle at steady state perfusion, we observe regularly a finite zero-flow pressure with values near 10 and 15 mmHg, while the venous pressure is maintained close to zero. During unsteady perfusion with oscillatory arterial pressure, the instantaneous zero-flow pressure rises to significantly higher values. Thus, there may be different mechanisms at steady and unsteady pressure perfusion. During steady perfusion pressures the finite zero-flow pressure has about the same value with plasma or whole blood cell perfusion suspension. It is present before and after dilation of the arterioles. Under these circumstances, the zero-flow pressure is due to a "hidden" arteriolar anastomosis between one of the main feeder arterioles and arterioles in the underlying skeletal muscle. The anastomosis is indeed hidden since only surgical removal of the arterioles reveals them. Reduction of the central blood pressure leads to instant reduction of zero-flow pressure to values less than 1 mmHg, accompanied by a shift of the pressure-flow curve towards the origin. In addition, if the arterial pressure is reduced to zero, i.e. below the zero-flow pressure, the flow in the feeder artery is actually reversed, yielding an outflow which clearly indicates the presence of an anastomosis. Theoretical network computations with anastomoses at different levels of the microvascular network show that in order to achieve a significant elevation of the zero-flow perfusion pressure, the vascular communication between the skeletal microcirculation must be located at the level of the arcade arterioles. Vascular communications from neighboring muscles into the capillary network or into the venules lead to negligible elevation of the zero-flow pressure in the feeder arteries of the gracilis muscle due to the inherent low venous resistance in the microcirculation (Schmid-Schönbein et al., 1989). This form of zero-flow pressure is similar to collateral pressures between different arterial inflows in other organs with interconnected arteriolar networks, such as in the canine coronary circulation or the hindlimb circulation.

Perfusion of rat skeletal muscle with aggregated red cells suspensions may lead to another form of zero-flow pressure. Under these circumstances the zero-flow pressure may reach higher values than the collateral pressure due to a yield stress (Cokelet et al., 1963). The pressure-flow curve at zero pressure does, however, not assume negative flow values, as in the case of the anastomosis, and instead stays near zero flows when the arterial pressure falls below the zero-flow pressure. In vasodilated rat gracilis muscle, the zero-flow pressure due to red cell aggregation is on average about 9 mmHg at 25-33% hematocrit with 3% 77KDa dextran (Sutton and Schmid-Schönbein, 1989).

Under oscillatory arterial pressures, the zero-flow pressure can assume significantly higher values than in the steady cases. The transient zero-flow pressure will shift in time and its magnitude depends on the frequency of pressure pulsations. It is due to dynamic viscous flow in distensible blood vessels under pulsatile blood pressure, which we discuss next.

Microcirculation under Pulsatile Arterial Pressure

In spite of the fact that inertia forces in the microcirculation are negligible, significant time dependent effects can be observed. They are due to arterial pressure pulsation and are independent of autoregulatory mechanisms. While the elastic stresses in the vessel walls and the viscous stress in the blood are in equilibrium under pulsatile pressure their interaction leads to time dependent flows. Hemodynamic parameters on the arterial side are not in synchrony with the venous side. Local flows are not only determined by input and output flow rates but also by local volume changes of blood vessels.

The axial velocity component dominates in skeletal microvessels since local radial distension is small. The equation of motion for a microvessel with circular crossection is given by Poiseuille's formula (equation 4). For a microvessel with impermeable wall and lumen crossection $A = \pi a^2$, the flow gradient along the length of the vessel is

$$\frac{\partial Q}{\partial z} = -\frac{\partial A}{\partial t} \tag{39}$$

where now Q(z, t) and A(z, t). The diameter strain E (equation 7) is

$$E = \frac{1}{2}\left(\frac{A}{A_o} - 1\right) \tag{40}$$

with $A_o = \pi a_o^2$ as reference crossection. The short term response for an elastic distension of a viscoelastic microvessel according to equation (6) is (Skalak and Schmid-Schönbein, 1986b)

$$P(z, t) = \frac{\alpha}{2}\left(\frac{A(z, t)}{A_o} - 1\right) \tag{41}$$

where $\alpha = \alpha_1 + \alpha_2$. Combining equations (39) and (40) yields

$$\frac{\partial Q}{\partial z} = -\frac{2\,A_o}{\alpha}\frac{\partial p}{\partial t} \tag{42}$$

and noting that Poiseuile's formula in terms of the lumen crossectional area is

$$Q = -\frac{A_o^2\,(2\,E + 1)^2}{8\,\mu\,l}\frac{\partial p}{\partial z} \tag{43}$$

we obtain the equation of motion in a microvessel under pulsatile pressure

$$\frac{\partial P}{\partial z} = C^2\,\frac{\partial^2 P^3}{\partial z^2} \tag{44}$$

for the normalized pressure $P(z, t) = ((2\ p/\alpha) + 1)$ and $C^2 = A_o\,\alpha/ 48\,\pi\,\mu$. Thus under pulsatile pressure, the flow in the microcirculation is determined by a differential equation which has the form of a nonlinear diffusion equation. Numerical solutions of equation (44) show, that after a step arterial pressure the arterial inflow exhibits an overshoot followed by a gradual return to steady flow, in line with experiments (Schmid-Schönbein et al., 1989). The flow overshoot is due to the fact that the blood volume pushed initially into the arteriolar side

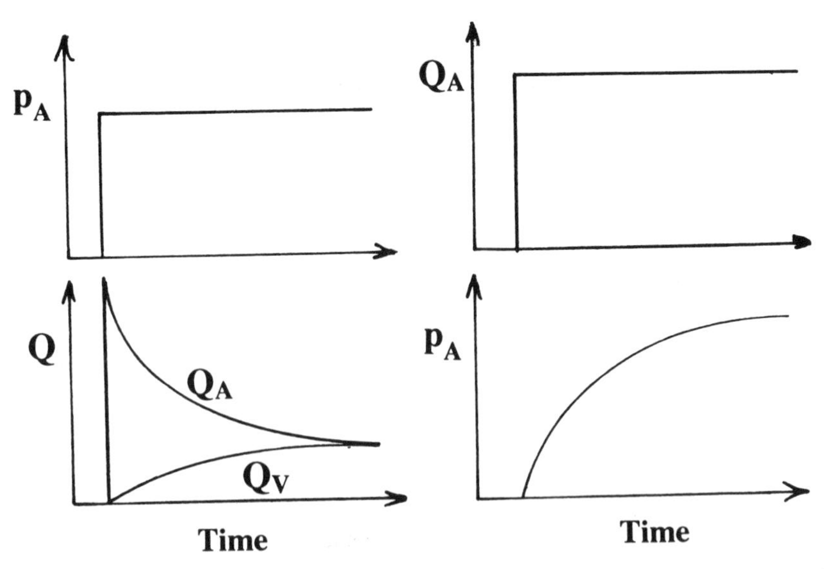

Figure 7. Schematic of dynamic viscous flow in the microcirculation for a step arterial pressure (left) and step arterial flow (right). During a step pressure the flow at the arterial inflow Q_A exhibits an overshoot due to expansion of the arterioles, while venous outflow Q_V increases gradually to steady state. After a step arterial flow, the arterial pressure, p_A rises gradually to steady state state value.

serves to distend the arterioles while in more distal vessels flow starts only after distension of the arterioles and rise of the pressure gradient along the microvascular network. The venular flow exhibits no overshoot and gradually increases to its steady state value at which it equals the arterial inflow. If in contrast a step arteriolar flow is applied, the arteriolar pressure rises gradually without overshoot towards steady state (Figure 7). If the arteriolar pressure is reduced in a step to zero pressure while venous pressure is maintained close to zero, flow *out* of the muscle is observed at both the arteriolar and venular side of the microcirculation. The magnitude of the outflow depends on distensibility of the microvessels, α, and the viscosity of the blood. In rat gracilis muscle, the transit time to reach less than 5% of the initial flow rate is about 20 sec in the presence of plasma and rises to about 40 sec at 27% hematocrit. The times required to reach steady state will increase with the size of the muscle and may be significantly longer in man. These results indicate that the microcirculation under physiological pressures is not at steady state so that equation (44) serves as the governing equation for microvascular flow. In the presence of oscillatory arteriolar pressures, the pressure-flow curves exhibit hysteresis. Different pressure-flow curves can be observed depending on the magnitude of the applied pressure or flow amplitudes. If, for example, the arteriolar inflow at constant venous pressure is oscillated between a physiological flow amplitude and zero flow, the hysteresis loop shifts towards higher arteriolar pressures as frequency of oscillation is increased. The arteriolar pressure at zero flow increases linearly with frequency (Sutton and Schmid- Schönbein, 1989). Consequently zero-flow pressures can be encountered which exceed the values under steady state

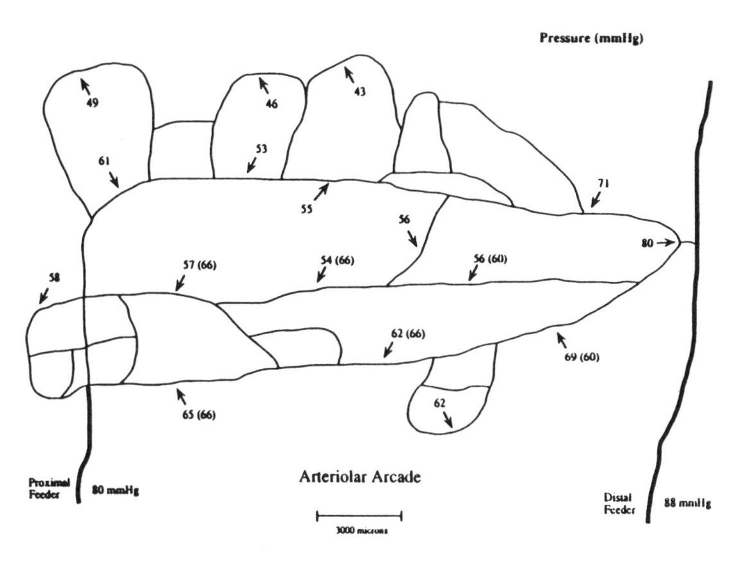

Figure 8. Pressure distribution in the arcade arterioles of rat gracilis muscle.

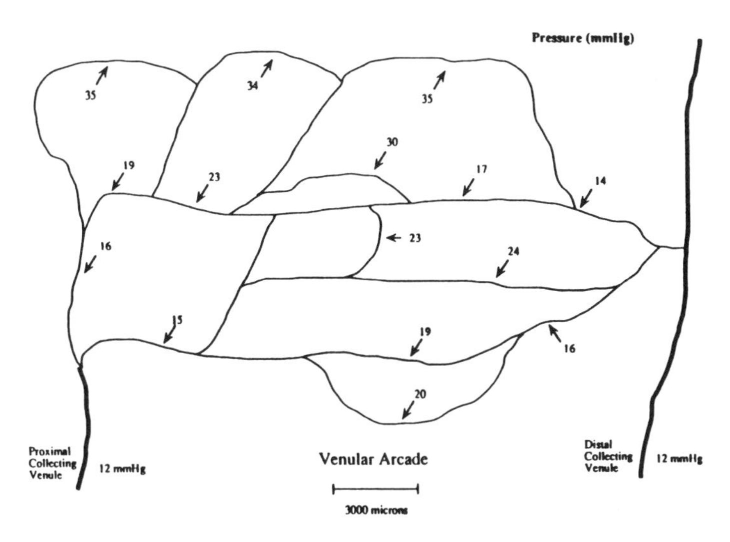

Figure 9. Pressure distribution in the arcade venules of rat gracilis muscle.

perfusion, discussed in the previous section.

There is one case of special interest. The hysteresis loop during pulsatile perfusion at the frequency of the heart rate oscillates clockwise. At lower frequencies the hysteresis loop turns counter-clockwise due to the viscoelastic vessel properties (Schmid-Schönbein, 1988). Therefore, at intermediate frequencies the artery pressure-flow curve undergoes a transition from a counter-clockwise hysteresis to a clockwise hysteresis. At the transition frequencies the pressure-flow curve forms a figure eight (Lee and Schmid-Schönbein, 1990).

Pressure (mmHg)

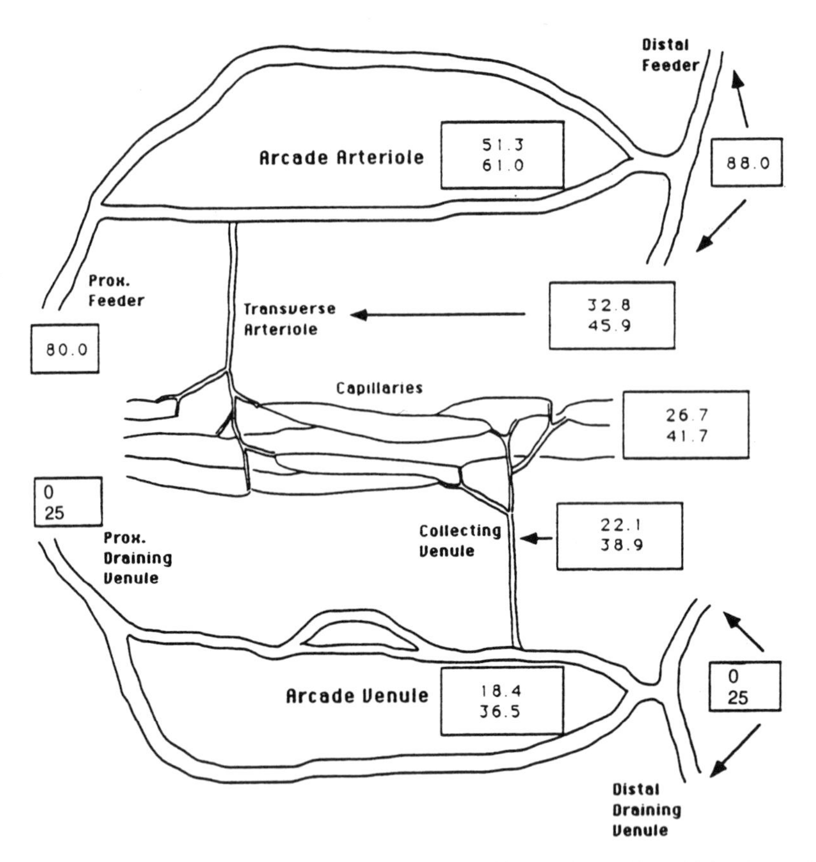

Figure 10. The average microvascular pressures in different hierarchies of blood vessels in rat skeletal microcirculation during steady perfusion with 45% hematocrit suspension (no white blood cells) under myogenic control. Pressures for two different venous outflow pressures are shown.

Microvascular Distribution of Pressure and Flow

With a reconstruction of the microvascular network, we are in a position to integrate the equations of motion over all vessels in the network. In skeletal muscle, previous network analyses have been presented for a capillary module in the cat sartorius muscle (Papenfuss and Gross, 1986) and the rat spinotrapezius muscle (Schmid-Schönbein et ål., 1989). Numerical network simulations in other organs have been summarized by Popel (1987). In the following we will illustrate the approach in abbreviated form for the rat gracilis muscle.

Let a single microvessel be defined by its position between two neighboring nodes, i and j, respectively. The flow, Q_{ij}, in the vessel is

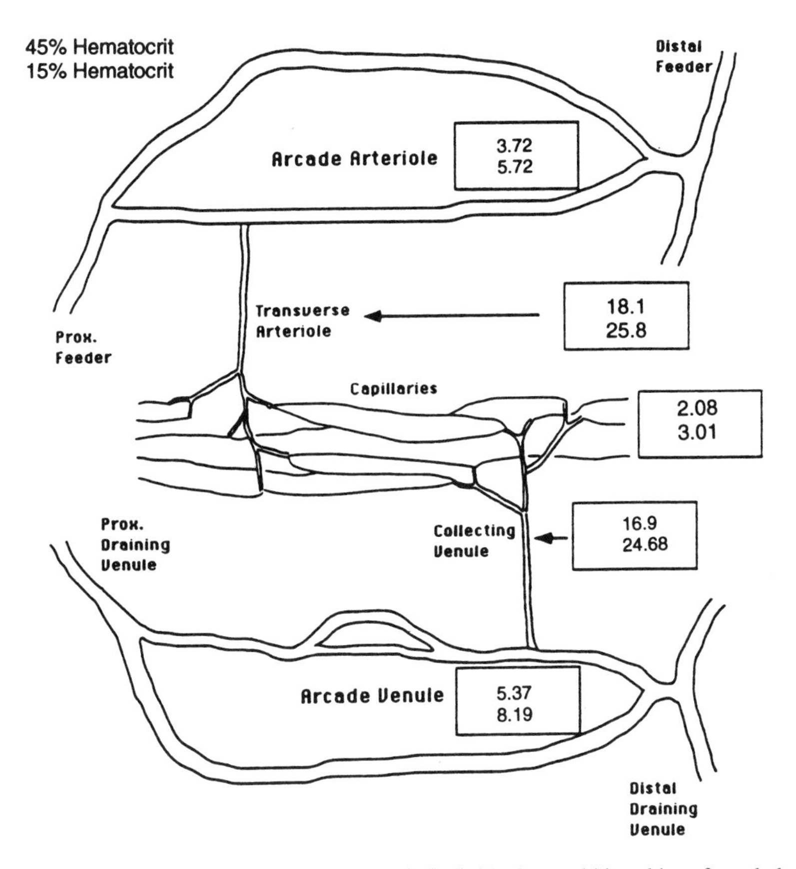

Figure 11. Average microvascular flow rates in individual vessel hirarchies of rat skeletal microcirculation during steady perfusion with 0 and 45% hematocrit (no white blood cells) under myogenic control. Flow rate units in the arcades are 10^{-2} mm^3/sec, in capillaries, transverse arterioles, and collecting venules they are 10^{-5} mm^3/sec.

$$Q_{ij} = (p_i - p_j)\ G_{ij} \frac{\pi}{128} \tag{45}$$

where $p_i - p_j$ is the pressure drop from node i to j. G_{ij} is the conductance (1/resistance) defined according to equation (4) by

$$G_{ij} = \frac{\pi\, d_{ij}^4}{8\, \mu_{ij}^{app}\ l_{ij}} \tag{46}$$

where d_{ij}, μ_{ij}^{app}, and l_{ij} refer to the diameter, apparent viscosity and length of the vessel between nodes i and j. Conservation of mass at each node requires that the total flow entering a node equals the flow exiting and therefore for all

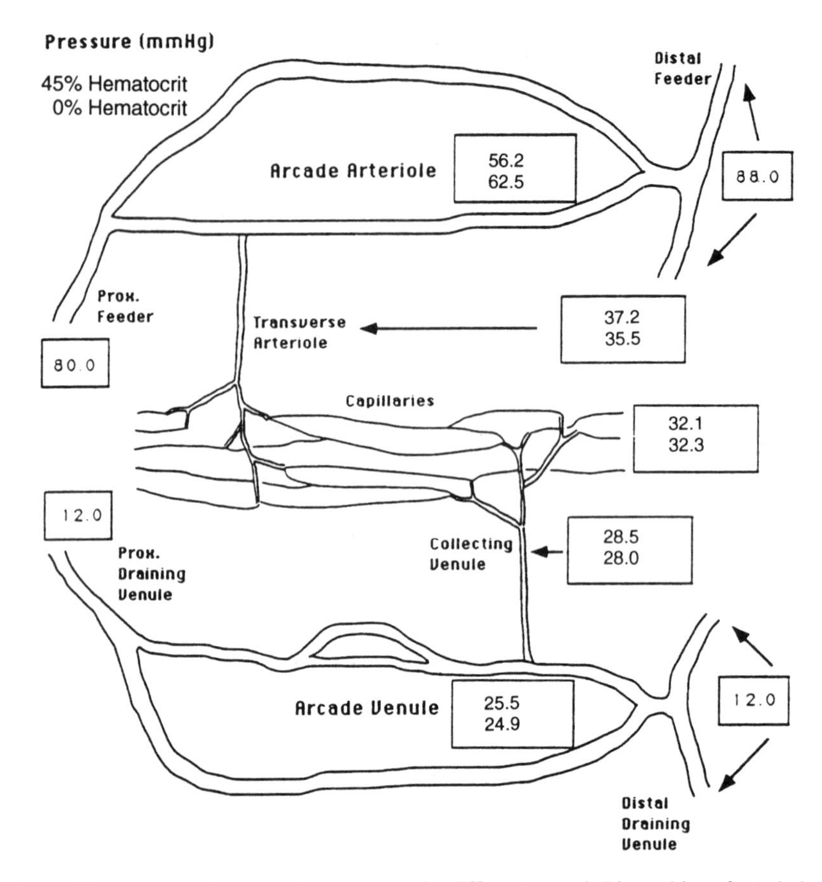

Figure 12. Average microvascular pressures in different vessels hierarchies of rat skeletal microcirculation during steady perfusion with 0 and 45% hematocrit (no white blood cells) under myogenic control.

nodes j connected to node i

$$\sum_j Q_{ij} = 0 \,. \tag{47}$$

Substitution of equation (45) into (47) for all nodes i in the network and application of boundary pressures at the arterial input and venous outflow vessels to the microcirculation yields a set of simultaneous equations

$$[\mathbf{G}](\mathrm{P}) = (\mathrm{B}) \tag{48}$$

which can be solved for the unknown nodal pressures (P) in the interior of the network (Skalak, 1984). The matrix [**G**] depends on the individual conductances and the network connectivity, and the vector (B) depends on the conductances of the boundary vessels and the pressures at the inflows and outflows to the microvascular network. The apparent viscosity is a function of vessel hematocrit, lumen diameter, and shear rate due to the non-Newtonian properties of blood, the Fahraeus effect and non-uniform cell separation at divergent vascular bifurcations. These effects have been subject to extensive previous explorations. In the simulations presented below an empirical approximation proposed by Pries et al. (1990) was utilized (for details see Fenster, 1992). The pressures at the feeder arteries were measured independently for the rat gracilis muscle by micropipette measurements (DeLano et al., 1991).

Since the network in rat gracilis muscle has several hundred thousand microvessels, the system of equations (48) becomes large, but can be solved numerically. Introduction of indefinite admittance parameters (Skalak, 1984) for repetitive microvascular subunits (e.g. the capillary modules) simplifies the computations. The hematocrit was determined in each vessel according to its diameter, flow rate, discharge hematocrit in the parent vessel, and the cell separation at bifurcations. To account for red cell distribution in all vessels of the network from instant to instant, a network traversal algorithm was developed. The computation of the individual vessel hematocrits is initiated at each feeder vessel with known hematocrit and progresses through the network in a non-recursive first search algorithm (Fenster, 1992). Several physiological effects have been investigated using such a network simulation.

One of the important consequences of arteriolar arcades is that the pressure at the entry into the transverse arterioles is relative uniform over the large extend of the organ. Two factors contribute to this uniformity, the presence of multiple inputs from the central arteries and the interconnected network topology of the arcades (Figure 8). Venular arcades exhibit a similar feature (Figure 9). The blood pressure distribution along different hierarchies of microvessels shows the typical steep pressure reduction in arterioles and more shallow pressure drops in venules. Local pressures are sensitive with respect to arterial and venous pressure feeding and draining the organ (Figure 10)

For any set of arterial and venous pressures, capillary flow is significantly influenced by the myogenic response. Myogenic contraction reduces the capillary and venular pressures by increasing the pressure drop in the arcade and transverse arterioles. The hematocrit has a significant influence on flow rates (Figure 11) while its influence on the microvascular pressure distribution is small in the presence of constant arterial and venous pressures (Figure 12).

The arcade arterioles are responsible for the notable uniformity of capillary flow throughout thousands of capillaries in muscle. Capillary modules in proximity to the feeder arteries have on average similar flow rates as capillary modules located away from feeder arteries and positioned deep inside the skeletal muscle. This feature is also reflected in a spatially uniform tube hematocrit in capillary modules. Average pressures in capillary modules are only mildly influenced by the exact branching pattern of the arterioles as long as arcades span all regions in the muscle.

The transverse arterioles form the inflow control into the capillary network. They have the largest pressure gradients in the network with the highest shear rates and lowest apparent viscosity. In contrast, the pressure drops in the collecting venules and arcade venules are lower than in their arteriolar counterparts, mostly due to the larger diameters and the higher network density in the venules. The high density of microvascular interconnections along the capillary bundles permits large variations of capillary flow during constriction of transverse arterioles (from physiological flows to stasis) while maintaining pressures at levels above the collecting venules and thereby reducing the fluid shift into the interstitium during vasomotion.

Synopsis

The interaction between non-linear distensibility of microvessels, active control mechanisms in the arterioles, and shear rate and cell concentration dependent properties of blood suspensions leads to a complex flow pattern in the microcirculation. But within this complexity lies a certain order, one of them is the level of capillary perfusion. The capillary perfusion is remarkably uniform in spite of the vastness of the microvascular network. Among many factors that influence capillary perfusion, the single most important one is the microvascular anatomy. In spite of network randomness, the main categories of microvessels, outlined here, can be detected in virtually all animals that have been investigated so far. The details in microvascular anatomy are different from animal to animal and even right and left muscle in the same animal differ at the microvascular level. Consequently the local flows will vary from one network to another. But in spite of this randomness, average capillary perfusion is robust with respect to details in network connectivity as long as the general pattern of the microvascular network is preserved.

The current analysis is a beginning to identify the mechanisms involved. It serves as a basis for future refinements with additional physiological control mechanisms and as point of departure for quantitative analysis of the microcirculation in disease.

Acknowledgement

This research was supported by NSF Grant DCB-88-19346 and in part by NIH Grant HL-10881. The authors wish to thank their colleague Dr. B.W. Zweifach for many valuable suggestions.

References

Apter, J. T.; Graessley, W. W.: A physical model for muscular behavior Biophysical J. 1970; 10: 539-555.

Bagge, U.: White blood cell deformability and plugging of skeletal muscle capillaries in hemorrhagic shock. Acta. Physiol. Scand. 1980; 108: 159-163.

Bohlen, H. G.; Harper, S. L.: Evidence of myogenic vascular control in the rat cerebral cortex. Circ. Res. 1984; 55: 554-559.

Borgström, P.; Grände, P.; Mellander, S.: A mathematical description of the myogenic response in the microcirculation Acta Physio. Scand. 1982; 116: 363 - 376.

Braakman, R.; Sipkema, P.; Westerhof, N.: A dynamic nonlinear lumped parameter model for skeletal muscle circulation. Ann. Biomed. Eng. 1989; 17: 593-616.

Braide, M.; Amundson, B.; Chien, S.; Bagge, U.: Quantitative studies on the influence of leukocytes on the vascular resistance in a skeletal muscle. preparation. Microvasc. Res. 1984; 27: 331-352.

Burton, A. C.: On the physical equilibrium of small blood vessels. Am. J. Physiol. 1951; 164: 319-329.

Chien, I. I. H.: A mathematical representation for vessel network. II J. Theoret. Biol. 1983; 104: 647-654.

Chien, I. I. H.: A mathematical representation for vessel network. III J. Theoret. Biol. 1984; 111: 115-121.

Cokelet, G. R.; Merrill, E. W.; Gilliland, E. R.; Shin, H.; Britten, A.; Wells, R. E.: The rheology of human blood measurement near and at zero shear rate. Trans. Soc. Rheol. 1963; 7: 303-317.

Davis, M. J.; Sikes, P. J.: Myogenic responses of isolated arterioles: test for a rate-sensitive mechanism. Am. J. Physiol. 1990; 259: 1890-1900.

DeLano, F. A.; Schmid-Schönbein, G. W.; Skalak, T. C.; Zweifach, B. W.: Penetration of the systemic blood pressure into the microvasculature of rat skeletal muscle. Microvasc. Res. 1991; 41: 92-110.

Engelson, E. T.; Schmid-Schönbein, G. W.; Zweifach, B. W.: The microvasculature in skeletal muscle. III. Venous network anatomy in normotensive and spontaneously hypertensive rats. Microvasc. Res. 1985; 4: 229-248.

Engelson, E. T.; Schmid-Schönbein, G. W.; Zweifach, B. W.: The microvasculature in skeletal muscle. II. Arteriolar network anatomy in normotensive and hypertensive rats. Microvasc. Res. 1986; 31: 356-374.

Engelson, E. T.; Skalak, T. C.; Schmid-Schönbein, G. W.: The microvasculature in skeletal muscle. I. Arteriolar network topology. Microvasc. Res. 1985; 30: 29-44.

Falcone, J. C.; Davis, M. J.; Meininger, G. A.: Endothelial independence of myogenic response in isolated skeletal muscle arterioles. Am. J. Physiol. 1991; 260: H130-H135.

Fenster, M. A.: A mathematical hemodynamic model of the microcirculation in skeletal muscle, including passive and active vessel properties, hematocrit and blood rheology. M.S. Thesis, University of California San Diego, 1992.

Fung, Y. C.: Biodynamics: Circulation. New York: Springer-Verlag; 1984.

Fung, Y. C.: Biomechanics: Mechanical Properties of Living Tissue New York: Springer-Verlag; 1993.

Granger, H.; Meininger, G. A.; Borders, J. L.; Morff, R. J.; Goodman, A. H.: Microcirculation of skeletal muscle. Phys. Pharm. Microcirc. 1984; 2: 181-265.

Iida, N.: Physical properties of resistance vessel wall in peripheral blood flow regulation - I. Mathematical model. J. Biomech. 1989; 22: 109-117.

Jackson, P. A.; Duling, B. R.: Myogenic response and wall mechanics of arterioles. Am. J. Physiol. 1989; 257: H1147-H1155.

Jerome, S. N.; Smith, C. W.; Korthuis, R. J.: CD18-dependent adherence reactions play an important role in the development of the no-reflow phenomenon. J. Appl. Physiol. 1993; 264: H479-H483.

Johnson, P. C.: The myogenic response. In: Bohr, D. F.; Somlyo, A. P.; Sparks, H. V. J.; Geiger, S. R. (eds.) Handbook of Physiology, Section 2, The

Cardiovascular System Bethesda, MD: American Physiological Society; 1980: p. 409-442.

Johnson, P. C.: The myogenic response. News Physiol. Sci.. 1991; 9: 41-42.

Johnson, P. C.; Intaglietta, M.: Contributions of pressure and flow sensitivity to autoregulation in mesenteric arterioles. Am. J. Physiol. 1976; 231: 1686-1698.

Koch, A. R.: Some mathematical forms of autoregulatory models. Circ. Res. 1964; 14 and 15: I269-I278.

Koller, A.; Dawant, B.; Liu, A.; Popel, A. S.; Johnson, P. C.: Quantitative analysis of arteriolar network architecture in cat sartorius muscle. Am. J. Physiol. 1987; 253: H154-H164.

Koller, A.; Johnson, P. C.: Methods for in vivo mapping and classifying microvascular networks in skeletal muscle. In: Popel, A. S.; Johnson, P. C. (eds.) Microvascular networks: Experimental and theoretical studies. Basel: Karger; 1986: p. 27-37.

Koller, A.; Kaley, G.: Prostaglandins mediate arteriolar dilation to increased blood flow velocity in skeletal muscle microcirculation Circ. Res. 1990; 67: 529-534.

Koller, A.; Seyedi, N.; Gerritsen, M. E.; Kaley, G.: EDRF released from microvascular endothelial cells dilates arterioles in vivo. Am. J. Physiol. 1991; 261: H128-H133.

Kuo, L.; Chilian, W. M.; Davis, M. J.: Coronary arteriolar myogenic response is independent of endothelium. Circ. Res. 1990; 66: 860-866.

Kuo, L.; Davis, M. J.; Chilian, W. M.: Myogenic activity in isolated subepicardial and subendocardial coronary arterioles. Am. J. Physiol. 1988; 255: H1558-H1562.

Kuo, L.; Davis, M. J.; Chilian, W. M.: Endothelium-dependent, flow-induced dilation of isolated coronary arterioles. Am. J. Physiol. 1990; 259: H1063-H1070.

Kuo, L.; Davis, M. J.; Chilian, W. M.: Endothelial modulation of arteriolar tone. News Physiol. Sci.. 1992; 7: 5-9.

Kurthius, R. J.; Grisham, M. B.; Granger, D. N.: Leukocyte depletion attenuates vascular injury in postischemic skeletal muscle. Am. J. Physiol. 1988; 254: H823-H827.

Lee, J.: The morphometry and mechanical properties of skeletal muscle capillaries. Ph.D. Dissertation, University of California San Diego, 1990.

Lee, S.: A biomechanical model of the skeletal muscle microcirculation with pulsatile pressure and the myogenic response. Ph.D., University of California San Diego, 1993.

Lee, S. Y.; Schmid-Schönbein, G. W.: Pulsatile pressure and flow in the skeletal muscle microcirculation. J. Biomech. Eng. 1990; 112: 437-443.

Nobis, U.; Pries, A. R.; Cokelet, G. R.; Gaehtgens, P.: Radial distribution of white cells during blood flow in small tubes. Microvasc. Res. 1984; 29: 295-304.

Papenfuss, H. D.; Gross, J. F.: Mathematical simulation of blood flow in microcirculatory networks. In: Popel, A. S.; Johnson, P. C. (eds.) Microvascular Networks: Experimental and Theoretical Studies. Basel: Karger; 1986: p. 168-181.

Popel, A. S.: Network models of peripheral circulation. In: Skalak, R.; Chien, S. (eds.) Handbook of Bioengineering. New York: McGraw Hill; 1987: p. 20.1-20.24.

Popel, A. S.; Torres-Filho, I. P.; Johnson, P. C.; Bonskela, E.: A new schema for hierarchical classification of anastomosing vessels. Int. J. Microcirc. Clin. Exp. 1988; 7: 131-138.

Pries, A. R.; Secom, T. W.; Gaehtgens, P.; Gross, J. F.: Blood flow in microvascular networks: Experiments and simulation. Circ. Res. 1990; 67: 826-834.

Reinke, W.; Gaehtgens, P.; Johnson, P. C.: Blood viscosity in small tubes: effect of shear rate, red cell aggregation, and sedimentation. Am J. Physiol. 1987; 253: H540-H547.

Saltzman, D. J.: Adrenergic innervation of arterioles in normotensive and spontaneously hypertensive rats. M.S. Thesis, University of California, San Diego, 1993.

Saunders, R. L. d. C. H.; Lawrence, E. J.; Maciner, D. A.; Nementhy, N.: The anatomical basis of the peripheral circulation in man. On the concept of macromesh and micromesh as illustrated by the blood supply in man. In: Redish, L.; Tango, F. F.; Sauders, R. L. d. C. H. (eds.) Peripheral Circulation in Health and Disease. New York: Grune & Stratton; 1957: p.

Schmid-Schönbein, G. W.: Mechanisms of granulocyte-capillary-plugging. Prog. Appl. Microcirc. 1987; 12: 223-230.

Schmid-Schönbein, G. W.: A theory of blood flow in skeletal muscle. J. Biomech. Eng. 1988; 110: 20-26.

Schmid-Schönbein, G. W.; DeLano, F. A.; Chu, S.; Zweifach, B. W.: Wall structure of arteries and arterioles feeding the spinotrapezius muscle of normotensive and spontaneously hypertensive rats. Int. J. Microcirc: Clin. Exp. 1990; 9: 47-66.

Schmid-Schönbein, G. W.; Firestone, G.; Zweifach, B. W.: Network anatomy of arteries feeding the spinotrapezius muscle in normotensive and hypertensive rats. Blood Vessels 1986; 23: 34-39.

Schmid-Schönbein, G. W.; Lee, S. Y.; Sutton, D. W.: Dynamic viscous flow in distensible vessels of skeletal muscle microcirculation: Application to pressure and flow transient. Biorheology 1989; 26: 215-227.

Schmid-Schönbein, G. W.; Murakami, H.: Blood flow in contracting arterioles. Int. J. Microcirc. Clin. Exp. 1985; 4: 311-328.

Schmid-Schönbein, G. W.; Skalak, T. C.; Firestone, G.: The microvasculature in skeletal muscle. V. The arteriolar and venular arcades in normotensive and hypertensive rats. Microvasc. Res. 1987; 34: 385-393.

Schmid-Schönbein, G. W.; Skalak, T. C.; Sutton, D. W.: Bioengineering analysis of blood flow in resting skeletal muscle. In: Lee, J. S.; Skalak, T. C. (eds.) Microvascular Mechanics New York: Springer Verlag; 1989: p. 65-99.

Schmid-Schönbein, G. W.; Zweifach, B. W.; Delano, F. A.; Chen, P. C.: Microvascular tone in a skeletal muscle of spontaneously hypertensive rats. Hypertension 1987; 9: 164-171

Secomb, T. W.; Pries, A. R.; Gaehtgens, P.; Gross, J. F.: Theoretical and experimental analysis of hematocrit distribution in microvascular networks. In: Lee, J.-S.; Skalak, T. C. (eds.) Microvascular Mechanics. New York: Springer Verlag; 1989: p. 39-49.

Skalak, T. C.: A mathematical hemodynamic network model of the microcirculation in skeletal muscle, using measured blood vessel distensibility and topology. Ph.D. Dissertation, University of California, San Diego, 1984.

Skalak, T. C.; Schmid-Schönbein, G. W.: The microvasculature in skeletal muscle. IV. A model of the capillary network. Microvasc. Res. 1986a; 32: 333-347.

Skalak, T. C.; Schmid-Schönbein, G. W.: Viscoelastic properties of microvessels in rat spinotrapezius muscle. J. Biomech. Eng. 1986b; 108: 193-200.

Skalak, T. C.; Schmid-Schönbein, G. W.; Zweifach, B. W.: New morphological evidence for a mechanism of lymph formation in skeletal muscle. Microvasc. Res. 1984; 28: 95-112.

Sutton, D. W.; Schmid-Schönbein, G. W.: Hemodynamics at low flow in the resting, vasodilated rat skeletal muscle. Am. J. Physiol. 1989; 257: H1419-H1427.

Sutton, D. W.; Schmid-Schönbein, G. W.: The pressure-flow relation for plasma in whole organ skeletal muscle and its verification. J. Biomech. Eng. 1991; 113: 452-457.

Sutton, D. W.; Schmid-Schönbein, G. W.: Elevation of organ resistance due to leukocyte perfusion. Am. J. Physiol. 1992; 262: H1646-H1650.

Thompson, T. N.; La Celle, P. L.; Cokelet, G. R.: Perturbation of red cell flow in small tubes by white blood cells. Pflügers Arch. 1989; 413: 372-377.

Ursino, M.; Fabbri, G.: Role of the myogenic mechanism in the genesis of microvascular oscillations (vasomotion): analysis with a mathematical model. Microvasc. Res. 1992; 43: 156-177.

Warnke, K. C.; Skalak, T. C.: The effects of leukocytes on blood flow in a model skeletal muscle capillary network. Microvasc. Res. 1990; 40: 118-136.

Woldenberg, M. J.: Quantitative analysis of biological and fluvial networks. In: Popel, A. S.; Johnson, P. C. (eds.) Microvascular Networks: Experimental and Theoretical Studies. Karger Basel; 1986: p. 12-26.

Yen, R. T.; Zhuang, F. Y.; Fung, Y. C.; Tremer, H.; Sobin, S. S.: Morphometry of the cat's pulmonary arterial tree. J. Biomech. Eng. 1984; 106: 131-136.

Zweifach, B. W.; Kovalcheck, S.; DeLano, F.; Chen, P.: Micropressure-flow relationships in a skeletal muscle of spontaneously hypertensive rats. Hypertension 1981; 3: 601-614.